Organic Reactions

Organic Reactions

VOLUME 28

JOHN WILEY & SONS, INC.

New York · Chichester · Brisbane · Toronto · Singapore

Published by John Wiley & Sons. Inc.

Library of Congress Catalog Card Number 42-20265

ISBN 0-471-86141-3

Printed in the United States of America

10 9 8 7 6 5 4 3 2 1

PREFACE TO THE SERIES

In the course of nearly every program of research in organic chemistry the investigator finds it necessary to use several of the better-known synthetic reactions. To discover the optimum conditions for the application of even the most familiar one to a compound not previously subjected to the reaction often requires an extensive search of the literature; even then a series of experiments may be necessary. When the results of the investigation are published, the synthesis, which may have required months of work, is usually described without comment. The background of knowledge and experience gained in the literature search and experimentation is thus lost to those who subsequently have occasion to apply the general method. The student of preparative organic chemistry faces similar difficulties. The textbooks and laboratory manuals furnish numerous examples of the application of various syntheses, but only rarely do they convey an accurate conception of the scope and usefulness of the processes.

For many years American organic chemists have discussed these problems. The plan of compiling critical discussions of the more important reactions thus was evolved. The volumes of *Organic Reactions* are collections of chapters each devoted to a single reaction, or a definite phase of a reaction, of wide applicability. The authors have had experience with the processes surveyed. The subjects are presented from the preparative viewpoint, and particular attention is given to limitations, interfering influences, effects of structure, and the selection of experimental techniques. Each chapter includes several detailed procedures illustrating the significant modifications of the method. Most of these procedures have been found satisfactory by the author or one of the editors, but unlike those in *Organic Syntheses* they have not been subjected to careful testing in two or more laboratories.

Each chapter contains tables that include all the examples of the reaction under consideration that the author has been able to find. It is inevitable, however, that in the search of the literature some examples will be missed, especially when the reaction is used as one step in an extended synthesis. Nevertheless, the investigator will be able to use the tables and their accompanying bibliographies in place of most or all of the literature search so often required.

Because of the systematic arrangement of the material in the chapters and the entries in the tables, users of the books will be able to find information desired by reference to the table of contents of the appropriate chapter. In the interest of economy the entries in the indices have been kept to a minimum, and, in particular, the compounds listed in the tables are not repeated in the indices.

The success of this publication, which will appear periodically, depends upon the cooperation of organic chemists and their willingness to devote time and effort to the preparation of the chapters. They have manifested their interest already by the almost unanimous acceptance of invitations to contribute to the work. The editors will welcome their continued interest and their suggestions for improvements in *Organic Reactions*.

Chemists who are considering the preparation of a manuscript for submission to *Organic Reactions* are urged to write either secretary before they begin work.

CONTENTS

Organic Reactions

CHAPTER 1

THE REIMER–TIEMANN REACTION

HANS WYNBERG AND EGBERT W. MEIJER

The University, Groningen, The Netherlands

CONTENTS

INTRODUCTION

The Reimer–Tiemann reaction owes its name to two young German chemists, Karl Reimer and Ferdinand Tiemann. In 1876 they isolated and identified hydroxyaldehydes as the principal reaction products of phenol and chloroform in alkaline medium.[1–4] The scope of this reaction was enlarged in 1884 by von Auwers, who discovered the chlorine-containing cyclohexadienones as by-products in the formylation of alkylphenols.[5–12] The ring-expansion products, namely, chloropyridines, were first noted by Ciamician when he subjected pyrroles to Reimer–Tiemann reaction conditions.[13–15] Nearly half a century passed before Woodward recognized that the conversion of an alkylphenol to a substituted cyclohexadienone could lead to the synthesis of terpenes and steroids containing an angular methyl group.[16] Although the method failed as a preparatively useful approach to the synthesis of steroids,[17–19] an A/B *trans*-fused hexahydrophenanthrene was prepared using a Reimer–Tiemann reaction.[20] The reaction was last reviewed some 20 years ago.[21]

It is convenient to divide the Reimer–Tiemann reaction into a normal and abnormal transformation depending on the reaction products. A normal Reimer–Tiemann reaction is one in which a phenol (or electron-rich aromatic such as pyrrole) yields one or more aldehydes on treatment with chloroform and alkali.[1–4]

OH (phenol) —aq. NaOH, $CHCl_3$, 60°→ OH, CHO (salicylaldehyde) (20%) + OH, CHO (4-hydroxybenzaldehyde) (10%)

The abnormal Reimer–Tiemann reaction product can be subdivided further into cyclohexadienones and ring-expansion products:

1. When *ortho*- or *para*-substituted phenols are subjected to the Reimer–Tiemann reaction conditions, 2,2- or 4,4-disubstituted cyclohexadienones may be obtained in addition to the normal products.[7–10]

OH, CH_3 —aq. NaOH, $CHCl_3$, 60°→ OH, CHO, CH_3 (25%) + O, CH_3 $CHCl_2$ (26%)

OH, CH_3, CH_3 —aq. NaOH, $CHCl_3$, 60°→

OH, OHC, CH_3, CH_3 (11%) + O, CH_3, $CHCl_2$, CH_3 (trace) + O, CH_3, CH_3 $CHCl_2$ (30%)

Several alkylphenols, alkylnaphthols, and tetralols have been converted into cyclohexadienones, whereas some alkylpyrroles are converted to pyrrolines in this manner.

2. A variety of five-membered rings yield ring-expansion products when subjected to the Reimer–Tiemann reaction conditions. These products are formed in addition to the normal products.[13–15]

N, K —aq. KOH, $CHCl_3$, 60°→ CHO, N, H (31%) + Cl, N (13%)

In addition to aromatic aldehydes, cyclohexadienones, pyrrolines, and ring-expansion products, a variety of other products have been noted, isolated, and identified in a few cases. All these by-products are discussed in in the Experimental Conditions section.

MECHANISM

The classical work of Hine,[22–28] followed by that of von Doering,[29] Skell,[30,31] Parham,[32–35] Robinson,[36] and others,[37,38] clearly established that dihalocarbenes were formed when a haloform was treated with alkali; direct nucleophilic substitution of the phenolate carbanion on chloroform is thus ruled out.[37] The recognition that dichlorocarbene was the reactive intermediate set the stage for the formulation of a rational mechanism.[39] It is convenient to consider separately the two reactions, namely, the hydrolysis of chloroform and the reaction of a phenol with dichlorocarbene.

The first reaction involves the generation of the carbene in a rate-determining step by the unimolecular elimination of a chloride ion from the trichloro-

$$CHCl_3 + NaOH \xrightleftharpoons{\text{Fast}} \overset{+}{Na}\overset{-}{C}Cl_3 + H_2O$$

$$\overset{-}{C}Cl_3 \xrightarrow{\text{Slow}} :CCl_2 + Cl^-$$

$$:CCl_2 + H_2O \longrightarrow [H_2\overset{+}{O}-\overset{-}{C}Cl_2] \longrightarrow HOCHCl_2$$

$$HOCHCl_2 + 2NaOH \longrightarrow CO + 2NaCl + 2H_2O$$

$$CO + NaOH \xrightarrow{\text{Slow}} \overset{+}{Na}\overset{-}{O}\overset{O}{\overset{\|}{C}}H \qquad \text{(Eq. 1)}$$

methyl anion.[21,22] Although Hine's original mechanism assumed simultaneous formation of carbon dioxide and formate,[22] Robinson's work clarified this point.[36] Dichlorocarbene reacts rapidly with water to generate carbon monoxide, while the latter slowly hydrolyzes to sodium formate in the alkaline medium. Superficially these last two steps appear unimportant to the main reaction, which involves bond formation between the carbene and the phenol. However, the hydrolysis of carbene consumes alkali and competes with the formylation reaction. The low conversions and subsequent recovery of starting phenol noticed by numerous investigators are often due to the fact that most if not all of the alkali becomes neutralized as the reaction proceeds. Unfortunately only limited use can be made of the apparently ideal phase-transfer conditions under which this reaction might be run, since the phenolic substrate remains in the aqueous alkaline layer while the carbene is largely present in the chloroform layer. The reaction of the dihalocarbene with the phenolate ion comprises the product-forming reaction. No significant changes in the overall mechanism have been proposed since its formulation 25 years ago.[39]

Dihalocarbene Reaction with Phenoxide Anion

Electrophilic attack of the dihalocarbene on each of the resonance forms of the phenoxide anion gives the anion having resonance forms **I**, **II**, and **III** (Eq. 2):

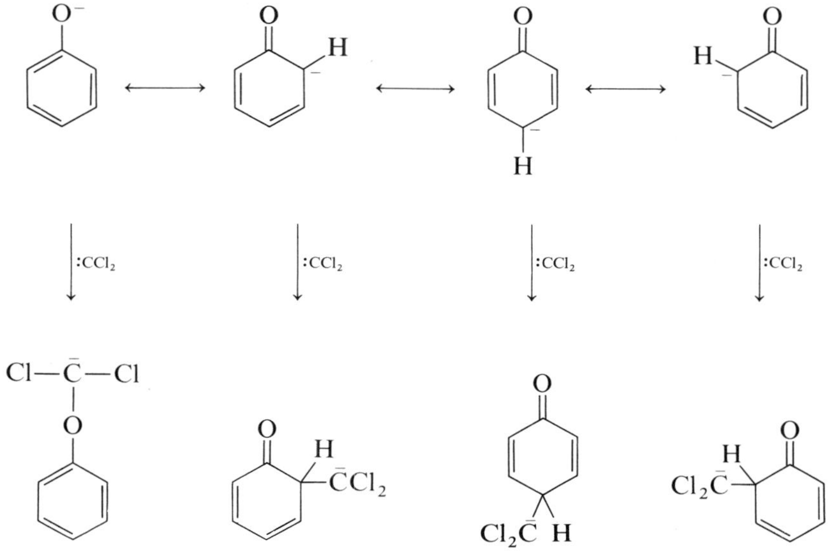

I II III II

(Eq. 2)

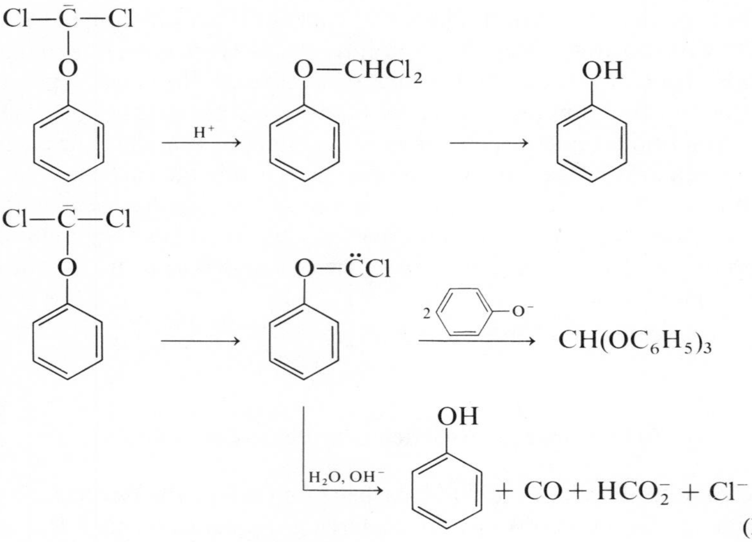

(Eq. 3)

The O-alkylation product, namely, anion **I**, either decomposes or reacts further to form small amounts of orthoformic esters (Eq. 3). The details of the transformation of anions **II** and **III** (Eq. 4 is given only for the *ortho*-substituted phenoxide anion) to the corresponding hydroxyaldehydes have been the subject of considerable speculation.

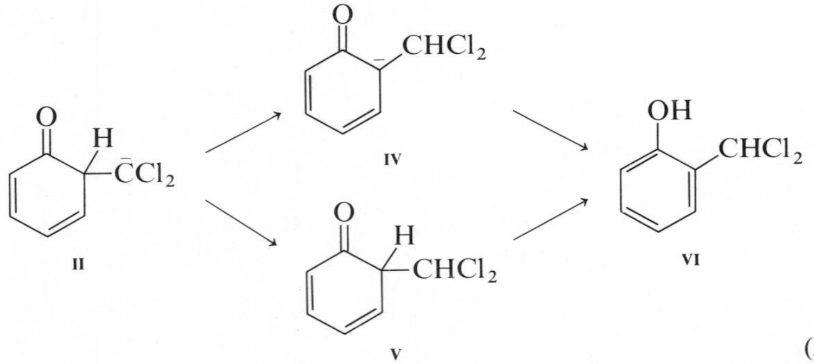

(Eq. 4)

Neither the anion **IV** nor one of the neutral products **V**, 2-dichloromethyl-3,5-cyclohexadienone and **VI**, 2-dichloromethylphenol, has ever been isolated from a Reimer–Tiemann reaction mixture despite claims to the contrary.[40] The elimination sequence suggested in Eq. 5 seems more reasonable.

The 1,2-proton shift of the anion **II** to the anion **IV** (Eq. 4) appears unlikely in view of Hückel and orbital-symmetry calculations.[41] Tritium isotope experiments show complete transfer of the isotope from tritium oxide to the formyl group.[41,42] Proton exchange between water and the aldehydic proton of salicylaldehyde is absent under the reaction conditions. These experiments favor a mechanism involving the neutral intermediate **V**.

The neutral intermediate **VI** is hydrolyzed to the aldehyde **VII** in alkaline medium (Eq. 5). Without speculating on the details of the intermediate steps, an explanation must be given for the fact that *ortho*- and *para*-hydroxybenzal halides (such as **VI**) have not been isolated, while benzal halides without *ortho*- or *para*-hydroxy functions are isolable. It is easily seen from Eq. 5 that a driving force exists that clearly aids the decomposition of **VI** into the aldehyde.

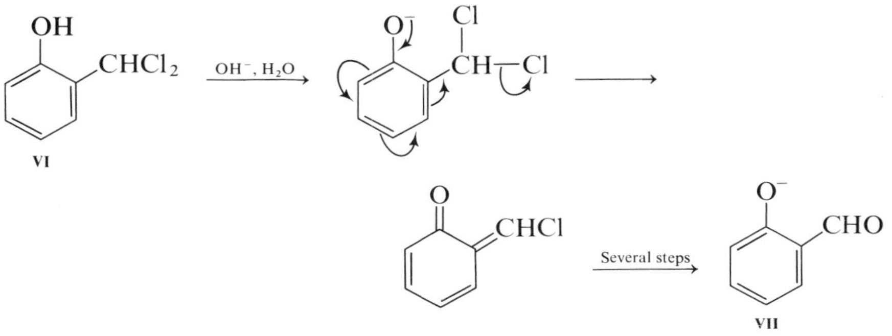

OH, $CHCl_2$ (VI) $\xrightarrow{OH^-, H_2O}$ O^-, CH(Cl)–Cl $\longrightarrow$ O, CHCl $\xrightarrow{\text{Several steps}}$ O^-, CHO (VII)

(Eq. 5)

Note that the abnormal products are not tautomeric with their phenols since they lack the needed proton on the α- or γ-carbon atom.

Dihalocarbene Reaction with *Ortho*- and/or *Para*-Substituted Phenoxide Anions

The first two steps of the mechanism of the abnormal reaction and the normal reaction are the same. After proton abstraction from the solvent, the intermediate does not hydrolyze to the corresponding aldehyde. It is worthy of note that both the *ortho* and *para* abnormal products represent a dihaloalkyl structure similar to that of a neopentyl halide, whose hydrolysis is exceedingly slow.

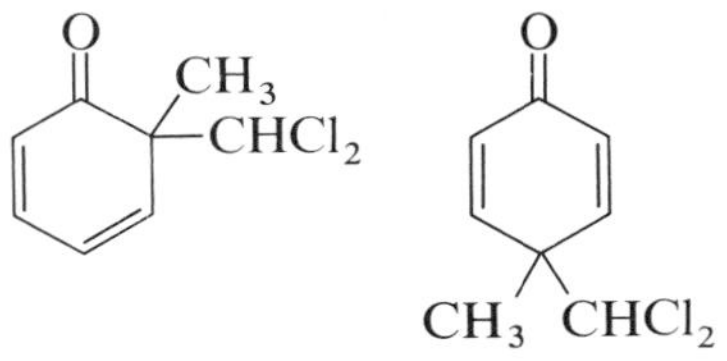

O, CH_3, $CHCl_2$ O, CH_3 $CHCl_2$

(Eq. 6)

Dihalocarbene Reaction with Substrates Other than Phenoxide Anion

In the absence of a large excess of strongly nucleophilic reagents (phenoxide, hydroxide), dichlorocarbene can add to a double bond. The mechanism of the addition of dihalocarbene to double bonds has been well described in a review.[43]

SCOPE AND LIMITATIONS

Phenols and Alkylphenols

Virtually all the phenols and alkylphenols that have been subjected to the action of chloroform and base have been found to give *ortho*- and/or *para*-aldehydes. Consistent exceptions are 2,4,6-trialkylphenols, which form alkyl dichloromethylcyclohexadienones.[19]

OH, CH_3 $\xrightarrow[60°]{\text{aq. NaOH, } CHCl_3}$ OH, OHC, CH_3 (20%) + OH, CH_3, CHO (8–10%) + abnormal product

HO $\xrightarrow[C_2H_5OH,\ 75°]{\text{aq. NaOH, } CHCl_3}$ HO, CHO (38–48%)

Although early isolation and identification procedures may not have been ideal compared to present-day methods, no serious discrepancies with the early work have been uncovered.

A recent report states that traces of hydroxyaldehydes are formed when 1-naphthol is treated with chloroform and base.[44] Earlier studies also mentioned the failure of this phenol to yield Reimer–Tiemann reaction products.[45] Even more astonishing, and in need of careful verification, is the report by the same workers that in addition to the well-known 2-hydroxy-1-naphthalenecarboxaldehyde, a trace (2%) of 2-hydroxy-4-naphthalenecarboxaldehyde, the *meta* product, is formed from 2-naphthol.[44] It has been proven conclusively that the major product from the formylation of 2-hydroxy-5,6,7,8-tetrahydronaphthalene is 2-hydroxy-5,6,7,8-tetrahydro-1-naphthalenecarboxaldehyde, as previously reported.[16] Traces of the 3-naphthalenecarboxaldehyde are also formed.[46]

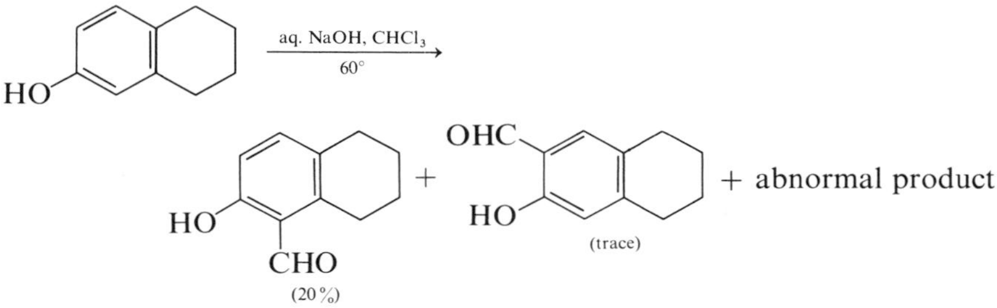

aq. NaOH, $CHCl_3$, 60°

(20%) + (trace) + abnormal product

Halophenols

All the halophenols studied yield normal Reimer–Tiemann reaction products.[40,47–57] Noteworthy is a patent describing the formylation of 2-fluorophenol in the presence of dimethylformamide to furnish the *ortho*-substituted hydroxybenzaldehyde to the exclusion of the *para*-isomer.[56] In a few cases dialdehydes have been isolated in unspecified low yield.[52–54] Chloral instead of chloroform has been found effective in some cases.[58] For example, a 41% yield of 5-chloro-2-hydroxybenzaldehyde could be isolated when 4-chlorophenol was treated with chloral in the presence of 50% aqueous sodium hydroxide.[58]

Hydroxy- and Alkoxyphenols

Catechol, resorcinol, and hydroquinone as well as their monomethyl ethers have been formylated successfully.[59–63]

In the reaction of 3-methoxyphenol with chloroform and aqueous alkali, 2-hydroxy-6-methoxybenzaldehyde has never been isolated, although it is one of the possible products.[60] The evidence is questionable for the formation of dialdehydes in the Reimer–Tiemann reaction of resorcinol and resorcinol monomethyl ether.[60,64]

The yield of vanillin from guaiacol is said to be affected favorably by the addition of ethanol.[62,63]

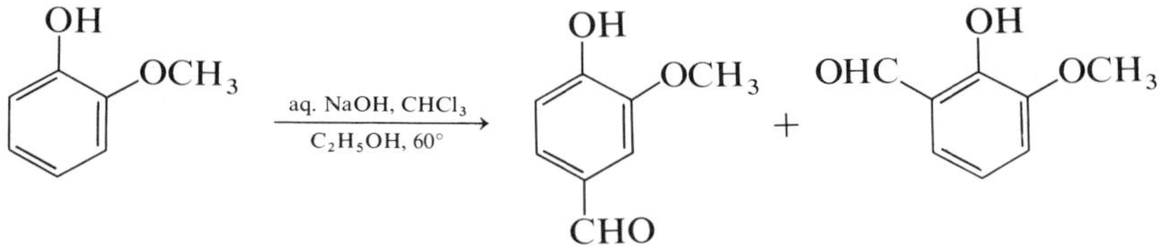

aq. NaOH, $CHCl_3$, C_2H_5OH, 60°

Carboxy and Sulfonic Acid Substituted Phenols

In view of the accepted mechanism (electrophilic carbene attack on the phenolate anion) of the Reimer–Tiemann reaction, it is not surprising that phenols containing electron-withdrawing groups furnish hydroxyaldehydes in lower yields.

Formylation of salicyclic acid, one of the earliest Reimer–Tiemann reactions, yields hydroxyaldehydes lacking the carboxyl group in addition to the expected products.[3,65–67]

OH
CO_2H
aq. NaOH, $CHCl_3$
70°

OH
OHC CO_2H
(20%)
\+
OH
CO_2H
CHO
(trace)
\+
OH
CHO
(1–2%)

Attack by dichlorocarbene at the 1 position of the salicylate dianion followed by decarboxylation of the resulting dienone intermediate is a possible mechanism. Even two carboxyl groups do not prevent nuclear formylation; both 4- and 5-hydroxyisophthalic acid yield the corresponding aldehydes.[65] In a series of patents[68,69] the successful formylation of naphtholmonosulfonic acid and naphtholdisulfonic acid was reported. No analytical data, yields, or structure proofs are available for these products.

Heterocyclic Phenols

Phenolic properties associated with hydroxy-substituted aromatics are exhibited only by certain nitrogen-containing heteroaromatic phenols. Thus 3-hydroxypyridine, hydroxyquinolines, and hydroxypyrimidines represent the important classes of heterocyclic phenols that undergo the Reimer–Tiemann reaction.

The reported conversion[70] of 8-hydroxyquinoline to the hydroxyaldehydes in high yield could not be verified.[71] This important conversion has also been reported using chloral instead of chloroform, and this report has been duplicated.[71,72]

Pyrroles and Related Compounds

A number of electron-rich heteroaromatics not containing a phenolic hydroxy group are nevertheless subject to formylation. Pyrrole and 2,5-dimethylpyrrole give the 2-aldehyde and the 3-aldehyde, respectively, and indole gives the 3-aldehyde. In addition to the heterocyclic aldehydes the ring-expansion products are also formed. No detailed studies have been published to explain these phenomena.

Miscellaneous Compounds

In view of the hundred-and-fifth anniversary of the discovery of the Reimer–Tiemann reaction at the date of this writing it is surprising that only about a dozen examples of the formation of hydroxyaldehydes are known to fall outside the six categories described thus far. The high yield of product from phenolphthalein is noteworthy (corrected from the original data).[73]

HO OH O O $\xrightarrow[C_2H_5OH,\ 95^\circ]{aq.\ NaOH,\ CHCl_3}$ CHO HO OH O O (59%)

It seems reasonable that an increasing number of hydroxyaldehydes of varying structure will become accessible using the Reimer–Tiemann reaction under phase-transfer conditions[74] as employed for some nitrogen heterocycles.[75] The introduction of a formyl group is unsuccessful when benzaldehyde,[76,77] thiophenol,[78] or hydroxyphenylarsonic acid[79] are used under Reimer–Tiemann reaction conditions.

Abnormal Products

Appropriately substituted phenols subjected to Reimer–Tiemann reaction conditions are transformed into cyclohexadienones.[5–12] Although history has ordained that the hydroxyaldehydes are called *normal* products, whereas the dihalomethyl cyclohexadienones and ring-expansion compounds are called *abnormal* products, an increasing number of high-yield abnormal reactions have been reported in the literature.[80] Several aspects of the abnormal reaction have contributed to its present use. The reaction with certain alkylphenols leads to nonaromatic products under basic conditions. The introduction of potential *gem*-dialkyl groups or an angular methyl group via the abnormal reaction has attracted interest in steroid and terpene synthesis.[16,20] Many abnormal products are formed in good to excellent yields, in contrast to the yields of normal products.

Cyclohexadienones and Related Nonaromatics. The conversion of appropriately substituted alkyl (or aryl) phenols, pyrroles, and indoles to angularly alkylated cyclohexadienones, pyrrolines, and indolines is well known.[20,80–83] A classic example is the transformation of 2-hydroxy-5,6,7,8-tetrahydronaphthalene, which is of potential use in steroid synthesis.[16]

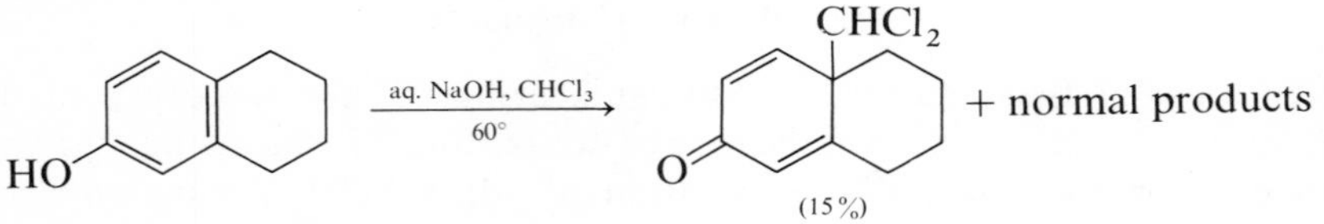

HO (tetrahydronaphthol) —aq. NaOH, $CHCl_3$, 60°→ $CHCl_2$ (dienone) (15%) + normal products

Examples of reactions that proceed in high yield with and without regioselectivity are shown below.[19,80]

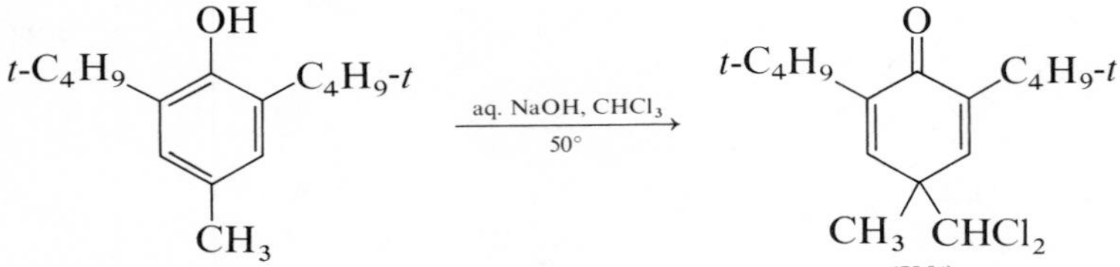

2,6-di-t-C_4H_9-4-CH_3-phenol (OH) —aq. NaOH, $CHCl_3$, 50°→ t-C_4H_9, C_4H_9-t, O, CH_3 $CHCl_2$ (78%)

OH, CH_3 (bridged phenol) —aq. NaOH, $CHCl_3$, 50°→ O, $CHCl_2$, CH_3 (76%)

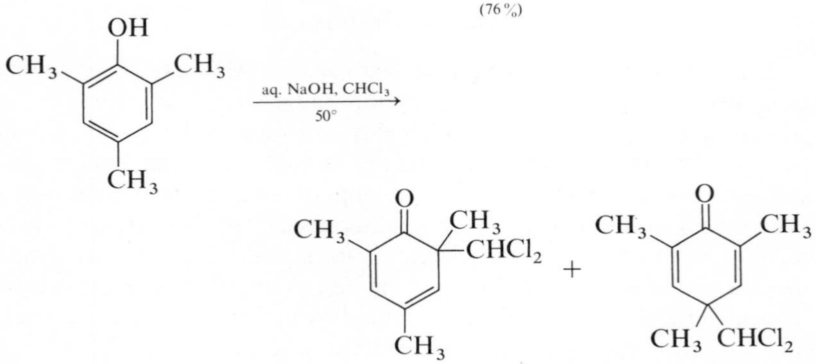

OH, CH_3, CH_3, CH_3 (mesitol) —aq. NaOH, $CHCl_3$, 50°→

CH_3, O, CH_3, $CHCl_2$, CH_3 + CH_3, O, CH_3, CH_3 $CHCl_2$

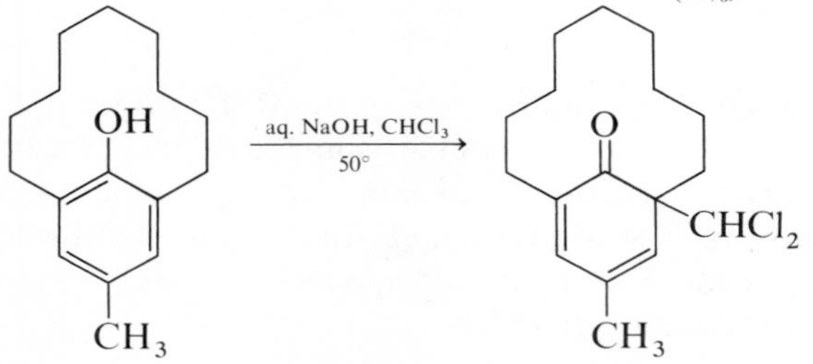

(total yield 67–77%)

The original papers should be consulted for a discussion of regioselectivity. The ratio of *ortho* to *para* isomers in the conversion of mesitol is dependent on the reaction conditions, varying from 1:1 to 2:1 (*ortho*:*para*);[19,80] with cyclodextrins, only *para* product was found.[185] A careful study has been published of normal, abnormal, and ring-expansion products resulting from the Reimer–Tiemann reaction with a series of 4-alkyl guaiacols.[84,85]

Ring-Expansion Products. The discussion on ring-expansion products is limited to those in which the normal Reimer–Tiemann reaction products are also formed. Carbene-insertion reactions on substrates other than phenols or electron-rich heteroaromatics have been treated extensively in other reviews.[43,86]

The first and only example of a phenol yielding a ring-expansion product has been uncovered by careful work of a group that was able to isolate a tropolone (in less than 1% yield).[84,85]

OH, CH_3O, R — aq. NaOH, $CHCl_3$, 60° → O, OH, Cl, R (1%) + other products

The Reimer–Tiemann reaction on pyrroles and indoles to furnish pyridines and quinolines is of limited utility; the yields are moderate, even when the reaction is catalyzed by a phase-transfer catalyst.[75]

COMPARISON WITH OTHER METHODS

Direct introduction of an aldehyde group into an aromatic nucleus is possible with the Gattermann,[87] Gattermann–Koch,[88] Vilsmeier,[89,90] Duff,[90] and Reimer–Tiemann reactions. All except the Reimer–Tiemann reaction are conducted under acidic and/or anhydrous conditions. Only the Gattermann and Duff reactions are applicable to phenols. Since in the Gattermann reaction the entering aldehyde group usually occupies the position *para* to the hydroxyl group and the Duff reaction fails with polyhydric phenols and phenols carrying electron-donating substituents, the Reimer–Tiemann reaction is occasionally the only method for the direct formylation of phenols. Thus 2-nitrophenol, pyrrole, and indole, which do not furnish aldehydes in a Gattermann reaction, are formylated successfully under Reimer–Tiemann reaction conditions. For example, the yields of 2-hydroxy-1-naphthalenecarboxaldehyde obtained when 2-hydroxynaphthalene is formylated using the Reimer–Tiemann, Gattermann, Vilsmeier, and Duff reactions are respectively 66, 45, 85, and 20%.[91] Clearly in terms of ease and safety of operations the Reimer–Tiemann reaction is the reaction of choice in this case.

EXPERIMENTAL CONDITIONS

Introduction

Even 105 years after its discovery, conditions for the Reimer–Tiemann reaction cannot be said to have been optimized. This is not too surprising for a reaction in which a quantitative yield has never been reported, and in which useful yields (of abnormal products) of 3–10% are not unusual. The discussion that follows concerns itself with the choice of the base, the solvent, and other variables, and must therefore be judged mainly from the point of view of complete literature coverage.

Effect of the Base

Some differences in the yields of *ortho*- and *para*-hydroxyaldehydes have been observed depending on the alkali hydroxide used. The *ortho–para* ratio of products obtained from phenol was determined in the presence of sodium, potassium, and cesium hydroxides as well as with triethylmethylammonium hydroxide.[92] Although the reported *ortho–para* ratio of 2 : 1 with 15 *N* sodium hydroxide appears at variance with an earlier ratio of 6 : 10,[51] the unmistakable trend toward increased *para* substitution with increasing size of the cation appears significant. The effect is attributed to a decrease in the coordination of the cation with the phenoxide ion as the cation increases in size. In the case of ring-expansion products it appears advantageous to use the potassium salts exclusively.

Effect of the Solvent

Normally the phenol is dissolved in 10–40% aqueous alkali, a large excess of chloroform is added, and the resulting two-phase system is stirred and/or refluxed for some time.

With the increase in understanding of the mechanism of the Reimer–Tiemann reaction and the role of the dichlorocarbene, more attention has been paid to the role of the solvent. Nevertheless, trivial factors such as the insolubility of the alkali salt of the phenol in the solvent or cosolvent may significantly change the yield.[93,94] No general pattern appears at present, although a patent claims that *para* substitution is favored by the addition of ethanol.[63]

Phase-Transfer Catalysis

High-yield dichlorocarbene reactions have recently been reported under phase-transfer conditions.[75,80] Attempts to increase the yield of normal Reimer–Tiemann reaction products using such phase-transfer conditions have been somewhat disappointing thus far.[95] Possibly the solubility of the phenolate ion and of the product in the aqueous phase is a contributing factor to the lack of immediate success. Nevertheless, high concentrations of carbene can be generated at room temperature when chloroform and aqueous alkali are treated with a quaternary ammonium salt.[75] Alkali concentration can be kept near 10% and, although the yields are not greatly improved, the reaction is somewhat cleaner.[71] Cyclodextrins influence the *ortho-para* ratio, giving high yield of *para* product.[183–185]

Other Reagents

Trichloroacetic acid,[96,97] chloral,[58] trichloronitromethane,[98] and other dichlorocarbene precursors yield Reimer–Tiemann reaction products. Recently phenolic aldehydes have been prepared by irradiation of a mixture of phenols, chloroform, and diethylamine in acetonitrile.[99] Trichloromethyl radicals were suggested as the active species. Whether the normal Reimer–Tiemann reaction is concurrently operative is not evident from the evidence presented.

Carbon Tetrachloride under Basic Conditions

Reimer and Tiemann showed hydroxy acids could be obtained when carbon tetrachloride was substituted for chloroform.[4,100–105] Although this is superfically reminiscent of the Kolbe reaction, fundamental differences must exist, since 2-nitrophenol, which is unreactive in the Kolbe carboxylation reaction,[106] is reported to furnish the two isomeric carboxylic acids when treated with carbon tetrachloride in alkali, with the *ortho* isomer predominating.[100]

By-products

By-products that are formed in yields not exceeding 3% each are the triphenylmethane-type dyes (**A**), their tautomers, and the orthoformic esters (**B**).[37,107–109] In the case of phenol itself "tars" amounting to 10% yield are probably mixtures of **A** and **B**.[94] Contrary to one report, acetals of the phenolic aldehydes have never been definitely identified as reaction products.[37,40]

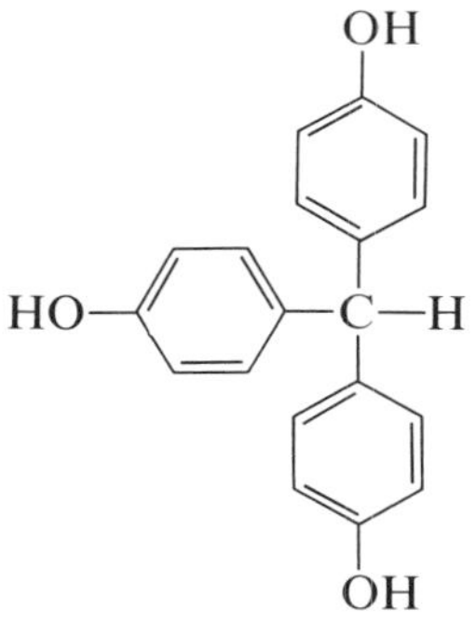

OH HO—C—H OH

A

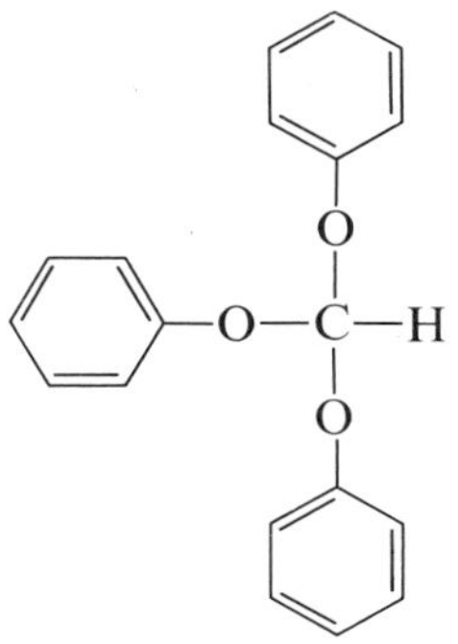

O—C—H O O

B

EXPERIMENTAL PROCEDURES

Procedures for the preparation of 2-hydroxy-1-naphthalenecarboxaldehyde in 38–40% yield can be found in *Organic Syntheses.*[93] The preparation of *p*-hydroxybenzaldehyde (8–10% yield) and salicylaldehyde (20% yield) are found in a standard text.[110]

2-Hydroxy-5-methoxybenzaldehyde.[111] In a 2-L, three-necked, round-bottomed flask, fitted with a mercury seal stirrer, a long reflux condenser, a separatory funnel, and a thermometer, were placed 125 g (1 mol) of *p*-methoxyphenol and a hot solution of 320 g of sodium hydroxide in 400 mL of water. It is convenient to add the sodium hydroxide solution to the phenol shortly after the alkali has been dissolved in the water. In this way the hot alkaline solution readily dissolves the phenol and excess carbonate formation is avoided. Chloroform (240 g, 160 mL, 2 mol) was added dropwise when the solution was brought to a temperature of 70°, and this temperature was maintained throughout the addition, which required 3–4 hours. When all the chloroform had been added, the reaction mixture, which had become dark brown and filled with a thick sludge, was stirred at 70–72° for another 15–20 minutes. The mixture was cooled, transferred to a 5-L round-bottomed flask with the aid of 400–500 mL of hot water, cooled again, and acidified to litmus by the careful addition of 150–200 mL of 10 *N* sulfuric acid. The 2-hydroxy-5-methoxybenzaldehyde was steam-distilled; from this mixture 5–6 L of distillate was collected. The aldehyde, a yellow oil, was extracted from the distillate with ether. The ether was dried over anhydrous sodium sulfate and the ether was removed by distillation. The residual light-brown oil (106–120 g, 70–79%) was distilled under diminished pressure in an atmosphere of nitrogen. From 166 g of the above product there was obtained 98 g of pure 2-hydroxy-5-methoxybenzaldehyde, bp 133°/15 mm.

Indole-3-carboxaldehyde.[112] Indole (20 g) was dissolved in a mixture of chloroform (150 mL) and 96% ethanol (400 mL) contained in a 2-L, three-necked flask fitted with a rubber stopper carrying a reflux condenser, a stirrer, and a dropping funnel. The mixture was kept gently boiling and stirred while a solution of potassium hydroxide (250 g) in water (300 mL) was gradually added over a period of 4–5 hours. The mixture was boiled for another 30 minutes after the last addition of potassium hydroxide. When the contents of the flask had cooled, the potassium chloride was collected on a Büchner funnel and washed with ethanol. The combined filtrate and washings were then steam-distilled, the receiver being changed when the chloroform and most of the alcohol had passed over. The distillation was continued for 30 minutes after the distillate was no longer turbid owing to the presence of 6-chloroquinoline. The hot aqueous liquid in the flask was decanted from the tarry residue and set aside to cool. The tarry material remaining was dissolved in the minimum quantity of hot ethanol, the alcoholic solution was poured into 1 L of hot water, and the whole solution was again boiled until the tarry globules had coalesced (an action that can be hastened by the addition of a little sodium chloride to the solution) and then filtered through a fluted filter paper using a heated funnel. The tar remaining on the filter was extracted once again in the same manner. Aldehyde separated from all three aqueous solutions. This was filtered and the combined filtrates amounting to some 2.5–3.0 L were concentrated to *ca.* 300 mL. This concentrated solution yielded a further crop of aldehyde on cooling. The total yield of aldehyde thus obtained amounted to 7.5 g (31 %). The crude material melted at 194°, and at 198° after recrystallization from water (lit. 194–196°).[113]

1-Dichloromethyl-1-methyl-2(1H)-naphthalenone.[17] A solution of 33.0 g of 1 methyl-2-naphthol and 66.0 g of sodium hydroxide in 660 mL of water was heated to 75°. Chloroform (132 g) was added over 3 hours. The resulting mixture was heated for an additional hour. The organic layer was separated from the aqueous layer, which was then extracted with 100 mL of chloroform. The combined chloroform solution was washed first with a dilute sodium hydroxide solution, then with water, and dried with magnesium sulfate. The chloroform was then removed by distillation. Distillation of the residual yellow oil under vacuum gave 1-dichloromethyl-1-methyl-2(1H)-naphthalenone (38.5 g, 77%), bp 131–134°/2.5 mm. Most of the material distilled at 131–132°/2.5 mm. On standing, the yellow distillate slowly crystallized. After one crystallization from ethanol–water (1:1) the product (34.5 g) melted at 64–65°.

3-Chloro-4-methylquinoline (phase-transfer catalysis).[75] A 33 % solution of sodium hydroxide (5 mL) was added to a vigorously stirred solution of 3-methylindole (1.0 g) and triethylbenzylammonium chloride (173 mg) in chloroform (10 mL) under ice-cooling. Stirring was continued under ice-cooling for 6 hours and then at room temperature for 24 hours. The aqueous layer was separated and extracted with chloroform. The combined organic layer was extracted with 20 % hydrochloric acid (3 × 30 mL). The aqueous layer was made alkaline with 10 % sodium hydroxide and extracted with chloroform. The extract was dried over sodium sulfate and concentrated to give crystals of 3-chloro-4-methylquinoline (720 mg, 53 %, mp 53.5–54.5° from *n*-hexane).

TABULAR SURVEY

The following tables summarize data in the literature through November 1981. Tables I–IX are compiled on the basis of the type of compound that is formylated. Tables X–XIII list compounds that give abnormal reaction products. All are tabulated according to increasing number of carbon atoms in the substrate.

Yield. The yield, listed in parentheses after the product, refers to product formed with the conditions cited; in most cases the highest value is reported. A dash indicates that the yield is not stated or is unavailable in the references cited.

References. The first reference cited refers to the conditions listed, which lead to the highest yield.

Reactions. Since most reactions have been carried out under closely similar conditions, no details are given in the tables. An exception is Table XIII, in which the source is recorded for the dihalocarbene. An asterisk indicates that the reaction in question is also listed in Table XIV, which gives those reactions that have been carried out under unusual conditions.

TABLE I. FORMYLATION OF PHENOL AND ALKYLPHENOLS

No. of C Atoms	Substrate	Product(s) and Yield(s) (%)	Refs.
C_6	Phenol	2-Hydroxybenzaldehyde (20), 4-hydroxybenzaldehyde (10)	110,40,70,97,109,114–124*
C_7	2-Methylphenol (*o*-cresol)	2-Hydroxy-3-methylbenzaldehyde (20), 4-hydroxy-3-methylbenzaldehyde (8)	125,40,49,51*
	3-Methylphenol (*m*-cresol)	2-Hydroxy-4-methylbenzaldehyde (—), 2-hydroxy-6-methylbenzaldehyde (20), 4-hydroxy-2-methylbenzaldehyde (8)	125,40,49,51,126,127
	4-Methylphenol (*p*-cresol)	2-Hydroxy-4-methylbenzaldehyde (25)	125,40*
C_8	2,4-Dimethylphenol	3,5-Dimethyl-2-hydroxybenzaldehyde (11)	126,9,122,128,129
	2,5-Dimethylphenol	3,6-Dimethyl-2-hydroxybenzaldehyde (—), 2,5-dimethyl-4-hydroxybenzaldehyde (—)	128,9
	3,4-Dimethylphenol	2,3-Dimethyl-6-hydroxybenzaldehyde (—), 3,4-dimethyl-6-hydroxybenzaldehyde (—)	126,8,11
	3,5-Dimethylphenol	4,6-Dimethyl-2-hydroxybenzaldehyde (—), 2,6-dimethyl-4-hydroxybenzaldehyde (—)	128,130
C_9	2,4,5-Trimethylphenol	2-Hydroxy-3,5,6-trimethylbenzaldehyde (5)	5,129,131
C_{10}	2-Methyl-5-*i*-propylphenol	2-Hydroxy-3-methyl-6-*i*-propylbenzaldehyde (—), 4-hydroxy-3-methyl-6-*i*-propylbenzaldehyde (—)	132–134
	5-Methyl-2-*i*-propylphenol	2-Hydroxy-6-methyl-3-*i*-propylbenzaldehyde (17), 4-hydroxy-6-methyl-3-*i*-propylbenzaldehyde (11)	70,135
	4-*i*-Butylphenol	5-*i*-Butyl-2-hydroxybenzaldehyde (—)	136
	4-*t*-Butylphenol	5-*t*-Butyl-2-hydroxybenzaldehyde (—)	122
C_{11}	4-*i*-Amylphenol	5-*i*-Amyl-2-hydroxybenzaldehyde (—)	137
C_{12}	4-Phenylphenol	2-Hydroxy-5-phenylbenzaldehyde (—)	138
C_{13}	4-Benzylphenol	2-Hydroxy-5-benzylbenzaldehyde (—)	138,139

TABLE II. FORMYLATION OF NAPHTHOLS AND ALKYLNAPHTHOLS

No. of C Atoms	Substrate	Product(s) and Yield(s) (%)	Refs.
C_{10}	1-Naphthol	1-Hydroxy-2-napththalenecarboxaldehyde (1), 4-hydroxy-2-naphthalenecarboxaldehyde (1)	44,45*
	2-Naphthol	2-Hydroxy-1-naphthalenecarboxaldehyde (38–48), 3-hydroxy-1-naphthalenecarboxaldehyde (1)	93,44,45, 140,141*
	5,6,7,8-Tetrahydro-1-naphthol	1-Hydroxy-5,6,7,8-tetrahydro-2-naphthalenecarboxaldehyde (2), 4-hydroxy-5,6,7,8-tetrahydro-1-naphthalenecarboxaldehyde (9)	19
	5,6,7,8-Tetrahydro-2-naphthol	2-Hydroxy-5,6,7,8-tetrahydro-1-naphthalenecarboxaldehyde (20–50), 3-hydroxy-5,6,7,8-tetrahydro-2-naphthalenecarboxaldehyde (—)	46,16
C_{11}	1-Methyl-5,6,7,8-tetrahydro-2-naphthol	3-Hydroxy-4-methyl-5,6,7,8- tetrahydro-2-naphthalenecarboxaldehyde (20)	138

TABLE III. FORMYLATION OF HALOPHENOLS

No. of C Atoms	Substrate	Product(s) and Yield(s) (%)	Refs.
C_6	2-Fluorophenol	3-Fluoro-2-hydroxybenzaldehyde (13), 3-fluoro-4-hydroxybenzaldehyde (15)	142,56
	3-Fluorophenol	4-Fluoro-2-hydroxybenzaldehyde (19), 2-fluoro-4-hydroxybenzaldehyde (20)	50
	4-Fluorophenol	5-Fluoro-2-hydroxybenzaldehyde (8)	142
	2-Chlorophenol	3-Chloro-2-hydroxybenzaldehyde (17), 3-chloro-4-hydroxybenzaldehyde (17)	40,45, 51*
	3-Chlorophenol	4-Chloro-2-hydroxybenzaldehyde (—), 2-chloro-4-hydroxybenzaldehyde (19), 6-chloro-2-hydroxybenzaldehyde (19)	40,47,45, 51,55
	4-Chlorophenol	5-Chloro-2-hydroxybenzaldehyde (41)	40,70*
	3,5-Dichlorophenol	2,6-Dichloro-4-hydroxybenzaldehyde (3)	57
	2-Bromophenol	3-Bromo-2-hydroxybenzaldehyde (13), 3-bromo-4-hydroxybenzaldehyde (3)	70
	3-Bromophenol	4-Bromo-2-hydroxybenzaldehyde (19), 2-bromo-4-hydroxybenzaldehyde (18), 6-bromo-2-hydroxybenzaldehyde (2)	48,49,51, 54,55
	4-Bromophenol	5-Bromo-2-hydroxybenzaldehyde (—)	122
	2-Iodophenol	3-Hydroxy-3-iodobenzaldehyde (—), 4-hydroxy-3-iodobenzaldehyde (—)	49
	3-Iodophenol	2-Hydroxy-4-iodobenzaldehyde (15), 4-hydroxy-2-iodobenzaldehyde (15), 2-hydroxy-6-iodobenzaldehyde (1)	48–52, 55
C_{10}	4-Chloro-2-naphthol	4-Chloro-2-hydroxy-1-naphthalenecarboxaldehyde (22)	143
	1-Bromo-5,6,7,8-tetrahydro-2-naphthol	4-Bromo-3-hydroxy-5,6,7,8-tetrahydro-2-naphthalenecarboxaldehyde (12)	138

TABLE IV. FORMYLATION OF HYDROXY- AND ALKOXYPHENOLS

No. of C Atoms	Substrate	Product(s) and Yield(s) (%)	Refs.
C_6	2-Hydroxyphenol (catechol)	2,3-Dihydroxybenzaldehyde (7), 3,4-dihydroxybenzaldehyde (11)	144
	3-Hydroxyphenol (resorcinol)	2,4-Dihydroxybenzaldehyde (6), 2,6-dihydroxybenzaldehyde (15), 2,4-dihydroxy-x-formylbenzaldehyde (—)	64,59,145
	4-Hydroxyphenol (hydroquinone)	2,5-Dihydroxybenzaldehyde (19)	145,61
C_7	2-Methoxyphenol (guaiacol)	4-Hydroxy-3-methoxybenzaldehyde (vanillin) (—), 2-hydroxy-3-methoxybenzaldehyde (low)	59,1,62,63
	3-Methoxyphenol	4-Hydroxy-2-methoxybenzaldehyde (25), 2-hydroxy-4-methoxybenzaldehyde (—), 4-hydroxy-x-formyl-2-methoxybenzaldehyde (—)	60
	4-Methoxyphenol	2-Hydroxy-5-methoxybenzaldehyde (50–65)	61,111,146
C_8	4-Methoxy-2-methylphenol	2-Hydroxy-5-methoxy-3-methylbenzaldehyde (—)	83
	2-Methoxy-4-methylphenol	2-Hydroxy-3-methoxy-4-methylbenzaldehyde (—)	85,84
	2,6-Dimethoxyphenol	3,5-Dimethoxy-4-hydroxybenzaldehyde (syringaldehyde) (—)	144
C_9	4-Ethyl-2-methoxyphenol	4-Ethyl-2-hydroxy-3-methoxybenzaldehyde (—)	85,84
C_{10}	2-Methoxy-4-*n*-propylphenol	2-Hydroxy-3-methoxy-4-*n*-propylbenzaldehyde (—)	85,84

TABLE V. FORMYLATION OF CARBOXY AND SULFONIC ACID SUBSTITUTED PHENOLS

No. of C Atoms	Substrate	Product(s) and Yield(s) (%)	Refs.
C_7	2-Hydroxybenzoic acid	5-Formyl-2-hydroxybenzoic acid (17), 3-formyl-2-hydroxybenzoic acid (—), 2-hydroxybenzaldehyde (2)	40,3,49,51,67,147
	3-Hydroxybenzoic acid	x-Formyl-3-hydroxybenzoic acid (7)	40
	4-Hydroxybenzoic acid	3-Formyl-4-hydroxybenzoic acid (15)	40,3,67
C_8	4-Hydroxyisophthalic acid	5-Formyl-4-hydroxyisophthalic acid (—)	65
	2-Hydroxyisophthalic acid	5-Formyl-2-hydroxyisophthalic acid (—)	65
	3-Hydroxy-4-methoxybenzoic acid	2-Formyl-3-hydroxy-4-methoxybenzoic acid (—), 6-formyl-3-hydroxy-4-methoxybenzoic acid (—)	66
C_{11}	2-Hydroxy-3-naphthoic acid	3-Hydroxy-4-formyl-2-naphthalenecarboxylic acid (—)	68,69

TABLE VI. FORMYLATION OF HETEROCYCLIC PHENOLS

No. of C Atoms	Substrate	Product(s) and Yield(s) (%)	Refs.
C_4	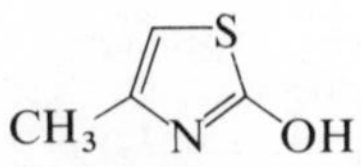 S, CH_3, N, OH (4-methylthiazol-2-ol structure)	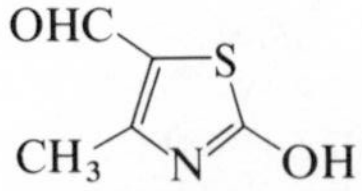 OHC, S, CH_3, N, OH (trace)	148
C_5	3-Pyridinol	3-Hydroxypyridine-2-carboxaldehyde (—)	149
C_9	4-Quinolinol	4-Hydroxyquinoline-3-carboxaldehyde (—)	150
	6-Quinolinol	6-Hydroxyquinoline-5-carboxaldehyde (—)	151
	7-Quinolinol	7-Hydroxyquinoline-8-carboxaldehyde (32)	152
	8-Quinolinol	8-Hydroxyquinoline-5-carboxaldehyde (30), 8-hydroxyquinoline-7-carboxaldehyde (10)	70,101,153
	2,4-Quinolinediol	2,4-Dihydroxyquinoline-3-carboxaldehyde (—)	154
C_{10}	2-Methyl-4-quinolinol	2-Methyl-4-hydroxyquinoline-3-carboxaldehyde (—)	155
	7-Methyl-8-quinolinol	7-Methyl-8-hydroxyquinoline-5-carboxaldehyde (6)	156
C_{13}	2,5,6,8-Tetramethyl-4-quinolinol	2,5,6,8-Tetramethyl-4-hydroxyquinoline-3-carboxaldehyde (—)	155
C_{14}	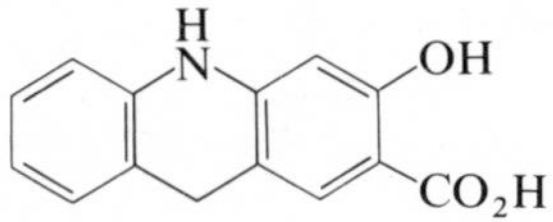 H, N, OH, CO_2H (acridan structure)	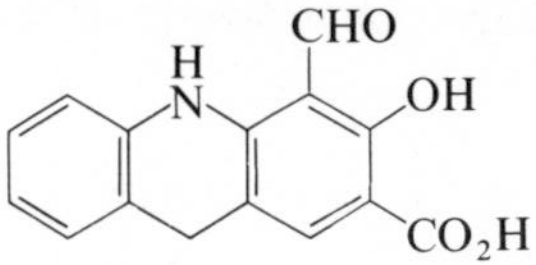 CHO, H, N, OH, CO_2H (—)	122

TABLE VII. FORMYLATION OF SUBSTITUTED PYRIMIDINES

$$\text{2-}R_2\text{-4-}R_4\text{-6-}R_6\text{-pyrimidine} \xrightarrow[\text{KOH, 80°, 1 hr}]{H_2O,\ C_2H_5OH,\ CHCl_3} \text{2-}R_2\text{-4-}R_4\text{-6-}R_6\text{-pyrimidine-5-CHO}$$

Formula	R_2	R_4	R_6	Yield(s) (%)	Refs.
$C_4H_4N_2O_2$	OH	OH	H	18	157
$C_4H_4N_2O_3$	OH	OH	OH	42	157
$C_5H_6N_2OS$	SH	OH	CH_3	17	157
$C_5H_6N_2O_2$	CH_3	OH	OH	29	157
	OH	OH	CH_3	14	157,158
$C_5H_7N_3O$	NH_2	OH	CH_3	38	158
$C_6H_8N_2O$	OH	CH_3	CH_3	26	157
	CH_3	OH	CH_3	13	157
$C_6H_8N_2OS$	SCH_3	OH	CH_3	14	157
$C_7H_{11}N_3O$	$N(CH_3)_2$	OH	CH_3	28	158
$C_9H_{13}N_3O_2$	piperidino (N-)	OH	OH	35	158
$C_{10}H_8N_2O_2$	C_6H_5	OH	OH	57[a]	158
$C_{10}H_{15}N_3O$	piperidino (N-)	OH	CH_3	18	158

[a] The yield of 57% is the yield of crude product.

TABLE VIII. FORMYLATION OF PYRROLES AND RELATED COMPOUNDS

No. of C Atoms	Substrate	Product(s) and Yield(s) (%)	Refs.
C_4	Pyrrole	Pyrrole-2-carboxaldehyde (31)	122,159,160
	2-Mercapto-4-methylglyoxaline	5-Formyl-2-mercapto-4-methylglyoxaline (—)	161
C_6	2,5-Dimethylpyrrole	2,5-Dimethylpyrrole-3-carboxaldehyde (—)	82
	2,4-Dimethylpyrrole	2,4-Dimethylpyrrole-3-carboxaldehyde (—)	162
C_8	Indole	Indole-3-carboxaldehyde (31)	112,163
C_9	2-Methylindole	2-Methylindole-3-carboxaldehyde (51)	112,164
	5-Methylindole	5-Methylindole-3-carboxaldehyde (27)	112
	7-Methylindole	7-Methylindole-3-carboxaldehyde (31)	112
	5-Methoxyindole	5-Methoxyindole-3-carboxaldehyde (29)	165,166
	6-Methoxyindole	6-Methoxyindole-3-carboxaldehyde (7)	167
	7-Methoxyindole	7-Methoxyindole-3-carboxaldehyde (18)	165,166
C_{14}	2-Phenylindole	2-Phenylindole-3-carboxaldehyde (52)	168

TABLE IX. FORMYLATION OF MISCELLANEOUS COMPOUNDS

No. of C Atoms	Substrate	Product(s) and Yield(s) (%)	Refs.
C_6	2-Nitrophenol	2-Hydroxy-3-nitrobenzaldehyde (9)	70
	3-Nitrophenol	2-Hydroxy-4-nitrobenzaldehyde (3)	40,70
	4-Nitrophenol	2-Hydroxy-5-nitrobenzaldehyde (8)	70
C_7	Tropolone	4-Formyltropolone (4)	169
C_8	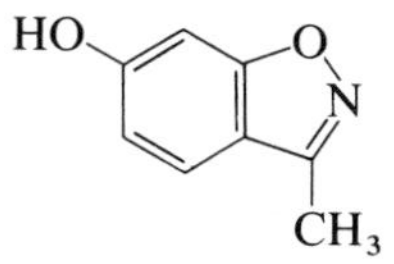 HO O N CH_3	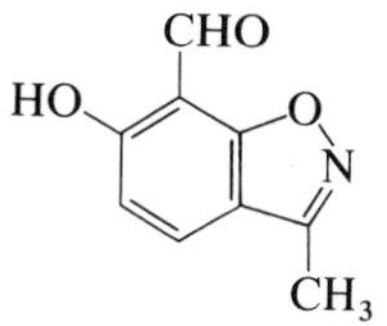 CHO HO O N CH_3 (40)	170
C_9	2-Hydroxycinnamic acid	3-Formyl-2-hydroxycinnamic acid (10)	70
	3-(4-Hydroxyphenyl)propionic acid	3-(3-Formyl-4-hydroxyphenyl)propionic acid (13)	19
	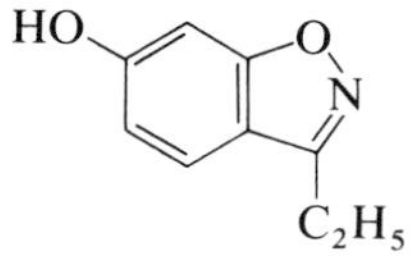 HO O N C_2H_5	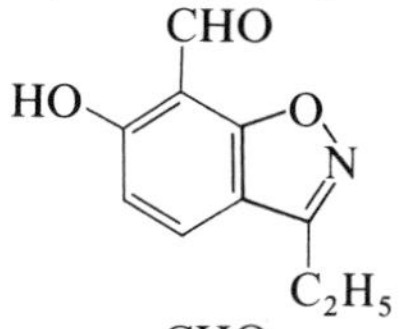 CHO HO O N C_2H_5 (30)	170
C_{10}	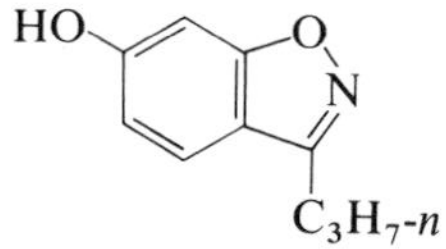 HO O N C_3H_7-*n*	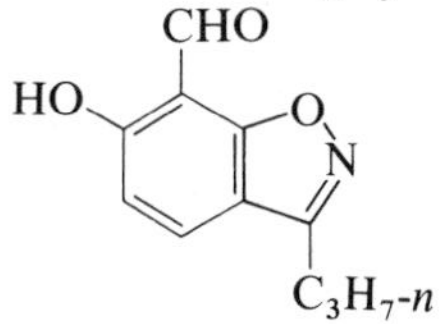 CHO HO O N C_3H_7-*n* (30)	170
C_{14}	2-Hydroxyanthraquinone	2-Hydroxyanthraquinone-1-carboxaldehyde (16)	70
C_{20}	Phenolphthalein	Phenolphthalein-3-carboxaldehyde (59)	73

TABLE X. CYCLOHEXADIENONES OBTAINED FROM SUBSTITUTED PHENOLS

No. of C Atoms	Substrate	Product(s) and Yield(s) (%)	Refs.
C_7	2-Methylphenol	2-Dichloromethyl-2-methyl-3,5-cyclohexadienone (8)	11
	4-Methylphenol	4-Dichloromethyl-4-methyl-2,5-cyclohexadienone (75)[a]	185,128,10,11,171
C_8	2,4-Dimethylphenol	2-Dichloromethyl-2,4-dimethyl-3,5-cyclohexadienone (—), 4-dichloromethyl-2,4-dimethyl-2,5-cyclohexadienone (30)	10,11
	2,5-Dimethylphenol	2-Dichloromethyl-2,5-dimethyl-3,5-cyclohexadienone (1)	11,172
	3,4-Dimethylphenol	4-Dichloromethyl-3,4-dimethyl-2,5-cyclohexadienone (28)	11
	2,6-Dimethylphenol	2-Dichloromethyl-2,6-dimethyl-3,5-cyclohexadienone (19)	83
	4-Methoxy-2-methylphenol	2-Dichloromethyl-4-methoxy-2-methyl-3,5-cyclohexadienone (10)	83
	2-Methoxy-4-methylphenol	4-Dichloromethyl-2-methoxy-4-methyl-2,5-cyclohexadienone (–)	85,84
C_9	2,4,5-Trimethylphenol	2-Dichloromethyl-2,4,5-trimethyl-3,5-cyclohexadienone (2), 4-dichloromethyl-2,4,5-trimethyl-2,5-cyclohexadienone (45)	129,5,6,126
	2,4,6-Trimethylphenol	2-Dichloromethyl-2,4,6-trimethyl-3,5-cyclohexadienone (59), 4-dichloromethyl-2,4,6-trimethyl-2,4-cyclohexadienone (17)	80,19
	3,4,5-Trimethylphenol	4-Dichloromethyl-3,4,5-trimethyl-2,5-cyclohexadienone (90)[a]	185
	4-Ethyl-2-methoxyphenol	4-Dichloromethyl-4-ethyl-2-methoxy-2,5-cyclohexadienone (—)	85,84
C_{10}	2-Methoxy-4-*n*-propylphenol	4-Dichloromethyl-2-methoxy-4-*n*-propyl-2,5-cyclohexadienone (—)	85,84
C_{12}	4-Phenylphenol	4-Dichloromethyl-4-phenyl-2,5-cyclohexadienone (85)[a]	185,85,84
C_{15}	2,6-Di-*t*-butyl-4-methylphenol	4-Dichloromethyl-2,6-di-*t*-butyl-4-methyl-2 5-cyclohexadienone (78)	80
C_{16}	OH, CH_3	O, $CHCl_2$, CH_3 (76)	80

[a] These results were obtained using cyclodextrin.

No. of C Atoms	Substrate	Product(s) and Yield(s) (%)	Refs.
C_{10}	5,6,7,8-Tetrahydro-1-naphthol	9-Dichloromethyl-5,6,7,8-tetrahydro-1(9H)-naphthalenone (3)	19,83
	5,6,7,8-Tetrahydro-2-naphthol	10-Dichloromethyl-5,6,7,8-tetrahydro-1(10H)-naphthalenone (15)	16
	1-Bromo-5,6,7,8-tetrahydro-2-naphthol	1-Bromo-4a-dichloromethyl-5,6,7,8-tetrahydro-2(4aH)-naphthalenone (5)	138
C_{11}	1-Methyl-2-naphthol	1-Dichloromethyl-1-methyl-2(1H)-naphthalenone (77)	17,172,173
	4-Methyl-1-naphthol	4-Dichloromethyl-4-methyl-1(4H)-naphthalenone (14)	176,20,173
	2-Methoxy-5,6,7,8-tetrahydro-1-naphthol	9-Dichloromethyl-2-methoxy-5,6,7,8-tetrahydro-1(9H)-naphthalenone (8)	86
	1-Methyl-5,6,7,8-tetrahydro-2-naphthol	10-Dichloromethyl-1-methyl-5,6,7,8-tetrahydro-2(10H)-naphthalenone (—), 1-dichloromethyl-1-methyl-5,6,7,8-tetrahydro-2(1H)-naphthalenone (15)	138,174
C_{12}	1-Ethyl-2-naphthol	1-Dichloromethyl-1-ethyl-2(1H)-naphthalenone (—)	172
	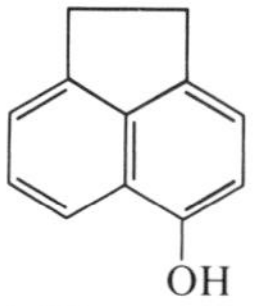 [structure: OH]	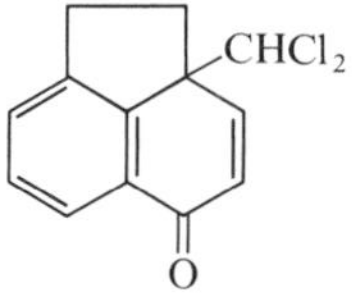 [structure: $CHCl_2$, O] (6)	176
C_{13}	1-Allyl-2-naphthol	1-Dichloromethyl-1-allyl-2(1H)-naphthalenone (47)	17,175
	1-(2-Cyanoethyl)-2-naphthol	1-Dichloromethyl-1-(2-cyanoethyl)-2(1H)-naphthalenone (41)	177
C_{14}	1-(3-Chloro-2-butenyl)-2-naphthol	1-Dichloromethyl-1-(3-chloro-2-butenyl)-2(1H)-naphthalenone (35)	17,175
	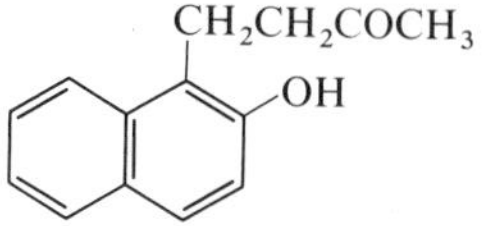 [structure: $CH_2CH_2COCH_3$, OH]	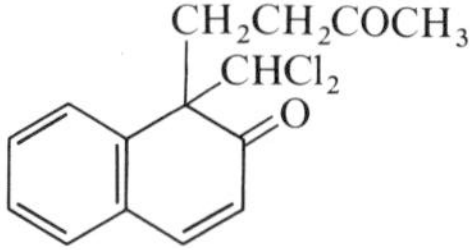 [structure: $CH_2CH_2COCH_3$, $CHCl_2$, O] (40)	177
C_{16}	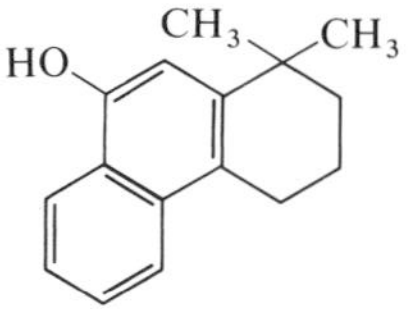 [structure: HO, CH_3, CH_3]	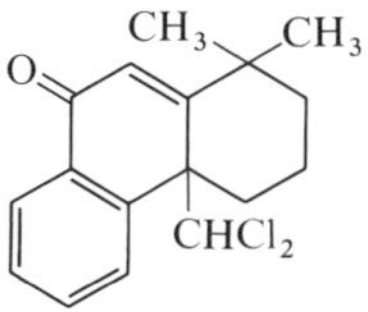 [structure: O, CH_3, CH_3, $CHCl_2$] (10)	18,173

TABLE XII. Abnormal Products Obtained from Miscellaneous Compounds

No. of C Atoms	Substrate	Product(s) and Yield(s) (%)	Ref.
C_6	2,5-Dimethylpyrrole	2-Dichloromethyl-2,5-dimethylpyrrolenine (—)	82
C_9	3-(4-Hydroxyphenyl)propionic acid	Ethyl 3-(1-dichloromethyl-2,5-cyclohexadienone-4)propionate (17)	19
C_{10}	2,3-Dimethylindole	3-Dichloromethyl-2,3-dimethylindolenine (—)	81
	CH_3 N OH	CH_3 $CHCl_2$ N O (—)	178
C_{13}	Tetrahydrocarbazole	11-Dichloromethyltetrahydrocarbazolenine (—)	81

TABLE XIII. RING-EXPANSION PRODUCTS OBTAINED FROM PYRROLES AND INDOLES

No. of C Atoms	Substrate	Carbene Precursor	Product(s) and Yield(s) (%)	Refs.
C_4	Pyrrole	$CHCl_3$	3-Chloropyridine (13)	37,14,15,81,179
		$CHBr_3$	3-Bromopyridine (9)	37,13,14,180
		$C_6H_5CHCl_2$	3-Phenylpyridine (1)	37,14
		CH_2Cl_2	Pyridine (0.5)	37
C_5	2-Methylpyrrole	$CHCl_3$	Chloromethylpyridines (—)	179
	3-Methylpyrrole	$CHCl_3$	Chloromethylpyridines (—)	179
C_6	2,3-Dimethylpyrrole	$CHBr_3$	Bromomethylpyridine (10)	181
	2,4-Dimethylpyrrole	$CHCl_3$	Chloromethylpyridine (—)	162
	2,5-Dimethylpyrrole	$CHCl_3$	3-Chloro-2,6-dimethylpyridine (—)	181
		$CHBr_3$	3-Bromo-2,6-dimethylpyridine (15)	181
C_8	Indole	$CHCl_3$	3-Chloroquinoline (9)	112,163
		$CHBr_3$	3-Bromoquinoline (10)	112
C_9	2-Methylindole	$CHCl_3$	3-Chloro-2-methylquinoline (45)	75,15,112,182
		$CHBr_3$	3-Bromo-2-methylquinoline (—)	182
	3-Methylindole	$CHCl_3$	3-Chloro-4-methylquinoline (53)	75,182
		$CHBr_3$	3-Bromo-4-methylquinoline (24)	75,182
	5-Methylindole	$CHCl_3$	3-Chloro-6-methylquinoline (—)	112
	7-Methylindole	$CHCl_3$	3-Chloro-8-methylquinoline (18)	112
	5-Methoxyindole	$CHCl_3$	3-Chloro-6-methoxyquinoline (21)	166,165
	6-Methoxyindole	$CHCl_3$	3-Chloro-7-methoxyquinoline (8)	167
	7-Methoxyindole	$CHCl_3$	3-Chloro-8-methoxyquinoline (5)	166,165
C_{10}	2,3-Dimethylindole	$CHCl_3$	3-Chloro-2,4-dimethylquinoline (55)	167
C_{14}	2-Phenylindole	$CHCl_3$	3-Chloro-2-phenylquinoline (68)	167,168
		$CHBr_3$	3-Bromo-2-phenylquinoline (37)	167

TABLE XIV. FORMYLATION UNDER UNUSUAL CONDITIONS

No. of C Atoms	Substrate	Reagents	Product(s) and Yield(s) (%)	Refs.
C_6		Chloral	[2-hydroxybenzaldehyde structure: OH, CHO] (23), [4-hydroxybenzaldehyde structure: OH, CHO] (10)	58
		Trichloroacetic acid	" (8), " (4)	97,96
		Trichloronitromethane	" (—), " (—)	98
		Irradiation, 2 hr, $(C_2H_5)_2NH$, CH_3OH, $CHCl_3$	" (22), " (31)	99
		β-Cyclodextrin, $CHCl_3$	" (—), " (—) (*o*/*p* ratio, 1:19)	183,184
	2-Chlorophenol	Chloral	3-Chloro-2-hydroxybenzaldehyde (17), 3-chloro-4-hydroxybenzaldehyde (17)	58
	4-Chlorophenol	Chloral	5-Chloro-2-hydroxybenzaldehyde (41)	58
C_7	2-Methylphenol	Chloral	2-Hydroxy-3-methylbenzaldehyde (10), 4-hydroxy-3-methylbenzaldehyde (7)	58
	4-Methylphenol	Chloral	2-Hydroxy-5-methylbenzaldehyde (5)	58
C_{10}	1-Naphthol	Chloral	1-Hydroxy-2-naphthalenecarboxaldehyde (3)	58
	2-Naphthol	Chloral	2-Hydroxy-1-naphthalenecarboxaldehyde (66)	58

REFERENCES

[1] K. Reimer, *Ber.*, **9**, 423 (1876).
[2] K. Reimer and F. Tiemann, *Ber.*, **9**, 824 (1876).
[3] K. Reimer and F. Tiemann, *Ber.*, **9**, 1268 (1876).
[4] K. Reimer and F. Tiemann, *Ber.*, **9**, 1285 (1876).
[5] K. Auwers, *Ber.*, **17**, 2976 (1884).
[6] K. Auwers, *Ber.*, **18**, 2655 (1885).
[7] K. Auwers, *Ber.*, **29**, 1109 (1896).
[8] K. Auwers, *Ber.*, **32**, 3598 (1899).
[9] K. Auwers and F. Winternitz, *Ber.*, **35**, 465 (1902).
[10] K. Auwers and G. Keil, *Ber.*, **35**, 4207 (1902).
[11] K. Auwers and G. Keil, *Ber.*, **36**, 1861 (1903).
[12] K. Auwers and G. Keil, *Ber.*, **36**, 3902 (1903).
[13] G. L. Ciamician and P. Silber, *Ber.*, **18**, 721 (1885).
[14] G. L. Ciamician and P. Silber, *Ber.*, **20**, 191 (1887).
[15] G. L. Ciamician, *Ber.*, **37**, 4201 (1904).
[16] R. B. Woodward, *J. Am. Chem. Soc.*, **62**, 1208 (1940).
[17] R. M. Dodson and W. P. Webb, *J. Am. Chem. Soc.*, **73**, 2767 (1951).
[18] M. S. Gibson, *Experientia*, **7**, 176 (1951).
[19] H. Wynberg and W. S. Johnson, *J. Org. Chem.*, **24**, 1424 (1959).
[20] E. Wenkert and T. E. Stevens, *J. Am. Chem. Soc.*, **78**, 5627 (1956).
[21] H. Wynberg, *Chem. Rev.*, **60**, 169 (1960).
[22] J. Hine, *J. Am. Chem. Soc.*, **72**, 2438 (1950).
[23] J. Hine and D. E. Lee, *J. Am. Chem. Soc.*, **73**, 22 (1951).
[24] J. Hine, E. L. Pollitzer, and H. Wagner, *J. Am. Chem. Soc.*, **75**, 5607 (1953).
[25] J. Hine and M. Dowell, *J. Am. Chem. Soc.*, **76**, 2688 (1954).
[26] J. Hine, *Physical Organic Chemistry*, McGraw-Hill, New York, 1956, p. 133, and 1962, p. 491.
[27] J. Hine and J. M. van der Veen, *J. Am. Chem. Soc.*, **81**, 6446 (1959).
[28] J. Hine and J. M. van der Veen, *J. Org. Chem.*, **26**, 1406 (1961).
[29] W. von E. Doering and A. K. Hoffmann, *J. Am. Chem. Soc.*, **76**, 6163 (1954).
[30] P. S. Skell and R. B. Woodward, *J. Am. Chem. Soc.*, **78**, 4496 (1956).
[31] P. S. Skell and S. R. Sandler, *J. Am. Chem. Soc.*, **80**, 2024 (1958).
[32] W. E. Parham and H. E. Reiff, *J. Am. Chem. Soc.*, **77**, 1177 (1955).
[33] W. E. Parham, H. E. Reiff, and P. Swartzentruber, *J. Am. Chem. Soc.*, **78**, 1437 (1956).
[34] W. E. Parham and R. R. Twelves, *J. Org. Chem.*, **22**, 730 (1957).
[35] W. E. Parham and C. D.Wright, *J. Org. Chem.*, **22**, 1473 (1957).
[36] E. A. Robinson, *J. Chem. Soc.*, **1961**, 1663.
[37] E. R. Alexander, A. B. Herrich, and J. M. Roder, *J. Am. Chem. Soc.*, **72**, 2760 (1950).
[38] I. Fells and E. A. Moelwyn-Hughes, *J. Chem. Soc.*, **1959**, 398.
[39] H. Wynberg, *J. Am. Chem. Soc.*, **76**, 4998 (1954).
[40] D. E. Armstrong and A. H. Richardson, *J. Chem. Soc.*, **1933**, 496.
[41] D. S. Kemp, *J. Org. Chem.*, **36**, 202 (1971).
[42] J. M. Al-Rawi, J. P. Bloxsidge, C. O'Brien, D. E. Caddy, J. A. Elvidge, J. R. Jones, and E. A. Evans, *J. Chem. Soc., Perkin Trans. II*, **1974**, 1635.
[43] W. E. Parham and E. E. Schweizer, *Org. Reactions*, **13**, 55 (1963).
[44] P. H. Deshpando, *Vikram*, **7**, 15 (1963) [*C.A.* **66**, 55265h (1967)].
[45] G. Kauffman, *Ber.*, **15**, 804 (1882).
[46] R. T. Arnold, H. E. Zaugg, and J. Sprung, *J. Am. Chem. Soc.*, **63**, 1314 (1941).
[47] H. H. Hodgson and T. A. Jenkinson, *J. Chem. Soc.*, **1927**, 1740.
[48] H. H. Hodgson and T. A. Jenkinson, *J. Chem. Soc.*, **1927**, 3041.
[49] H. H. Hodgson and T. A. Jenkinson, *J. Chem. Soc.*, **1929**, 469.
[50] H. H. Hodgson and J. Nixon, *J. Chem. Soc.*, **1929**, 1632.
[51] H. H. Hodgson and T. A. Jenkinson, *J. Chem. Soc.*, **1929**, 1639.
[52] S. Kobaysahi, S. Tagawa, and S. Nakajima, *Chem. Pharm. Bull.*, **11**, 123 (1963).
[53] S. Kobayashi, M. Azekawa, and H. Morita, *Chem. Pharm. Bull.*, **17**, 89 (1969).

[54] S. Kobayashi, M. Azekawa, and M. Taoka, *Chem. Pharm. Bull.*, **17**, 1279 (1969).
[55] K. A. Thakar and D. D. Goswami, *J. Indian Chem. Soc.*, **49**, 707 (1972).
[56] C. F. Albright, U.S. Pat. 3,972,945 [*C.A.*, **86**, 16421z (1977)].
[57] J. J. Baldwin, E. L. Engelhardt, R. Hirschmann, G. F. Lundell, G. S. Ponticello, C. T. Ludden, C. S. Sweet, A. Scriabine, N. N. Shane, and R. Hall, *J. Med. Chem.*, **22**, 687 (1979).
[58] C. Hamada, *Nippon Kagaku Zasshi*, **76**, 993 (1955) [*C.A.* **51**, 17859g (1957)].
[59] F. Tiemann and O. Koppe, *Ber.*, **14**, 2015 (1881).
[60] F. Tiemann and A. Parrisius, *Ber.*, **13**, 2354 (1880).
[61] F. Tiemann and W. H. M. Müller, *Ber.*, **14**, 1985 (1881).
[62] M. C. Traube, *Ber.*, **28R**, 524 (1895).
[63] M. C. Traube, Ger. Pat., 80,195 (1894); *Frdl.*, **4**, 1287 (1899).
[64] F. Tiemann and L. Lewy, *Ber.*, **10**, 2210 (1877).
[65] C. L. Reimer, *Ber.*, **11**, 793 (1878).
[66] F. Tiemann and B. Mendelsohn, *Ber.*, **9**, 1278 (1876).
[67] F. Tiemann and K. L. Reimer, *Ber.*, **10**, 1562 (1877).
[68] J. R. Geigy and Co., Ger. Pat., 98,466 (1898); *Frdl.*, **5**, 140 (1901).
[69] J. R. Geigy and Co., Ger. Pat., 97,934 (1898); *Frdl.*, **5**, 140 (1901).
[70] R. N. Sen and S. K. Ray, *J. Indian Chem. Soc.*, **9**, 173 (1931).
[71] E. W. Meijer, unpublished data.
[72] K. Matsumura and C. Sone, *J. Am. Chem. Soc.*, **53**, 1493 (1931).
[73] E. J. van Kampen, *Rec. Trav. Chim.*, **7**, 954 (1952).
[74] This technique has been used to demonstrate that estrone furnishes both isomeric *o*-hydroxyaldehydes (E. W. Meijer, unpublished data).
[75] S. Kwon, Y. Nishimura, and Y. Tamura, *Synthesis*, **1976**, 249.
[76] T. C. Chaudhuri, *J. Am. Chem. Soc.*, **64**, 315 (1942).
[77] W. S. Rapson, D. H. Saunders, and E. Stewart, *J. Chem. Soc.*, **1944**, 74.
[78] S. Gabriel, *Ber.* **10**, 185 (1877).
[79] I. E. Balaban, *J. Chem. Soc.*, **1931**, 885.
[80] T. Hiyama, Y. Ozaki, and H. Nozaki, *Tetrahedron*, **30**, 2661 (1974).
[81] G. Plancher and O. Carrasco, *Atti Accad. Nazl. Lincei*, **13**, I, 632 (1904) [*Chem. Zentr.*, **1904**, II, 341].
[82] G. Plancher and U. Ponti, *Atti Accad. Nazl. Lincei*, **18**, II, 469 (1909) [*C.A.* **4**, 2452 (1910)].
[83] H. Wynberg, Ph.D Dissertation, University of Wisconsin, 1952.
[84] R. S. McCredie, E. Ritchie, and W. C. Taylor, *Aust. J. Chem.*, **22**, 1011 (1969).
[85] E. R. Krajniak, E. Ritchie, and W. C. Taylor, *Aust. J. Chem.*, **26**, 1337 (1973).
[86] H. C. van der Plas, *Ring Transformations of Heterocycles*, Academic Press, London, 1973, Vol. 4, p. 210.
[87] W. E. Truce, *Org. Reactions*, **9**, 37 (1957).
[88] N. N. Crounse, *Org. Reactions*, **5**, 290 (1949).
[89] L. F. Fieser, J. L. Hartwell, J. E. Jones, J. H. Wood, and R. W. Bost, *Org. Syntheses*, Coll. Vol. **III**, 98 (1955).
[90] L. N. Ferguson, *Chem. Rev.*, **38**, 229 (1946).
[91] R. B. Wagner and H. D. Zook, *Synthetic Organic Chemistry*, Wiley, London, 1953, p. 307.
[92] O. L. Brady and J. Jakobovits, *J. Chem. Soc.*, **1950**, 767.
[93] A. Russell and L. B. Lockhart, *Org. Syntheses*, Coll. Vol. **III**, 463 (1955).
[94] D. F. Pontz, U.S. Pat. 3,365,500 (1968) [*C.A.*, **69**, 27041a (1968)].
[95] Y. Sasson and M. Yonovich, *Tetrahedron Lett.*, **1979**, 3753.
[96] G. R. de Almeida, *Anais Fac. Farm. Univ. Recife*, **1**, 55 (1958) [*C.A.*, **54**, 22464h (1960)].
[97] J. van Alphen, *Rec. Trav. Chim.*, **46**, 144 (1927).
[98] S. Berlingozzi and P. Badolotto, *Atti Accad, Nazl. Lincei*, **33**, I, 290 (1924) [*C.A.*, **18**, 3365 (1924)].
[99] K. Hirao, M. Ikegame, and O. Yonemitsu, *Tetrahedron*, **30**, 2301 (1974).
[100] G. Hasse, *Ber.*, **10**, 2185 (1877).
[101] E. Lippman and F. Fleisher, *Ber.*, **19**, 2467 (1886).
[102] M. S. Newman and L. L. Wood, *J. Org. Chem.*, **23**, 1236 (1958).
[103] E. J. Villani and J. Lang, *J. Am. Chem. Soc.*, **72**, 2301 (1950).
[104] M. Shimizu and M. Maki, *J. Pharm. Soc. Jpn.*, **71**, 961 (1951) [*C.A.*, **46**, 8058c (1952)].
[105] J. A. Tykal, Ph.D. Dissertation, Southern Illinois University, 1972; *Diss. Abstr. Int. B.*, **33**, 4206 (1973).
[106] F. Wessely, *Monatsh. Chem.*, **81**, 1071 (1950).

[107] H. Baines and J. E. Driver, *J. Chem. Soc.*, **123**, 1214 (1923).
[108] H. Baines and J. E. Driver, *J. Chem. Soc.*, **125**, 907 (1924).
[109] P. Bamfield and I. Cheetham, Brit. Pat., 1,490,350 [*C.A.*, **88**, 192757p (1978)].
[110] A. I. Vogel, *Textbook of Practical Organic Chemistry*, Longman, London, 1978, pp. 757 and 761.
[111] M. B. Gillespie, *Biochem. Prep.*, **3**, 79 (1973).
[112] W. J. Boyd and W. Robson, *Biochem. J.*, **29**, 555 (1935).
[113] A. C. Shabica, E. E. Howe, J. B. Ziegler, and M. Tishler, *J. Am. Chem. Soc.*, **68**, 1156 (1946).
[114] V. Migrdichian, *Organic Syntheses*, Chapman & Hall, London, 1957, p. 1336.
[115] O. Fernandes, *Anales Soc. Espan. Fis. Quim.*, **26**, 33 (1928) [*C.A.*, **22**, 1764 (1928)].
[116] L. Gattermann, *Die Praxis des Organische Chemikers*, 41st ed., Walter de Gruyter & Co., Berlin, 1962, p. 206.
[117] H. H. Hodgson, *J. Soc. Dyers Colourist*, **45**, 259 (1929) [*C.A.*, **24**, 1294 (1930)].
[118] K. Hatano and M. Matsui, *Agric. Biol. Chem.*, **37**, 2917 (1973) [*C.A.*, **80**, 70487q (1974)].
[119] H. H. Hodgson, *J. Soc. Dyers Colourist*, **46**, 39 (1930) [*C.A..*, **24**, 2952 (1930)].
[120] J. M. Montanes del Olmo and E. Ravina, *An. Quim.*, **1969**, 613 [*C.A.*, **71**, 101452v (1969)].
[121] M. Verzele and H. Sion, *Bull. Soc. Chim. Belges*, **65**, 627 (1956).
[122] E. Iwata, *J. Chem. Soc. Jpn., Ind. Chem. Sect.* **55**, 258 (1952) [*C.A.*, **48**, 7577b (1954)].
[123] P. H. Despando, *Vikram*, **7**, 11 (1963) [*C.A.*, **66**, 55174e (1967)].
[124] Houben-Weyl, *Methoden der Organische Chemie*, 4th ed., Sauerstoff Volume II, Part I, **1952**, 36.
[125] F. Tiemann and C. Schotten, *Ber.*, **11**, 767 (1878).
[126] E. Ravina Rubira, J. M. Montanes del Olmo, and M. T. Cobreros, *An. Quim.*, **1973**, 657 [*C.A.*, **79**, 78291w (1973)].
[127] Ph. Chuit and F. Bolsing, *Bull. Soc. Chim. Fr.*, **35**, 129 (1906).
[128] O. Anselmino, *Ber.*, **35**, 4108 (1902).
[129] E. Ravina and J. M. Montanes, *Chim. Ther.*, **8**, 290 (1973) [*C.A.*, **79**, 115541a (1973)].
[130] K. Auwers and E. Borche, *Ber.*, **48**, 1698 (1915).
[131] E. Ravina, J. M. Montanes, M. T. Cobreros, and F. Tato, *An. Quim.*, **1969**, 709 [*C.A.*, **72**, 54851h (1970)].
[132] F. Bell and T. A. Henry, *J. Chem. Soc.*, **1928**, 2215.
[133] S. Lustig, *Ber.*, **19**, 11 (1886).
[134] E. Nordmann, *Ber.*, **17**, 2632 (1884).
[135] M. Kobek, *Ber.*, **16**, 2096 (1883).
[136] F. B. Daines and I. R. Rothroch, *J. Am. Chem. Soc.*, **16**, 634 (1894).
[137] T. Koyama and T. Sadanga, *J. Pharm. Soc. Jpn.*, **69**, 538 (1949) [*C.A.*, **44**, 3916e (1950)].
[138] T. Haruhawa and H. Ishihawa, *J. Pharm. Soc. Jpn.*, **70**, 338 (1950).
[139] T. Koyama and T. Asou, *J. Pharm. Soc. Jpn.*, **71**, 31 (1951) [*C.A.*, **45**, 7044d (1951)].
[140] M. R. Fosse, *Bull. Soc. Chim. Fr.*, **25**, 371 (1901).
[141] M. G. Rousseau, *C.R.*, **44**, 133 (1882).
[142] L. N. Ferguson, J. C. Reid, and M. Calvin, *J. Am. Chem. Soc.*, **68**, 2502 (1946).
[143] H. Burton, *J. Chem. Soc.*, **1945**, 280.
[144] C. Graebe and E. Mantz, *Ber.*, **36**, 1031 (1903).
[145] P. H. Despando, *Vikram*, **7**, 19 (1963) [*C.A.*, **66**, 55175d (1967)].
[146] H. Günther, J. Prestien, and P. Joseph-Nathan, *Org. Magn. Res.*, **7**, 339 (1975).
[147] A. Fürth, *Ber.*, **16**, 2180 (1883).
[148] E. Ochiai and H. Nagasawa, *Ber.*, **72**, 1470 (1939).
[149] S. Ginsburg and I. B. Wilson, *J. Am. Chem. Soc.*, **79**, 481 (1957).
[150] B. Bobrański, *Ber.*, **69**, 1113 (1936).
[151] B. Bobrański, *J. Prakt. Chem.*, **134**, [2], 141 (1932).
[152] L. Kochańska and B. Bobrański, *Ber.*, **69**, 1807 (1936).
[153] C. Hamada, Y. Hirano, and T. Iida, *Nippon Kagaku Zasshi*, **77**, 1107 (1956) [*C.A.*, **53**, 5267f (1959)].
[154] Y. Asahina and F. Fuzikawa, *Ber.*, **65**, 61 (1932).
[155] M. Conrad and L. Limpach, *Ber.*, **21**, 1965 (1881).
[156] C. Hamada, K. Isogai, and Y. Nakajima, *Nippon Kagaku Zasshi*, **82**, 1284 (1961) [*C.A.*, **57**, 11163a (1962)].
[157] R. H. Wiley and Y. Yamamoto, *J. Org. Chem.*, **25**, 1906 (1960).
[158] R. Hull, *J. Chem. Soc.*, **1957**, 4845.
[159] E. Bamberger and G. Djierdjian, *Ber.*, **32**, 536 (1900).

[160] H. Fisher, H. Bellen, and A. Stern, *Ber.*, **61**, 1078 (1928).
[161] H. Health, A. Lawson, and C. Rimington, *J. Chem. Soc.*, **1951**, 2223.
[162] O. Pipoty, W. Krannich, and H. Will, *Ber.*, **47**, 2531 (1914).
[163] A. Ellinger, *Ber.*, **39**, 2515 (1906).
[164] A. Ellinger, *H-Z. Physiol. Chem.*, **91**, 49 (1914).
[165] K. G. Blaikie and W. H. Perkin, *J. Chem. Soc.*, **125**, **296** (1924).
[166] R. M, Marchant and P. G. Harvey, *J. Chem. Soc.*, **1951**, 1808.
[167] W. O. Kermach, W. H. Perkin, and R. Robinson, *J. Chem. Soc.*, **121**, 1882 (1922).
[168] R. G. Blume and H. G. Lindwall, *J. Org. Chem.* **10**, 255 (1945).
[169] J. W. Cook, R. A. Raphael, and A. I. Scott, *J. Chem. Soc.*, **1952**, 4416.
[170] K. A. Thakar and B. M. Bhawal, *Ind. J. Chem.*, **15B**, 1056 (1977).
[171] J. Schreiber, M. Pesaro, W. Leimgruber, and A. Eschenmoser, *Helv. Chim. Acta*, **41**, 2103 (1958).
[172] F. Bell and H. W. Hunter, *J. Chem. Soc.*, **1950**, 2903.
[173] M. S. Gibson, *J. Chem. Soc.*, **1961**, 2251.
[174] C. Ukita, S. Nojima, and K. Nagasawa, *J. Pharm. Soc. Jpn.*, **72**, 1327 (1952) [*C.A.*, **47**, 7470i (1953)].
[175] W. P. Webb, Ph.D. Dissertation, University of Minnesota, 1951; *Diss. Abstr. Int. B.*, **12**, 820 (1952).
[176] R. C. Fuson and T. G. Miller, *J. Org. Chem.*, **17**, 316 (1952).
[177] C. G. Krespan, Ph.D. Dissertation, University of Minnesota (1952); *Diss. Abstr. Int. B.* **12**, 812 (1952).
[178] C. Hamada, K. Isozaki, and M. Kaizu, *Nippon Kagaku Zasshi*, **79**, 1446 (1958) [*C.A.*, **54**, 5644g (1960)].
[179] G. L. Ciamician and M. Dennstedt, *Ber.*, **14**, 1153 (1881).
[180] G. L. Ciamician and M. Dennstedt, *Ber.*, **15**, 1172 (1882).
[181] O. Bocchi, *Gazz. Chim. Ital.*, **30**, 8994 (1900).
[182] G. Magnanini, *Ber.*, **20**, 2608 (1887).
[183] M. Ohara and J. Fukuda, *Pharmazie*, **33**, 467 (1978) [*C.A.*, **89**, 146548u (1978)].
[184] M. Komiyama and H. Hirai, *Bull. Chem. Soc. Jpn.*, **54**, 2053 (1981).
[185] M. Komiyama and H. Hirai, *Makromol. Chem., Rapid Commun.*, **2**, 177 (1981).

CHAPTER 2

THE FRIEDLÄNDER SYNTHESIS OF QUINOLINES

CHIA-CHUNG CHENG AND SHOU-JEN YAN

Mid-America Cancer Center Program,
University of Kansas Medical Center,
Kansas City, Kansas

CONTENTS

ACKNOWLEDGMENTS

The authors thank Mrs. Caroline F. Sidor of E. I. du Pont de Nemours and Company and Miss Kathy Delker of Oak Ridge National Laboratory for their assistance in surveying the literature; Professor Edward A. Fehnel, Professor Donald E. Pearson, and Dr. Eugene G. Podrebarac for examining the manuscript; and Ms. Lanora Moore for typing this review. They are especially indebted to Dr. Robert M. Joyce for careful editorial revision of the final manuscript.

INTRODUCTION

Friedländer prepared quinoline in 1882 by the condensation of *o*-aminobenzaldehyde with acetaldehyde in the presence of sodium hydroxide.[1] This type of reaction has since been extensively explored and, in its most general form, can be defined as an acid- or base-catalyzed condensation followed by a cyclodehydration between an *o*-amino-substituted aromatic aldehyde, ketone, or derivative thereof with an appropriately substituted aldehyde, ketone, or other carbonyl compound containing a reactive α-methylene group (Eq. 1). Some quinolines can also be prepared simply by heating a mixture of the reactants with or without a solvent.

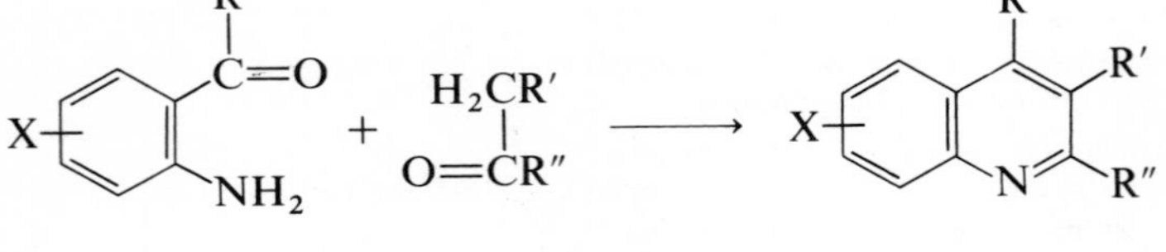

(Eq. 1)

Because both reactants contain a carbonyl group, certain quinolines can be prepared from a single starting compound. Of historic interest is the formation of the dye flavaniline (**1**) from acetanilide.[2] When acetanilide is heated with zinc chloride, the acetyl group of some molecules migrates to the *ortho* position, and that of others migrates to the *para* position. The resulting mixture of *o*-acetylaniline and *p*-acetylaniline then undergoes Friedländer condensation to form flavaniline (Eq. 2).[3]

$$2\,C_6H_5NHCOCH_3 \longrightarrow o\text{-}H_2NC_6H_4COCH_3 + p\text{-}H_2NC_6H_4COCH_3$$

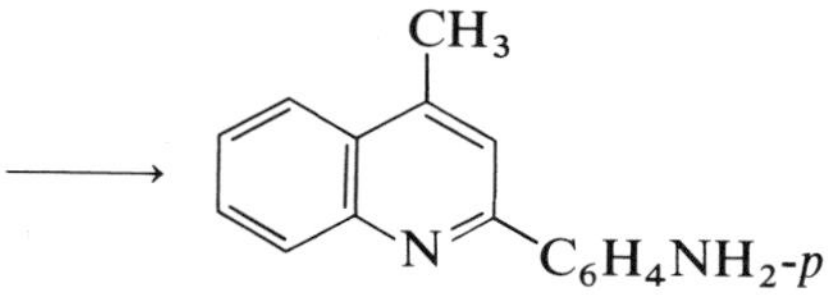

$$\longrightarrow \quad \text{(4-methyl-2-}(C_6H_4NH_2\text{-}p)\text{quinoline)} \quad \textbf{1} \qquad \text{(Eq. 2)}$$

Two related quinoline syntheses, the Pfitzinger reaction[4] and the Niementowski reaction,[5] can be considered as extensions of the Friedländer synthesis. The Pfitzinger reaction uses an isatic acid or isatin (Eq. 3) and the Niementowski reaction uses an anthranilic acid (Eq. 4) for condensation to form a 4-quinolinecarboxylic acid and a 4-hydroxyquinoline, respectively.

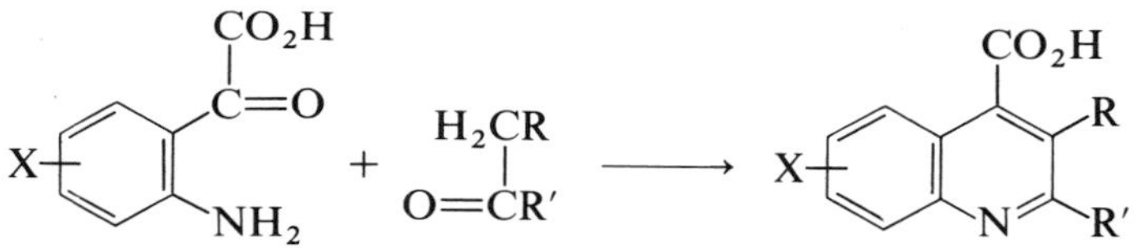

$$X\text{-}C_6H_3(COCO_2H)(NH_2) + H_2CR{-}C(=O)R' \longrightarrow X\text{-quinoline}(4\text{-}CO_2H, 3\text{-}R, 2\text{-}R') \qquad \text{(Eq. 3)}$$

$$X\text{-}C_6H_3(CO_2H)(NH_2) + H_2CR{-}C(=O)R' \longrightarrow X\text{-quinoline}(4\text{-}OH, 3\text{-}R, 2\text{-}R') \qquad \text{(Eq. 4)}$$

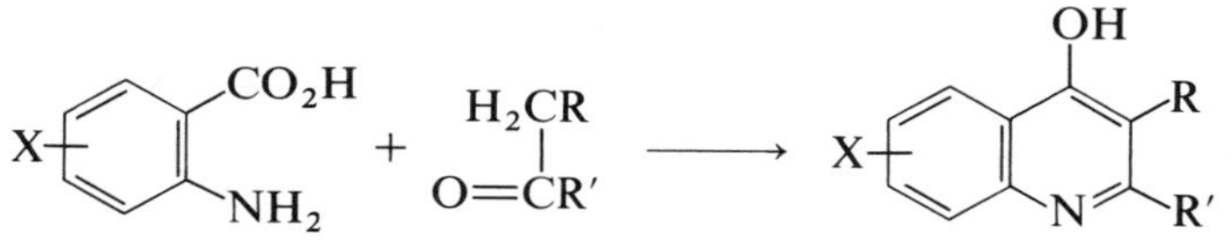

Several other related reactions, such as the Combes quinoline synthesis,[6] the Doebner–Miller quinoline synthesis,[7] and the Skraup quinoline synthesis,[8,9] use different types of starting compounds. It is of interest that all of the six aforementioned syntheses of quinolines were discovered within a period of 14 years in the latter part of the nineteenth century, undoubtedly reflecting many scientific interests and commercial activities at that time, including the isolation of quinolines from coal tar, the manufacture of dyes, and studies of natural products such as alkaloids.

MECHANISM

Despite the fact that the Friedländer synthesis of quinolines has been known for one hundred years, the mechanism or mechanisms of the reaction are still not completely understood. Two different reaction paths, each involving different uncyclized intermediates, have been postulated. These are illustrated by considering the original Friedländer synthesis of quinoline from *o*-aminobenzaldehyde and acetaldehyde. The two possible initial reactions are (*a*) Schiff base formation or (*b*) Claisen condensation to an α,β-unsaturated carbonyl compound;[10–14] Friedländer himself proposed that *o*-amino-cinnamaldehyde was the intermediate in this reaction. In either case the second reaction is cyclodehydration to quinoline.

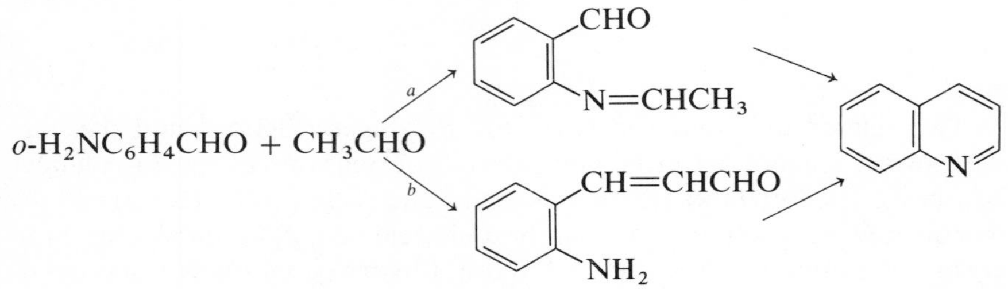

o-$H_2NC_6H_4CHO$ + CH_3CHO —*a*→ [o-(N=CHCH$_3$)C$_6$H$_4$CHO] → quinoline (N)

o-$H_2NC_6H_4CHO$ + CH_3CHO —*b*→ [o-(CH=CHCHO)C$_6$H$_4$NH$_2$] → quinoline (N)

The reaction between *o*-aminoacetophenone and bicyclo[2.2.2]octanone to yield the normal Friedländer product **2** (a condensed quinoline) and a side product **3** suggests the involvement of an intermediate Schiff base (Eq. 5).[15] A reported two-step Friedländer reaction that proceeds by way of Schiff base **4** also supports this postulate (Eq. 6).[16] Nevertheless, the alternative sequence of events has also been suggested on the basis of the two-step Friedländer reaction in which the aldol intermediate **5** was isolated (Eq. 7).[17] However, the reaction conditions for the isolation of intermediates **4** and **5** and for the subsequent cyclization reactions are not those usually employed in the one-step Friedländer syntheses, and these observations do not by themselves constitute rigorous proof for either of the proposed mechanisms in the more typical reactions.

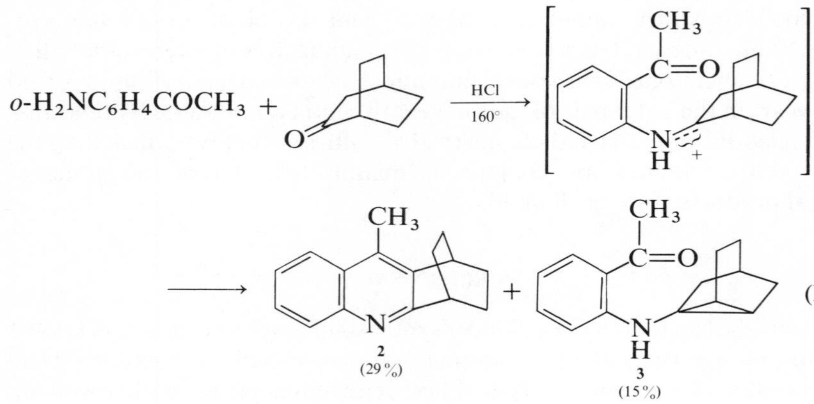

o-$H_2NC_6H_4COCH_3$ + bicyclo[2.2.2]octanone (O) —HCl, 160°→ [iminium intermediate: CH$_3$, C=O, N, H, +]

→ **2** (29%) (CH$_3$, N) + **3** (15%) (CH$_3$, C=O, N, H) (Eq. 5)

o-$H_2NC_6H_4COR$ + [3-oxo-γ-butyrolactone] $\xrightarrow{100°}$ **4**

R = CH_3, C_6H_5

$\xrightarrow[\text{Room temp}]{\text{Conc } H_2SO_4}$ (>70%) (Eq. 6)

[2-aminonicotinaldehyde] + [1,3-cyclohexanedione] $\xrightarrow[\text{EtOH, room temp}]{\text{Piperidine}}$ **5** (59%)

$\xrightarrow[\text{Heat}]{C_6H_5CH_3}$ (94%) (Eq. 7)

In addition, in some reactions the formation of a Schiff-base intermediate and a Claisen-condensation intermediate may be competitive.[18] In Friedländer syntheses of 2-hydroxyquinolines (carbostyrils) from β-ketoesters, another alternative reaction path has been suggested that involves an intermediate anilide of the type o-$RCOC_6H_4NHCOCH_2COR'$.[16]

It is likely, therefore, that the mechanism of the Friedländer synthesis may vary, depending on the nature of reactants, catalysts, solvents, and reaction conditions such as temperature and pH. The possibility that either or both of the previously postulated intermediates can be formed in reversible side reactions and are not directly involved as intermediates in the usual one-step Friedländer synthesis has not been ruled out.

SCOPE AND LIMITATIONS

Because the o-aminocarbonyl reactants may carry a variety of substituents (X, R), and because a wide choice of groups (R′, R″) may be selected for the active methylene reactants (Eq. 1), the Friedländer synthesis is useful for the preparation of quinolines with a wide range of substituents at positions 2–8. The substituent X may be hydrogen, alkyl, alkoxy, halogen, or nitro; R may be hydrogen, alkyl, aryl, dimethoxymethyl, hydroxy (Niementowski), or carboxy

(Pfitzinger); R′ may be hydrogen, alkyl, aryl, nitro, acyl, carboxy, carbalkoxy, carboxamido, cyano, hydroxy, sulfonyl, or related functions; and R″ may be hydrogen, alkyl, aryl, alkoxy, acetoxy, hydroxy, or amino.

Catalysts

Friedländer syntheses can be catalyzed by bases or acids, or may take place without a catalyst. In some instances identical reactants may yield different products in the presence or absence of a catalyst[19] or when catalyzed by an acid or a base.[20] Classical Friedländer syntheses are generally carried out either in the presence of basic catalysts or simply by heating a mixture of reactants in the absence of catalysts and solvents. The most frequently used basic catalysts are sodium hydroxide, potassium hydroxide, and piperidine.[21] Others include sodium alkoxides, alkali carbonates, pyridine, and ion-exchange resins. Most *o*-aminobenzaldehydes and some *o*-aminoacetophenones condense readily with active methylene compounds in the presence of basic catalysts. However, *o*-aminobenzophenone fails to undergo base-catalyzed Friedländer condensation with many typical active methylene compounds such as acetaldehyde,[22] cyclohexanone,[23] deoxybenzoin,[23] and *β*-ketoesters.[19,20] This problem can often be overcome by the use of acidic catalysts.

Uncatalyzed Friedländer syntheses usually require more drastic reaction conditions, with temperatures in the range 150–220°.[23] At lower temperatures many condensed intermediates fail to undergo the final cyclodehydration to form the quinoline ring.[24] Nevertheless, the uncatalyzed Friedländer condensations often give higher yields of quinolines than those catalyzed by base when the active methylene reactant is an aldehyde. For example, condensation of *o*-aminobenzaldehyde and propionaldehyde with potassium hydroxide as catalyst gives a modest yield of 3-methylquinoline;[25] with the same reactants at 220° without a catalyst, 80–85% yields of the same product can be obtained.[26,27] *β*-Ketoesters can be induced to react under these conditions, but the products are 3-acylcarbostyrils (2-hydroxyquinolines) rather than the 3-carbalkoxyquinolines that might have been expected.[16,20]

Acids have been widely employed as catalysts in Friedländer syntheses in the last 35 years. Hydrochloric acid, sulfuric acid, *p*-toluenesulfonic acid, and polyphosphoric acid are the usual agents.[20,28–30a,b] In the few reactions in which acetic acid has been used alone, it probably functioned simply as a convenient solvent and not as a catalyst. A modification by Kempter and co-workers, which involves the use of hydrochlorides of *o*-aminobenzaldehydes, *o*-aminoacetophenones, or related *o*-aminoaryl ketones, is also satisfactory.[31–34] For example, the hydrochloride of *o*-aminobenzophenone reacts smoothly with the furan ketone **6** at 140° to give a 63% yield of the expected quinoline (Eq. 8); the free base gives only a 16% yield at the same

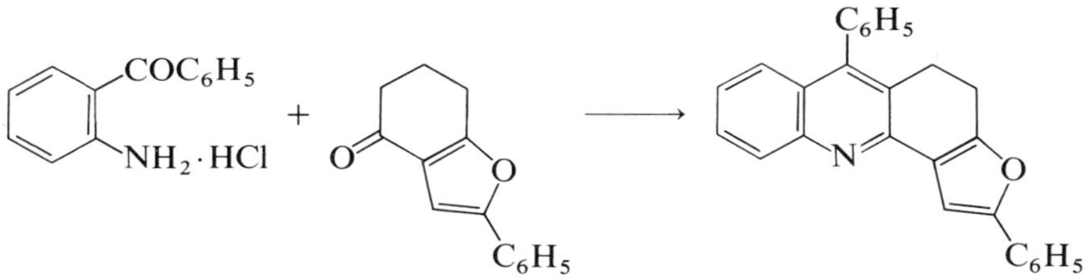

6

(Eq. 8)

temperature. When the reaction is conducted in ethanolic potassium hydroxide, no quinoline can be isolated.[33] The Kempter modification, however, is not suitable for the synthesis of quinolines from sterically hindered reactants.[15,35]

Another useful procedure, developed by Fehnel, in which the condensation is carried out in acetic acid under reflux in the presence of a small amount of sulfuric acid, often provides the expected product in good yields.[19,20,36] Under these conditions *β*-ketoesters react "normally" with *o*-aminoaryl ketones to give 3-carbalkoxyquinolines instead of the carbostyrils that are obtained in uncatalyzed reactions at high temperature.[19] In many instances acid-catalyzed Friedländer syntheses have been found to be more effective than those catalyzed by bases, especially when one of the reactants is an *o*-aminoaryl ketone.[19,20,33,34,37] However, the converse is observed in the reaction of *o*-aminobenzaldehyde with cyclobutanone.[37a]

The *o*-Aminocarbonyl Component

The majority of compounds used as the *o*-aminocarbonyl component are *o*-aminobenzaldehydes, *o*-aminoacetophenones, and *o*-aminobenzophenones. Benzene rings that contain two sets of *o*-aminocarbonyl functions, such as 4,6-diaminoisophthalaldehyde[38,39] and 2,5-diaminoterephthalaldehyde,[38] can also undergo the normal condensation without complication. Several types of polymers containing quinoline rings have been prepared by condensations of compounds that have two *o*-aminobenzophenone functions with bifunctional active methylene compounds.[39–39f] In extensions of the Friedländer synthesis the benzenoid ring may be replaced by polynuclear aromatic,[33] nonbenzenoid aromatic,[40,41] or heteroaromatic rings.[42–44] The aromatic ring may even be eliminated completely, as in the preparation of 3-acylpyridines and ethyl nicotinates from 3-aminoacroleins.[45,45a] The *o*-aminophenyl alkyl ketones, which contain an active methyl or methylene group, may undergo self-condensation even in the presence of another *α*-methylenecarbonyl compound in the reaction mixture. This is especially likely to occur under more vigorous reaction conditions (Eq. 9).[15]

$COCH_3$ / $NH_2 \cdot HCl$ + O / $CO_2C_2H_5$ / O / $CO_2C_2H_5$ $\xrightarrow{<175^\circ}$ No reaction

$\xrightarrow{>175^\circ}$ CH_3 / NH_2 / N

(Eq. 9)

The instability of *o*-aminobenzaldehyde and its derivatives, as well as the tendency of *o*-aminoaryl ketones to undergo self-condensation, cause difficulties when these compounds are used in carrying out Friedländer reactions with less reactive α-methylenecarbonyl compounds, *e.g.*, simple esters, amides, nitriles, or sterically hindered ketones, which usually require more drastic reaction conditions or prolonged reaction times. These drawbacks particularly limit larger-scale preparations of certain quinoline derivatives.

A modification developed by Borsche circumvents these difficulties by employing the azomethines (arylimines) of *o*-aminobenzaldehydes in place of their parent compounds.[46–49] Such reactions are often conducted in an alkaline ethanolic solution or in the presence of piperidine. As shown in the following example, the required azomethine **7** can be conveniently prepared by the reaction of *p*-toluidine with the appropriate *o*-nitroaromatic aldehyde, followed by sodium sulfide reduction. Condensation of **7** with an active methylene compound in the presence of base forms the desired quinoline.[47] In most cases the Borsche modification of the Friedländer synthesis gives quinolines in moderate to good yields (60–90%).

O / O / CHO / NO_2 + $p\text{-}H_2NC_6H_4CH_3$ $\longrightarrow$

O / O / $CH{=}NC_6H_4CH_3\text{-}p$ / NO_2 (75%) $\xrightarrow{Na_2S}$

O / O / $CH{=}NC_6H_4CH_3\text{-}p$ / NH_2 **7** (54%) $\xrightarrow{p\text{-}ClC_6H_4COCH_3}$

O / O / N / $C_6H_4Cl\text{-}p$ (90%)

About the only instance in which the Borsche modification compares unfavorably with the original Friedländer synthesis was reported by Borsche himself. *o*-Aminobenzaldehyde is smoothly condensed with phenylpyruvic acid in ethanolic sodium hydroxide to yield 3-phenylquinaldic acid, whereas only an amorphous resin can be isolated when N-(*o*-aminobenzylidene)-*p*-toluidine is used in place of the free aldehyde under the same reaction conditions.[46,50] On the other hand, the azomethines of alkoxy-substituted *o*-aminobenzaldehydes, such as compound **7** and the corresponding dimethoxy analog, are condensed readily with phenylpyruvic acid to form 6,7-methylenedioxy-3-phenylquinaldic and 6,7-dimethoxy-3-phenylquinaldic acid, respectively.[46]

The original Borsche modification uses bases as catalysts. In the absence of catalysts quinolines can also be obtained, although in lower yields.[46] A more recent variation that uses an acid catalyst during the condensation between an azomethine and a ketone can be conveniently applied to large-scale syntheses of a variety of quinolines.[51]

A recently reported extension of the Friedländer synthesis involves the use of an anthranilic ester as the *o*-aminocarbonyl component.[51a] Substituted quinolines are prepared by condensation of anthranilic esters with active methylene compounds, using hexamethylphosphoramide–polyphosphoric acid as the condensing agent.

The Active Methylene Component

The Friedländer condensation is applicable to aldehydes, ketones, carboxylic acids, esters, amides, nitriles, and aldoximes that have an adjacent enolizable group, In general, the reactivity of this methylene group is the principal factor that influences the condensation. For less reactive α-methylene derivatives longer reaction times or more drastic reaction conditions are usually required.[18,42,48,52,53] Steric hindrance in the vicinity of the methylene-carbonyl group is also an important factor. A single aldehydic or ketonic carbonyl group is often sufficient to activate the neighboring methylene group for the desired condensation. Esters, amides, and nitriles usually require the presence of another electron-withdrawing group at the other side of the methylene linkage to provide additional activation. Electron-withdrawing groups that furnish activation in the Knoevenagel condensation should in principle also be effective in the Friedländer reaction. The order of activation by these groups is comparable in both reactions.[18,42,52,53]

Unsymmetrical methyl ketones, such as 2-butanone, 2-heptanone, and phenylacetone, may undergo cyclization in two different ways, depending on whether the condensation with the carbonyl group of the *o*-aminoaromatic carbonyl compound occurs at C_1 or C_3 of the methyl ketone. For methyl ketones of the type CH_3COCH_2X, where X is nitro,[30a] hydroxy,[19] phthalimido,[29a] arylsulfonyl,[54] or aryl,[55] the methylene rather than the methyl carbon is usually involved. Base-catalyzed condensation between 2-butanone and *o*-aminobenzaldehyde in sodium hydroxide gives exclusively 2,3-dimethylquinoline.[56,57] Similarly, condensation between 2-butanone and 4-aminonicotinaldehyde in ethanolic sodium hydroxide forms only 2,3-dimethyl-1,6-naphthyridine (**8**).[52] However, almost equal amounts of the naphthyridine **8** and 2-ethyl-1,6-naphthyridine (**9**) are obtained with piperidine as the catalyst.[18] Condensation between 2-butanone and *o*-aminobenzophenone gives a mixture of two products (Eq. 10); the ratio of the products depends on the nature of the catalysts used,[20] with C–C bond formation

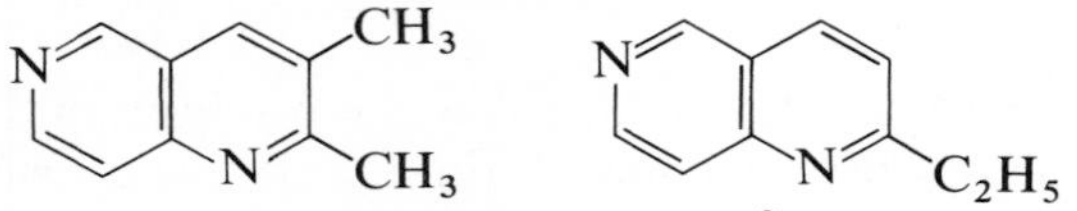

8 9

occurring predominantly at the α-methylene carbon in the presence of acid and at the α-methyl carbon in the presence of base. Similarly, reactions between

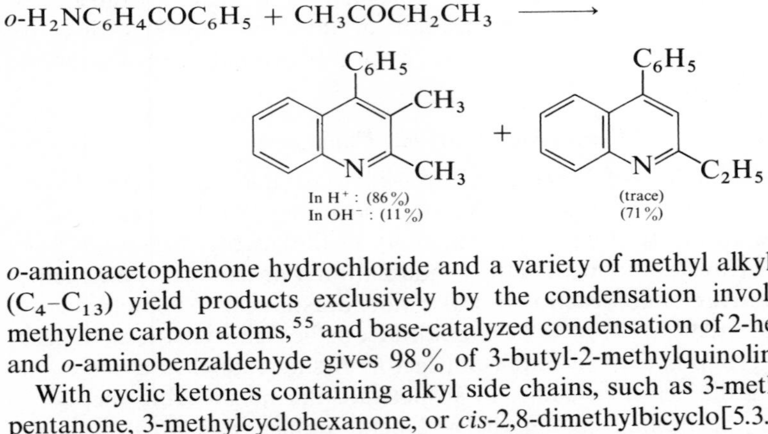

$$o\text{-}H_2NC_6H_4COC_6H_5 + CH_3COCH_2CH_3 \longrightarrow$$

In H^+ : (86%) (trace)
In OH^- : (11%) (71%)

(Eq. 10)

o-aminoacetophenone hydrochloride and a variety of methyl alkyl ketones (C_4–C_{13}) yield products exclusively by the condensation involving the methylene carbon atoms,[55] and base-catalyzed condensation of 2-heptanone and *o*-aminobenzaldehyde gives 98% of 3-butyl-2-methylquinoline.[57]

With cyclic ketones containing alkyl side chains, such as 3-methylcyclopentanone, 3-methylcyclohexanone, or *cis*-2,8-dimethylbicyclo[5.3.0]decan-5-one, condensation with *o*-aminoacetophenone hydrochloride takes place preferentially at the active methylene carbon farther from the alkyl substituent, *i.e.*, at the less sterically hindered active methylene group (*e.g.*, Eq. 11).[31] With

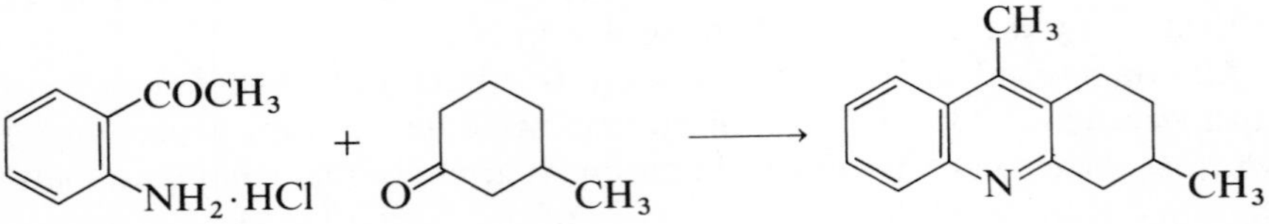

(Eq. 11)

3,4-benzocyclohex-3-en-1-one or 3,4-benzocyclohept-3-en-1-one,[58] the benzyl carbon adjacent to the ketone group is favored in condensation (Eq. 12), much like phenylacetone.[20]

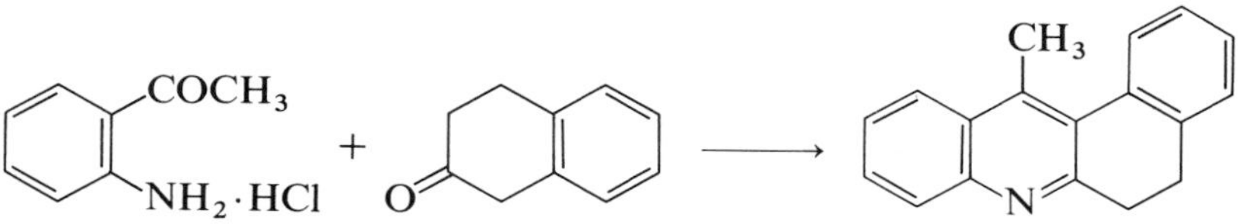

$COCH_3$ $NH_2 \cdot HCl$ + O ⟶ CH_3 N

(Eq. 12)

Bifunctional compounds containing ketone groups, such as pyruvic acids,[19,59,60] levulinic acid,[19] isonitrosoacetone,[46,62] α-nitroacetophenone,[62] and related compounds,[61] generally react like simple ketones to yield the corresponding quinolines. The reaction of *o*-aminobenzaldehyde is claimed not to proceed with a few compounds, *e.g.*, dihydroxyacetophenone and benzyl phenyl ketone;[56] the latter, however, does condense with *o*-aminobenzophenone,[20] *o*-aminoacetophenone hydrochloride,[55] and substituted N-(*o*-aminophenylmethylene)-*p*-toluidines.[47]

With bifunctional compounds that have a methylene group attached directly to two functional groups (such as keto, aldehyde, cyano, or carbalkoxy) capable of reacting with the amino group of the *o*-aminocarbonyl compound, the amino group usually attacks the more electrophilic carbonyl group. Thus ethyl acetoacetate undergoes condensation with *o*-aminobenzaldehyde to give 3-carbethoxy-2-methylquinoline (**10**),[57,64] α-cyanoacetophenone reacts with *o*-aminobenzophenone to form 3-cyano-2,4-diphenylquinoline (**11**),[19] and formylacetophenone condenses with 2-amino-6-phenylnicotinaldehyde to yield 2-phenyl-6-benzoyl-1,8-naphthyridine (**12**).[42] Similar condensations

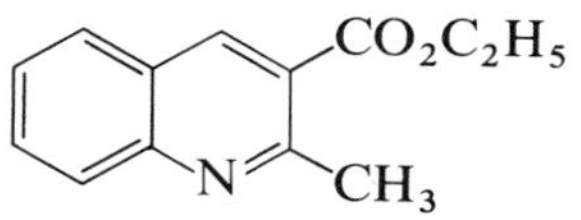

$CO_2C_2H_5$ N CH_3

10

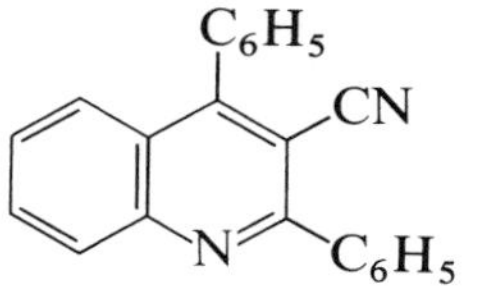

C_6H_5 CN N C_6H_5

11

COC_6H_5 C_6H_5 N N

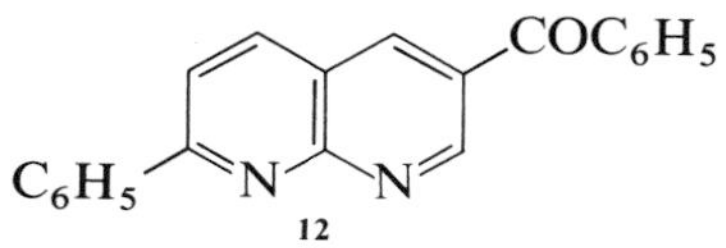

12

have been reported between α-cyanoacetophenone and other *o*-aminocarbonyl compounds.[52,65–67] In some reactions the mode of cyclization may depend on reaction conditions, the nature of reactants, and/or catalysts (Eqs. 13,[19,23] 14,[18] and 15[52,53]).

o-$H_2NC_6H_4COC_6H_5$ + $CH_3COCH_2CO_2C_2H_5$

AcOH, H_2SO_4 → 4-C_6H_5-3-$CO_2C_2H_5$-2-CH_3-quinoline (85%)

150° → 4-C_6H_5-3-$COCH_3$-2-OH-quinoline (60%)

Base → No reaction

(Eq. 13)

4-amino-3-pyridinecarboxaldehyde (CHO, NH_2) + $(CH_3)_2NCOCH_2CN$

Piperidine → CN, OH (Yield not reported)

NaOH → $CON(CH_3)_2$, NH_2 (63%)

(Eq. 14)

4-amino-3-pyridinecarboxaldehyde (CHO, NH_2)

$NCCH_2CO_2C_2H_5$, Piperidine, EtOH → CN, OH (92%)

$NCCH_2CO_2H$, Piperidine, EtOH → CO_2H, NH_2 (52%)

(Eq. 15)

The α-diketone biacetyl reacts readily with two molecules of *o*-aminobenzaldehyde to form biquinoline;[68] cyclic α-diketones, such as 1,2-cyclohexanedione, usually require higher reaction temperatures (>160°) for the formation of the corresponding biquinolines (Eq. 16).[24]

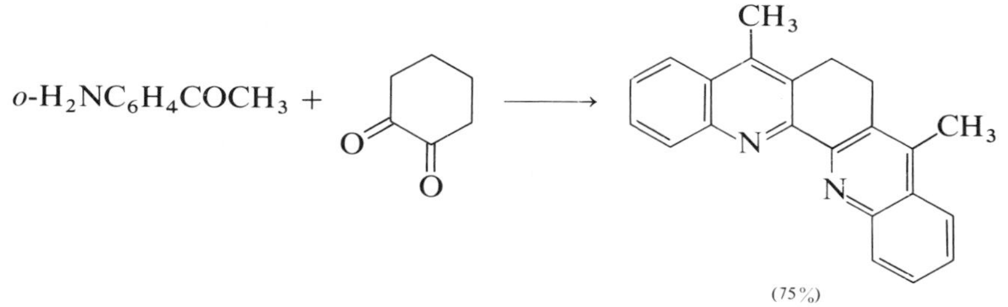

o-$H_2NC_6H_4COCH_3$ + [1,2-cyclohexanedione] ⟶ [product, CH_3, CH_3, N, N]

(75%)

(Eq. 16)

1,3-Diketones generally react with 1 mol of *o*-aminoaromatic aldehydes and ketones to give the corresponding quinolines.[18,19,21,69,70] Phloroglucinol has been shown to react with one, two, or three molecules of *o*-aminobenzaldehyde, depending on the ratio of reactants and the reaction conditions.[48,56,71] With unsymmetrical aryl 1,3-diketones in which one of the keto groups is conjugated with the aromatic ring, the keto group that condenses with the amine is usually the one not attached to the aromatic ring (Eq. 17).[19,72,73]

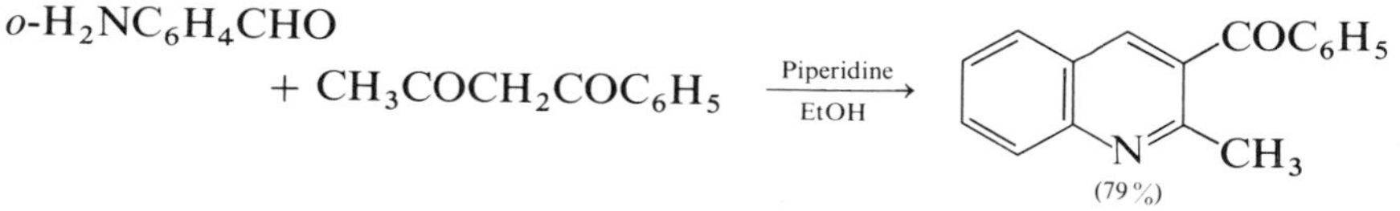

o-$H_2NC_6H_4CHO$ + $CH_3COCH_2COC_6H_5$ $\xrightarrow[\text{EtOH}]{\text{Piperidine}}$ [quinoline, COC_6H_5, CH_3, N]

(79%)

(Eq. 17)

Diketones with two keto groups separated by more than one carbon atom usually condense with two molecules of the *o*-aminoarylcarbonyl compound[31,48] except when steric hindrance becomes important.[15,74a,b]

Acetic anhydride facilitates the cyclization of azomethines of 6-aminopiperonal and 6-aminoveratraldehyde, but not the azomethines of *o*-aminobenzaldehydes, in the Borsche modification of the Friedländer synthesis.[46] The oxime of acetaldehyde has also been used to condense with the azomethine of 6-aminoveratraldehyde in refluxing ethanol.[46] The oxime of nitroacetaldehyde condenses readily with various *o*-aminobenzaldehydes even at room temperature to form the corresponding 3-nitroquinolines (Eq. 18).[28]

$$\text{X-}C_6H_3(CHO)(NH_2) + O_2NCH_2CH{=}NOH \xrightarrow[\text{EtOH}]{\text{HCl}} \text{X-(3-nitroquinoline)}$$

(Eq. 18)

COMPARISON WITH OTHER METHODS

Although numerous syntheses leading to the formation of the quinoline ring system are available,[10,11] two of the most important are (1) condensation of an aromatic amine with another reactant or reactants that provide the three-carbon unit required to complete the quinoline ring, and (2) condensation of an *ortho*-C-substituted aniline with a reactant that provides the two-carbon unit for completing the quinoline ring. The first type is represented by the Skraup synthesis,[8,9] the Combes synthesis,[6] and the Doebner–Miller synthesis,[7] and the second type by the Friedländer synthesis as well as the Pfitzinger[4] and the Niementowski[5] syntheses. Another variation of the second type involves a Michael addition of the amino group of either 6-aminopiperonal or *o*-aminoacetophenone to an acetylenic carbon of acetylenedicarboxylic esters followed by cyclization (Eq. 19).[75] A comparison of these two major types is therefore of interest.

$$\text{6-aminopiperonal (CHO, } NH_2\text{)} + CH_3O_2C{\equiv}CCO_2CH_3 \xrightarrow{\text{MeOH}}$$

$$\text{piperonal-6-}NHC(CO_2CH_3){=}CHCO_2CH_3 \xrightarrow[H_2SO_4]{\text{MeOH/CHCl}_3} \text{dimethyl 6,7-methylenedioxyquinoline-2,3-dicarboxylate } (64\%)$$

(Eq. 19)

The Skraup synthesis (Eq. 20) involves condensation of an aniline that has at least one vacant *ortho* position with an α,β-unsaturated carbonyl compound (or a suitable precursor) in the presence of an acidic condensing agent (such as concentrated sulfuric acid or phosphoric acid) and an oxidizing agent (such as arsenic acid or arsenic pentoxide) at relatively high temperature (*ca.* 150°).

$$X\text{-}C_6H_4NH_2 + R_1CH{=}CHCOR_2 \longrightarrow \text{X-quinoline (2-}R_1\text{, 4-}R_2\text{)}$$

(Eq. 20)

Because substituted anilines are generally more readily available than substituted *o*-aminobenzaldehydes, the Skraup synthesis and related reactions are often more useful than the Friedländer synthesis for the preparation of quinolines with substituents on the benzene portion of the quinoline ring, *i.e.*, substituents at positions 5, 6, 7, and 8. This limitation of the Friedländer synthesis can sometimes be overcome by replacement of the starting substituted *o*-aminobenzaldehydes with the more accessible substituted isatins[76,77] or isatic acids [Pfitzinger reaction (Eq. 3)], followed by decarboxylation of the resulting quinoline-4-carboxylic acid derivatives to the desired quinolines.

Only those substituted anilines that are not affected by hot concentrated acids can be used as starting materials in the Skraup synthesis. Nevertheless, this method finds frequent application in the preparation of 2- and/or 4-substituted quinolines. On the other hand, the Friedländer synthesis is the preferred route to quinolines substituted at the 3 position, and particularly those with electron-attracting substituents such as $-NO_2$, $-SO_3H$, $-CO_2H$, and $-CN$, which are not easily made by the Skraup method.

One drawback of the Skraup-type syntheses is isomer formation when a substituent is present *meta* to the amino group in the aniline reactant. The ratio of the resulting 5- and 7-substituted quinolines depends primarily on the nature of that substituent, but can also be influenced by the strength of the sulfuric acid used as condensing agent.[9] Isomer formation of this type cannot occur in the Friedländer synthesis because of the regiochemistry of the condensation. As discussed earlier, the Friedländer synthesis can yield a mixture of isomers (or different isomers under different conditions) when the other reactant, the active methylene compound, is either unsymmetrical or polyfunctional (see Eqs. 10, 13–15). Similar possibilities for isomer formation also exist in the aniline condensation reactions, as exemplified by the Conrad–Limpach[78–80] (Eq. 21) and the Knorr[81–83] (Eq. 22) syntheses.

$C_2H_5O_2C$, CH_2, C, N, CH_3 —260°→ OH, N, CH_3

(Eq. 21)

< 100°

$$C_6H_5NH_2 + CH_3COCH_2CO_2C_2H_5$$

> 100°

CH_3CO, CH_2, CO, N, H —Heat, H^+→ CH_3, N, OH

(Eq. 22)

The appropriate method for the preparation of a particular quinoline is, therefore, dependent on a number of factors: the availability of starting materials, the reactivity and position of functional groups in the reactants, reaction conditions, and possible contamination of the final product by isomers or other side-reaction products. Synthesis of 4-hydroxyquinolines, for example, can be carried out by most of the aforementioned reactions as well as by the Camps synthesis[84] (from *o*-acylaminoalkyl ketones), the Niementowski synthesis[5] (Eq. 4), and the Gould–Jacobs synthesis[85] (from anilines and ethoxymethylenemalonic esters). The type and position of the functional groups desired in the quinoline play a key role in selecting the route.

EXPERIMENTAL CONDITIONS

Selection of catalysts for the Friedländer synthesis depends on the nature of both reactants. In general, acid- or base-catalyzed condensations are better than uncatalyzed reactions.

For reactions between *o*-aminobenzaldehydes and ketones, catalysts such as sodium and potassium hydroxide are used widely. The use of strongly basic anion-exchange resins, (*e.g.*, Amberlite IRA-400 and Dowex-2) has also been reported.[69] With methyl ketones or *β*-keto acids better yields are often realized when the reaction is conducted at room temperature for several days. For less accessible *o*-aminobenzaldehydes, such as the alkoxy-substituted compounds, the use of their azomethine derivatives in place of the parent aldehydes (the Borsche modification[46–49]) is of value because undesired side reactions of the aldehydes can usually be minimized. This is especially useful in large-scale preparations.[51,86,87]

For condensations involving *o*-aminoacetophenone, the readily accessible hydrochloride rather than the free base (the Kempter modification[15,31,32]) is often the preferred reactant. This modification is also applicable to *o*-aminobenzophenones.[15,33,37] Condensations of *o*-aminoaryl ketones can often be carried out most conveniently in refluxing acetic acid in the presence of a small amount of sulfuric acid (the Fehnel modification), particularly when highly volatile active methylene compounds are involved.[19,20,36,88]

Piperidine or sodium hydroxide can be used as the catalyst for condensations involving 2- or 4-aminonicotinaldehyde with most reactive methylene compounds. When the latter reactants contain a cyano group rather than a carbonyl (acetonitrile or other alkyl cyanides), a stronger base, such as sodium methoxide, is usually employed. For highly enolic active methylene compounds such as 1,3-diketones or 3-keto esters, piperidine is the catalyst most frequently used in condensation reactions with *o*-aminobenzaldehyde.[48,72] With *o*-aminoaryl ketones, on the other hand, acid catalysis is generally preferred.[19]

It should be emphasized that with reactions involving unsymmetrical ketones such as 2-butanone[20] and N-carbethoxy-3-pyrrolidone,[89] two modes of cyclization are possible. Selection of catalysts can greatly influence the course of the reaction and the yield of the desired product. As mentioned previously, an acidic catalyst may favor the formation of one isomeric product whereas a basic catalyst may favor the other.[20] In some systems even the choice of different basic catalysts, such as sodium hydroxide or piperidine, may result in the formation of different proportions of two isomeric products from the same reactants.[18,52]

The selection of solvents generally depends on the solubility of reactants and catalysts. Ethanol is most commonly used for both catalyzed and uncatalyzed reactions. Aqueous ethanol or water has frequently been used in the more conventional types of alkali hydroxide–catalyzed condensations with *o*-aminobenzaldehyde as one of the reactants. The use of alkali hydroxide in an aqueous reaction medium, however, has some disadvantages, one being that *o*-aminobenzaldehyde is only sparingly soluble in water, and undesirable side reactions may thus occur.[69] This problem has been avoided in one case by generating an aqueous solution of *o*-aminobenzaldehyde *in situ* from *o*-dichloromethylphenyl isocyanate and proceeding immediately with a Friedländer condensation with chloroacetone.[89a] Other alcohols such as methanol (for sodium methoxide catalyst), *n*-pentanol, and glycerol (for piperidine catalyst[48]) have also been used. Acetic acid is frequently the solvent for mineral acid catalysts.[19,20,36] Less often exploited, less polar solvents such as toluene (for catalysis by *p*-toluenesulfonic acid[51]), xylene (for piperidine[90]), and mineral oil (for uncatalyzed reactions[91]) may offer advantage in condensations that require higher temperatures. Hexamethylphosphoramide (an animal carcinogen) has been used as a solvent for the preparation of polymers containing anthrazoline units.[39]

The most common reaction temperature for acid- or base-catalyzed reactions is the reflux temperature of the solvent. When the hydrochloride of the *o*-aminoaryl carbonyl compound is the reactant, the reaction temperature ranges from 110° to 190° without the use of solvent. Uncatalyzed condensations usually require more drastic reaction conditions, with temperatures ranging between 150° and 220°. The reaction time required for these condensations depends on the activity of the reactants. Some are complete in 30 minutes, whereas others take up to 24 hours. Reactions conducted at room temperature usually require one to several days.

Stoichiometric amounts of both reactants, or a slight excess of the active methylene component, is normally employed. When the latter is readily accessible (*e.g.*, acetone and 2-butanone), it can be used in excess as both reactant and solvent to minimize losses due to self-condensation of the *o*-aminocarbonyl compound. The latter reaction can also be minimized by slow addition of the *o*-aminoaryl carbonyl component to the active methylene component, as in the Kempter modification.[15,31,32]

EXPERIMENTAL PROCEDURES

4,7-Dimethyl-2,3,8,9-dibenzo-5,6-dihydro-1,10-phenanthroline.[24] A mixture of 5.4 g (0.040 mol) of *o*-aminoacetophenone and 2.2 g (0.0196 mol) of 1,2-cyclohexanedione was heated in a metal bath at 160° for 8 hours. The brown syrupy mass was mixed with a little methanol and stirred. The resulting precipitate was recrystallized twice from ethanol to give 4.6 g (76%) of white needles, mp 271°.

6H-[1]Benzopyrano[4,3-*b*]quinoline (Chromeno[4,3-*b*]quinoline).[92] To an ice-cold solution of 1.5 g (0.010 mol) of 2,3-dihydrobenzopyran-4-one (chroman-4-one) and 1.2 g (0.010 mol) of *o*-aminobenzaldehyde in 15 mL of methanol was added, with stirring, 3 mL of 2*N* sodium hydroxide. After a short time the condensation product began to precipitate. The reaction mixture was left overnight and the crude product (1.7 g, 73%) was collected by filtration. It was recrystallized from either benzene or methanol to give white needles, mp 121.5°.

2-Amino-3-(*p*-bromophenyl)-1,6-naphthyridine.[53] A mixture of 0.366 g (0.0030 mol) of 4-aminonicotinaldehyde, 0.71 g (0.0060 mol) of phenylacetonitrile, and 0.4 mL (0.01 mol) of 10% aqueous sodium hydroxide in 5 mL of ethanol was heated under reflux on a steam bath for 2 hours. The solvent was evaporated under reduced pressure and the residue recrystallized from benzene to give 0.774 g (86%) of the naphthyridine, mp 195–197°. Proton magnetic resonance (d_6-dimethyl sulfoxide) δ: 7.94 (*s*, 1 H, H-4), 8.92 (*s*, 1 H, H-5), 8.42 (*s*, 1 H, H-7), and 7.36 (*s*, 1 H, H-8).

2-(4-Chlorophenyl)-6,7-methylenedioxyquinoline. (Borsche modification).[47] A mixture of 2.54 g (0.010 mol) of N-[(2-amino-4,5-methylenedioxy)benzylidene]-*p*-toluidine[93] and 1.54 g (0.010 mol) of 4-chloroacetophenone in 25 mL of ethanol and 8 mL of 2 *N* sodium hydroxide was heated on a water bath for 8 hours. The resulting *p*-toluidine and ethanol were removed by steam distillation. The product, which solidified on cooling, was collected by filtration. It was recrystallized from methanol to give 2.55 g (90%) of the quinoline as green-yellow flakes, mp 183°.

4-Dimethoxymethyl-2-methylquinoline.[94] To a solution of 0.6 g (0.026 g-atom) of sodium in 40 mL of absolute ethanol was added 3.9 g (0.02 mol) of *o*-aminophenylglyoxal dimethylacetal and 1.62 mL (0.022 mol) of acetone. The mixture was refluxed for 15 minutes. The reaction mixture was evaporated under reduced pressure to near dryness, and to the residue was added 150 mL of water. The resulting solid was collected by filtration and recrystallized from ethanol to give 4.1 g (95%) of the quinoline as needles, mp 47°.

Ethyl 2-Methyl-1,8-naphthyridine-3-carboxylate.[42] A mixture of 3.66 g (0.03 mol) of 2-aminonicotinaldehyde, 7.8 g (0.060 mol) of ethyl acetoacetate, and 0.75 mL (0.0075 mol) of piperidine in 5 mL of ethanol was heated under reflux on a water bath for 1 hour. The reaction mixture was evaporated under reduced pressure and the residue triturated with ligroin (bp 60–80°). The product was collected by filtration and recrystallized from ethanol to give 5.25 g (81%) of the 1,8-naphthyridine, mp 85–86°. Infrared (Nujol) cm^{-1}: 1720 (C=O), 1630, 1610, 1560, 1430, 1370, 1260, 1180, and 820.

10-Methyl-11 H-indeno[1,2-*b*]quinoline. (Kempter Modification).[31] Equimolar quantities of *o*-aminoacetophenone hydrochloride and 1-indanone were used for this preparation. The indanone was placed in an open vessel and heated to 100° on a metal bath. About one-tenth of the *o*-aminoacetophenone hydrochloride was added, and the reaction temperature was raised to 120°. Every 10 minutes another one-tenth portion of the amine salt was added while the temperature was kept at 120°. The progress of the reaction was indicated by the effervescence of the reaction mixture upon each incremental addition. Water formed during the condensation was removed by directing a stream of nitrogen over the surface of the mixture. Gradual formation of crystalline products in the hot mixture was an indication of the completion of reaction. The mixture was cooled, and the solid was collected by filtration and washed with benzene. Recrystallization from cyclohexane–petroleum ether gave an 88% yield of the indenoquinoline, mp 120°.

3 - Acetyl - 4 - phenylquinaldine (3 - Acetyl - 2 - methyl - 4 - phenylquinoline). (Fehnel modification).[36] A solution of 1.97 g (0.01 mol) of *o*-aminobenzophenone and 1 g (0.01 mol) of acetylacetone in 10 mL of glacial acetic acid containing 0.1 mL of concentrated sulfuric acid was heated under reflux for 2 hours. The reaction mixture was cooled and poured slowly, with stirring, into an ice-cold solution of 15 mL of concentrated ammonium hydroxide in 40 mL of water. The resulting suspension was allowed to stand in an ice bath until the gummy precipitate had hardened, after which the crude product was collected, washed with water, and recrystallized from aqueous ethanol to give 2.19 g (84%) of off-white needles, mp 110–112°. Further recrystallization from aqueous ethanol raised the melting point to 113–114°. Ultraviolet (ethanol), nm max (log ε): 237 (4.50), 285 (3.78), and sh 320 (3.54); infrared (Nujol) cm^{-1}: 1690 (C=O), 1560, 1210, 1155, 765, 750, 715, and 705; proton magnetic resonance ($CDCl_3$) δ: 2.00 (*s*, 3H), 2.70 (*s*, 3H), and 7.30–8.30 (m, 9H).

Ethyl 2 - Ethoxycarbonyl - 1,3 - dihydro - 2H - pyrrolo[3,4 - *b*]quinoline - 3 - acetate.[51] A solution of 630 g (3.00 mol) of N-(2-aminobenzylidene)-*p*-toluidine,[49] 730 g (3.00 mol) of ethyl N-carbethoxy-3-pyrrolidinone-2-acetate,[95] and 15 g of *p*-toluenesulfonic acid in 3 L of toluene was heated under reflux, with mechanical stirring, until no additional water was collected by a Dean-Stark apparatus (*ca.* 3.5 hours of reflux). The reaction mixture was filtered while hot to remove a small amount of insoluble solid. The filtrate was evaporated under reduced pressure to remove most of the solvent. The residue was then cooled to 0°, and 750 mL of ether was added. The product, which precipitated from the mixture, was collected by filtration, washed with ether, and dried to give 775 g (76.3%) of the quinoline, mp 108–110°. Recrystallization from ethyl acetate furnished white crystals, mp 111–113°. Infrared (Nujol) cm^{-1}: 1750 (ester carbonyl) and 1675 (amide carbonyl).

2-[(2-Acetamido)phenyl]-4-methylquinoline.[30b] Equimolar quantities (20 mmol) of *o*-aminoacetophenone (2.7 g) and *o*-acetamidoacetophenone (3.54 g) were heated in 62 g of polyphosphoric acid (10 times the amount of reactants) at 100° for 2 hours with stirring. The resulting mixture was decomposed with 200 mL of ice-water, and the aqueous solution was extracted with chloroform (3 × 30 mL). The extract was washed with dilute sodium carbonate solution, dried over anhydrous sodium sulfate, and evaporated. The residual syrup solidified upon heating with 30 mL of ethanol. Recrystallization from ethanol gave 4.69 g (85%) of the quinoline, mp 138°.

Ethyl 4-(2-Fluorophenyl)-2-methylquinoline-3-carboxylate.[164] A mixture of 45.4 g (0.211 mol) of 2-(2-fluorobenzoyl)aniline, 45.4 g (0.349 mol) of ethyl acetoacetate, 4.5 g of zinc chloride, and 500 mL of benzene was refluxed for 24 hours with separation of water. The benzene solution was washed with water, dried, and evaporated. Two crystallizations from ethanol gave 54.1 g (83 %) of white product, mp 123–125°. A third recrystallization from ethanol raised the melting point to 126–128°. Ultraviolet, nm max (log ε): 208 (4.61), 236 (4.67), sh 263 (3.71), sh 271 (3.75), 282 (3.76), infl 305 (3.59), and 320 (3.52); infrared ($CHCl_3$): 1730 cm^{-1} ($CO_2C_2H_5$); proton magnetic resonance ($CDCl_3$): δ 1.0 ppm (t, 3, $J = 7$ Hz, CH_3), 2.85 (s, 3, CH_3), 4.13 (q, 2, $J = 7$ Hz, OCH_2), and 7–8.3 (m, 8, aromatic H).

1,2,3,4-Tetrahydroacridine.[69] A mixture of 1.8 g (0.015 mol) of *o*-aminobenzaldehyde, 2.0 g (0.02 mol) of cyclohexanone, 15 mL of ethanol, and 0.5 mL of Amberlite IRA-400 ion-exchange resin was heated under reflux with stirring for 4 hours. The reaction mixture was filtered to remove the resin, and ethanol was evaporated from the filtrate. The residue was fractionally distilled and the fraction boiling in the range 180–180.5° (15–15.5 mm) was collected as an oil, which readily crystallized on standing to give 2.5 g (92 %) of product, mp 54°.

TABULAR SURVEY

The tables are arranged in order of condensations of active methylene compounds or their derivatives with (I) *o*-aminobenzaldehyde, (II) ring-substituted *o*-aminobenzaldehydes, (III) substituted N-(*o*-amino-substituted phenylmethylene)-*p*-toluidines, (IV) *o*-aminoacetophenones, (V) *o*-aminophenylglyoxal dimethylacetals, (VI) *o*-aminobenzophenone, (VII) substituted *o*-aminobenzophenones, (VIII) aza analogs of *o*-aminobenzaldehyde, (IX) other *o*-aminoaromatic and *o*-aminoheterocyclic carbonyl compounds and their dervatives, and (X) 3-aminoacroleins. The last two tables contain reactions that may be considered as extensions of the Friedländer synthesis. Within each table compounds are usually listed in the order of increasing numbers of carbon atoms in the α-methylenecarbonyl reactants and related derivatives. Compounds with the same number of carbon atoms are arranged by increasing complexity of the reactants and the products.

The literature survey has been conducted through December 1979. While the authors believe that the data collected are reasonably complete, some publications and examples of the Friedländer synthesis subordinated to other topics may have been overlooked. Yields indicated by a dash (—) are not specified in the reference(s) cited.

Abbreviations for solvents and catalysts are as follows: Me_2CO, acetone; AcOH, acetic acid; EtOH, ethanol; HMPA, hexamethylphosphoramide; MeOH, methanol; pTSA, *p*-toluenesulfonic acid; PPA, polyphosphoric acid; and THF, tetrahydrofuran.

TABLE I. Condensations with o-Aminobenzaldehyde

No. of C Atoms	Reactant	Catalyst	Reaction Conditions	Product(s) and Yield(s) (%)	Refs.
C_2	CH_3CHO	NaOH	40–50°, H_2O	Quinoline (—)	1
	$ClCH_2CHO$[a]	"	20°, 2–4 hr	3-Hydroxyquinoline (90)	89a
C_3	CH_3COCH_2Cl[a]	"	"	3-Hydroxy-2-methylquinoline (90)	89a
	$O_2NCH_2CH{=}CHNOH$	HCl	2 d; EtOH	3-Nitroquinoline (48)	28
	C_2H_5CHO	—	220°, 1 hr	3-Methylquinoline (80–85)	26
		KOH	25°, 20 hr	" ("Satisfactory")	25
	CH_3COCH_3	NaOH	H_2O	2-Methylquinoline (—)	96
		(pH 12)	25°, 7 d; H_2O	2-Methylquinoline (11)	57
		(pH 13)	25°, 7 d; H_2O	2-Methylquinoline (86)	57
	$CH_3COCH_2NO_2$	(pH 3)	24 hr; H_2O	2-Methyl-3-nitroquinoline (69)	30a
	CH_3COCO_2H	NaOH	Heat, 4 hr; EtOH	2-Quinolinecarboxylic acid (—)	46
	$CH_2(CO_2H)_2$	—	120°	2-Hydroxy-3-quinolinecarboxylic acid (—)	97
	$CH_3COCH_2SO_3Na$	NaOH	24 hr; H_2O	2-Methyl-3-quinolinesulfonic acid (59)	63
	$CH_3COCH_2SO_3H$[a]	"	20°, 2–4 hr	" (~90)	89a
C_4	$CH_3(CH_2)_2CHO$	—	120–130°, 24 hr	3-Ethylquinoline (80)	27
	$CH_3COC_2H_5$	NaOH	Cold; H_2O	2,3-Dimethylquinoline (—)	56,99
	$CH_3COC_2H_5$	KOH	Reflux, 1 hr; EtOH	" (58)	170
	$C_2H_5COCO_2H$	NaOMe	Reflux, 12 hr; MeOH, then H_2SO_4, reflux, 24 hr	CH_3, CO_2CH_3, N (87)	59
	$CH_3COCH_2CO_2H$	(pH 9)	25°, 8 d	2-Methylquinoline (66)	57
		(pH 9)	25°, 16 d (excess reactant)	" (90)	57
		(pH 13)	25°	CO_2H, CH_3, N (~100)	57
	$CH_3COCOCH_3$	KOH	Heat, 6 hr; EtOH	N, N (59)	68

		—	Reflux, 1 hr; EtOH	(73)	16
		KOH	3 d; EtOH	(55)	37a
		H_2SO_4	AcOH	" (25)	37a
	$CH_3C(=NH)CH_2CN$	—	120°	2-Methyl-3-cyanoquinoline (—)	100
		—	100°, 1hr; H_2O	(—)	101
		NaOAc	100°, AcOH then Ac_2O, boil, 90 min	(—)	102
C_5	$C_2H_5COC_2H_5$	NaOH	Cold; H_2O	2-Ethyl-3-methylquinoline (—)	56
	$CH_3COCH_2COCH_3$	Piperidine	Reflux, 5 hr; EtOH	3-Acetyl-2-methylquinoline (91)	21
		Dowex-2	Reflux, 4 hr; EtOH	" (69)	69
	$CH_3COCH_2COCH_3$[a]	NaOH	20°, 4–6 hr	3-Acetyl-2-methylquinoline[b] (94)	89a
	$CH_3COCH_2CO_2CH_3$[a]	"	"	(~90)	89a
	$CH_3COC(OH)(CH_3)_2$	NaOH	8 d; H_2O–EtOH	(60)	103
	$NCCH_2CO_2C_2H_5$	—	140–190°	3-Cyano-2-hydroxyquinoline (—)	104
		NaOH	20°, 48 hr; EtOH	(51)	32
		—	Heat, 1.5 hr; AcOH	(56)	102

TABLE I. CONDENSATIONS WITH *o*-AMINOBENZALDEHYDE (*Continued*)

No. of C Atoms	Reactant	Catalyst	Reaction Conditions	Product(s) and Yield(s) (%)	Refs.
C_6	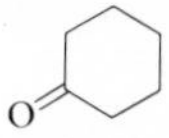 [cyclohexanone]	Amberlite IRA-400	Reflux, 4 hr; EtOH	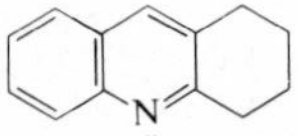 [1,2,3,4-tetrahydroacridine] (92)	69
	[a]	NaOH	20°, 4–6 hr	" (~90)	89a
	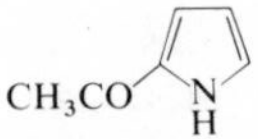 CH_3CO–[2-pyrrolyl, N–H]	KOH	Heat, 1 hr; EtOH	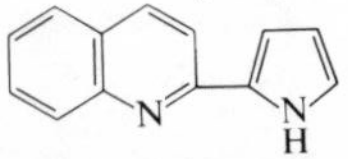 [2-(2-pyrrolyl)quinoline, N–H] (90)	68
	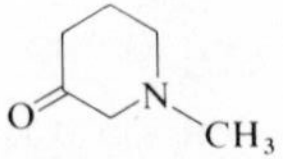 [1-CH_3-3-piperidone]	NaOH	20°, 48 hr; EtOH	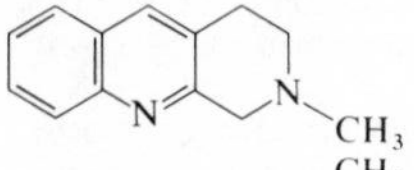 [2-CH_3-1,2,3,4-tetrahydrobenzo[b][1,6]naphthyridine] (52)	32
	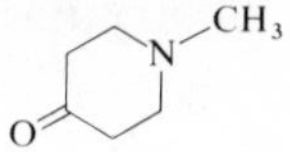 [1-CH_3-4-piperidone]	"	"	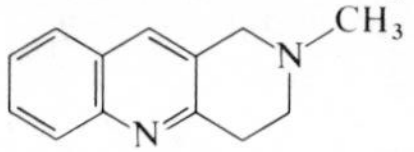 [2-CH_3-1,2,3,4-tetrahydrobenzo[b][1,6]naphthyridine isomer] (61)	32
	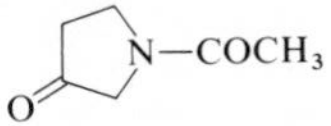 [1-acetyl-3-pyrrolidinone, N–$COCH_3$]	NaOH	45 hr; EtOH	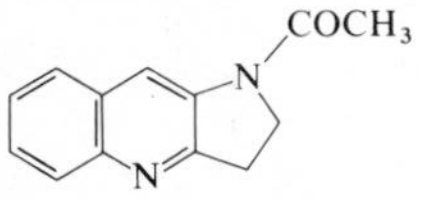 [1-$COCH_3$-2,3-dihydro-1H-pyrrolo[2,3-b]quinoline] (5) 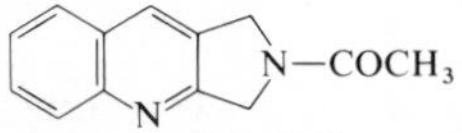 + [2-$COCH_3$-2,3-dihydro-1H-pyrrolo[3,4-b]quinoline, N–$COCH_3$] (Trace)	89
	$CH_3COCH_2CO_2C_2H_5$	NaOH		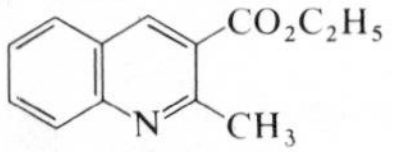 [3-$CO_2C_2H_5$-2-CH_3-quinoline] (—)	96
		"		" (86)	69
		Piperidine	100°, 4 hr; EtOH	" (93)	64
		Amberlite IRA-400	1. 40–50°, 2 hr; EtOH 2. 78°, 30 min	" (89)	69
	[a]	NaOH	20°, 4–6 hr	" (~90)	89a

	$CH_3COCH_2CO_2C_2H_5$	—	160°	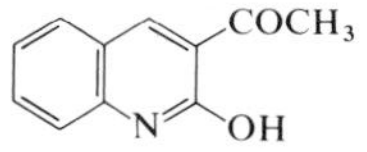 $COCH_3$, N, OH (—)	96
	$NCCH_2COCO_2C_2H_5$	HCl	25°, 7 d; EtOH	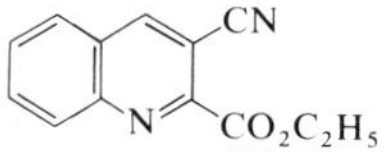 CN, N, $CO_2C_2H_5$ (32)	105
	$CH_3O_2CCHNaCOCO_2CH_3$	—	100°, 3 hr	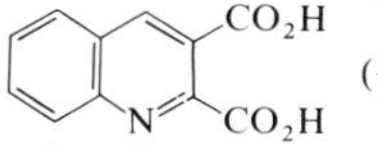 CO_2H, N, CO_2H (—)	106
	$CH_3COCH_2CH_2COCH_3$	NaOH	Cold, 12 hr; H_2O	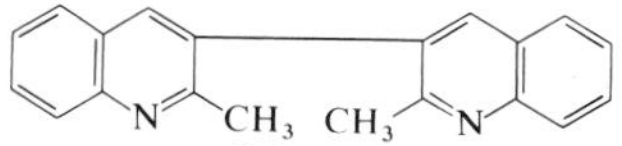 N, CH_3, CH_3, N (—)	56
	1,3,5-$C_6H_3(OH)_3$	NaOH	1 eq. reactant, 100°, H_2O	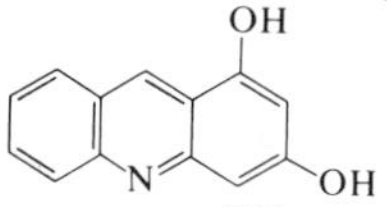 OH, N, OH (—)	56
		—	0.5 eq. reactant, 115–150°, 2 hr	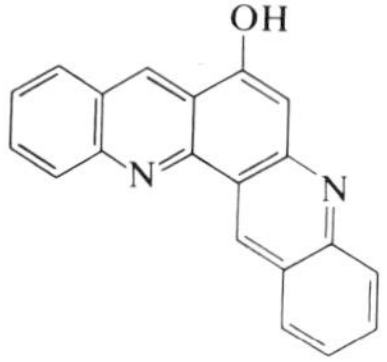 OH, N, N (60)	71
		—	0.33 eq. reactant, 115–120°	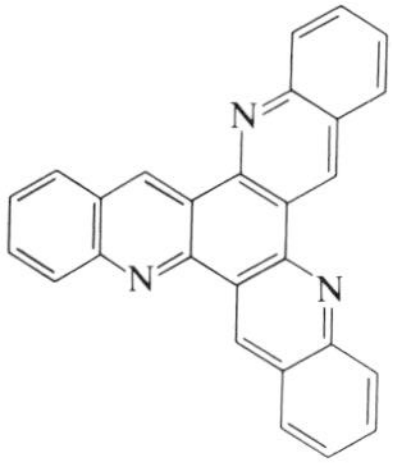 N, N, N (—)	71
C_7	$CH_3(CH_2)_5CHO$	—	180°, 5 hr	3-*n*-Amylquinoline (~100)	27
	$CH_3CO(CH_2)_4CH_3$	(pH 13)	25°, 7 d; H_2O	2-Methyl-3-*n*-butylquinoline (98)	57

TABLE I. CONDENSATIONS WITH *o*-AMINOBENZALDEHYDE (*Continued*)

No. of C Atoms	Reactant	Catalyst	Reaction Conditions	Product(s) and Yield(s) (%)	Refs.
C_7 (*Contd.*)	$CH_3COCH_2COCO_2C_2H_5$	KOH	Several d; EtOH	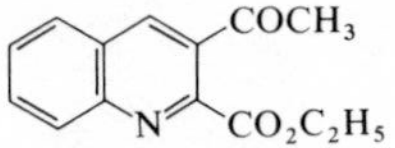 $COCH_3$, $CO_2C_2H_5$, N (44)	107,108
	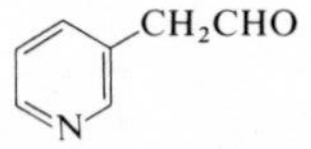 CH_2CHO, N	NaOH	Water bath, 2 hr; H_2O–EtOH	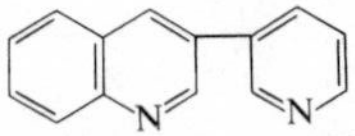 N, N (—)	109
	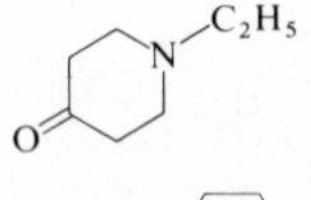 C_2H_5, N, O	NaOH	20°, 2 d; EtOH	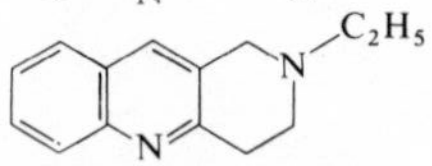 C_2H_5, N, N (36)	32
	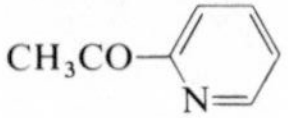 CH_3CO–, N	KOH	Reflux, 1 hr; EtOH	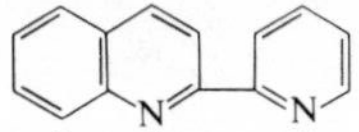 N, N (87)	68 109
	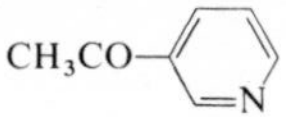 CH_3CO–, N	"	Water bath, 1 hr; EtOH	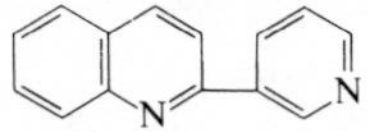 N, N (—)	109
	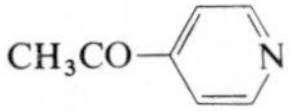 CH_3CO–, N	"	"	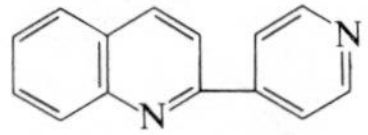 N, N (—)	109
	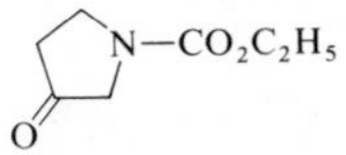 N–$CO_2C_2H_5$, O	10% NaOH	24 hr; EtOH	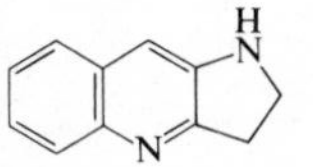 H, N, N (65) 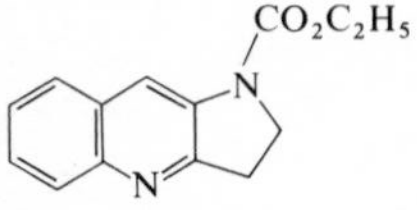 + $CO_2C_2H_5$, N, N (20) 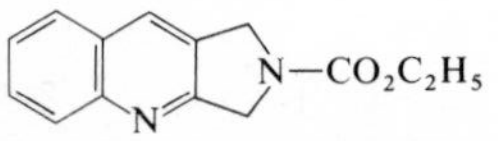 + N–$CO_2C_2H_5$, N (15)	89

	1-Ethoxycarbonyl-3-pyrrolidinone ($N-CO_2C_2H_5$)	85% NaOH	24 hr; EtOH	2,3-Dihydro-1H-pyrrolo[3,2-b]quinoline (79) + 2-($CO_2C_2H_5$)-2,3-dihydro-1H-pyrrolo[3,4-b]quinoline (18)	89
		pTSA	190–195°, 5 min	2-($CO_2C_2H_5$)-2,3-dihydro-1H-pyrrolo[3,4-b]quinoline (90) + 1-($CO_2C_2H_5$)-2,3-dihydro-1H-pyrrolo[3,2-b]quinoline (10)	98
		H_2SO_4	Steam bath, 1 hr; AcOH	2-($CO_2C_2H_5$)-2,3-dihydro-1H-pyrrolo[3,4-b]quinoline (29) + 1-($CO_2C_2H_5$)-2,3-dihydro-1H-pyrrolo[3,2-b]quinoline (36)	89
C_8	$C_6H_5CH_2CHO$	NaOH	H_2O–EtOH	3-Phenylquinoline (—)	96
	$C_6H_5COCH_3$	(pH 13)	25°, 7 d; H_2O	2-Phenylquinoline (100)	57
		Amberlite IRA-400	Reflux, 3.5 hr; EtOH	" (47)	69
		NaOH	"	" (59)	69
	$C_6H_5COCH_2Br$	NaOH	Few d; H_2O	3-Hydroxy-2-phenylquinoline (—)	110
	$C_6H_5CH_2CN$[c]	NaOEt	Heat; EtOH	2-Amino-3-phenylquinoline (—)	111
	$CH_3(CH_2)_4COCH_2CO_2H$	(pH 7–9)	25°	2-*n*-Amylquinoline (73)	57
		(pH 13)	25°	2-$(CH_2)_4CH_3$-quinoline-3-CO_2H (—)	57
	$C_6H_5CH_2CO_2Na$	$ZnCl_2$	Water bath, 5 hr; Ac_2O	2-Hydroxy-3-phenylquinoline (—)	111

TABLE I. CONDENSATIONS WITH o-AMINOBENZALDEHYDE (*Continued*)

No. of C Atoms	Reactant	Catalyst	Reaction Conditions	Product(s) and Yield(s) (%)	Refs.
C_8 (*Contd.*)	$C_2H_5OCOCH{=}C(ONa)CO_2C_2H_5$	KOH	Boil, 90 min; EtOH–H_2O	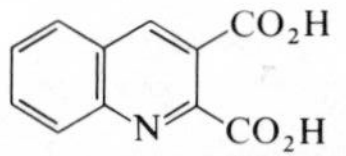 CO_2H, N, CO_2H (—)	106
	$C_6H_5SO_2CH_2CONH_2$	—	150–160°, 3 hr	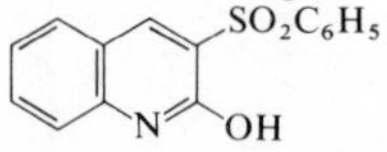 $SO_2C_6H_5$, N, OH (—)	112
	$C_2H_5OCOCH_2COCO_2C_2H_5$	KOH	7 d; EtOH	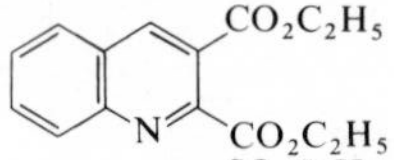 $CO_2C_2H_5$, N, $CO_2C_2H_5$ (37)	113
	$C_6H_5SO_2CH_2CN$	NaOH	Water bath, 3 hr; EtOH	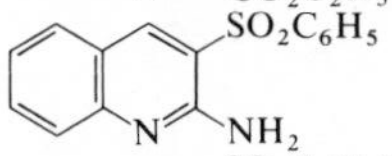 $SO_2C_6H_5$, N, NH_2 (—)	112
	p-$ClC_6H_4SO_2CH_2CN$	"	"	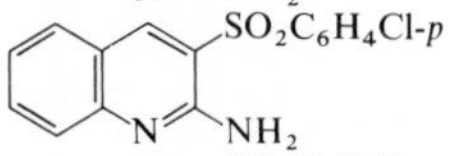 $SO_2C_6H_4Cl$-p, N, NH_2 (—)	112
	p-$BrC_6H_4SO_2CH_2CN$	"	Reflux, 3 hr; EtOH	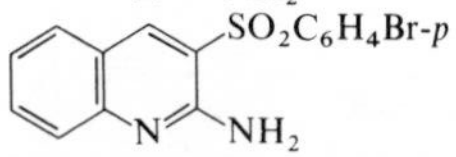 $SO_2C_6H_4Br$-p, N, NH_2 (—)	114
	$CH_3COC_6H_4NO_2$-m	Amberlite IRA-400	Reflux, 5 hr; EtOH	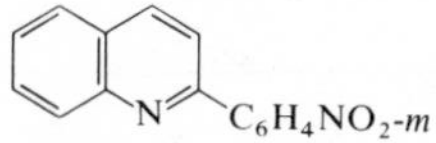 N, $C_6H_4NO_2$-m (81)	69
	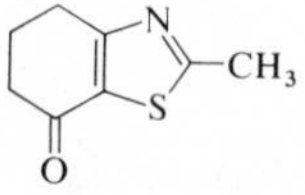 N, S, CH_3, O	$NaOC_2H_5$	1 hr; EtOH	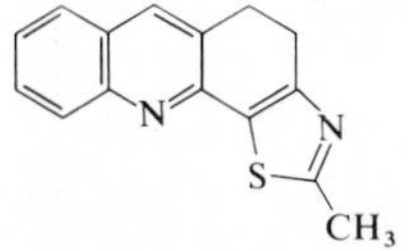 N, N, S, CH_3 (75)	115

	Reactant	Catalyst	Conditions	Product (Yield %)	Refs.
	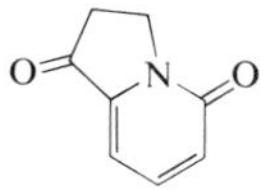 [structure: pyrrolo-pyridinedione, O, N, O]	NaOH	10–21°, 24 hr; EtOH	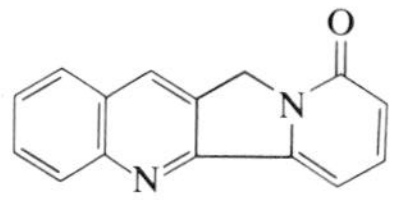 [structure: O, N, N] (32)	116
	$CH_3O_2CCH_2C(NH_2){=}C(CN)CO_2CH_3$	Piperidine	AcOH–C_6H_6	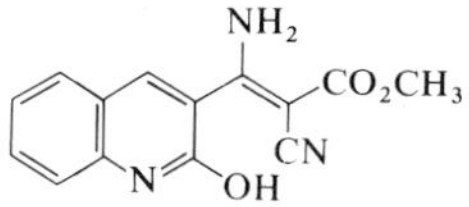 [structure: NH_2, CO_2CH_3, CN, N, OH] (56)	117
	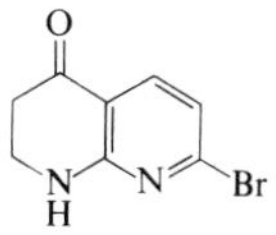 [structure: O, N, H, N, Br]	HCl	Cold; EtOH	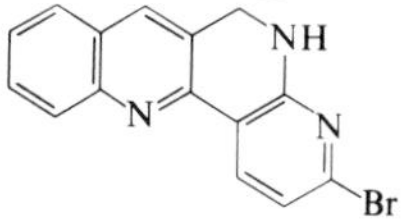 [structure: NH, N, N, Br] (59)	117a
		KOH	Reflux, 15 hr; EtOH	" (47) 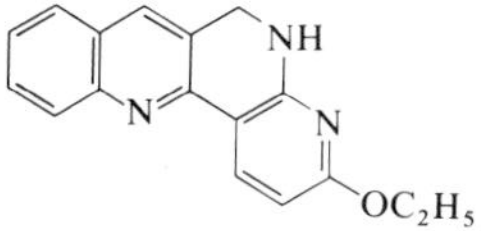 + [structure: NH, N, N, OC_2H_5] (25)	117a
C_9	$C_6H_5COCH_2CO_2H$	(pH 9)	25°	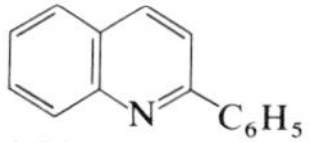 [structure: N, C_6H_5] (95)	57
	$C_6H_5COCH_2CO_2H$	(pH 13)	25°	2-Phenyl-3-carboxyquinoline ("v. good")	57
	$C_6H_5CH_2COCO_2H$	NaOH	100°, 12 hr; H_2O–EtOH	2-Carboxy-3-phenylquinoline (90)	50
	$C_6H_5C({=}NH)CH_2CN$	—	180°; EtOH	2-Phenyl-3-cyanoquinoline (—)	100
	p-$CH_3OC_6H_4COCH_2Cl$	NaOH	3–4 d; H_2O	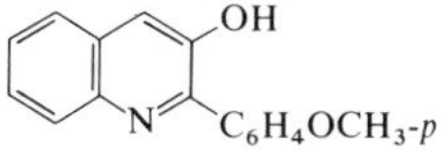 [structure: OH, N, $C_6H_4OCH_3$-p] (—)	110
	$C_2H_5OCOCH_2COCH_2CO_2C_2H_5$	NaOH	10 d; EtOH	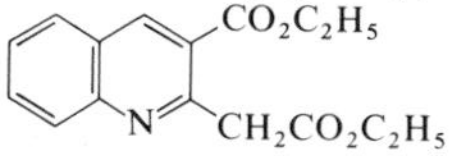 [structure: $CO_2C_2H_5$, N, $CH_2CO_2C_2H_5$] (70)	107
		NH_4OH	Several hr; H_2O	" (—)	118

TABLE I. CONDENSATIONS WITH *o*-AMINOBENZALDEHYDE (*Continued*)

No. of C Atoms	Reactant	Catalyst	Reaction Conditions	Product(s) and Yield(s) (%)	Refs.
C_9 (*Contd.*)	p-$CH_3C_6H_4SO_2CH_2CN$	NaOH	Water bath, 3 hr; EtOH	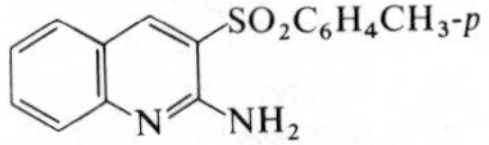 $SO_2C_6H_4CH_3$-p, N, NH_2 (95)	112
	$CH_3COCH_2SO_2C_6H_5$	NaOH	Water bath, long time; EtOH	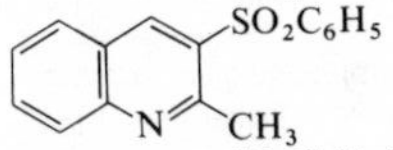 $SO_2C_6H_5$, N, CH_3 (—)	54
	$CH_3COCH_2SO_2C_6H_4Cl$-p	NaOH	Reflux, 4 hr; H_2O–EtOH	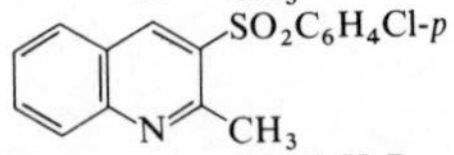 $SO_2C_6H_4Cl$-p, N, CH_3 (—)	54
	$CH_3COCH_2SO_2C_6H_4Br$-p	NaOH	Reflux, 4 hr; EtOH	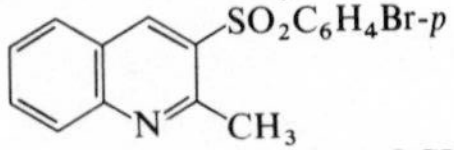 $SO_2C_6H_4Br$-p, N, CH_3 (80)	119
	o-$CH_3OC_6H_4SO_2CH_2CN$	NaOH	Water bath, 1 hr; EtOH	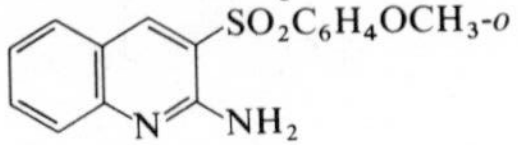 $SO_2C_6H_4OCH_3$-o, N, NH_2 (—)	114
	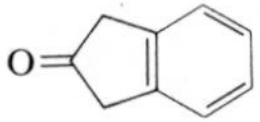 O=	HCl	Reflux, 90 min	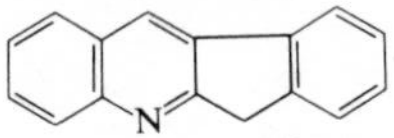 N (—)	29
	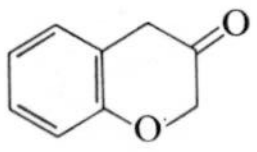 O, O	HCl	140°, 30 min	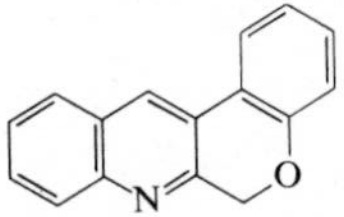 N, O (65)	120
	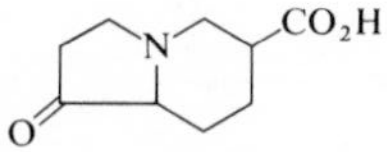 N, CO_2H, O	KOH	Reflux, 90 min; EtOH	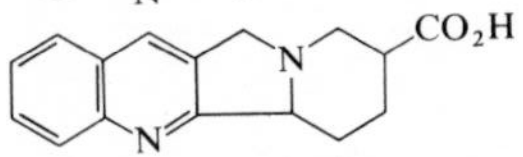 N, N, CO_2H (52)	87
	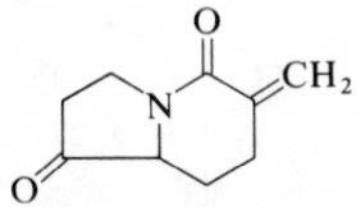 O, N, CH_2, O	KOH	Reflux, 24 hr; EtOH	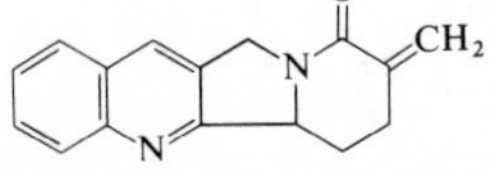 O, N, CH_2, N (33)	87

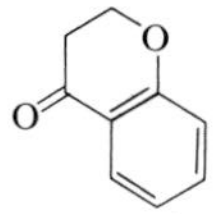	NaOEt	20°, 2 hr; EtOH	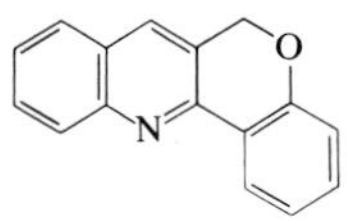	(25)	121
	NaOH	H_2O–MeOH	”	(73)	92
	HCl	o-$HCOC_6H_4NH_2 \cdot HCl$, 140°, 30 min	”	(60)	34
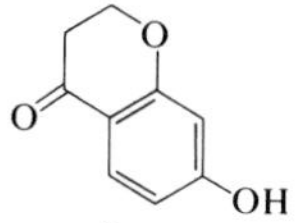	NaOMe	Reflux, 1 hr; MeOH	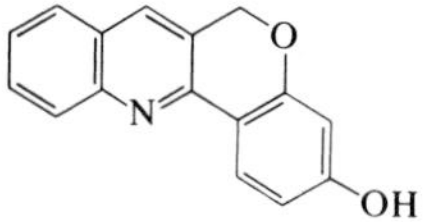	(—)	92
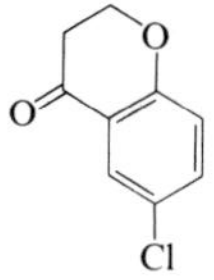	—	o-$HCOC_6H_4NH_2 \cdot HCl$, 140°, 30 min	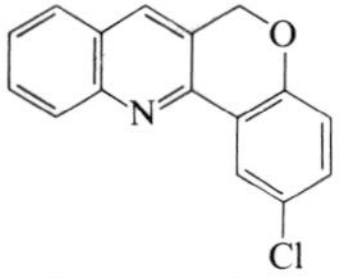	(60)	34, 121
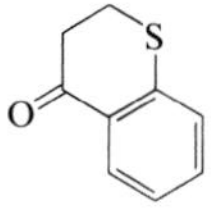	—	”	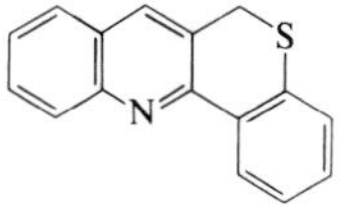	(60)	34
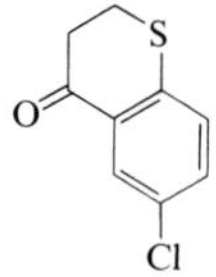	—	”	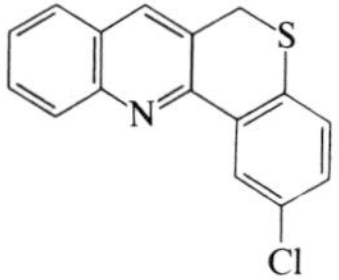	(60)	34
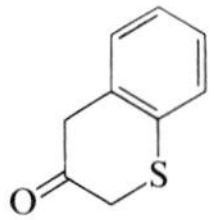	—	”	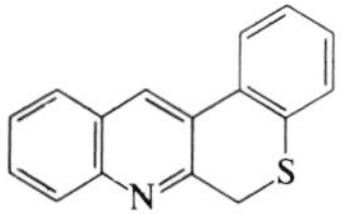	(—)	121a

TABLE I. CONDENSATIONS WITH *o*-AMINOBENZALDEHYDE (*Continued*)

No. of C Atoms	Reactant	Catalyst	Reaction Conditions	Product(s) and Yield(s) (%)	Refs.
C_9 (*Contd.*)		HCl	0°; EtOH	(83)	117a
		KOH	Reflux, 15 hr; EtOH	" (59)	117a
		—	Heat, 1.5 hr; AcOH	(—)	102
		NaOH	Reflux, 40 hr; H_2O	(29)	122
		$(C_2H_5)_2NH$	EtOH	(—)	123
		HCl	AcOH	" (—)	123
		Piperidine	Reflux, 10 hr; EtOH	(57)	123a

C_{10}	$C_6H_5(CH_2)_2COCO_2H$	NaOH	100°, 3 hr; H_2O	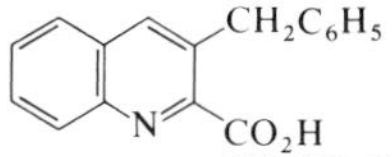 $CH_2C_6H_5$, CO_2H, N (75)	50
	$C_6H_5CO(CH_2)_2CO_2H$	NaOH	100°, 6 hr; MeOH	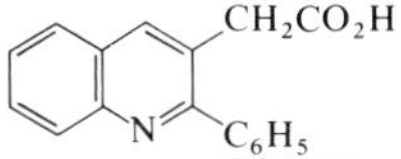 CH_2CO_2H, C_6H_5, N (62)	124
	$C_6H_5COCH_2COCH_3$	Piperidine	Reflux, 24 hr; EtOH	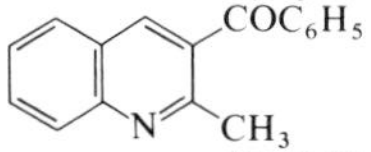 COC_6H_5, CH_3, N (79)	72
	$CH_3COCH_2SO_2C_6H_4CH_3$-*p*	NaOH	Water bath, 24 hr; EtOH	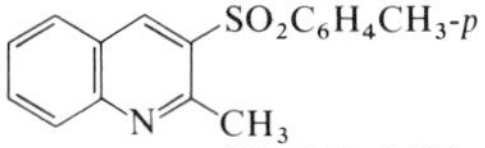 $SO_2C_6H_4CH_3$-*p*, CH_3, N ("Good")	54
	$CH_3COCH_2SO_2C_6H_4OCH_3$-*p*	"	Reflux, 1 hr; EtOH	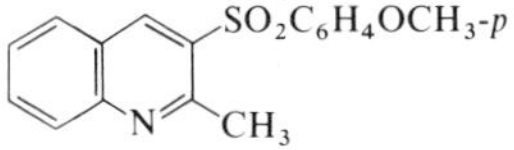 $SO_2C_6H_4OCH_3$-*p*, CH_3, N (—)	125
	$CH_3COCH_2SO_2C_6H_4OCH_3$-*o*	"	"	$SO_2C_6H_4OCH_3$-*o*, CH_3, N (—)	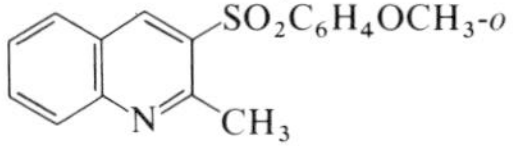 125
	$C_6H_5SO_2CH_2CO_2C_2H_5$	—	170°, 3 hr	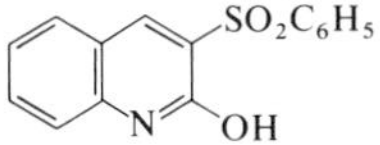 $SO_2C_6H_5$, OH, N (—)	112
	p-$ClC_6H_4SO_2CH_2CO_2C_2H_5$	—	"	$SO_2C_6H_4Cl$-*p*, OH, N (—)	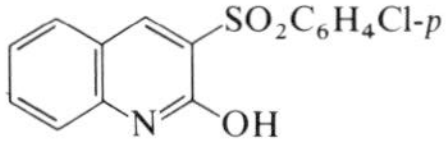 112
	p-$C_2H_5OC_6H_4SO_2CH_2CN$	NaOH	Water bath, several hr; EtOH 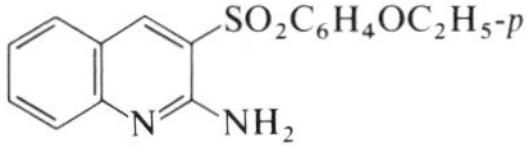	$SO_2C_6H_4OC_2H_5$-*p*, NH_2, N (—)	114
	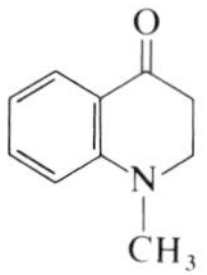 O, N, CH_3	NaOH	3 hr; EtOH 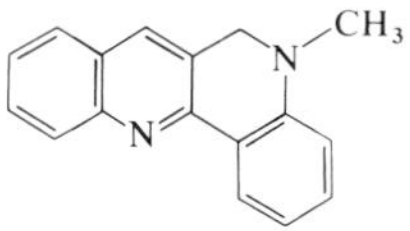	N, N, CH_3 (87)	74b

TABLE I. CONDENSATIONS WITH *o*-AMINOBENZALDEHYDE (*Continued*)

No. of C Atoms	Reactant	Catalyst	Reaction Conditions	Product(s) and Yield(s) (%)	Refs.
C_{10} (*Contd.*)	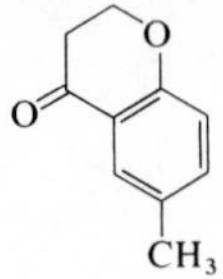 O, O, CH_3	—	*o*-$HCOC_6H_4NH_2\cdot HCl$, 140°, 30 min	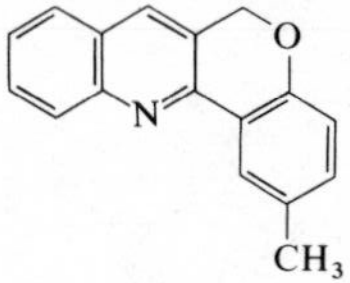 N, O, CH_3 (60)	34, 121
	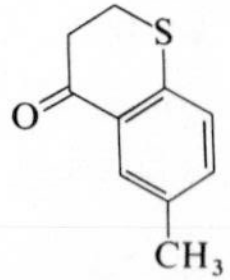 S, O, CH_3	—	"	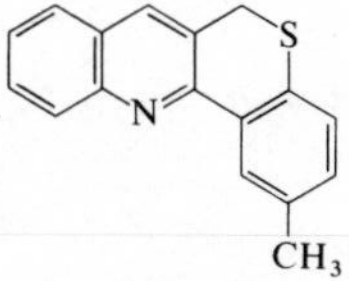 N, S, CH_3 (60)	34
	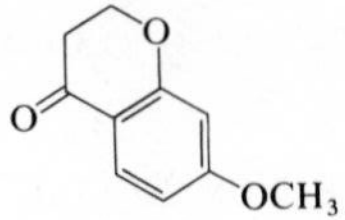 O, O, OCH_3	NaOMe	Reflux, 30 min; MeOH	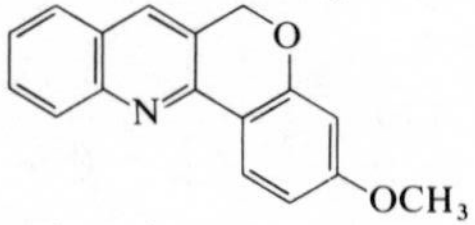 N, O, OCH_3 (76)	92
	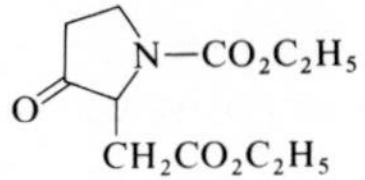 O, $N—CO_2C_2H_5$, $CH_2CO_2C_2H_5$	NaOH	—	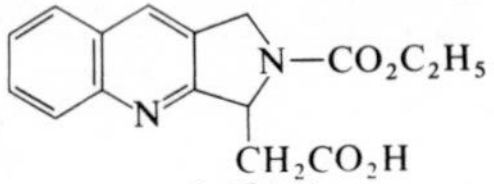 N, $N—CO_2C_2H_5$, CH_2CO_2H (50)	126
		NaH	Overnight; C_6H_6	" (31.5)	61
	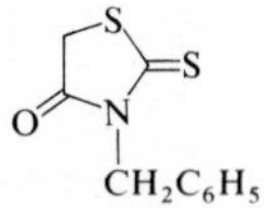 S, =S, O, N, $CH_2C_6H_5$	NaOAc	Heat; AcOH	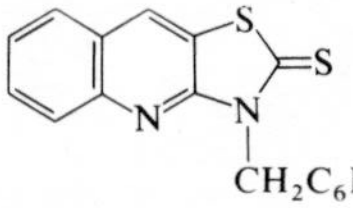 N, S, =S, N, $CH_2C_6H_5$ (—)	102
	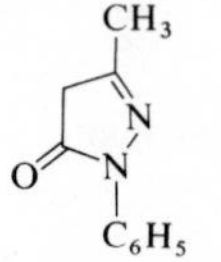 CH_3, N, N, O, C_6H_5	—	140°	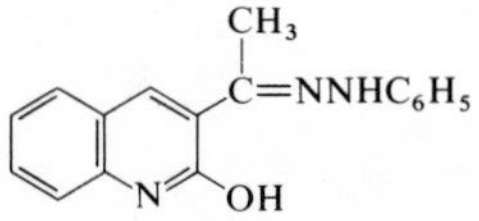 CH_3, $C=NNHC_6H_5$, N, OH (—)	127

CH_3, O, N, N, $C_6H_4Cl\text{-}o$	—	265°	C_6H_5, N, N, CH_3, N, OH (—) + N, N, OH (—)	127
	—	150°	OH, CH_3, $C{=}NNHC_6H_4Cl\text{-}o$, N, OH (80) + CH_3, $C{=}NNHC_6H_4Cl\text{-}o$, N, OH (10)	127
	—	180–190°, 3 hr	OH, CH_3, $C{=}NNHC_6H_4Cl\text{-}o$, N, OH (70) + $C_6H_4Cl\text{-}o$, N, N, CH_3, N (25)	127

TABLE I. CONDENSATIONS WITH o-AMINOBENZALDEHYDE (*Continued*)

No. of C Atoms	Reactant	Catalyst	Reaction Conditions	Product(s) and Yield(s) (%)	Refs.
C_{10} (*Contd.*)	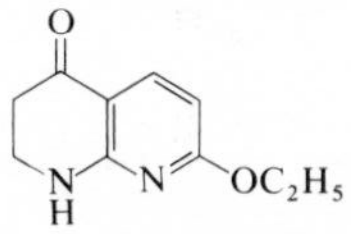 [structure: O; N, H; N; OC_2H_5]	HCl	EtOH	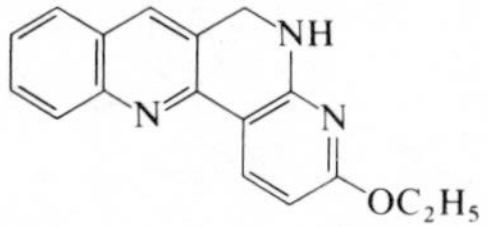 [structure: NH; N; N; OC_2H_5] (74)	117a
		KOH	Reflux, 15 hr; EtOH	" (34)	117a
C_{11}	$C_6H_5CO(CH_2)_3CO_2H$	NaOH	Heat, 24 hr; H_2O	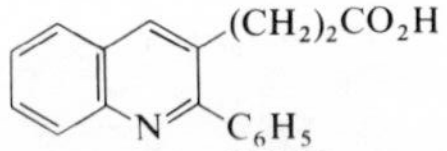 [structure: N; $(CH_2)_2CO_2H$; C_6H_5] (58)	23
	$C_6H_5COCH_2CO_2C_2H_5$	"	6 d; H_2O	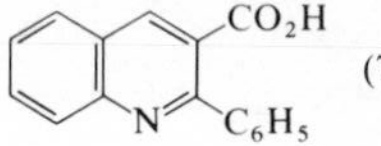 [structure: N; CO_2H; C_6H_5] (70)	124
	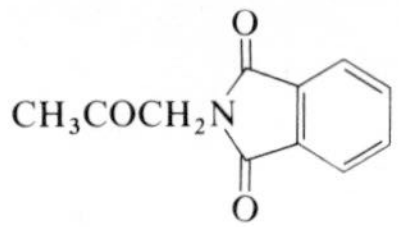 CH_3COCH_2N [structure: O; O]	"	Few d; H_2O	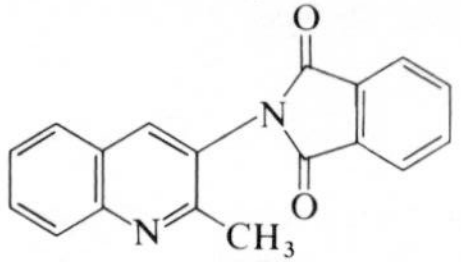 [structure: N; N; O; O; CH_3] (—)	110
	$CH_3COCH_2SO_2C_6H_4OC_2H_5$-*p*	"	Reflux, 6 hr; EtOH	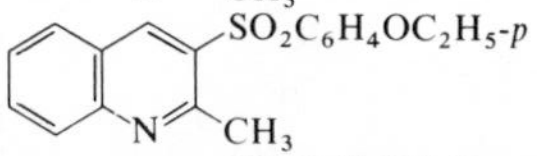 [structure: N; $SO_2C_6H_4OC_2H_5$-*p*; CH_3] (—)	125
	$CH_3COCH_2SO_2C_6H_4OC_2H_5$-*o*	"	"	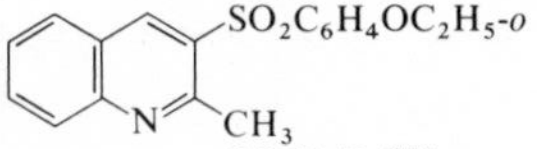 [structure: N; $SO_2C_6H_4OC_2H_5$-*o*; CH_3] (—)	125
	p-$CH_3C_6H_4SO_2CH_2CO_2C_2H_5$	—	160–170°, 3–4 hr	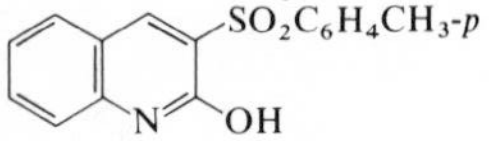 [structure: N; $SO_2C_6H_4CH_3$-*p*; OH] (—)	112
	$C_2H_5O_2CCH_2CONHC_6H_4Br$-*p*	Piperidine	Steam bath, 3 hr	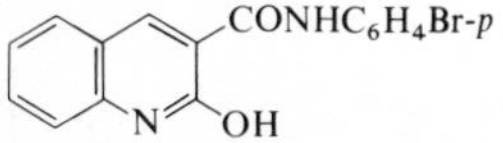 [structure: N; $CONHC_6H_4Br$-*p*; OH] (88)	128

Reactant	Base	Conditions	Product (Yield, %)	Refs.
$(C_2H_5OCOCOCH_2)_2CO$	KOH	100°, 1 hr; H_2O	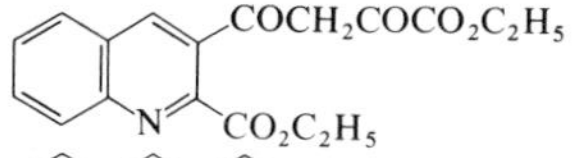 $COCH_2COCO_2C_2H_5$, $CO_2C_2H_5$ (—)	130
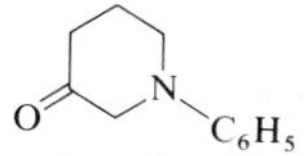	NaOH	20°, EtOH	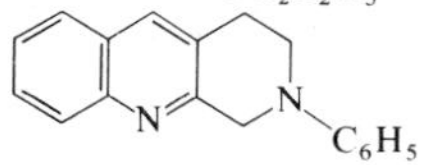 (15)	32
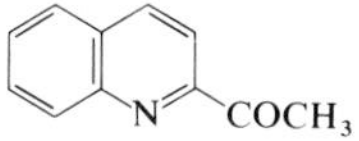	KOH	Reflux, 1 hr; EtOH	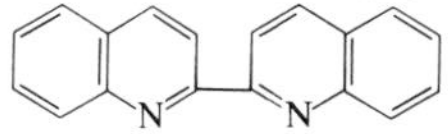 (59)	68
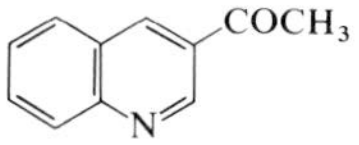	"	Several d; EtOH	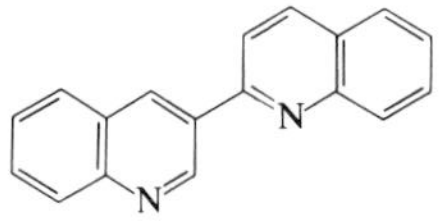 (53)	130
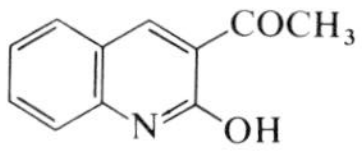	NaOH	Reflux; EtOH	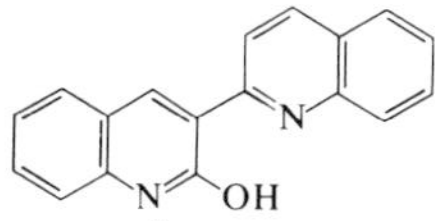 (—)	127
	—	265°	" (—)	127
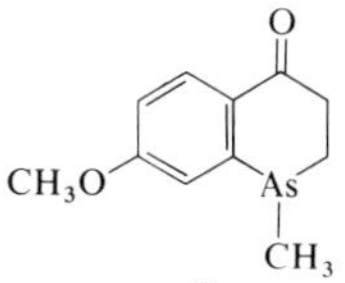	NaOH	3 d; EtOH	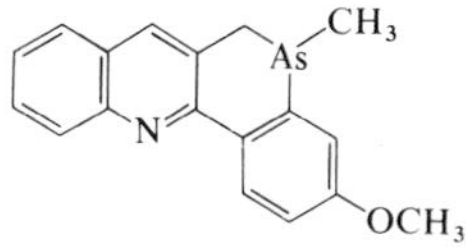 (39)	74a
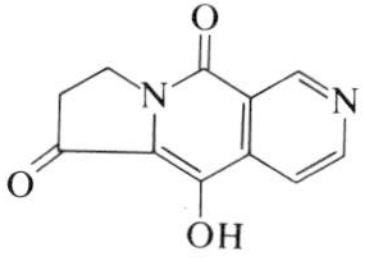	"	Reflux, 5 hr; H_2O–EtOH	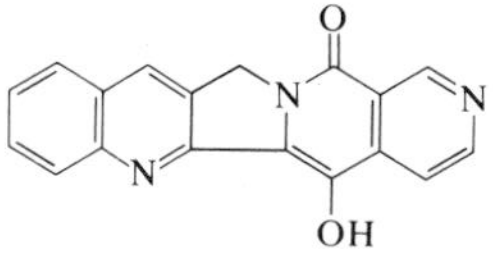 (—)	132
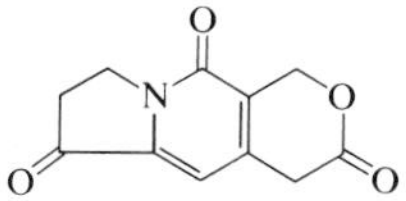	"	16 hr; EtOH	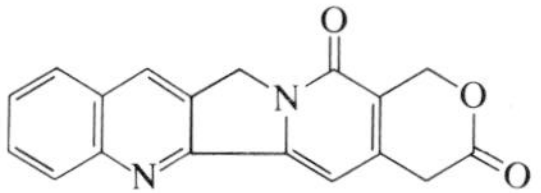 (71)	133

TABLE I. CONDENSATIONS WITH *o*-AMINOBENZALDEHYDE (*Continued*)

No. of C Atoms	Reactant	Catalyst	Reaction Conditions	Product(s) and Yield(s) (%)	Refs.
C_{12}	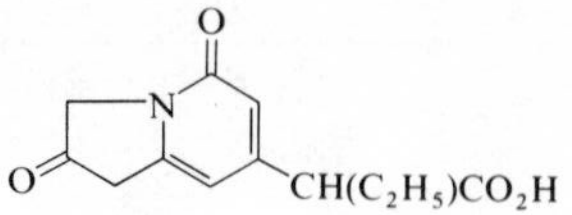 O, N, O, $CH(C_2H_5)CO_2H$	NaOH	MeOH	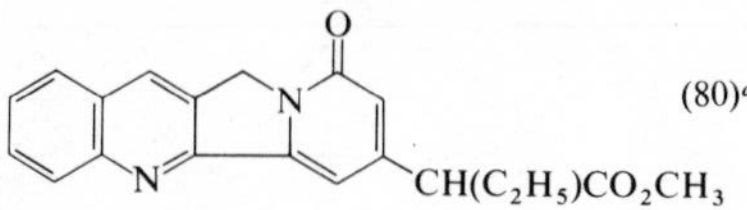 O, N, N, $CH(C_2H_5)CO_2CH_3$ (80)[d]	131
	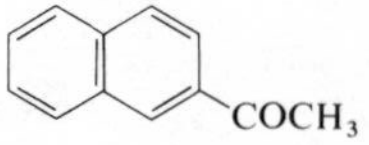 $COCH_3$	"	Reflux; EtOH	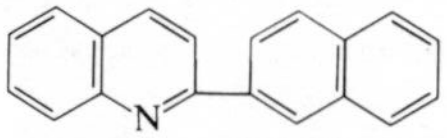 N (—)	56
	$C_2H_5O_2CCH_2CONHC_6H_4OCH_3$-*p*	Piperidine	Steam bath, 3 hr	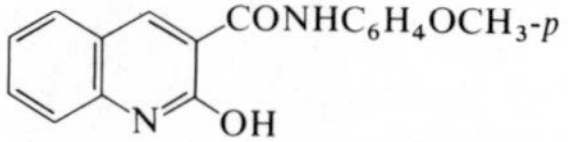 $CONHC_6H_4OCH_3$-*p*, N, OH (91)	128
	$C_2H_5O_2CCH_2CONHC_6H_4OCH_3$-*m*	"	"	 $CONHC_6H_4OCH_3$-*m*, N, OH (90)	128
	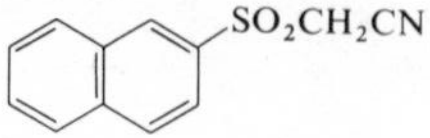 SO_2CH_2CN	NaOH	100°, 3 hr; EtOH	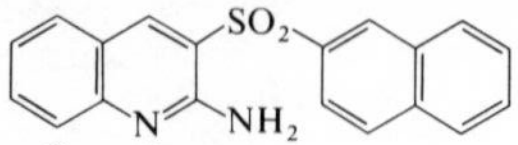 SO_2, N, NH_2 (—)	114
	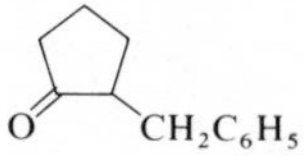 O, $CH_2C_6H_5$	Piperidine	150°; xylene	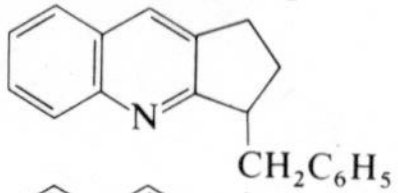 N, $CH_2C_6H_5$ (60)	90
	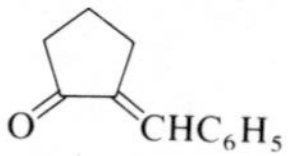 O, CHC_6H_5	"	"	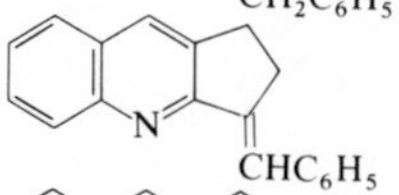 N, CHC_6H_5 (60)	90
	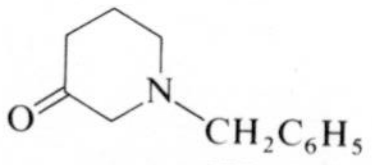 O, N, $CH_2C_6H_5$	NaOH	20°, 2 d; EtOH	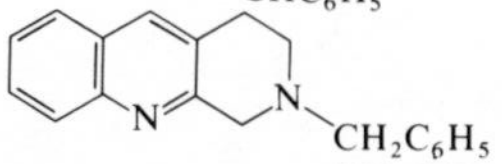 N, N, $CH_2C_6H_5$ (69)	32
	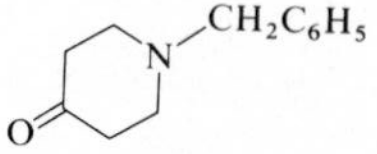 O, N, $CH_2C_6H_5$	"	"	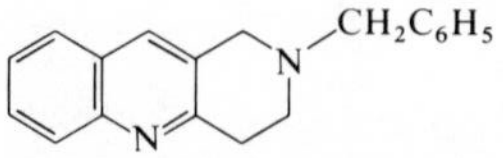 N, N, $CH_2C_6H_5$ (47)	32

	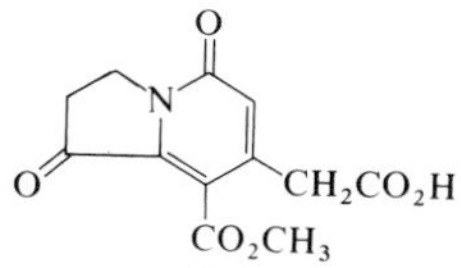 O; N; O; CH_2CO_2H; CO_2CH_3	"	Reflux, 36 hr; H_2O	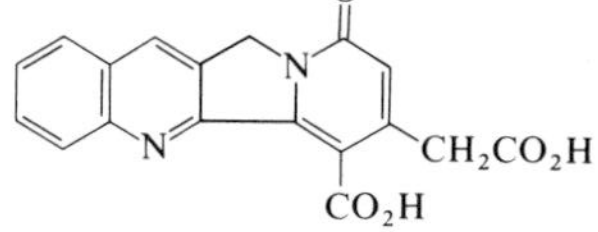 O; N; N; CH_2CO_2H; CO_2H (—)	134
	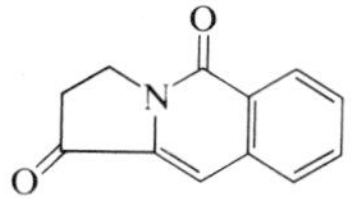 O; N; O	Triton B	Reflux, 1 hr; EtOH	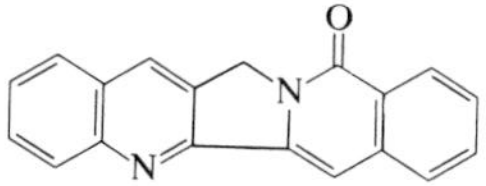 O; N; N (57)	132
	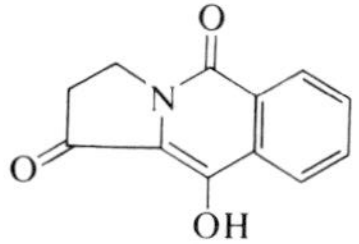 O; N; O; OH	"	Reflux, 6 hr; EtOH	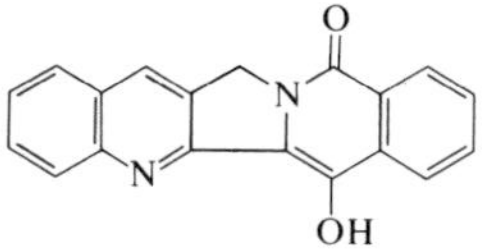 O; N; N; OH (37)	132
	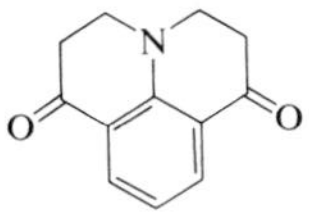 N; O; O	NaOH	36 hr; EtOH	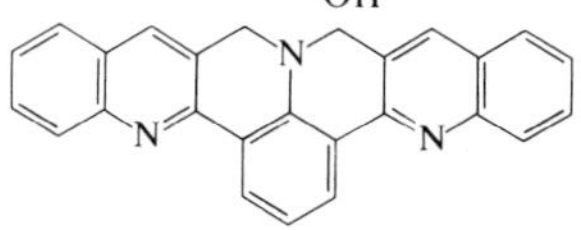 N; N; N (81)	74b
C_{13}	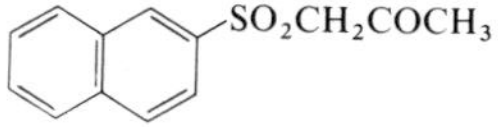 $SO_2CH_2COCH_3$	NaOH	100°, long time; EtOH	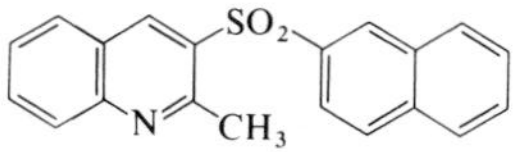 SO_2; N; CH_3 (—)	54
	$C_2H_5O_2CCH_2CONHC_6H_4OC_2H_5$-*o*	Piperidine	Steam bath, 3 hr	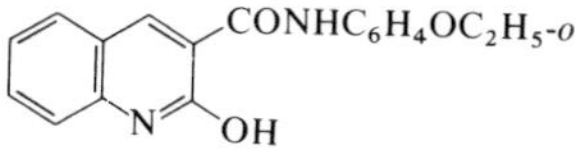 $CONHC_6H_4OC_2H_5$-*o*; N; OH (83)	128
	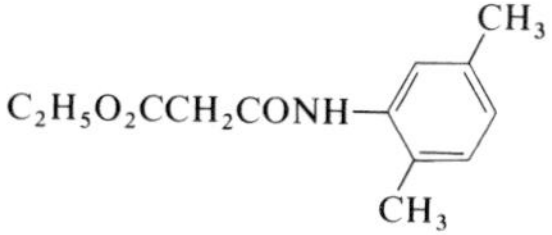 $C_2H_5O_2CCH_2CONH$—; CH_3; CH_3	"	"	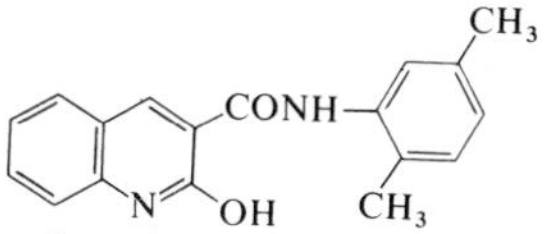 CONH—; CH_3; CH_3; N; OH (90)	128
	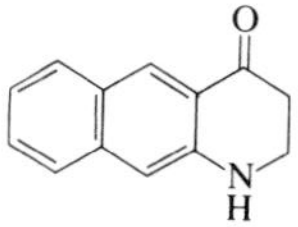 O; N; H	NaOH	2 d	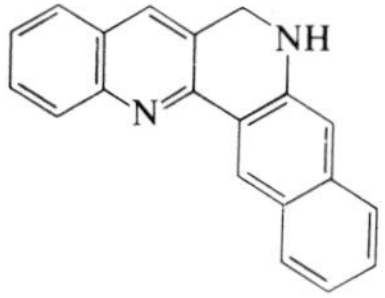 NH; N (82)	129

TABLE I. CONDENSATIONS WITH *o*-AMINOBENZALDEHYDE (*Continued*)

No. of C Atoms	Reactant	Catalyst	Reaction Conditions	Product(s) and Yield(s) (%)	Refs.
C_{13} (*Contd.*)	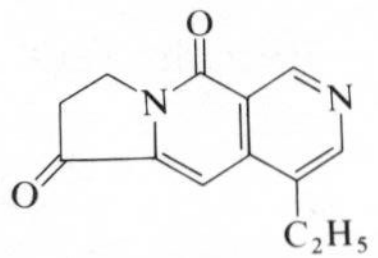 [structure: pyrrolo-naphthyridinedione with C_2H_5]	NaOH	24 hr; EtOH–H_2O	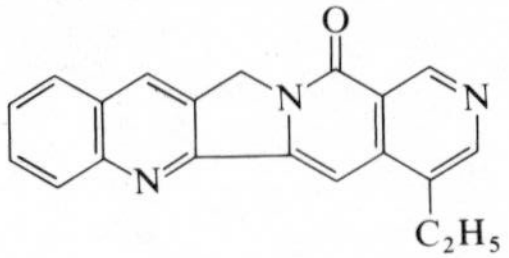 [structure with C_2H_5] (48)	135
	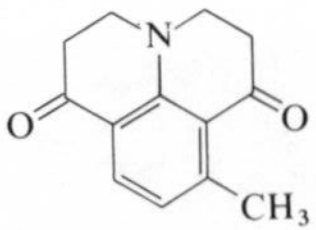 [structure: julolidine-dione with CH_3]	"	36 hr; EtOH	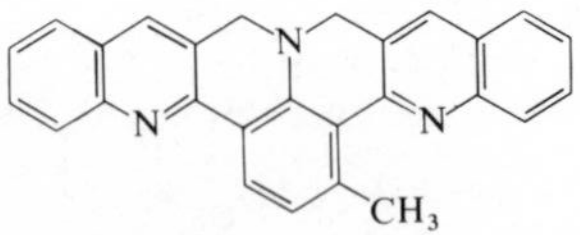 [structure with CH_3] (83)	74b
	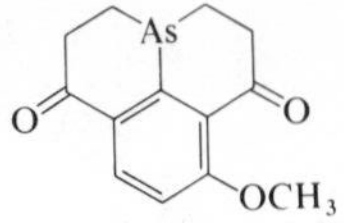 [structure: As-containing dione with OCH_3]	"	5 d; EtOH	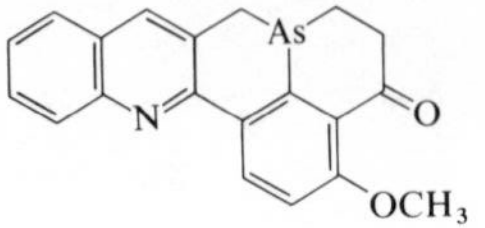 [structure with OCH_3] (—)	74a
C_{14}	$C_6H_5COCH_2SO_2C_6H_5$	—	200°, 2 hr	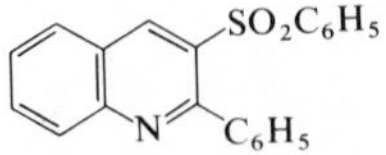 3-$SO_2C_6H_5$-2-C_6H_5-quinoline (70)	136
	$C_6H_5COCH_2SO_2C_6H_4Cl$-*p*	—	240°, 2 hr	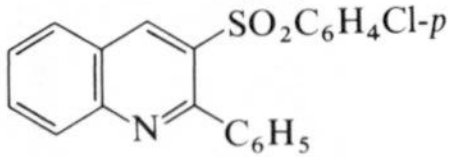 3-$SO_2C_6H_4Cl$-*p*-2-C_6H_5-quinoline (—)	136
	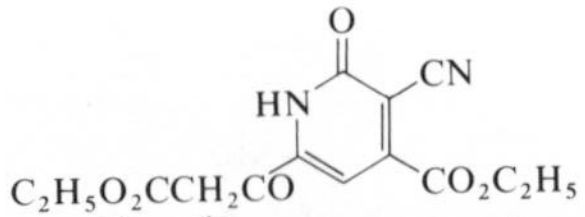 [structure: 6-($C_2H_5O_2CCH_2CO$)-pyridone with CN, $CO_2C_2H_5$]	Amberlite IRA-400	Reflux, 24 hr; EtOH	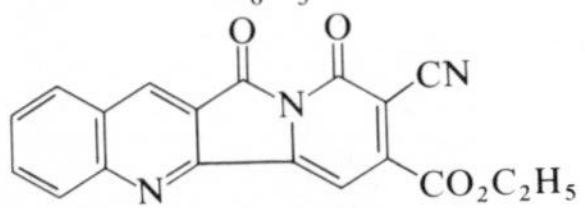 [structure with CN, $CO_2C_2H_5$] (10)	137
	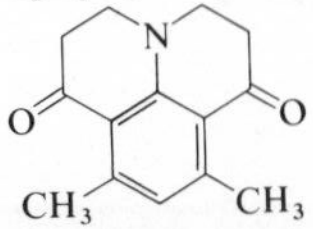 [structure: julolidine-dione with two CH_3]	NaOH	36 hr; EtOH	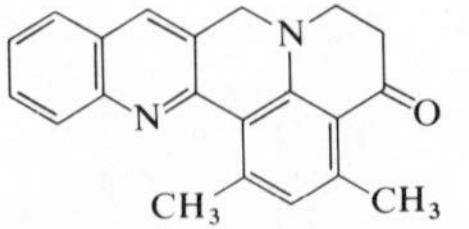 [structure with two CH_3] (50)	74b

C_{15}	$C_6H_5COCH_2COC_6H_5$	—	200–210°, 5 hr	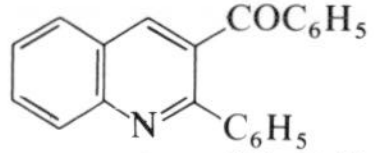 COC_6H_5, N, C_6H_5 (78)	72
	$C_6H_5COCH_2SO_2C_6H_4CH_3$-*p*	—	200°, 1.5 hr	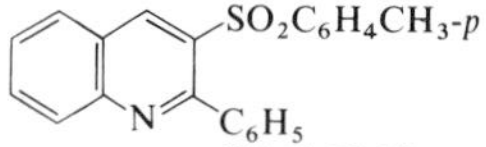 $SO_2C_6H_4CH_3$-*p*, N, C_6H_5 (—)	136
	p-$ClC_6H_4SO_2CH_2COCH_2SO_2C_6H_5$	—	130–150°, 20 min	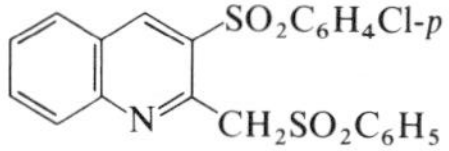 $SO_2C_6H_4Cl$-*p*, N, $CH_2SO_2C_6H_5$ (—)	136
	$(C_6H_5SO_2CH_2)_2CO$	—	160°, 2.5 hr	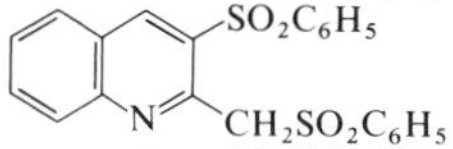 $SO_2C_6H_5$, N, $CH_2SO_2C_6H_5$ (—)	136
	(*p*-$ClC_6H_4SO_2CH_2)_2CO$	—	165°, 15 min	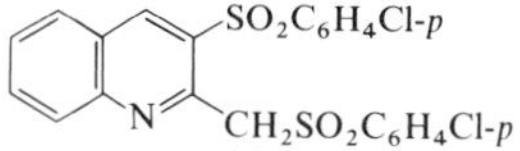 $SO_2C_6H_4Cl$-*p*, N, $CH_2SO_2C_6H_4Cl$-*p* (—)	136
	$C_6H_5COCH_2SO_2C_6H_4OCH_3$-*o*	—	6–8 hr, 220°	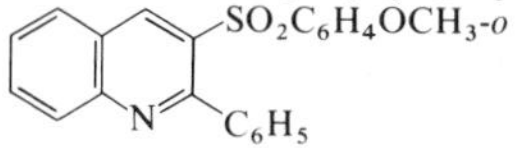 $SO_2C_6H_4OCH_3$-*o*, N, C_6H_5 (—)	98
	$C_6H_5COCH_2SO_2C_6H_4OCH_3$-*p*	—	Heat, 190–205°	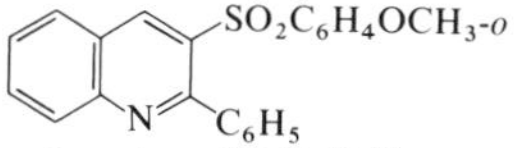 $SO_2C_6H_4OCH_3$-*o*, N, C_6H_5 (—)	98
	(*p*-$BrC_6H_4SO_2CH_2)_2CO$	—	Water bath, 3 d 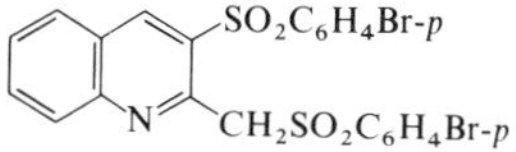	$SO_2C_6H_4Br$-*p*, N, $CH_2SO_2C_6H_4Br$-*p* (—)	138
	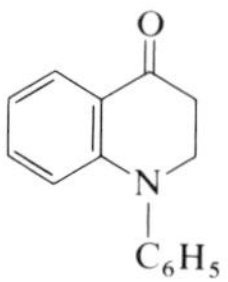 O, N, C_6H_5	NaOH	3 hr, EtOH	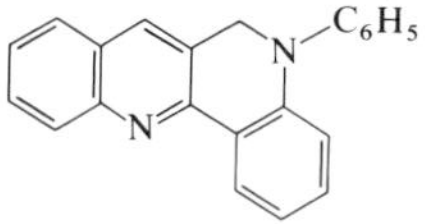 N, N, C_6H_5 (96)	74b
C_{16}	$C_6H_5CH_2COCH_2COC_6H_5$	Piperidine	Reflux, 48 hr; EtOH	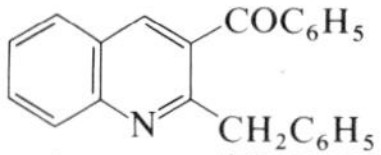 COC_6H_5, N, $CH_2C_6H_5$ (43)	73
	p-$ClC_6H_4SO_2CH_2CO$ / *p*-$CH_3C_6H_4SO_2CH_2$	—	165°, 10 min	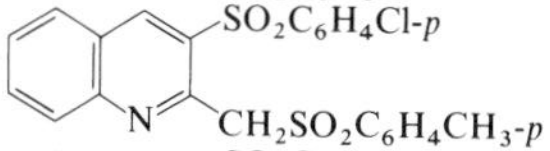 $SO_2C_6H_4Cl$-*p*, N, $CH_2SO_2C_6H_4CH_3$-*p* (—)	136
	p-$BrC_6H_4SO_2CH_2CO$ / *p*-$CH_3C_6H_4SO_2CH_2$	—	150°, 25 min 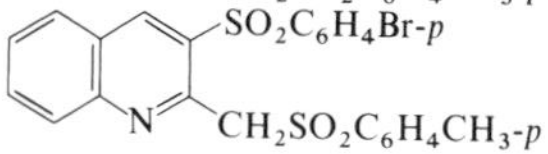	$SO_2C_6H_4Br$-*p*, N, $CH_2SO_2C_6H_4CH_3$-*p* (—)	138

TABLE I. CONDENSATIONS WITH *o*-AMINOBENZALDEHYDE (*Continued*)

No. of C Atoms	Reactant	Catalyst	Reaction Conditions	Product(s) and Yield(s) (%)	Refs.
C_{16} (*Contd.*)	$p\text{-}BrC_6H_4SO_2CH_2CO$ / $o\text{-}CH_3OC_6H_4SO_2CH_2$	—	100°, 2 d	$SO_2C_6H_4Br\text{-}p$; N; $CH_2SO_2C_6H_4OCH_3\text{-}o$ (—)	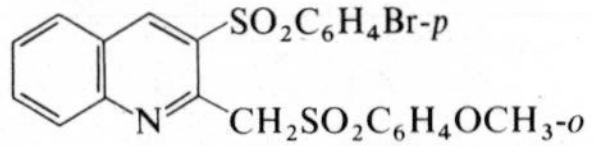 138
	$C_6H_5COCH_2SO_2C_6H_4OC_2H_5\text{-}o$	—	Heat, 180–200°	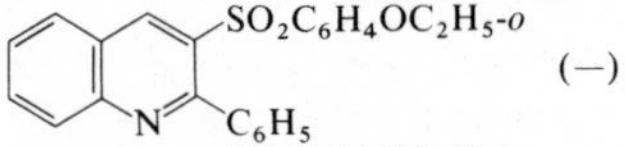 $SO_2C_6H_4OC_2H_5\text{-}o$; N; C_6H_5 (—)	98
	$C_6H_5COCH_2SO_2C_6H_4OC_2H_5\text{-}p$	—	10 hr, 200°	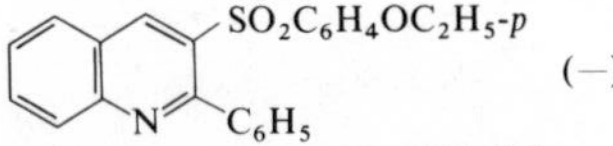 $SO_2C_6H_4OC_2H_5\text{-}p$; N; C_6H_5 (—)	98
	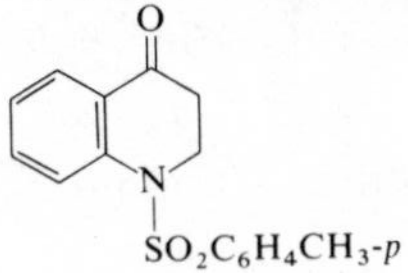 O; N; $SO_2C_6H_4CH_3\text{-}p$	NaOH	3 hr; EtOH	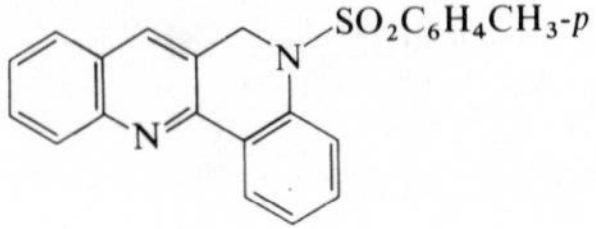 N; N; $SO_2C_6H_4CH_3\text{-}p$ (47)	74b
C_{17}	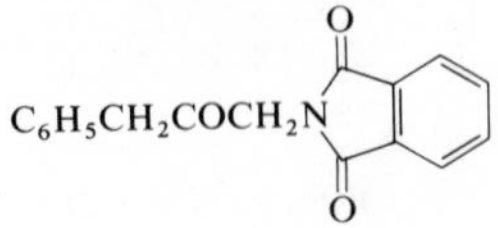 $C_6H_5CH_2COCH_2N$ (O; O)	NaOH	Few d, H_2O	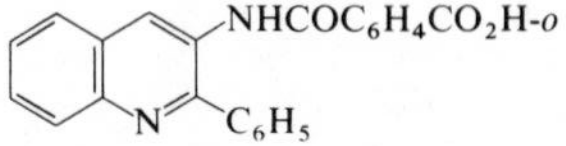 $NHCOC_6H_4CO_2H\text{-}o$; N; C_6H_5 (—)	110
	$(p\text{-}CH_3C_6H_4SO_2CH_2)_2CO$	—	155°, 15 min	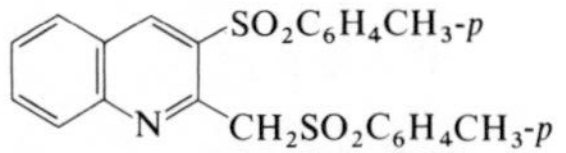 $SO_2C_6H_4CH_3\text{-}p$; N; $CH_2SO_2C_6H_4CH_3\text{-}p$ (—)	136
	$(o\text{-}CH_3OC_6H_4SO_2CH_2)_2CO$	—	180°, 2 hr	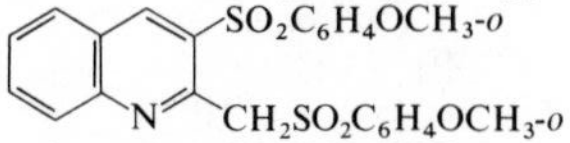 $SO_2C_6H_4OCH_3\text{-}o$; N; $CH_2SO_2C_6H_4OCH_3\text{-}o$ (—)	138
	$p\text{-}CH_3OC_6H_4SO_2CH_2CO$ / $p\text{-}CH_3C_6H_4SO_2CH_2$	—	150°, 2 hr	$SO_2C_6H_4OCH_3\text{-}p$; N; $CH_2SO_2C_6H_4CH_3\text{-}p$ (—)	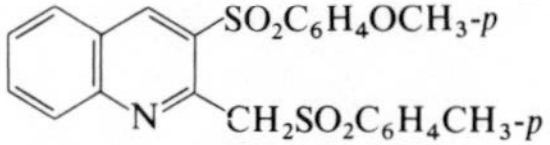 138
	$o\text{-}CH_3OC_6H_4SO_2CH_2CO$ / $p\text{-}CH_3C_6H_4SO_2CH_2$	—	160°, 1 hr	$SO_2C_6H_4OCH_3\text{-}o$; N; $CH_2SO_2C_6H_4CH_3\text{-}p$ (—)	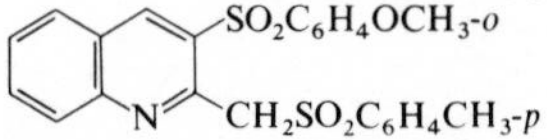 138

C_{18}	p-$C_2H_5OC_6H_4SO_2CH_2CO$ / p-$CH_3C_6H_4SO_2CH_2$	—	130°, 30 min	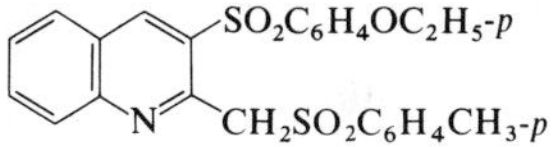 $SO_2C_6H_4OC_2H_5$-p, N, $CH_2SO_2C_6H_4CH_3$-p (—)	138
	o-$C_2H_5C_6H_4SO_2CH_2CO$ / p-$CH_3C_6H_4SO_2CH_2$	—	150°, 20 min 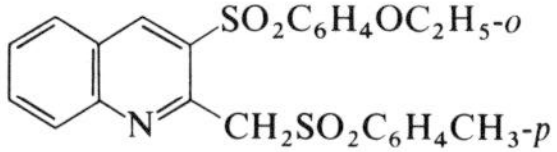	$SO_2C_6H_4OC_2H_5$-o, N, $CH_2SO_2C_6H_4CH_3$-p (—)	138
	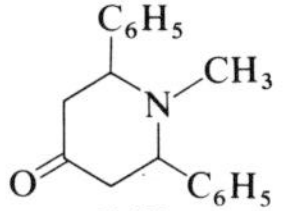 C_6H_5, N, CH_3, C_6H_5, O	NaOH	20°, 48 hr; EtOH 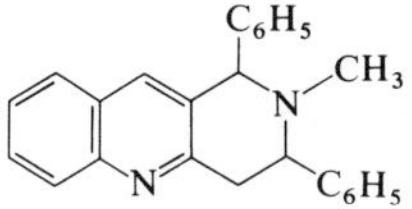	C_6H_5, N, CH_3, C_6H_5, N (66)	32
	C_6H_5, NH, C_6H_5, O, CH_3	”	”	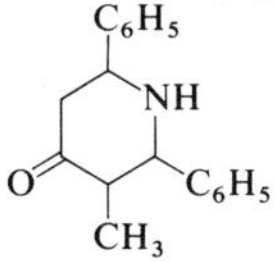 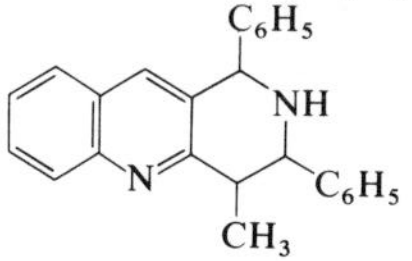 C_6H_5, NH, C_6H_5, CH_3, N (64)	32
C_{19}	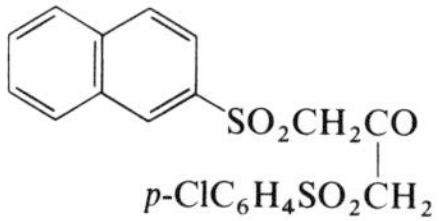 SO_2CH_2CO / p-$ClC_6H_4SO_2CH_2$	—	170°, 20 min 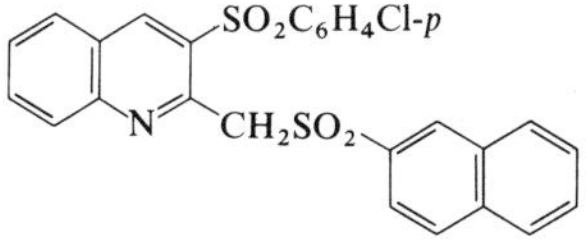	$SO_2C_6H_4Cl$-p, N, CH_2SO_2 (—)	136
	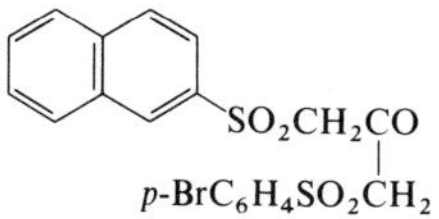 SO_2CH_2CO / p-$BrC_6H_4SO_2CH_2$	—	200°, 2 hr	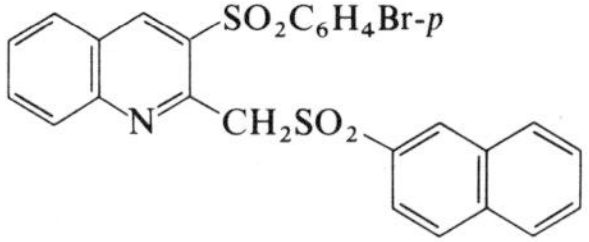 $SO_2C_6H_4Br$-p, N, CH_2SO_2 (—)	138

[a] o-Aminobenzaldehyde was generated *in situ* from o-dichloromethylphenyl isocyanate with (1) aq $Ba(OH)_2$ at 15–18°, and then (2) aq HCl at 0–5°.

[b] The reference gives the product as "3-acetyl-3-hydroxyquinaldine."

[c] The other reactant was o-acetamidobenzaldehyde.

[d] The product was isolated after Fischer esterification.

TABLE II. CONDENSATIONS WITH SUBSTITUTED *o*-AMINOBENZALDEHYDES

A. 5-Methoxy

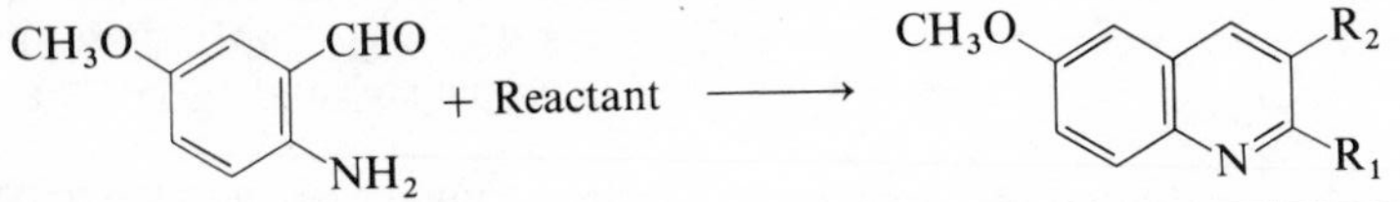

CH_3O, CHO, NH_2 + Reactant ⟶ CH_3O, R_2, N, R_1

Reactant	Catalyst	Reaction Conditions	Product(s) and Yield(s) (%) R_1	R_1R_2	R_2	Refs.
$CH_2(CO_2H)_2$	Pyridine	Reflux; EtOH	OH		CO_2H (—)	139
$CH_2(CN)_2$	"	Reflux 4 hr; EtOH	NH_2		CN (—)	139
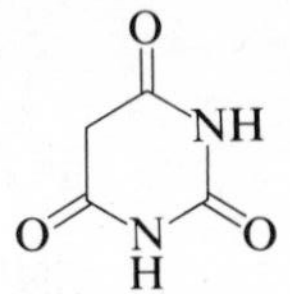 O, NH, O, N, H, O	NaOH	Reflux 2 hr; EtOH		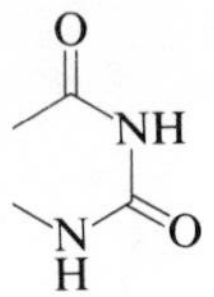 O, NH, N, H, O	(—)	139
$NCCH_2CO_2C_2H_5$	Pyridine	Reflux; EtOH	OH		CN (—)	139
$CH_3COCH_2CO_2C_2H_5$	NaOH	Reflux 3 hr; EtOH	CH_3		$CO_2C_2H_5$ (—)	139
		160°, 5 hr	OH		$COCH_3$ (—)	139
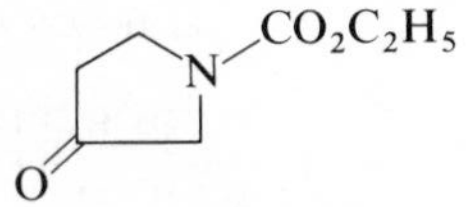 N—$CO_2C_2H_5$, O	pTSA	190–195°, 5 min	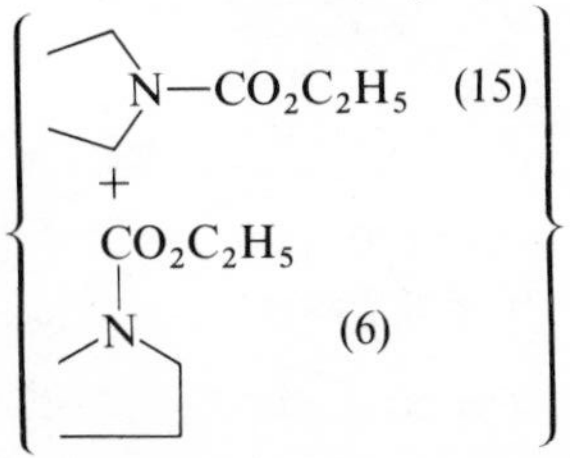 N—$CO_2C_2H_5$ (15) + $CO_2C_2H_5$ N (6)			89

$C_6H_5CH_2CN$	NaOH	Reflux 4 hr; EtOH	NH_2	C_6H_5 (—)	139
p-$BrC_6H_4SO_2CH_2CN$	"	Reflux; EtOH	NH_2	$SO_2C_6H_4Br$-p (—)	139
$C_6H_5COCH_2CN$	"	Reflux 2 hr; EtOH	C_6H_5	CN (—)	139
$C_6H_5COCH_2CO_2C_2H_5$	"	100°, 2 hr; H_2O	C_6H_5	$CO_2C_2H_5$ (—)	139
	—	160°, 4 hr	OH	COC_6H_5 (—)	139

B. 4-Methoxy

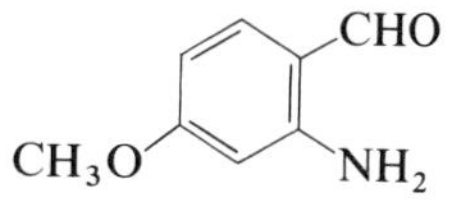

CH_3O, CHO, NH_2 + Reactant

Reactant	Catalyst	Reaction Conditions	Product(s) and Yield(s) (%)	Refs.
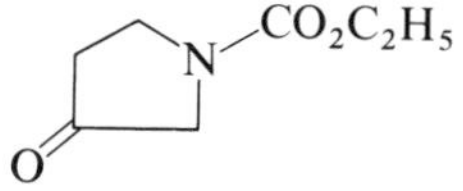 N-$CO_2C_2H_5$, O	pTSA	190–195°, 5 min	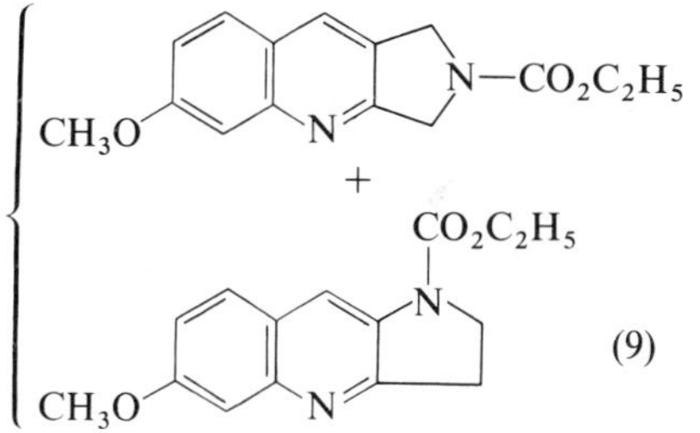 CH_3O, N, N—$CO_2C_2H_5$ (31) + $CO_2C_2H_5$, N, CH_3O, N (9)	89

TABLE II. CONDENSATIONS WITH SUBSTITUTED *o*-AMINOBENZALDEHYDES (*Continued*)

C. 3-Methoxy

2-Amino-3-methoxybenzaldehyde (CHO, NH_2, OCH_3) + Reactant ⟶ 8-methoxyquinoline (R_1 at C-2, R_2 at C-3, OCH_3 at C-8)

Reactant	Catalyst	Reaction Conditions	Product(s) and Yield(s)(%) R_1	R_2	Refs.
CH_3COCH_3	NaOH	Reflux; EtOH	CH_3	H (—)	140
$CH_2(CN)_2$	NaOH or Pyridine	Reflux 6 hr; EtOH	NH_2	CN (—)	65
$CH_2(CO_2H)_2$	Pyridine	100°	OH	CO_2H (—)	140
$NCCH_2CO_2C_2H_5$	"	Reflux; EtOH	OH	CN (—)	140
$C_6H_5COCH_3$	NaOH	Reflux 1 hr; EtOH	C_6H_5	H (—)	140
$C_6H_5CH_2CN$	"	Reflux 2 hr; EtOH	NH_2	C_6H_5 (—)	140
$C_6H_5COCH_2CN$	"	Reflux 1 hr; EtOH	C_6H_5	CN (—)	65
$C_6H_5COCH_2CO_2C_2H_5$	"	Reflux; EtOH	C_6H_5	$CO_2C_2H_5$ (—)	65
	—	160°, 3 hr	OH	COC_6H_5 (—)	65

D. 4,5-Dimethoxy 2-amino-4,5-dimethoxybenzaldehyde (CH_3O, CH_3O, CHO, NH_2) + Reactant ⟶ 6,7-dimethoxyquinoline (CH_3O, CH_3O, R_1, R_2)

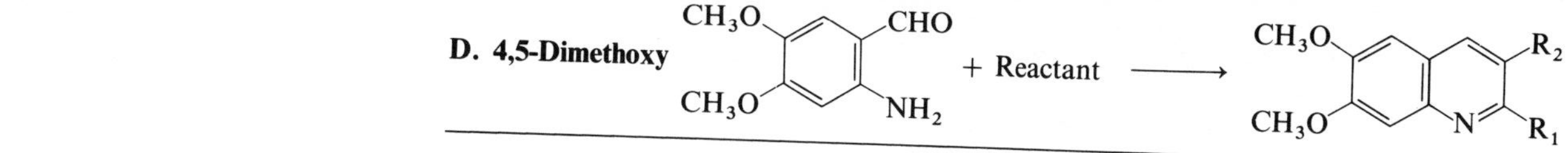

Reactant	Catalyst	Reaction Conditions	Product(s) and Yield(s) (%) R_1	R_2	Refs.
$O_2NCH_2CH{=}NOH$[a]	HCl	20°, 3 d; Me_2CO	H	NO_2 (57)	140a
"	"	2 d; EtOH	H	NO_2 (26)	28
CH_3COCH_3	Na_2CO_3	100°; H_2O	CH_3	H (—)	157
$CH_3COC_2H_5$	"	"	CH_3	CH_3 (—)	157
$C_6H_5COCH_3$	"	"	C_6H_5	H (—)	157
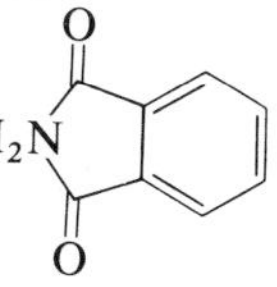 CH_3COCH_2N-phthalimido	KOH	3 d	CH_3	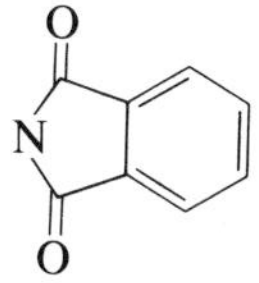 N-phthalimido (72)[b]	28

TABLE II. CONDENSATIONS WITH SUBSTITUTED *o*-AMINOBENZALDEHYDES (*Continued*)

E. Other Methoxy-Substituted *o*-Aminobenzaldehydes

R_1, R_2, R_3, R_4, CHO, NH_2

R	Reactant	Catalyst	Reaction Conditions	Product(s) and Yield(s) (%)	Refs.
$R_1 = R_2 = H$ $R_3 = R_4 = OCH_3$	$CH_3COCH_2CO_2C_2H_5$	Piperidine	100°, 6 hr	$CO_2C_2H_5$, CH_3O, N, CH_3, OCH_3 (90)	141
$R_1 = R_3 = H$[c] $R_2 = OCH_3$ $R_4 = OH$	CH_3CO, N, CO_2CH_3, O_2N, CH_3	KOH	1. $Na_2S_2O_4$ 2. 0°, 6 hr; $MeOH-H_2O$	CH_3O, N, N, CO_2CH_3, OH, H_2N, CH_3 (9)[d]	141a
$R_1 = R_3 = H$[c] $R_2 = R_4 = OCH_3$	"	KOH	0°, 3 hr; $MeOH-H_2O$	CH_3O, N, N, CO_2CH_3, CH_3O, O_2N, CH_3 (33)	141a

Substituents	Ketone	Base	Conditions	Product(s) (Yield, %)	Refs.
$R_1 = OSO_2C_6H_5$ $R_2 = OCH_3$ $R_3 = R_4 = H$	2-Acetylpyridine	Triton B	15 hr; THF	5-$OSO_2C_6H_5$, 6-CH_3O-2-(2-pyridyl)quinoline (8)[d] + 5-OH, 6-CH_3O-2-(2-pyridyl)quinoline (29)[d]	146
$R_1 = R_2 = OCH_3$[c] $R_3 = H$ $R_4 = OH$	"	KOH	0°, 6 hr; MeOH	5-OCH_3, 6-CH_3O, 8-OH-2-(2-pyridyl)quinoline (26)[d]	141b

TABLE II. CONDENSATIONS WITH SUBSTITUTED *o*-AMINOBENZALDEHYDES (*Continued*)

E. Other Methoxy-Substituted *o*-Aminobenzaldehydes (*Continued*)

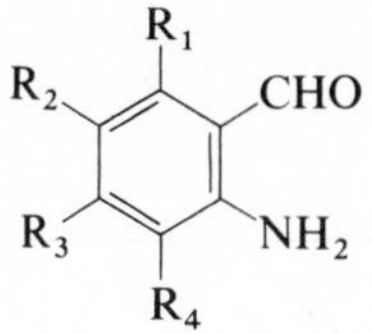

R	Reactant	Catalyst	Reaction Conditions	Product(s) and Yield(s) (%)	Refs.
$R_1 = R_2 = OCH_3$ $R_3 = H$ $R_4 = NHCOCH_3$	2-Acetylpyridine	KOH	0°, 6 hr; THF	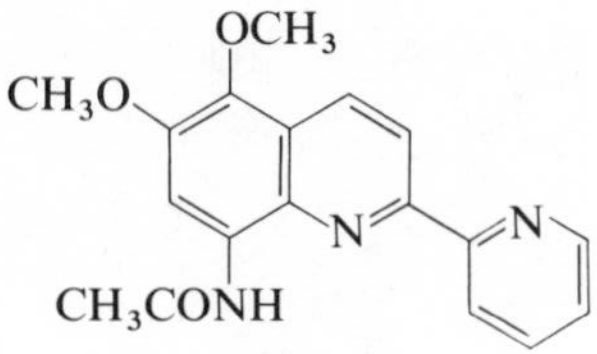 [structure: 5-OCH₃, 6-CH₃O, 8-CH₃CONH-2-(2-pyridyl)quinoline] (60)	141b
	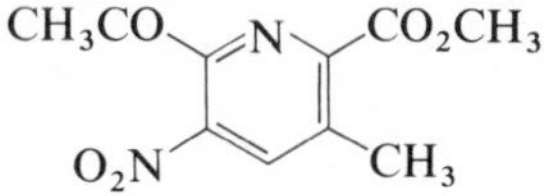 [structure: 2-CH₃CO-6-CO₂CH₃-5-CH₃-3-O₂N-pyridine]	KOH	0°, 4 hr; THF	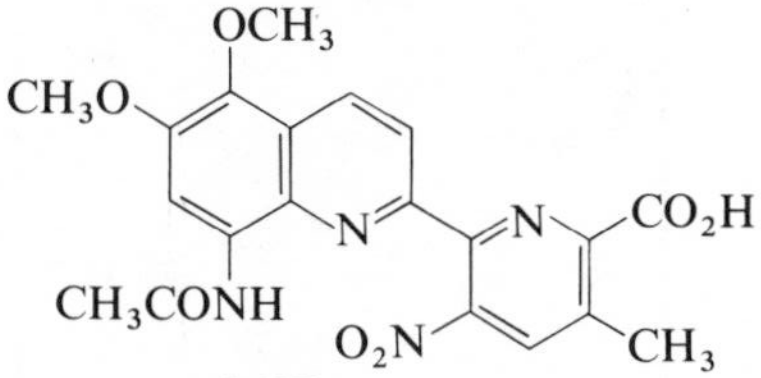 [structure: 5-OCH₃, 6-CH₃O, 8-CH₃CONH-quinolin-2-yl pyridine bearing CO₂H, CH₃, O₂N] (65)	141b
$R_1 = R_2 = OCH_3$[c] $R_3 = Br$ $R_4 = NH_2$	"	1. KOH 2. CH_2N_2	0°, 4 hr; THF	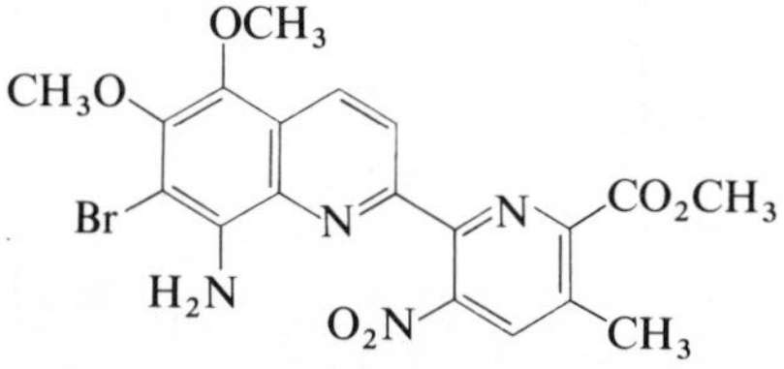 [structure: 5-OCH₃, 6-CH₃O, 7-Br, 8-H₂N-quinolin-2-yl pyridine bearing CO₂CH₃, CH₃, O₂N] (60)[d]	141b

F. 4,5-Methylenedioxy

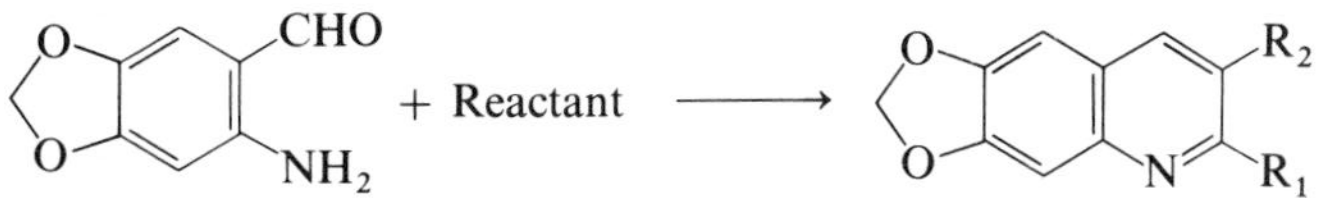

(O)(O) CHO NH_2 + Reactant ⟶ (O)(O) N R_1 R_2

Reactant	Catalyst	Reaction Conditions	Product(s) and Yield(s) (%) R_1	R_1R_2	R_2	Refs.
$O_2NCH_2CH{=}NOH$	HCl	2 d; EtOH	H		NO_2 (68)	28
CH_3COCH_3	Na_2CO_3	100°; H_2O	CH_3		H (~100)	93
$CH_3COCH_2NO_2$	AcOH	46 hr; Me_2CO	CH_3		NO_2 (51)	30a
$CH_3COC_2H_5$	Na_2CO_3	100°; H_2O	CH_3		CH_3 (—)	93
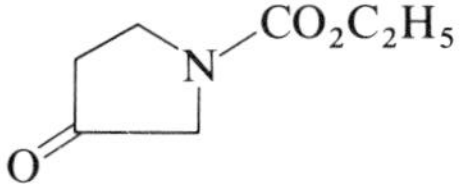 O, N–$CO_2C_2H_5$	pTSA	190–195°; 5 min		 N—$CO_2C_2H_5$ (13)		89
$CH_3COC_6H_5$	K_2CO_3	EtOH	C_6H_5		H (—)	93
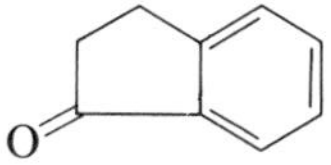 O	KOH	1. Few min; EtOH 2. Boil; AcOH		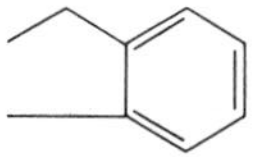 (60)		142a
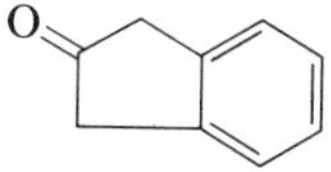 O	"	Few min; EtOH		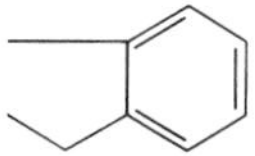 (84)		142a

TABLE II. CONDENSATIONS WITH SUBSTITUTED *o*-AMINOBENZALDEHYDES (*Continued*)

F. 4,5-Methylenedioxy (*Continued*)

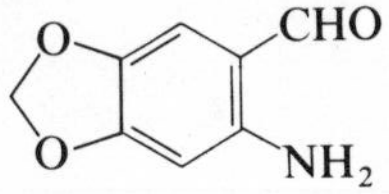

CHO, NH_2 + Reactant ⟶ R_2, N, R_1

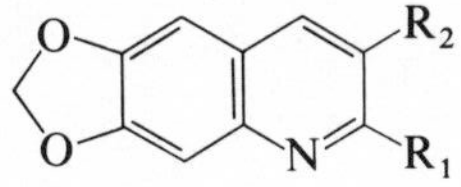

Reactant	Catalyst	Reaction Conditions	Product(s) and Yield(s) (%)			Refs.
			R_1	R_1R_2	R_2	
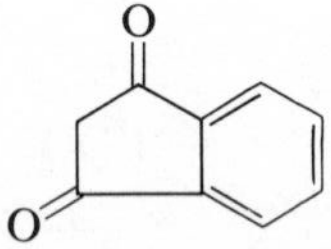	AcOH	Reflux, 10 min		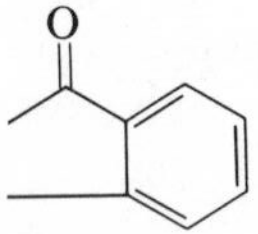 (69)		142a
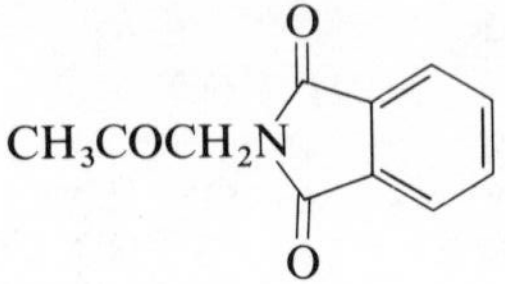 CH_3COCH_2N	NaOH	1. 24 hr; H_2O 2. Acid	CH_3		$NHCOC_6H_4CO_2H$-*o* (65)	143
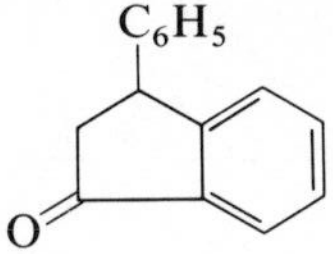 C_6H_5	KOH	Cold; EtOH		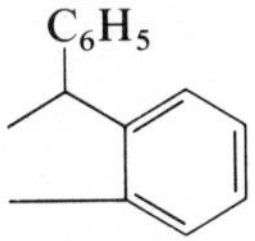 C_6H_5 (83)		142b
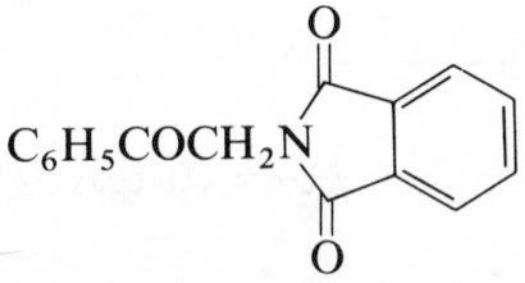 $C_6H_5COCH_2N$	NaOH	1. 24 hr; $EtOH-H_2O$ 2. Acid	C_6H_5		$NHCOC_6H_4CO_2H$-*o* (70)	143

G. Halosubstituted *o*-Aminobenzaldehydes

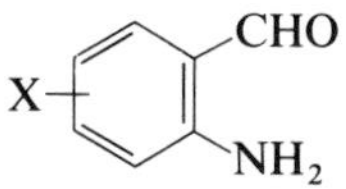

X	Reactant	Catalyst	Reaction Conditions	Product(s) and Yield(s) (%)	Refs.
5-Cl	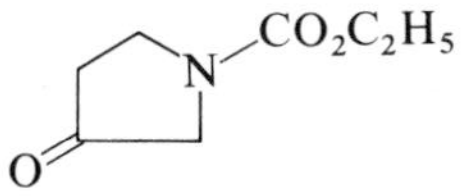 1-(ethoxycarbonyl)pyrrolidin-3-one (N–$CO_2C_2H_5$)	pTSA	190–195°, 5 min	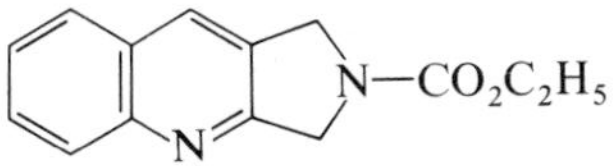 pyrrolo[3,4-b]quinoline, N–$CO_2C_2H_5$ (26)	89
	R_1–R_4-substituted 2-(dichloromethyl)phenyl isocyanate ($CHCl_2$, NCO) $\xrightarrow{\text{1. } Ba(OH)_2\text{, 15–18°; 2. HCl, 0–5°}}$ [R_1–R_4-2-aminobenzaldehyde (CHO, NH_2)] + CH_3COCH_2Cl $\xrightarrow[H_2O]{NaOH}$ R_1–R_4-3-hydroxy-2-methylquinoline (OH, CH_3)				

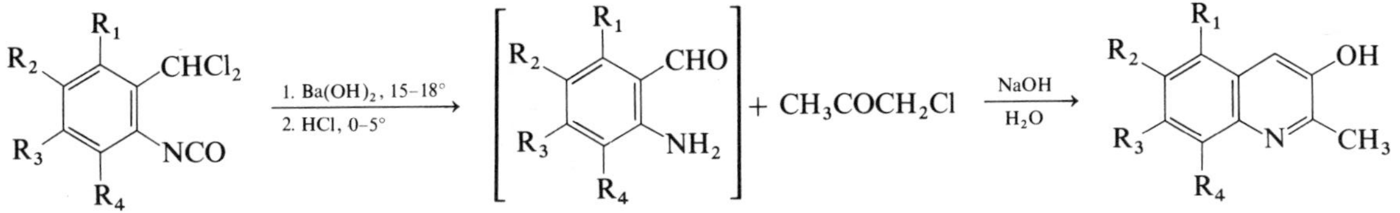

Temperature	R_1	R_2	R_3	R_4	Yield (%)	Refs.
20°	Cl	H	H	H	(~90)	89a
"	H	Cl	H	H	(~90)	89a
"	H	H	Cl	H	(~90)	89a
"	H	H	H	Cl	(~90)	89a
"	H	H	F	H	(~90)	89a
"	H	H	CF_3	H	(~90)	89a
"	Cl	Cl	H	H	(~90)	89a
"	Cl	H	H	Cl	(~90)	89a
"	Cl	Cl	H	Cl	(~90)	89a
50°	Cl	Cl	Cl	Cl	(85)	89a

TABLE II. CONDENSATIONS WITH SUBSTITUTED *o*-AMINOBENZALDEHYDES (*Continued*)

H. 2,4-Diaminoisophthalaldehyde (OHC, CHO, H_2N, NH_2) + Reactant ⟶ (R_2, R_1, N, N, R_2, R_1)

Reactant	Catalyst	Reaction Conditions	Product(s) and Yield(s) (%) R_1	R_2	Refs.
$CH_3COCH_2COCH_3$	Piperidine	180–190°; 1.5 hr	CH_3	$COCH_3$ (86)	145
$CH_3COCH_2CO_2C_2H_5$	NaOH	1. 2 d; EtOH 2. 100°	CH_3	$CO_2C_2H_5$ (75)	144
	"	3 d; EtOH	CH_3	$CO_2C_2H_5$ (—)	145
2-Acetylpyridine	KOH	Reflux, 8 hr; EtOH	2-Pyridyl	H (90)	39
$C_2H_5O_2CCH_2COCO_2C_2H_5$	Piperidine	Reflux, 30 min; xylene	$CO_2C_2H_5$	$CO_2C_2H_5$ (57)	145
$C_6H_5CH_2CHO$	KOH	Reflux, 8 hr; EtOH	H	C_6H_5 (30)	39
$C_6H_5COCH_3$	"	100°, 10 min	C_6H_5	H (70)	144, 39
p-$CH_3COC_6H_4OCH_3$	"	150°, 5 min; MeOH	$C_6H_4OCH_3$-*p*	H (82)	145
2,6-Diacetylpyridine	KOH or LiCl-LiOH	120°, 4 hr; HMPA	[polymer: N, N, N]$_n$	(93–99)	39
p-$CH_3COC_6H_4COCH_3$	KOH or LiCl-LiOH	120°, 4 hr; HMPA	[polymer: N, N]$_n$	(91–94)	39

Reactant	Catalyst	Reaction Conditions	Product(s) and Yield(s) (%)	Refs.
$C_6H_5COCH_2CO_2C_2H_5$	Piperidine	150°	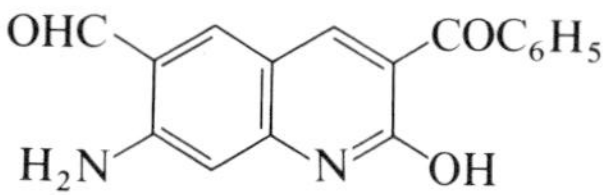 OHC, COC_6H_5, H_2N, N, OH (95)	145
p-$CH_3COC_6H_4OC_6H_5$	"	"	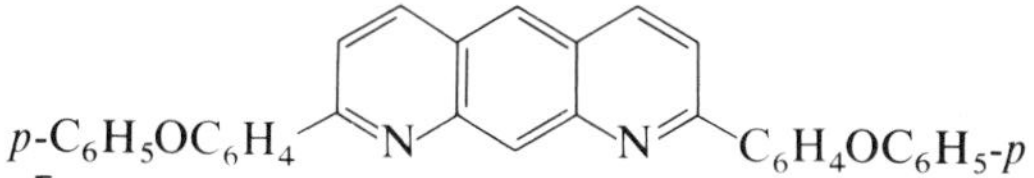 p-$C_6H_5OC_6H_4$, N, N, $C_6H_4OC_6H_5$-p (91)	39
$(p\text{-}CH_3COC_6H_4)_2O$	KOH or LiCl–LiOH	120–140°, 4 hr; HMPA	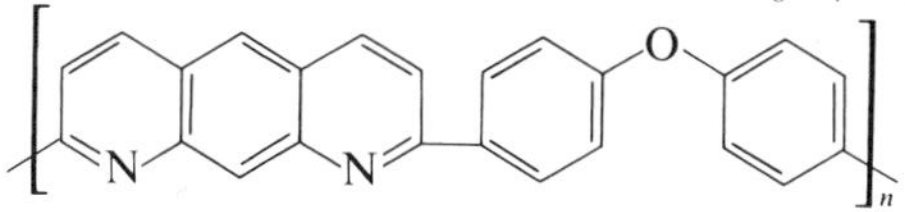 N, N, O, n (95–98)	39

I. 2,5-Diaminoterephthalaldehyde[e]

Reactant	Catalyst	Reaction Conditions	Product(s) and Yield(s) (%)	Refs.
$C_6H_5COCH_3$	—	190–197°, 90 min	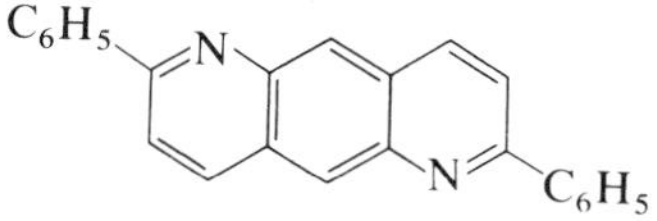 C_6H_5, N, N, C_6H_5 (33)	38

[a] The aminoaldehyde reactant was CH_3O, CH_3O, CH, O, O, CH_2, CH_2, N=CH, H_2N, OCH_3, OCH_3

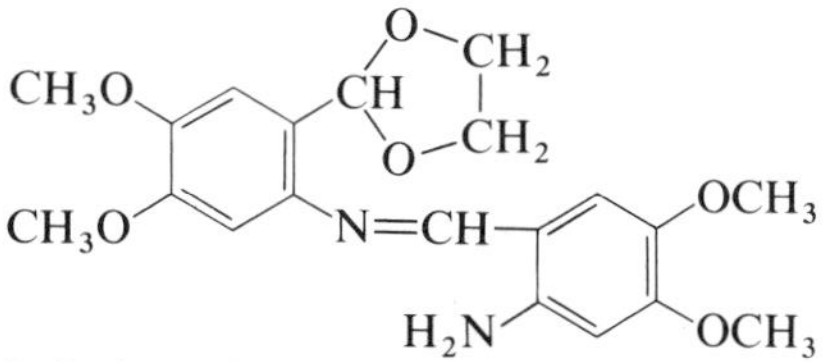

[b] The product was isolated after hydrolysis to the 3-NH_2 derivative.

[c] The aminoaldehyde was generated *in situ* by sodium dithionite reduction of the nitroaldehyde.

[d] Yields were calculated from the starting nitroaldehyde rather than the aminoaldehyde.

[e] The bis-p-toluenesulfonamide of the aldehyde was used in this reaction.

TABLE III. CONDENSATIONS WITH SUBSTITUTED *N*-(*o*-AMINOPHENYLMETHYLENE)-*p*-TOLUIDINES

[Structure: 4-R_1-5-R_2-2-aminobenzylidene-p-toluidine, R_1, R_2 on ring; $CH{=}N{-}C_6H_4{-}CH_3$; NH_2]

	Reactant	Catalyst	Reaction Conditions	Product(s) and Yield(s) (%)	Refs.
$R_1, R_2 = H$	$CH_3COCH{=}NOH$	KOH	Reflux, 6 hr; EtOH	2-Quinolyl-$CH{=}NOH$ (87)	46
	Barbituric acid (O, NH, O, N–H, O)	Piperidine	Gentle boiling, 6 hr; glycerol	Pyrimido[4,5-b]quinoline-2,4-dione (O, NH, N, N–H, O) (—)	48
	$CH_3COCH_2COCH_3$	"	Heat, 4 hr	3-$COCH_3$-2-CH_3-quinoline (—)	46
	$CH_3COCH_2CO_2C_2H_5$	"	Water bath, 2 hr	3-$CO_2C_2H_5$-2-CH_3-quinoline (—)	96
	"	"	Water bath, 1 d	" (88)	49

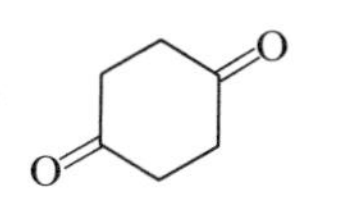	Piperidine	Water bath, 12 hr	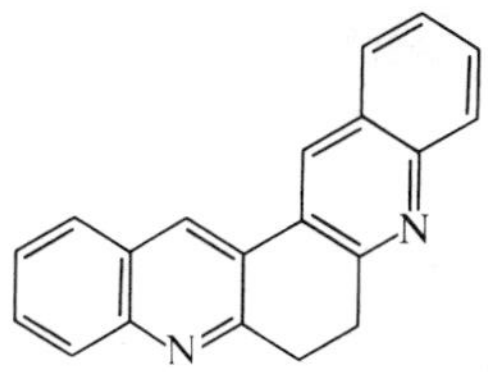 (55)	48
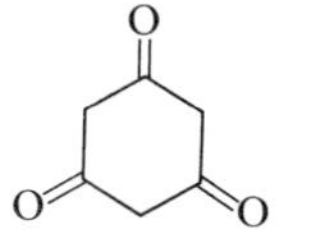	"	Water bath, 10 hr; EtOH	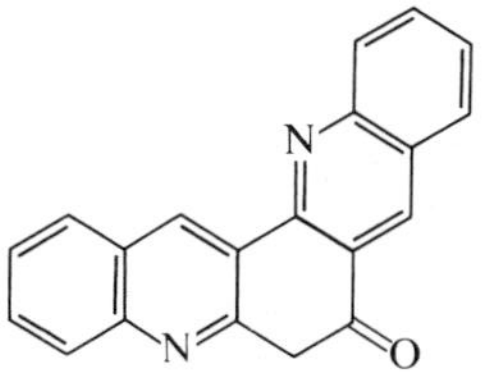 (60)	48
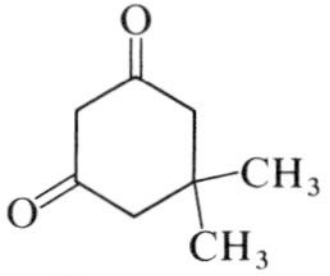	"	100°, 8 hr	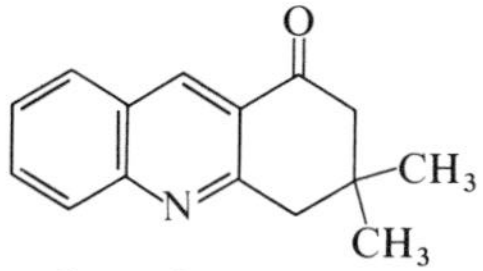 (80)	48
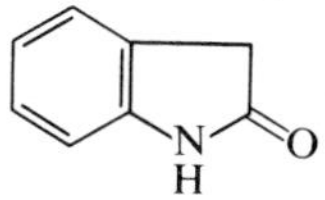	"	150°, 3 hr	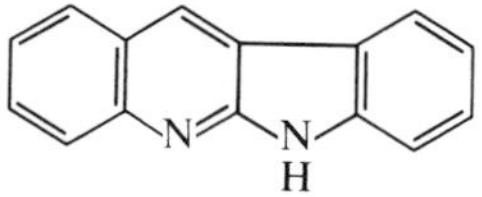 (70)	48
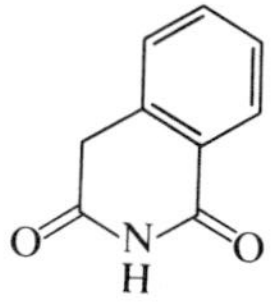	Piperidine	Heat, 6 hr; 1-pentanol	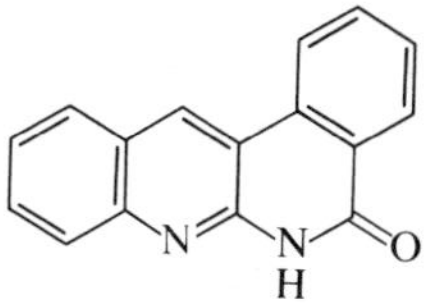 (95)	48

TABLE III. CONDENSATIONS WITH SUBSTITUTED *N*-(*o*-AMINOPHENYLMETHYLENE)-*p*-TOLUIDINES (*Continued*)

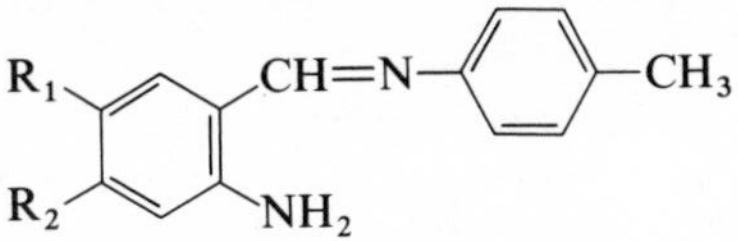

R_1, R_2, $CH{=}N$, CH_3, NH_2

	Reactant	Catalyst	Reaction Conditions	Product(s) and Yield(s) (%)	Refs.
$R_1, R_2 = H$ (*Contd.*)	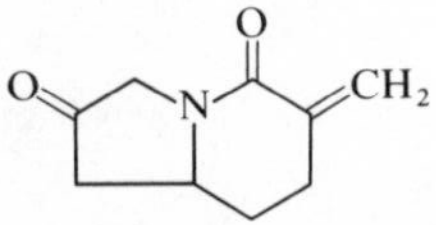 O, N, O, CH_2	pTSA	Reflux; $C_6H_5CH_3$	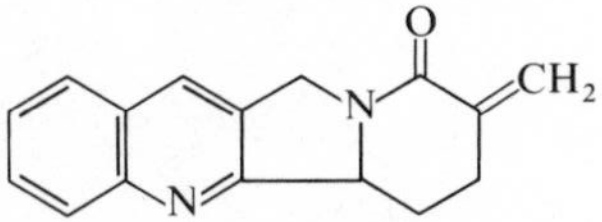 N, N, O, CH_2 (76)	87
	$C_6H_5COCH_2CO_2C_2H_5$	Piperidine	100°; 12 hr	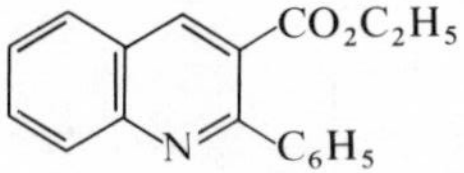 $CO_2C_2H_5$, N, C_6H_5 (90)	46
	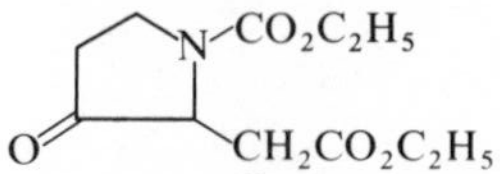 N, $CO_2C_2H_5$, O, $CH_2CO_2C_2H_5$	pTSA	Reflux, 4 hr; $C_6H_5CH_3$	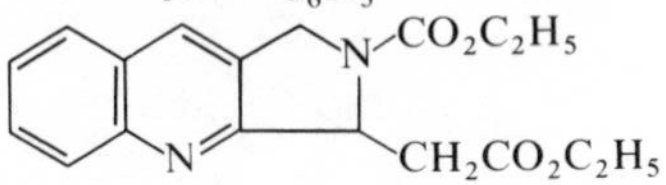 N, $CO_2C_2H_5$, N, $CH_2CO_2C_2H_5$ (76)	51
	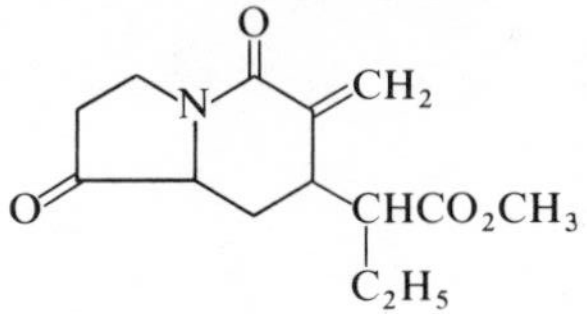 O, N, CH_2, O, $CHCO_2CH_3$, C_2H_5	"	Reflux, 3 hr; $C_6H_5CH_3$	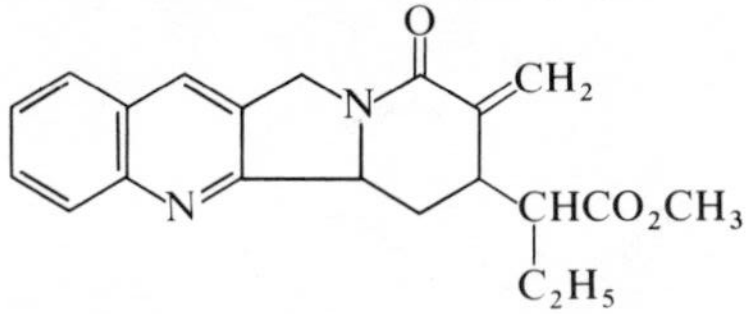 O, N, CH_2, N, $CHCO_2CH_3$, C_2H_5 (75)	86
$R_1, R_2 = OCH_2O$	$CH_3COCH{=}NOH$	KOH	Reflux, 6 hr; EtOH	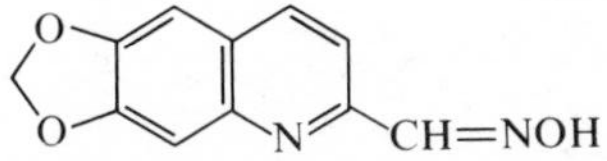 O, O, N, $CH{=}NOH$ (—)	46
	CH_3COCO_2H	NaOH	Reflux, 6 hr; EtOH	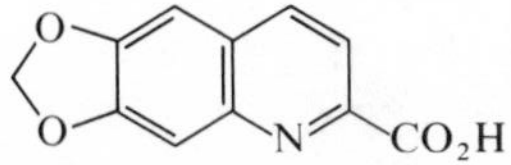 O, O, N, CO_2H (55)	46

$(CH_3CO)_2O$	—	Et_2O	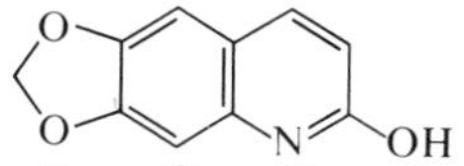 OH	(—)	46
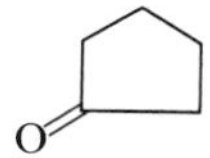	NaOH	Water bath, 7 hr; EtOH	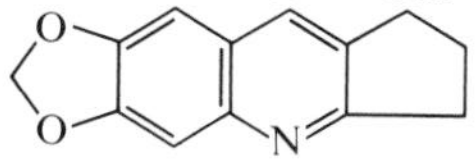	(90)	47
$CH_3COCH_2COCH_3$	Piperidine	Water bath, 4 hr	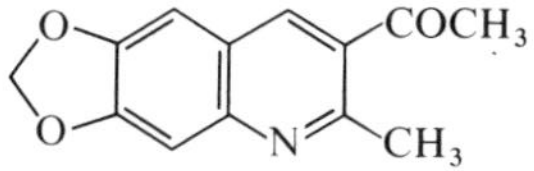 $COCH_3$, CH_3	(—)	46
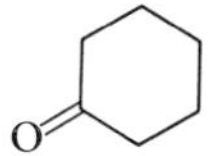	NaOH	Water bath, 7 hr; EtOH	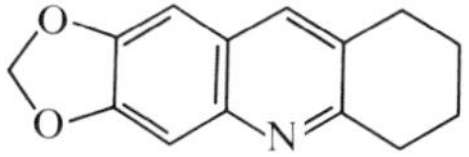	(75)	47
$CH_3COCH_2CO_2C_2H_5$	Piperidine	Water bath, 5 hr	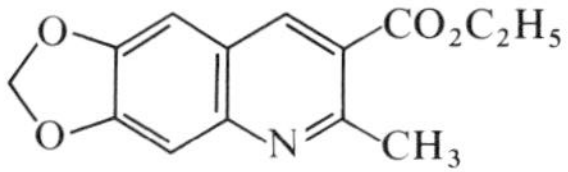 $CO_2C_2H_5$, CH_3	(70)	47
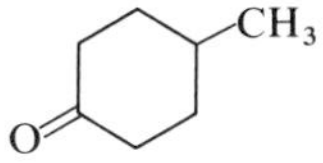 CH_3	NaOH	Water bath, 7 hr; EtOH	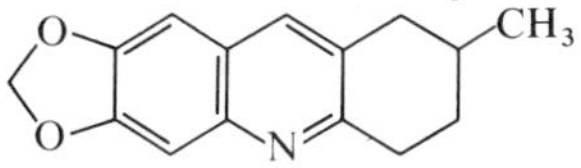 CH_3	(80)	47
$CH_3COCH_2COCO_2C_2H_5$	Piperidine	Water bath, 2 hr	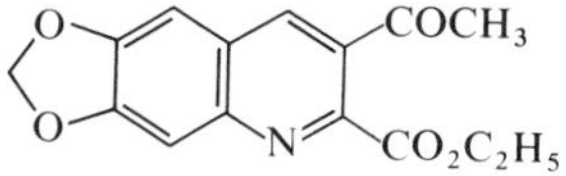 $COCH_3$, $CO_2C_2H_5$	(—)	46
p-$CH_3COC_6H_4Cl$	NaOH	Water bath, 7 hr; EtOH	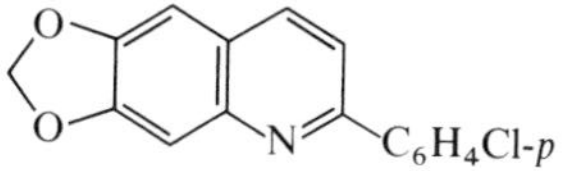 C_6H_4Cl-*p*	(90)	47
p-$CH_3COC_6H_4OCH_3$	"	"	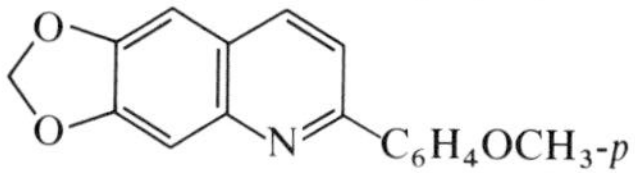 $C_6H_4OCH_3$-*p*	(75)	47
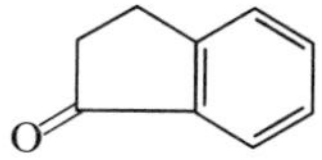	"	"	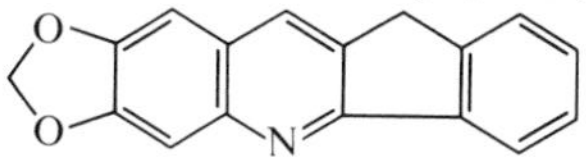	(90)	47

TABLE III. CONDENSATIONS WITH SUBSTITUTED *N*-(*o*-AMINOPHENYLMETHYLENE)-*p*-TOLUIDINES (*Continued*)

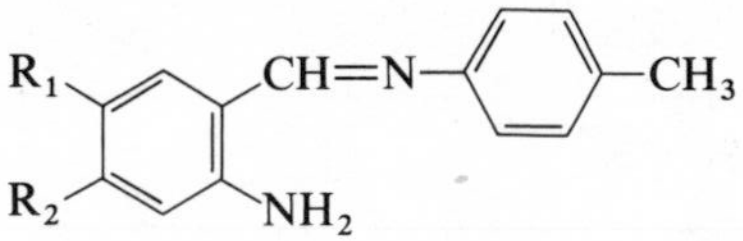

	Reactant	Catalyst	Reaction Conditions	Product(s) and Yield(s) (%)	Refs.
$R_1, R_2 = OCH_2O$ (*Contd.*)	$C_6H_5CH_2COCO_2H$	NaOH	Heat, 6 hr; EtOH	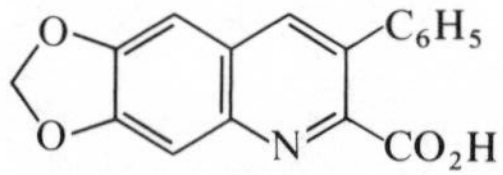 (—)	46
	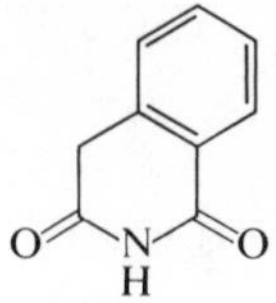	Piperidine	Heat, 6 hr; 1-pentanol	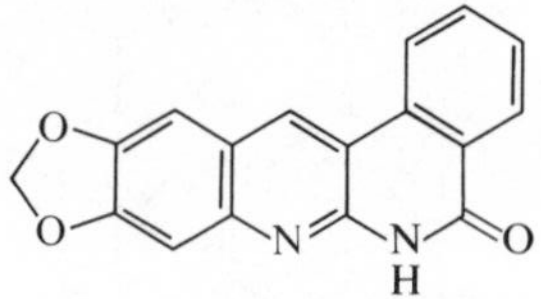 (87)	48
	$C_6H_5COCH_2CO_2C_2H_5$	Piperidine	—	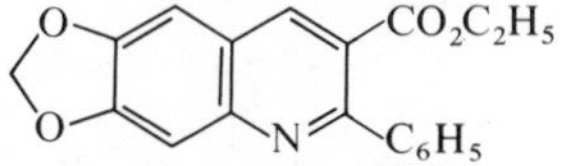 (—)	46
	$C_6H_5CH_2COC_6H_5$	NaOH	Water bath, 7 hr; EtOH	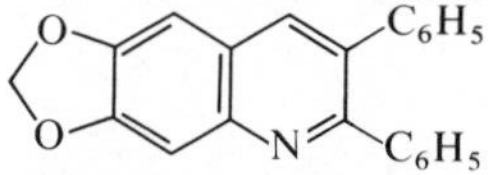 (—)	47
$R_1, R_2 = OCH_3$	$CH_3CH{=}NOH$	KOH	Reflux, 10 hr; EtOH	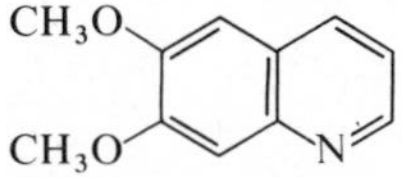 (36)	46
	$CH_3COCH{=}NOH$	KOH	Reflux, 6 hr; EtOH	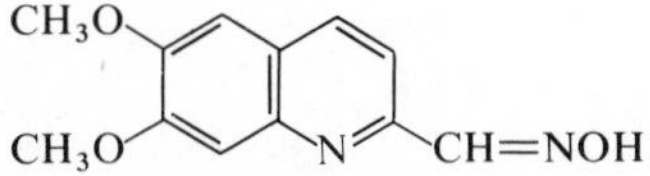 (65)	46

CH_3COCO_2H	NaOH	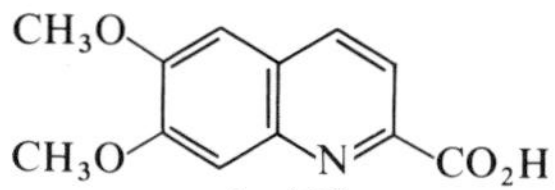 "	CH_3O, CH_3O, N, CO_2H (75)	46
"	Piperidine or no catalyst	Water bath, 6 hr; EtOH	" (63)	46
$(CH_3CO)_2O$		Et_2O 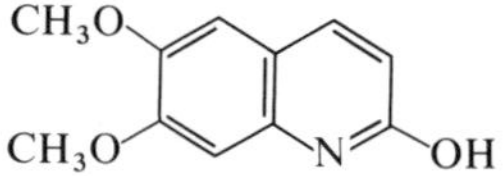	CH_3O, CH_3O, N, OH (68)	46
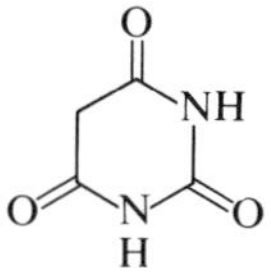 O, NH, O, N, H, O	Piperidine	Gentle boiling, 6 hr; glycerol 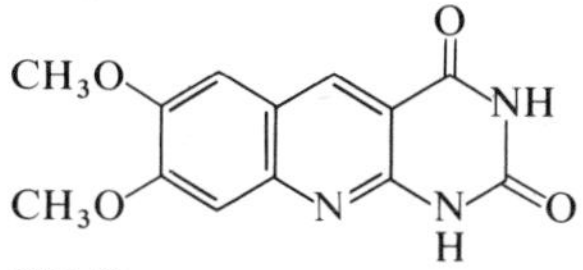	CH_3O, CH_3O, O, NH, N, N, H, O (40)	48
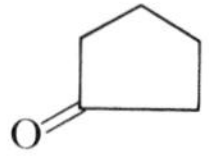 O	NaOH	Heat; EtOH	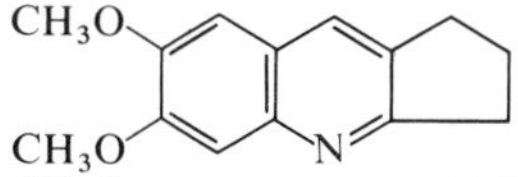 CH_3O, CH_3O, N (85)	47
$CH_3COCH_2COCH_3$	Piperidine	Heat, 4 hr	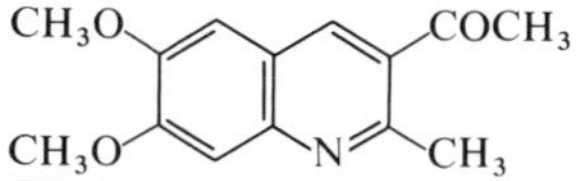 CH_3O, CH_3O, N, $COCH_3$, CH_3 (90)	46
$CH_3COCH_2CO_2C_2H_5$	"	EtOH 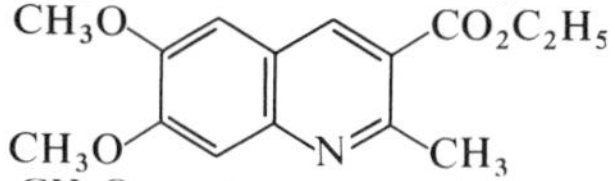	CH_3O, CH_3O, N, $CO_2C_2H_5$, CH_3 (85)	47
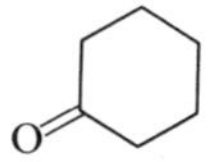 O	NaOH	Heat; EtOH 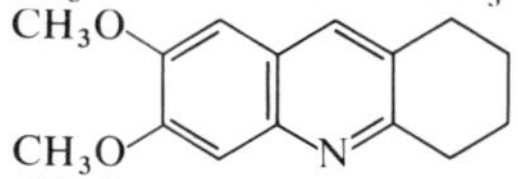	CH_3O, CH_3O, N (80)	47
O, CH_3	"	"	CH_3O, CH_3O, N, CH_3 (90)	47

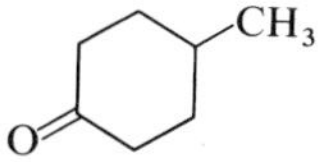

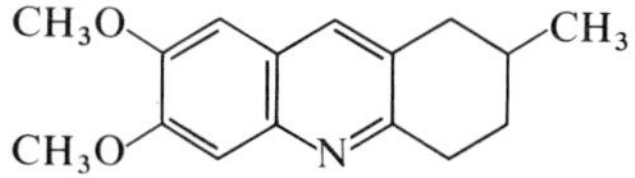

TABLE III. CONDENSATIONS WITH SUBSTITUTED *N*-(*o*-AMINOPHENYLMETHYLENE)-*p*-TOLUIDINES (*Continued*)

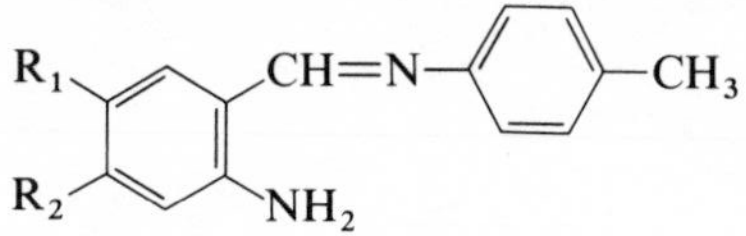

R_1, R_2, CH=N, CH_3, NH_2

	Reactant	Catalyst	Reaction Conditions	Product(s) and Yield(s) (%)	Refs.
$R_1, R_2 = OCH_3$ (*Contd.*)	$CH_3COCH_2COCO_2C_2H_5$	Piperidine	Water bath, 2 hr	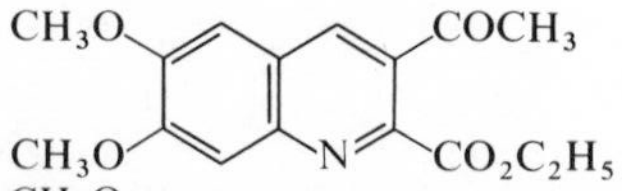 CH_3O, CH_3O, N, $COCH_3$, $CO_2C_2H_5$ (—)	46
	$CH_3COC_6H_5$	NaOH	Water bath, 7 hr; EtOH	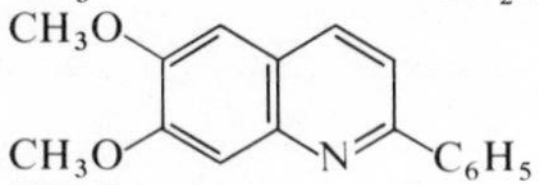 CH_3O, CH_3O, N, C_6H_5 (—)	47
	p-$CH_3COC_6H_4Cl$	"	"	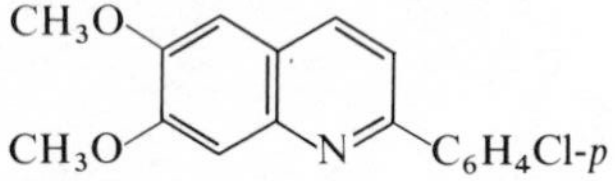 CH_3O, CH_3O, N, C_6H_4Cl-*p* (90)	47
	p-$CH_3COC_6H_4OCH_3$	"	"	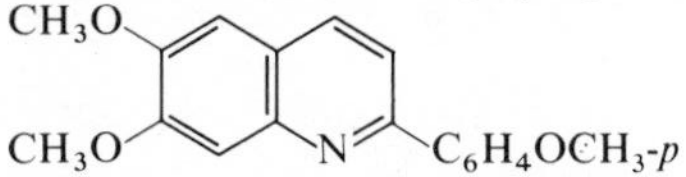 CH_3O, CH_3O, N, $C_6H_4OCH_3$-*p* (80)	47
	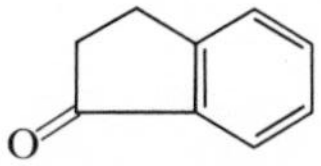 O	Heat; EtOH		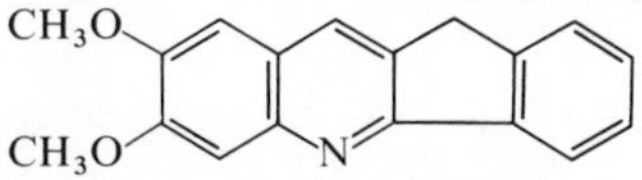 CH_3O, CH_3O, N (75)	47
	$C_6H_5CH_2COCO_2H$	"	Water bath, 6 hr; EtOH	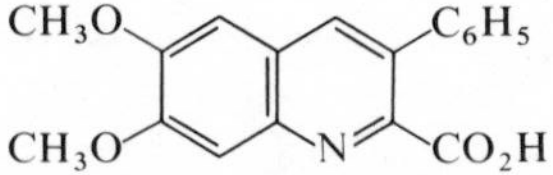 CH_3O, CH_3O, N, C_6H_5, CO_2H (—)	46

	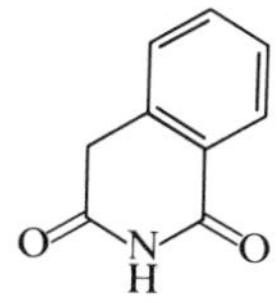 (structure: O, N, H, O)	Piperidine	Heat, 6 hr; 1-pentanol	(structure: CH_3O, CH_3O, N, N, H, O)	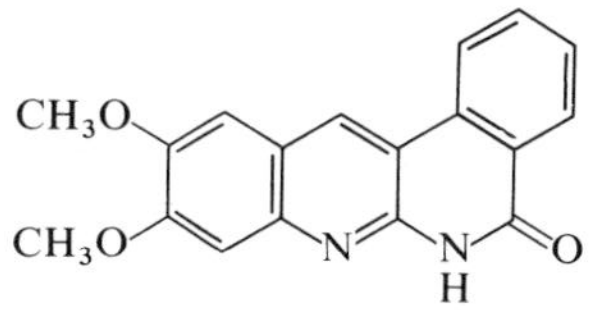 (94)	48
	$CH_3COCH_2COC_6H_5$	Piperidine	Water bath	(structure: CH_3O, CH_3O, N, COC_6H_5, CH_3)	(—)	46
	$C_6H_5COCH_2CO_2C_2H_5$	"	Water bath, 12 hr	(structure: CH_3O, CH_3O, N, $CO_2C_2H_5$, C_6H_5)	(—)	46
	$C_6H_5CH_2COC_6H_5$	NaOH	Water bath, 7 hr; EtOH	(structure: CH_3O, CH_3O, N, C_6H_5, C_6H_5)	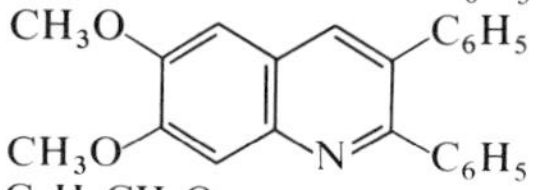 (80)	47
$R_1 = C_6H_5CH_2O$, $R_2 = H$	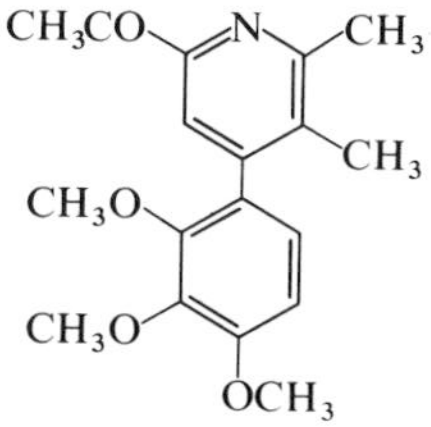 (structure: CH_3CO, N, CH_3, CH_3, CH_3O, CH_3O, OCH_3)	NaOH	Reflux, 24 hr; MeOH	(structure: $C_6H_5CH_2O$, N, N, CH_3, CH_3, CH_3O, CH_3O, OCH_3)	(72)	147

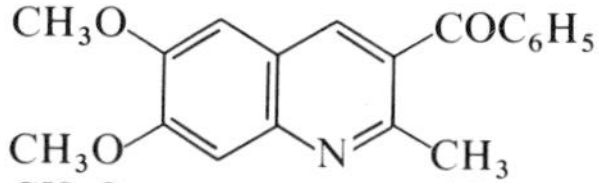

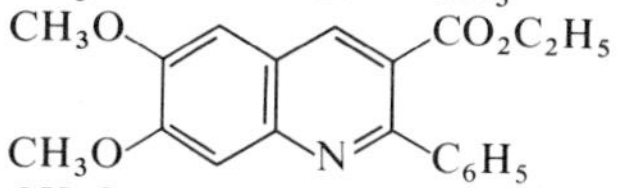

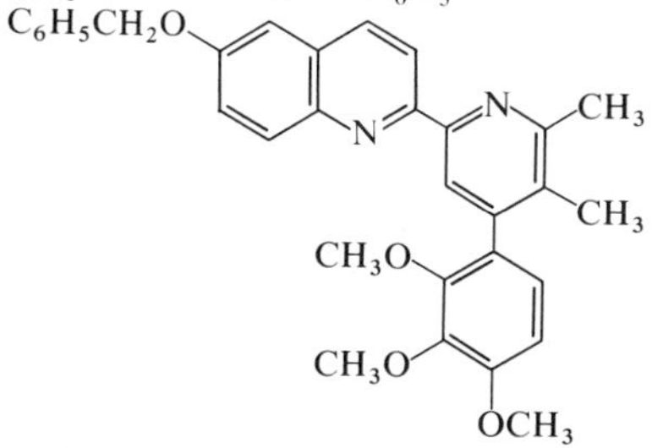

TABLE IV. CONDENSATIONS WITH *o*-AMINOACETOPHENONE HYDROCHLORIDE

No. of C Atoms	Reactant	Catalyst	Reaction Conditions	Product(s) and Yield(s) (%)	Refs.
C_4	$CH_3COCH_2CH_3$	—	79°, 30 hr	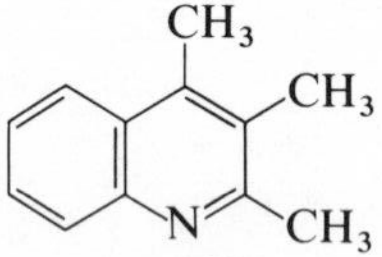 (77)	55
	$CH_3C(=NH)CH_2CN$	—	150°;[a] AcOH	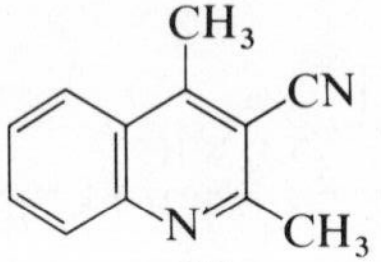 (—)	100
	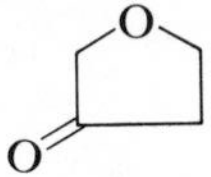	—	145°	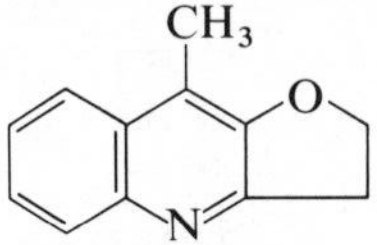 (48)	148
	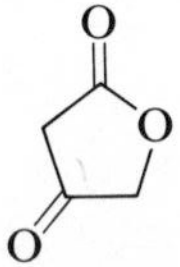	1. None 2. H_2SO_4	$EtOH$[a] 1. Steam bath 2. Cold, conc H_2SO_4	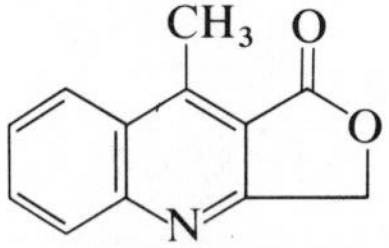 (76)	16
	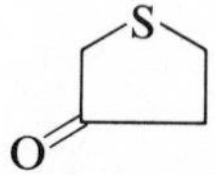	—	135°	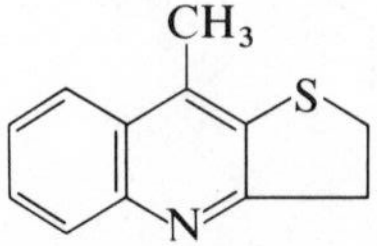 (59)	148

C_5	$CH_3COCH_2COCH_3$	—	110°	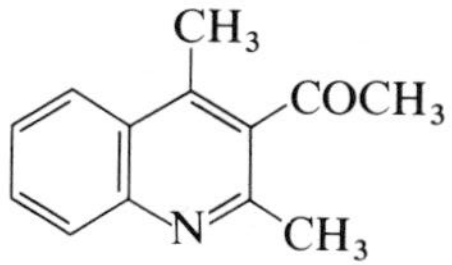 CH_3, $COCH_3$, N, CH_3 (60)	55
	$CH_3CO(CH_2)_2CH_3$	—	102°	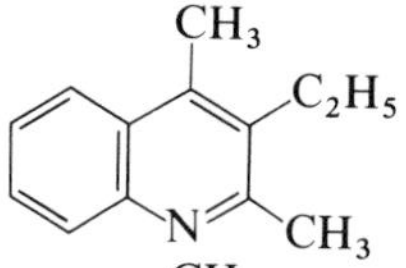 CH_3, C_2H_5, N, CH_3 (56)	55
	$NCCH_2CO_2C_2H_5$	—	200°,[a]	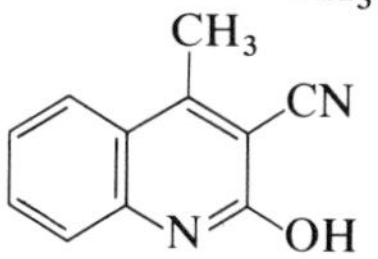 CH_3, CN, N, OH (96)	104,160
	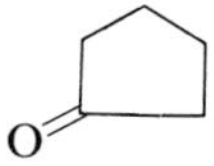 O	Piperidine	160°,[a] xylene	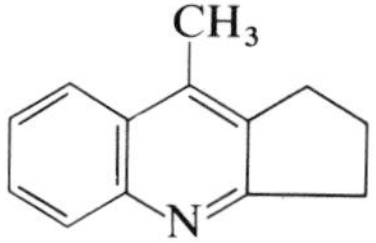 CH_3, N (70)	90
		—	115°	" (75)	31,149
	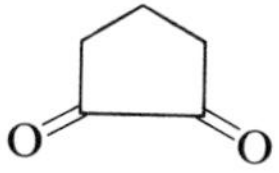 O, O	—	100°,[a] 3 hr	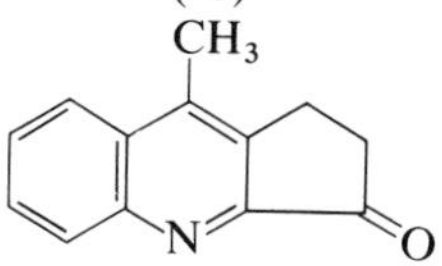 CH_3, N, O (92)	24,150
	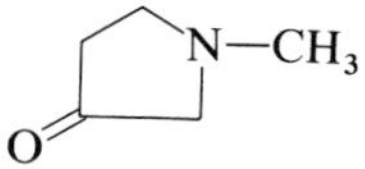 O, $N—CH_3$	—	150°	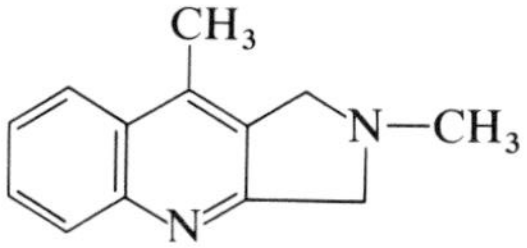 CH_3, $N—CH_3$, N (56)	32

TABLE IV. CONDENSATIONS WITH *o*-AMINOACETOPHENONE HYDROCHLORIDE (*Continued*)

No. of C Atoms	Reactant	Catalyst	Reaction Conditions	Product(s) and Yield(s) (%)	Refs.
C_5 (*Contd.*)	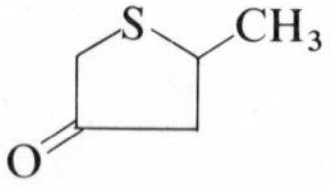 [structure: S, CH_3, O]	—	135°	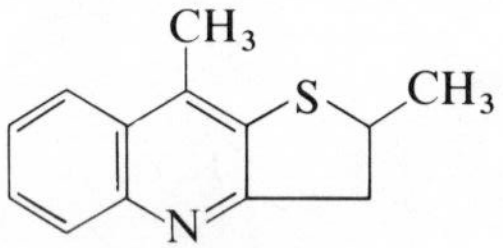 [structure: CH_3, S, CH_3, N] (70)	148
	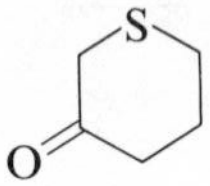 [structure: S, O]	—	130°	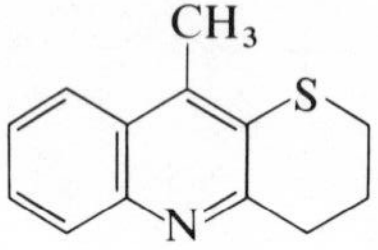 [structure: CH_3, S, N] (88)	148
	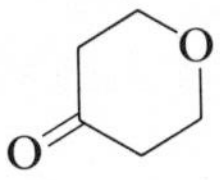 [structure: O, O]	—	120°	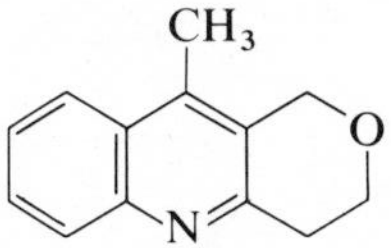 [structure: CH_3, O, N] (—)	148
	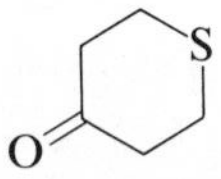 [structure: S, O]	—	140°	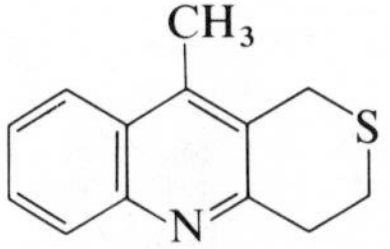 [structure: CH_3, S, N] (63)	148
C_6	$CH_3COC_4H_9$-*n*	—	127°	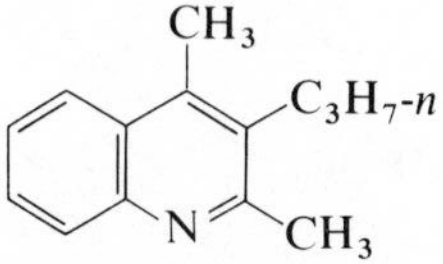 [structure: CH_3, C_3H_7-*n*, N, CH_3] (73)	55

O, CH_3	—	120°	CH_3, N, CH_3 (65)	31,149
O, CH_3	—	130°	CH_3, N, CH_3 (70)	31,149
O	—	110°	CH_3, N (90)	31,149
O, Cl	—	115°	CH_3, N, Cl (79)	31,149
O	—	120°	CH_3, N (58)	149
O, O	—	160°,[a] 8 hr	CH_3, N, N, CH_3 (76)	24,150

TABLE IV. CONDENSATIONS WITH *o*-AMINOACETOPHENONE HYDROCHLORIDE (*Continued*)

No. of C Atoms	Reactant	Catalyst	Reaction Conditions	Product(s) and Yield(s) (%)	Refs.
C_6 (*Contd.*)	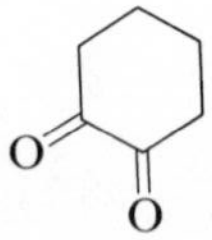 O, O	—	120–170°,[a] xylene	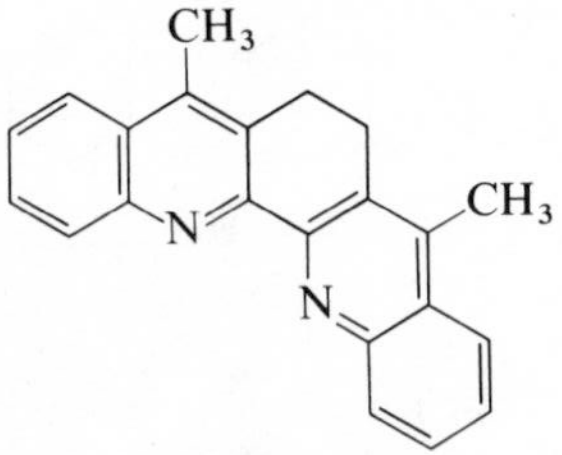 CH_3, CH_3, N, N (46)	151
		—	130°	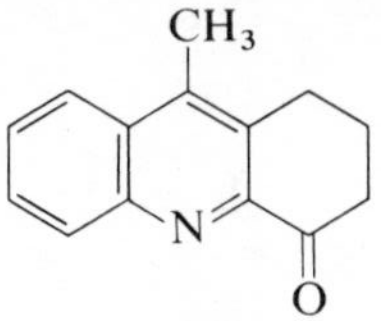 CH_3, N, O (61)	55
	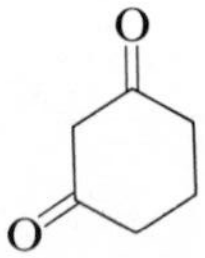 O, O	—	120°	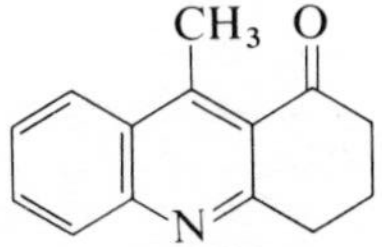 CH_3, O, N (90)	55
	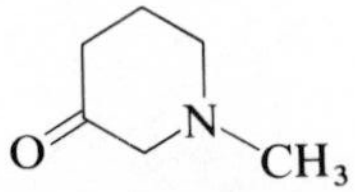 O, N, CH_3	—	120°	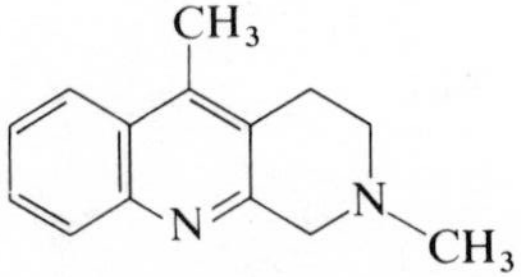 CH_3, N, N, CH_3 (48)	32

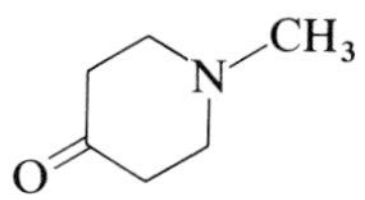

O, N, CH_3	—	130°	CH_3, N, N, CH_3 (62)	32
CH_3CO, O	—	130°	CH_3, N, O (34)	152
CH_3CO, S	—	125°	CH_3, N, S (35)	152
CH_3CO, N, H	—	110–130°	CH_3, N, N, H (14)	153
CH_3CO, S, N, CH_3	—	110–130°	CH_3, N, S, N, CH_3 (37)	153
CH_3, N, CH_3CO, S, NH_2	—	110–130°	CH_3, CH_3, N, N, S, NH_2 (20)	153

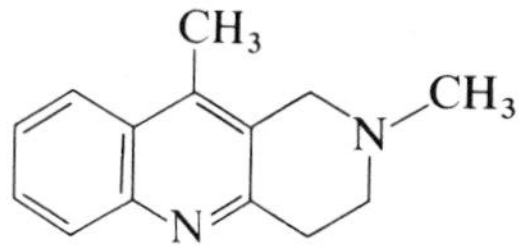

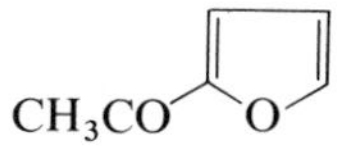

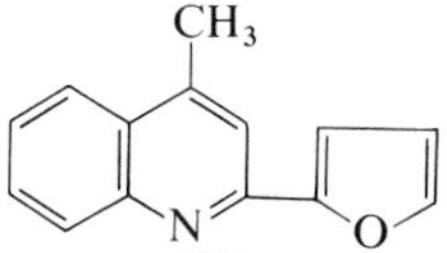

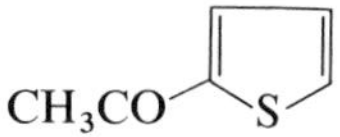

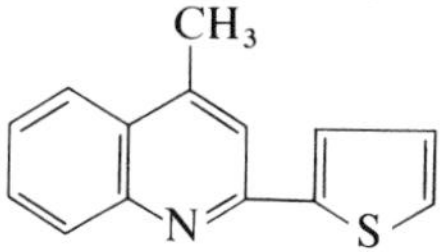

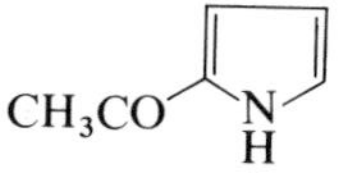

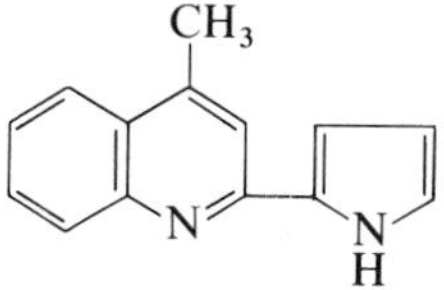

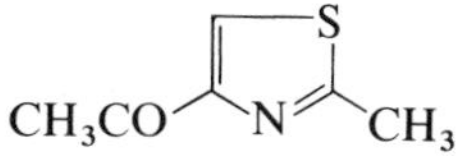

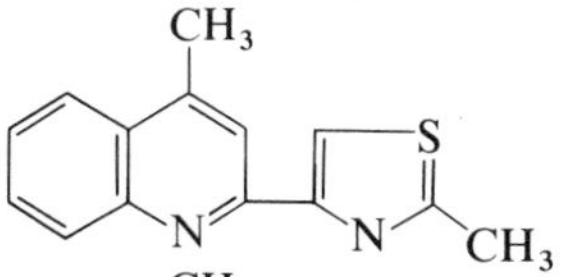

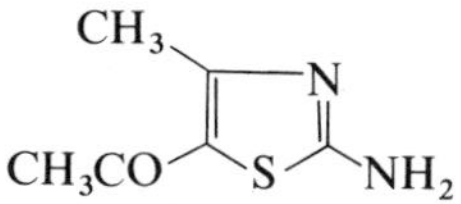

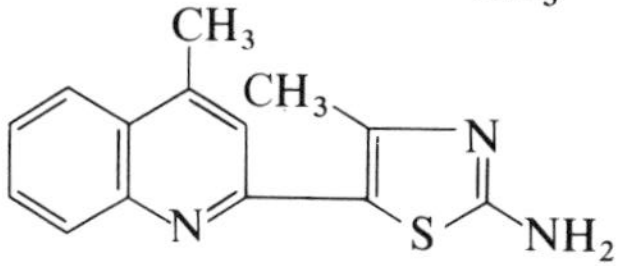

TABLE IV. Condensations with *o*-Aminoacetophenone Hydrochloride (*Continued*)

No. of C Atoms	Reactant	Catalyst	Reaction Conditions	Product(s) and Yield(s) (%)	Refs.
C_7	$CH_3COC_5H_{11}$-*n*	—	150°	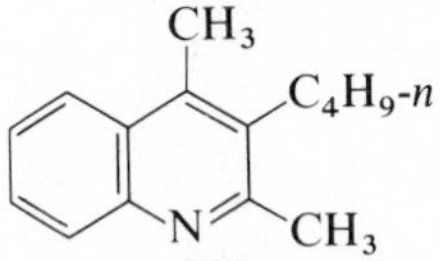 [quinoline structure: CH_3, C_4H_9-*n*, N, CH_3] (65)	55
	$CH_3CO(CH_2)_2CO_2C_2H_5$	—	110–130°	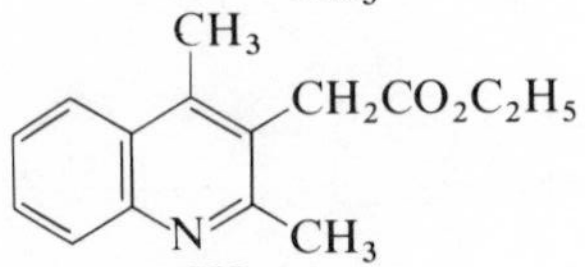 [quinoline structure: CH_3, $CH_2CO_2C_2H_5$, N, CH_3] (60)	154
	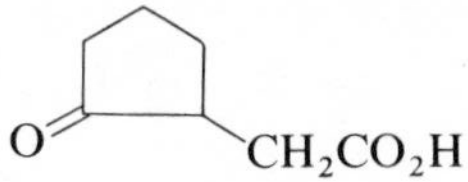 [cyclopentanone structure: O, CH_2CO_2H]	—	130°	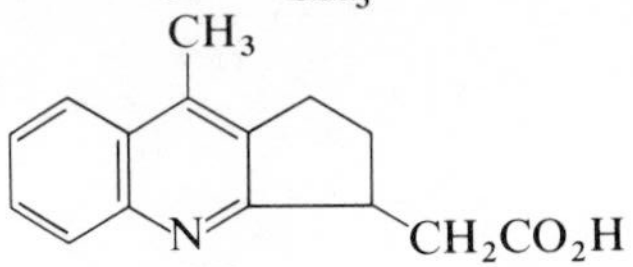 [structure: CH_3, N, CH_2CO_2H] (66)	31,149
	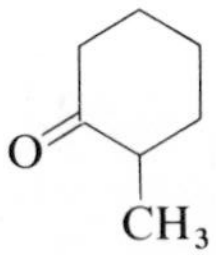 [cyclohexanone structure: O, CH_3]	—	110°	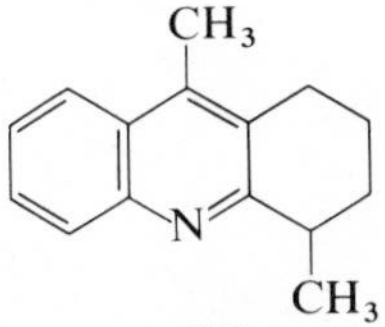 [structure: CH_3, N, CH_3] (51)	31,149
	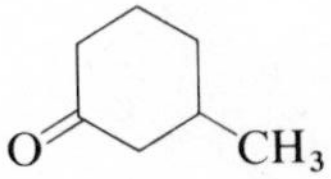 [cyclohexanone structure: O, CH_3]	—	115°	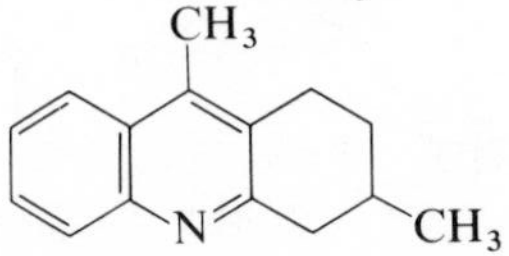 [structure: CH_3, N, CH_3] (75)	31,149

Cycloheptanone	—	140°	CH_3-substituted cyclohepta[b]quinoline (88)	31,149
2-Chlorocycloheptanone (Cl)	—	125°	CH_3-substituted chloro-cyclohepta[b]quinoline, Cl (51)	31,149
Norcamphor	—	155°	{ 4-CH_3-quinolin-2-yl aryl HN-(nortricyclyl) + CH_3CO-aryl NH-(norbornyl) } (37)	15,35
3-Quinuclidinone · HCl	—	175°	CH_3-quinoline fused quinuclidine · HCl (76)	15

TABLE IV. CONDENSATIONS WITH *o*-AMINOACETOPHENONE HYDROCHLORIDE (*Continued*)

No. of C Atoms	Reactant	Catalyst	Reaction Conditions	Product(s) and Yield(s) (%)	Refs.
C_7 (*Contd.*)	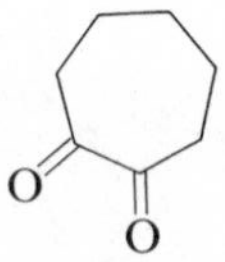 O, O	—	Water bath,[a] 15 hr, or 160°, 3 hr	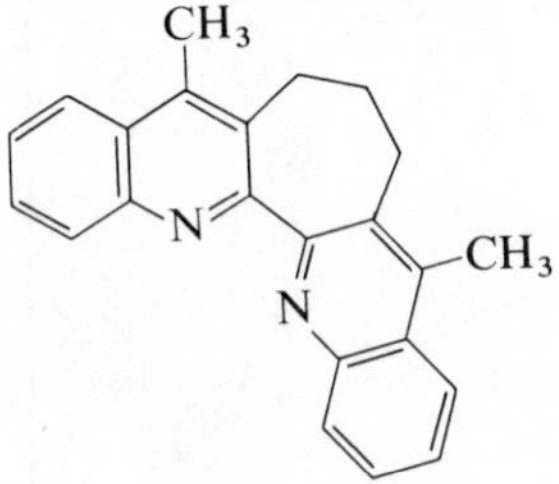 CH_3, N, N, CH_3 (60)	24
	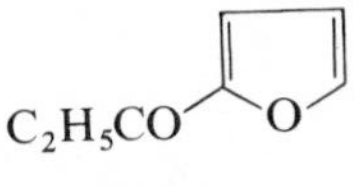 C_2H_5CO, O	—	140°	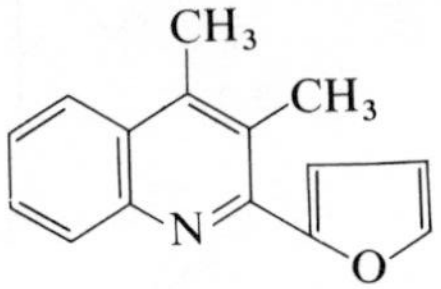 CH_3, CH_3, N, O (42)	152
	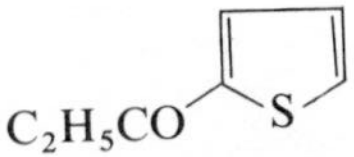 C_2H_5CO, S	—	135°	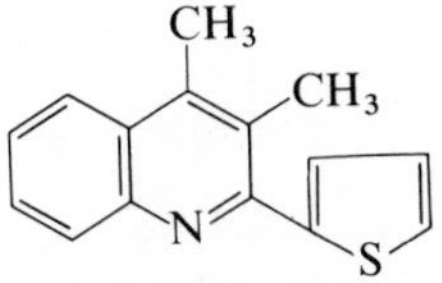 CH_3, CH_3, N, S (55)	152
	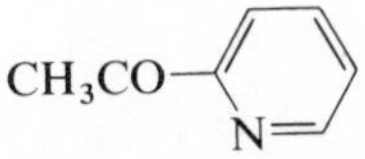 CH_3CO, N	NaOH	Reflux,[a] 18 hr; EtOH	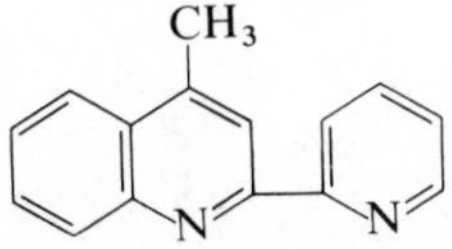 CH_3, N, N (63)	109

			145°	" (50)	152
	CH_3CO–(3-pyridyl)	NaOH	Reflux,[a] 18 hr; EtOH	4-CH_3-2-(3-pyridyl)quinoline (67)	109
	CH_3CO–(4-pyridyl)	"	"[a]	4-CH_3-2-(4-pyridyl)quinoline (61)	109
	1-C_2H_5-4-piperidone	—	135°	10-CH_3-2-C_2H_5-1,2,3,4-tetrahydrobenzo[b][1,6]naphthyridine (57)	32
	CH_3CO–(4-CH_3-2-CH_3-thiazol-5-yl)	—	110–130°	4-CH_3-2-(2,4-di-CH_3-thiazol-5-yl)quinoline (45)	153
C_8	$CH_3COC_6H_{13}$-*n*	—	155°	2,4-di-CH_3-3-C_5H_{11}-*n*-quinoline (75)	55
	$CH_3COC_6H_5$	—	165°	4-CH_3-2-C_6H_5-quinoline (55)	55

TABLE IV. CONDENSATIONS WITH *o*-AMINOACETOPHENONE HYDROCHLORIDE (*Continued*)

No. of C Atoms	Reactant	Catalyst	Reaction Conditions	Product(s) and Yield(s) (%)	Refs.
C_8 (*Contd.*)	$C_2H_5O_2CCOCH_2CO_2C_2H_5$	—	110–130°	4-CH_3-quinoline-2,3-di($CO_2C_2H_5$) (80)	155
	p-$CH_3COC_6H_4Cl$	—	165°	4-CH_3-2-(C_6H_4Cl-*p*)quinoline (64)	55
	p-$CH_3COC_6H_4Br$	—	165°	4-CH_3-2-(C_6H_4Br-*p*)quinoline (23)	55
	p-$CH_3COC_6H_4OH$	—	165°	4-CH_3-2-(C_6H_4OH-*p*)quinoline (33)	55
	o-$CH_3COC_6H_4OH$	—	165°	4-CH_3-2-(C_6H_4OH-*o*)quinoline (20)	55

Reactant	Catalyst	Conditions	Product (Yield, %)		Refs.
p-$CH_3COC_6H_4NO_2$	—	160°	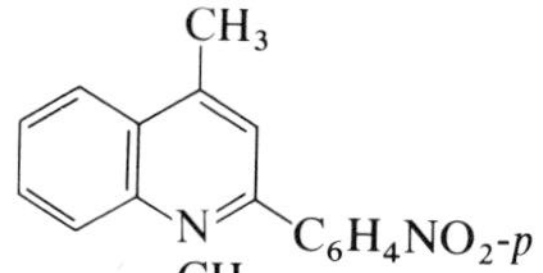 CH_3, N, $C_6H_4NO_2$-p	(52)	55
p-$CH_3COC_6H_4NH_2$	$ZnCl_2$	90–100°[a]	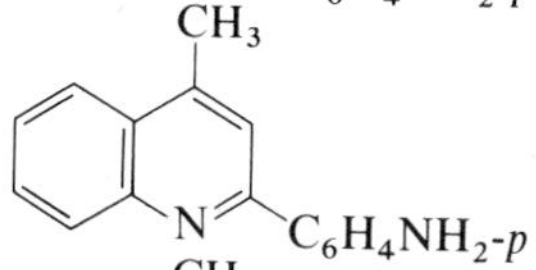 CH_3, N, $C_6H_4NH_2$-p	(—)	156
o-$CH_3COC_6H_4NH_2$	$ZnCl_2$	230°[a]	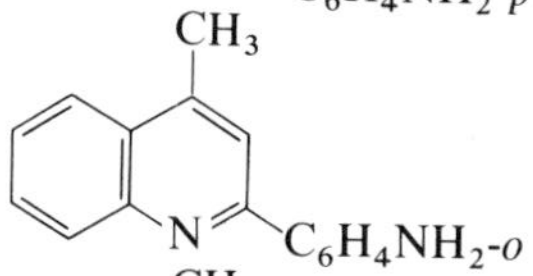 CH_3, N, $C_6H_4NH_2$-o	(—)	3
	NaOH	Heat;[a] H_2O–EtOH	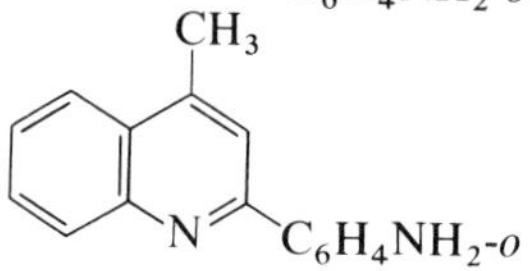 CH_3, N, $C_6H_4NH_2$-o	(—)	158
o-$CH_3COC_6H_4NH_2 \cdot HCl$	—	160°	"	(66)	55
 CH_3COCH_2CO–(2-thienyl, S)	—	160°	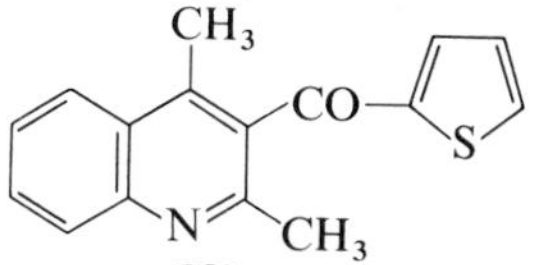 CH_3, CO–(2-thienyl, S), N, CH_3	(70)	55
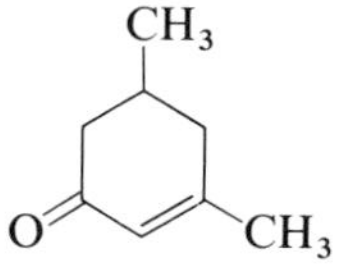 O, CH_3, CH_3	—	160°	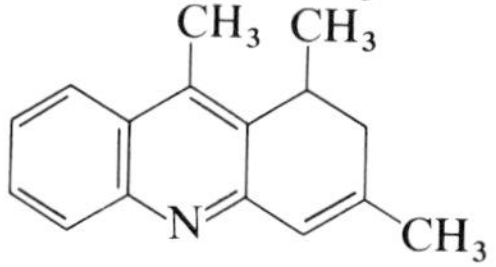 CH_3, CH_3, N, CH_3	(84)	55

TABLE IV. CONDENSATIONS WITH *o*-AMINOACETOPHENONE HYDROCHLORIDE (*Continued*)

No. of C Atoms	Reactant	Catalyst	Reaction Conditions	Product(s) and Yield(s) (%)	Refs.
C_8 (*Contd.*)	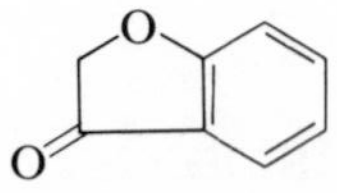 O; O	—	130°	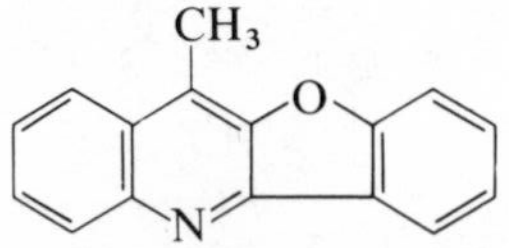 CH_3; O; N (90)	32,148
	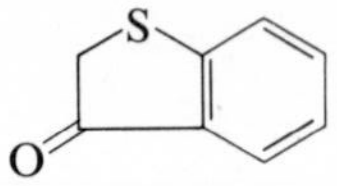 S; O	—	130°	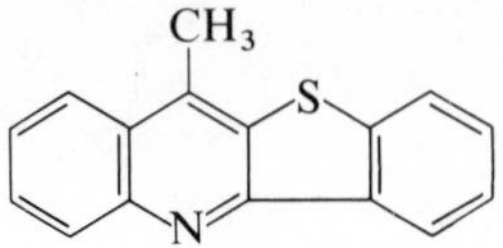 CH_3; S; N (—)	148
	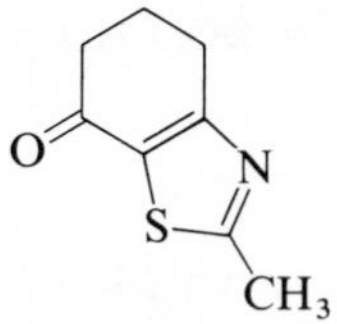 O; N; S; CH_3	—	150°	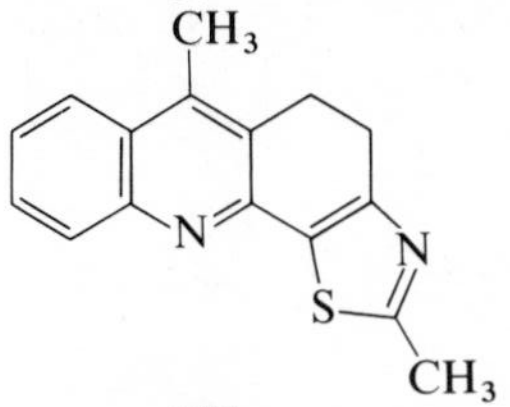 CH_3; N; N; S; CH_3 (51)	55
	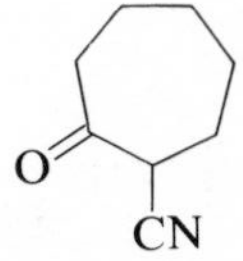 O; CN	—	160°	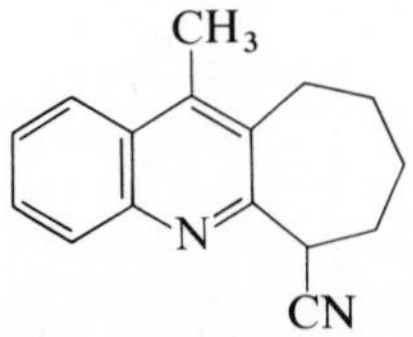 CH_3; N; CN (73)	31,149

Ketone	Catalyst	Conditions	Product (Yield, %)	Refs.
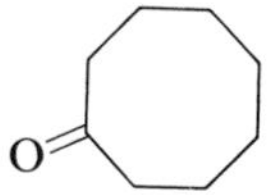	—	120°	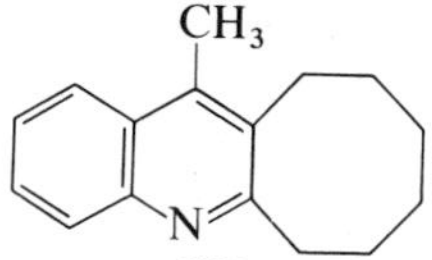 (70)	31,149
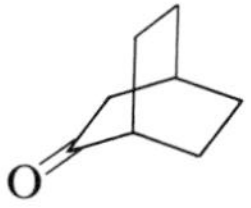	HCl	160°	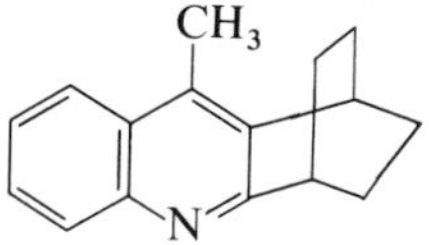 (29) + 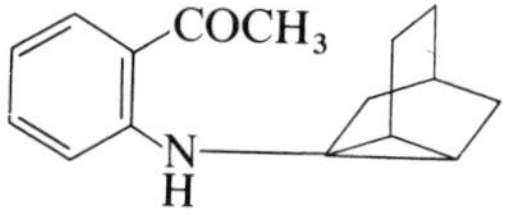 (15)	15,35
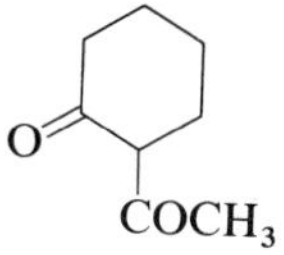	—	125°	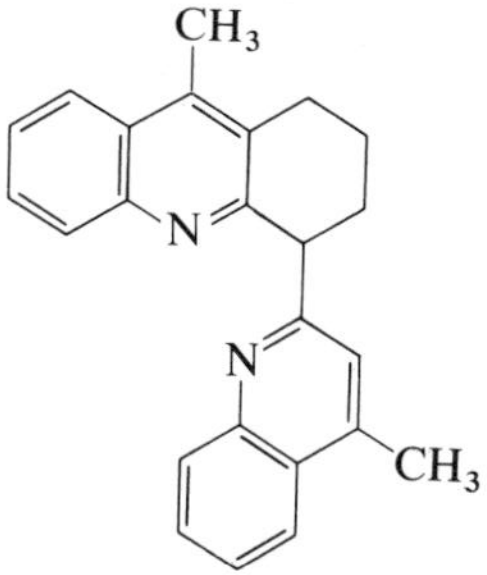 (43)	31,149
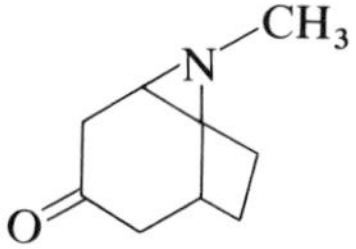	—	190°	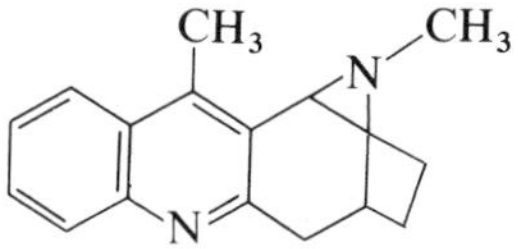 (16)	15
	NaOH	40–50°;[a] H_2O	" (17)	15

TABLE IV. CONDENSATIONS WITH *o*-AMINOACETOPHENONE HYDROCHLORIDE (*Continued*)

No. of C Atoms	Reactant	Catalyst	Reaction Conditions	Product(s) and Yield(s) (%)	Refs.
C_9	$CH_3COC_7H_{15}$-*n*	—	155°	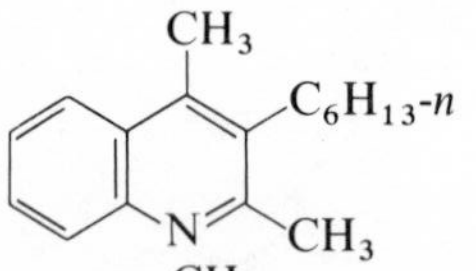 CH_3, C_6H_{13}-*n*, N, CH_3 (74)	55
	$CO(CH_2CO_2C_2H_5)_2$	—	110–130°	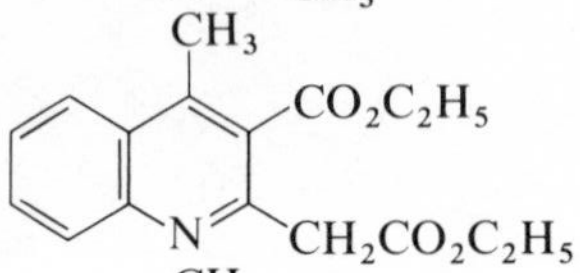 CH_3, $CO_2C_2H_5$, N, $CH_2CO_2C_2H_5$ (80)	155
	$CH_3CH_2COC_6H_5$	—	150°	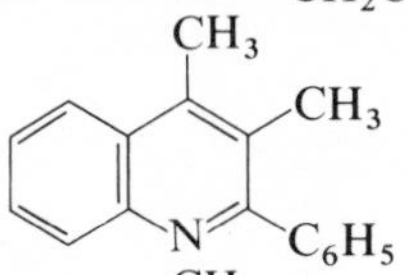 CH_3, CH_3, N, C_6H_5 (60)	55
	$C_6H_5C(=NH)CH_2CN$	—	150°;[a] AcOH	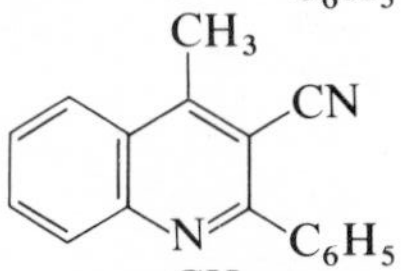 CH_3, CN, N, C_6H_5 (—)	100
	$CH_3COCH_2C_6H_5$	—	140°	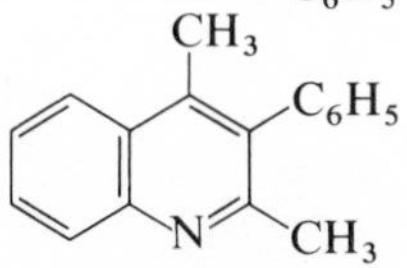 CH_3, C_6H_5, N, CH_3 (90)	55

C_2H_5, O, CH_3	—	165°	CH_3, C_2H_5, N, CH_3 (71)	55
O, $CH_2CO_2C_2H_5$	—	120°	CH_3, N, $CH_2CO_2C_2H_5$ (60)	31,149
O	—	120°	CH_3, N (88)	31,149
O, O	—	120°	CH_3, N, O (85)	31
O, $CO_2C_2H_5$	—	125°	CH_3, N, $CO_2C_2H_5$ (32)	31
O, O	—	140°, 30 min	CH_3, N, O (75)	120

TABLE IV. CONDENSATIONS WITH *o*-AMINOACETOPHENONE HYDROCHLORIDE (*Continued*)

No. of C Atoms	Reactant	Catalyst	Reaction Conditions	Product(s) and Yield(s) (%)	Refs.
C_9 (*Contd.*)	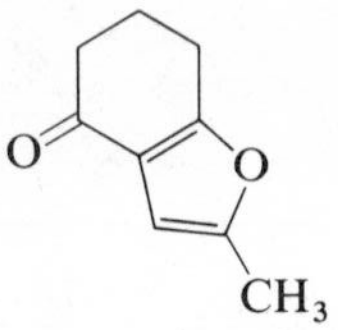 [structure: CH_3, O]	—	140°, 1 hr	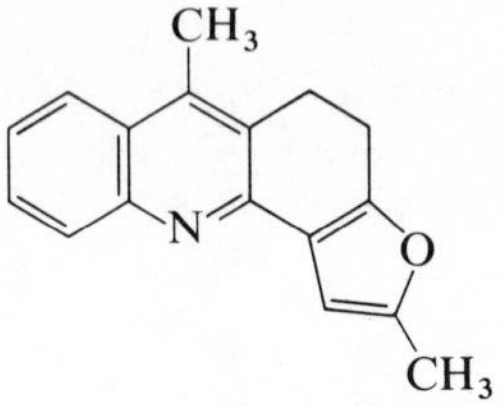 [structure: CH_3, N, O, CH_3] (40)	33
	o-$HO_2CC_6H_4CH_2CO_2H$	PPA	130–160° [a]	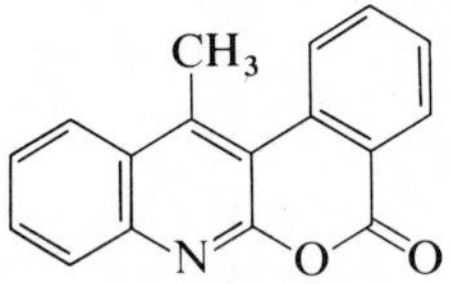 [structure: CH_3, N, O, O] (69)	161
	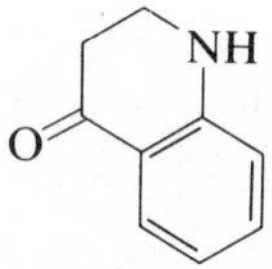 [structure: NH, O]	—	140°	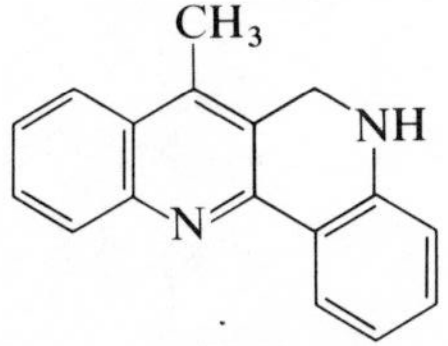 [structure: CH_3, N, NH] (61)	32

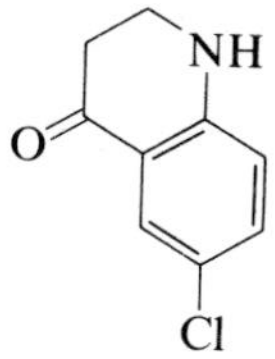	—	135°	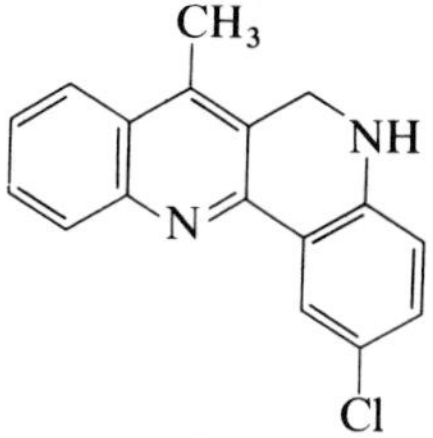 (49)	32
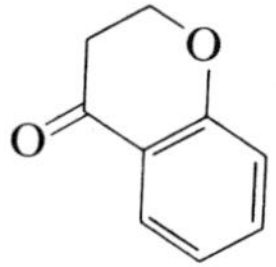	—	150°, 5 hr	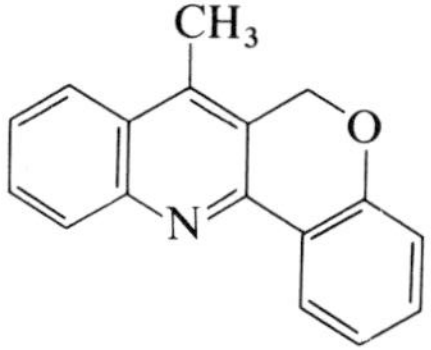 (77)	32,148
	—	140°, 30 min	" (75)	34
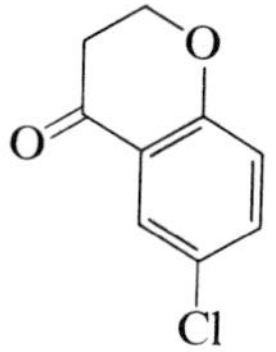	—	140°, 30 min	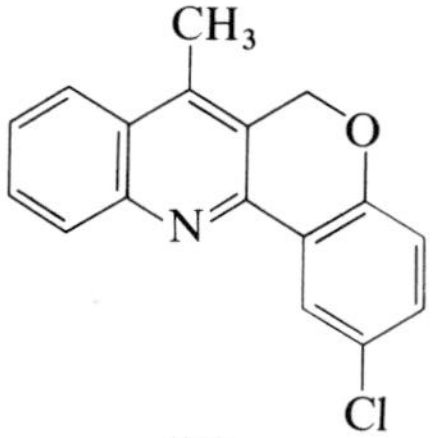 (75)	34
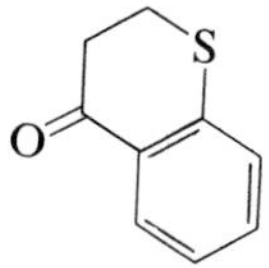	—	1. 150°, 10 min 2. 110°, 30 min	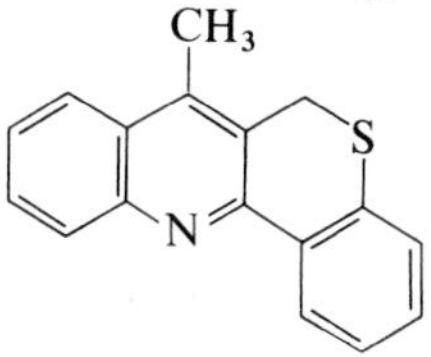 (81)	32
	—	140°, 30 min	" (75)	34
	—	145°	" (54)	148

TABLE IV. CONDENSATIONS WITH *o*-AMINOACETOPHENONE HYDROCHLORIDE (*Continued*)

No. of C Atoms	Reactant	Catalyst	Reaction Conditions	Product(s) and Yield(s) (%)	Refs.
C_9 (*Contd.*)	O, S, Cl	—	140°, 30 min	CH_3, S, N, Cl (75)	34
	O, S	—	"	CH_3, N, S (—)	121a
	O, CH_3, CH_3	—	175°	CH_3, N, CH_3, CH_3 (15)	15
	O, O	—	150°	CH_3, N, N, CH_3 (82)	15

Carbon count	Ketone	Catalyst	Conditions	Product	(Yield, %)	Ref.
	$p\text{-}CH_3COC_6H_4OCH_3$	—	160°	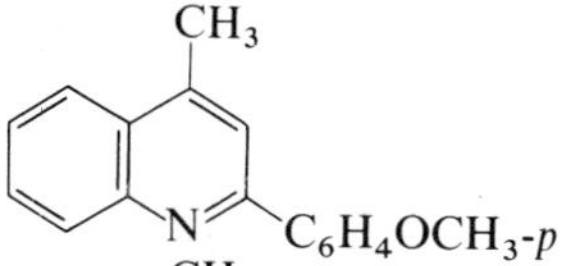 CH_3; N; $C_6H_4OCH_3\text{-}p$	(36)	55
	$o\text{-}CH_3COC_6H_4OCH_3$	—	160°	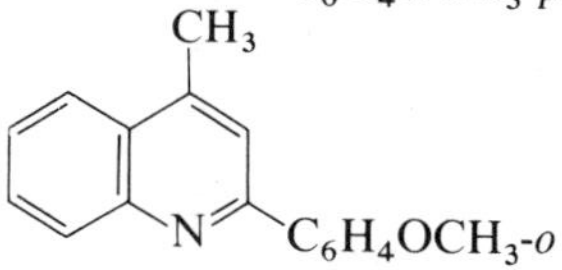 CH_3; N; $C_6H_4OCH_3\text{-}o$	(50)	55
C_{10}	$CH_3COC_8H_{17}\text{-}n$	—	155°	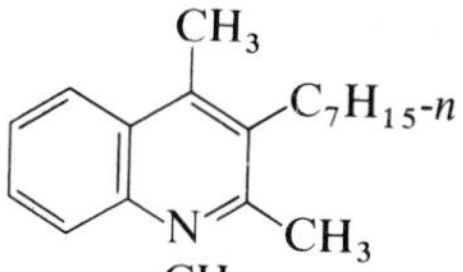 CH_3; $C_7H_{15}\text{-}n$; N; CH_3	(74)	55
	$CH_3COCH_2COC_6H_5$	—	140°	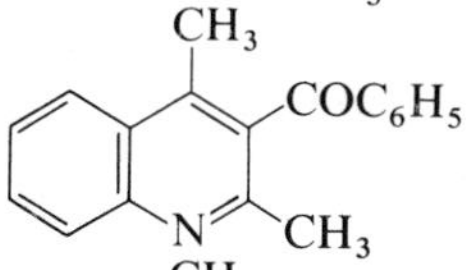 CH_3; COC_6H_5; N; CH_3	(80)	55
	$p\text{-}CH_3C_6H_4C({=}NH)CH_2CN$	—	150°;[a] AcOH	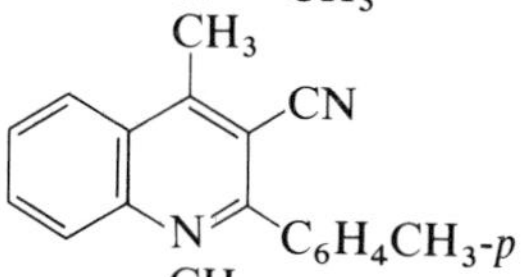 CH_3; CN; N; $C_6H_4CH_3\text{-}p$	(—)	100
	$o\text{-}CH_3COC_6H_4NHCOCH_3$	PPA	100°,[a] 2 hr	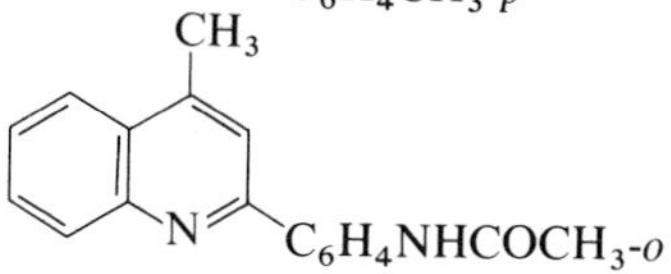 CH_3; N; $C_6H_4NHCOCH_3\text{-}o$	(85)	30b
	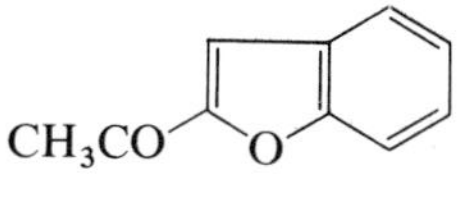 CH_3CO; O	—	130°	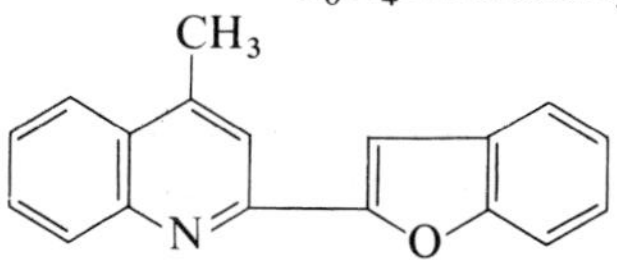 CH_3; N; O	(49)	152

TABLE IV. CONDENSATIONS WITH *o*-AMINOACETOPHENONE HYDROCHLORIDE (*Continued*)

No. of C Atoms	Reactant	Catalyst	Reaction Conditions	Product(s) and Yield(s) (%)	Refs.
C_{10} (*Contd.*)	4-C_4H_9-*t*-cyclohexanone (O; C_4H_9-*t*)	—	125°	CH_3; C_4H_9-*t*; N (80)	31,149
	CH_3; O; $C(CH_3)_2$	—	150°	CH_3 CH_3; N; $C(CH_3)_2$ (33)	31,149
	CH_3; O; C_3H_7-*i*	—	145°	CH_3 CH_3; N; C_3H_7-*i* (32)	31,149
	C_3H_7-*n*; O; CH_3	—	165°	CH_3 C_3H_7-*n*; N; CH_3 (71)	55
	C_3H_7-*i*; O; CH_3	—	165°	CH_3 C_3H_7-*i*; N; CH_3 (74)	55

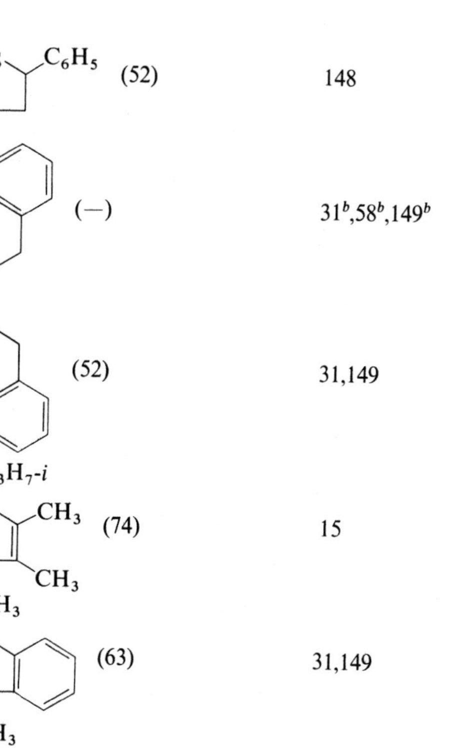

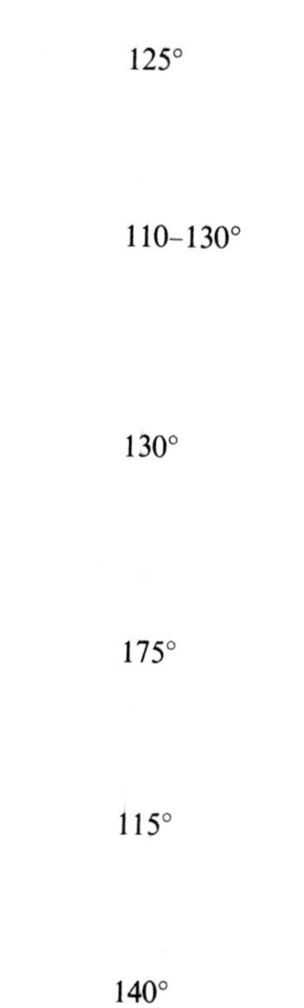

	—	125°	(52)	148
	—	110–130°	(—)	31[b],58[b],149[b]
	—	130°	(52)	31,149
	—	175°	(74)	15
	—	115°	(63)	31,149
	—	140°	(46)	31,149

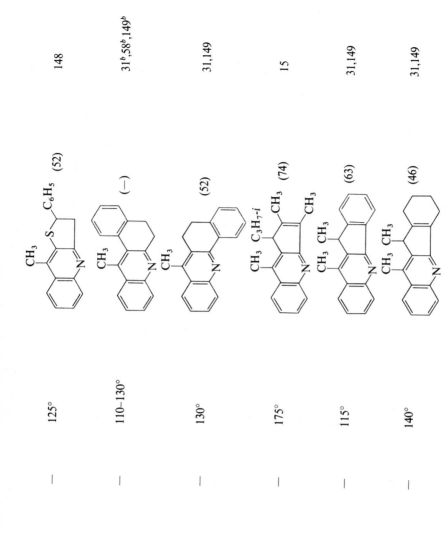

TABLE IV. CONDENSATIONS WITH *o*-AMINOACETOPHENONE HYDROCHLORIDE (*Continued*)

No. of C Atoms	Reactant	Catalyst	Reaction Conditions	Product(s) and Yield(s) (%)	Refs.
C_{10} (*Contd.*)	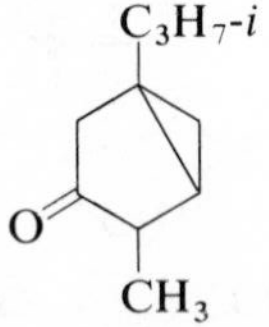 C_3H_7-*i*, O, CH_3	—	170°	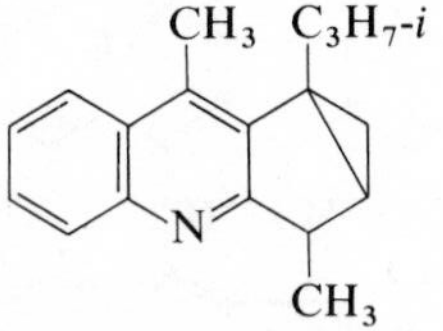 CH_3, C_3H_7-*i*, N, CH_3 (43)	15
	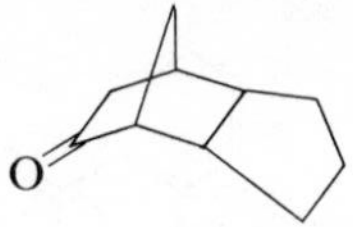 O	—	170°	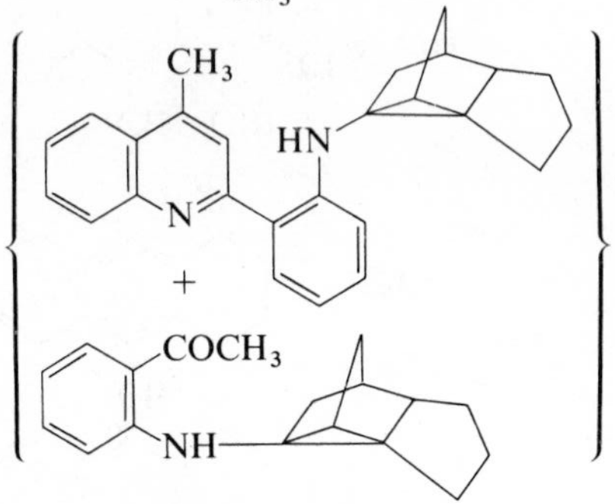 CH_3, N, HN + $COCH_3$, NH (26)	15,35
	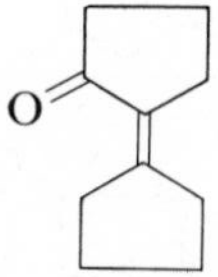 O	—	120°	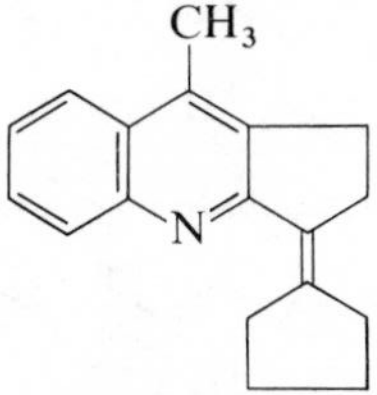 CH_3, N (80)	31,149

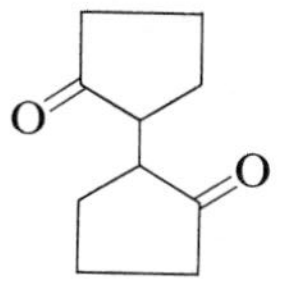

O, O	—	125°	CH_3, N, N, CH_3 (61)	31,149
NH, O, OCH_3	—	145°	CH_3, NH, N, OCH_3 (49)	32
N, CH_3, O	—	170°	CH_3, N, CH_3, N (53)	32
O, O, CH_3	—	140°, 30 min	CH_3, O, N, CH_3 (75)	34

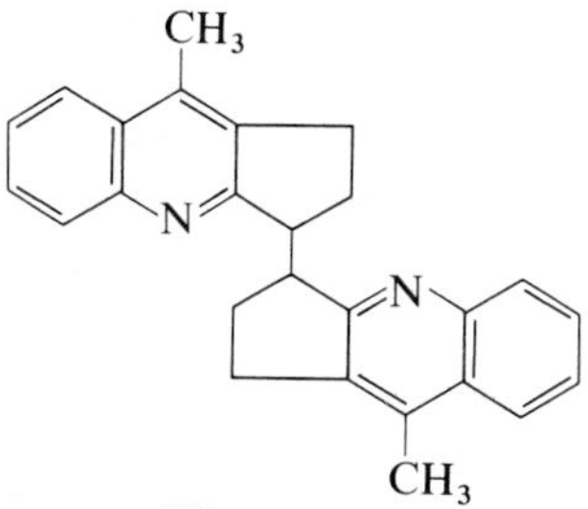

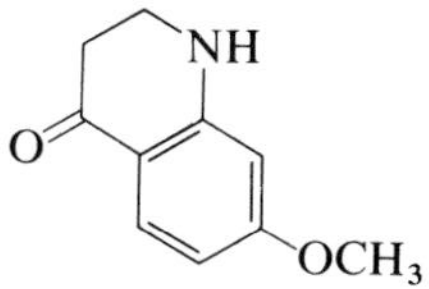

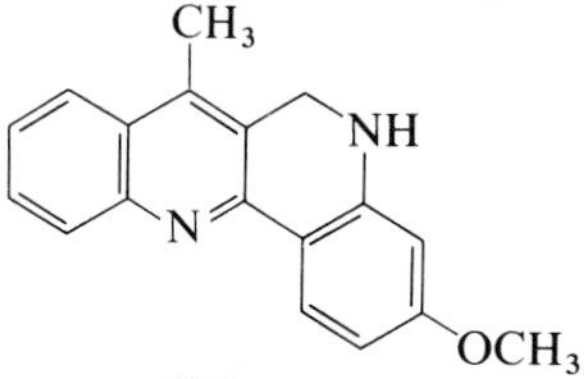

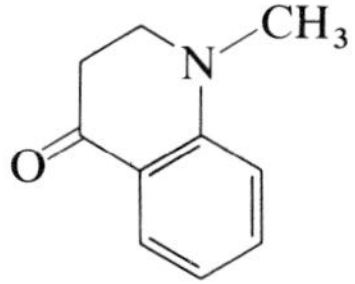

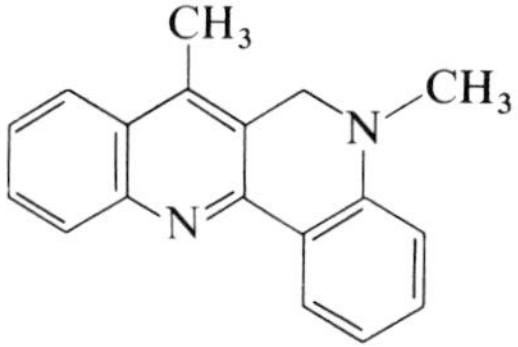

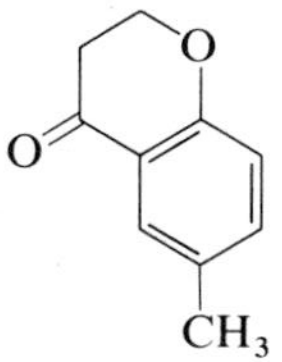

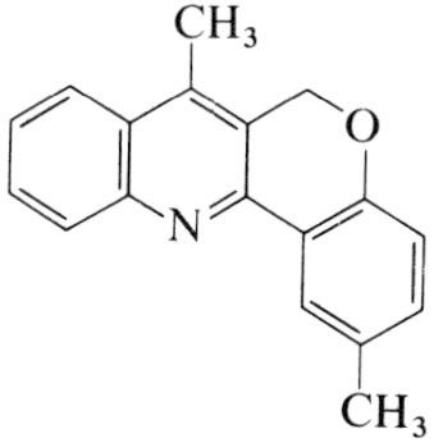

TABLE IV. CONDENSATIONS WITH *o*-AMINOACETOPHENONE HYDROCHLORIDE (*Continued*)

No. of C Atoms	Reactant	Catalyst	Reaction Conditions	Product(s) and Yield(s) (%)	Refs.
C_{10} (*Contd.*)	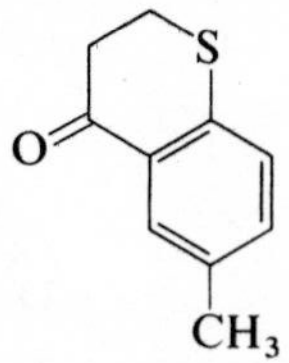	—	140°, 30 min	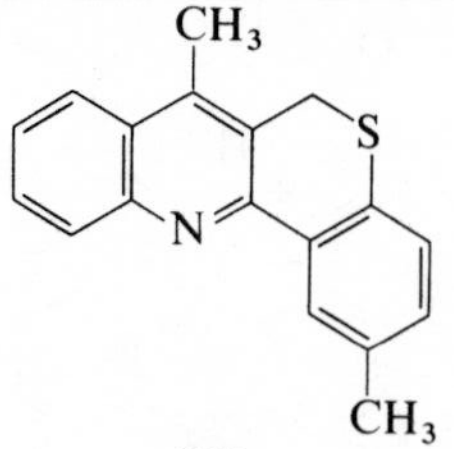 (75)	34
	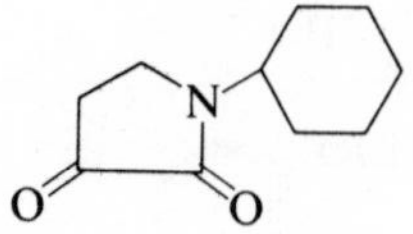	H_2SO_4	Reflux, 3 hr; AcOH	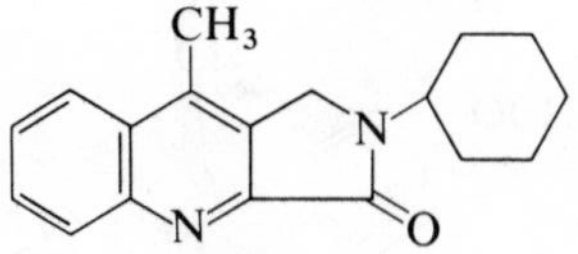 (100)	88
	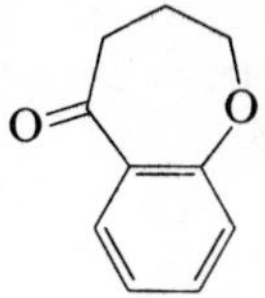	—	140°	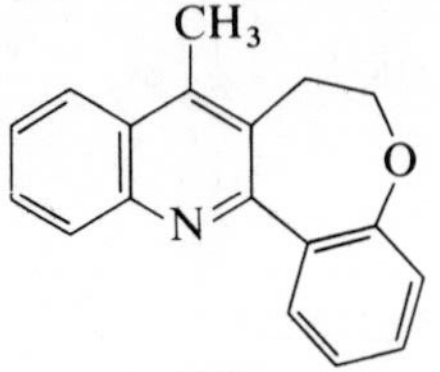 (56)	148
	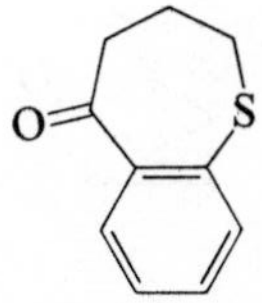	—	160°	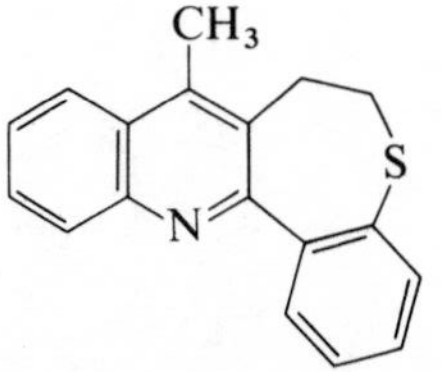 (54)	148

C_{11}	$CH_3COC_9H_{19}$-*n*	—	155°	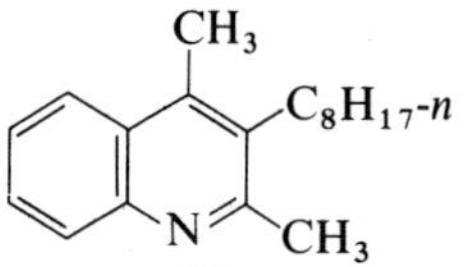 CH_3, C_8H_{17}-*n*, CH_3, N (72)	55
	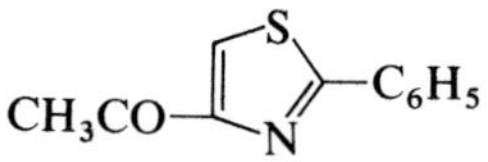 CH_3CO, S, N, C_6H_5	—	110–130°	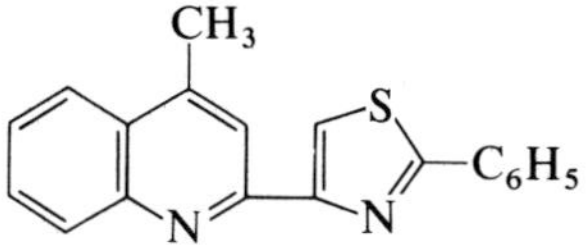 CH_3, N, S, N, C_6H_5 (58)	153
	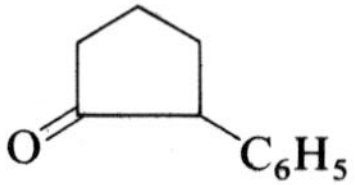 O, C_6H_5	Piperidine	180°;[a] xylene	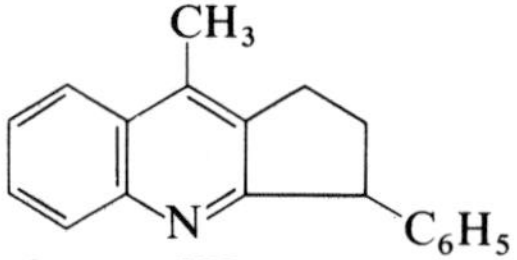 CH_3, N, C_6H_5 (80)	90
	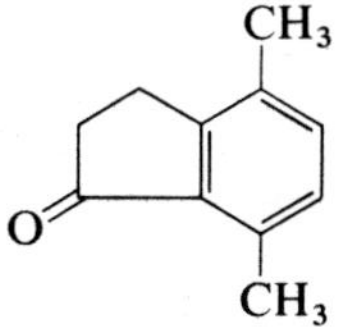 CH_3, O, CH_3	—	130°	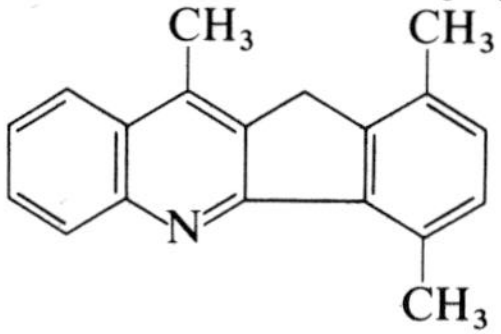 CH_3, CH_3, N, CH_3 (82)	31,149
	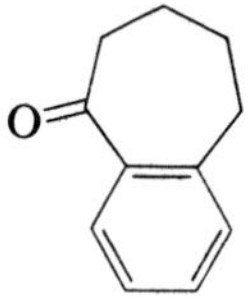 O	—	140°	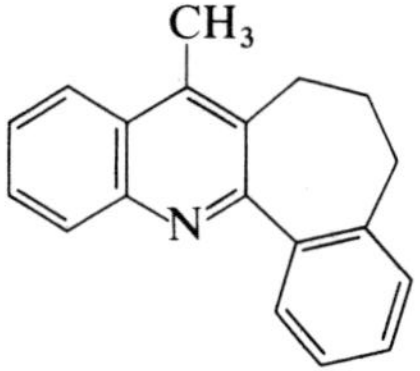 CH_3, N (95)	31,149

TABLE IV. CONDENSATIONS WITH *o*-AMINOACETOPHENONE HYDROCHLORIDE (*Continued*)

No. of C Atoms	Reactant	Catalyst	Reaction Conditions	Product(s) and Yield(s) (%)	Refs.
C_{11} (*Contd.*)	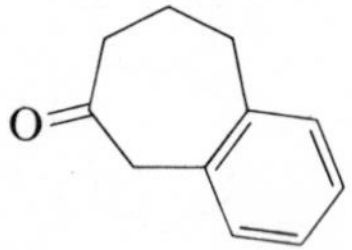 O	—	110–130°	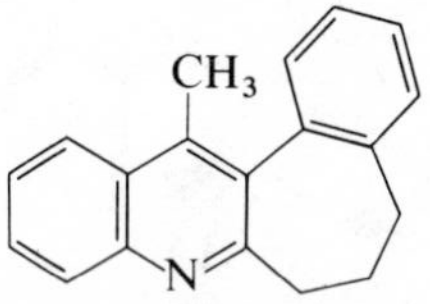 CH_3, N (—)	31[c],58[c]
	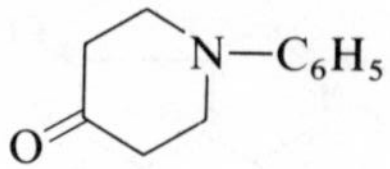 $N-C_6H_5$, O	NaOH	20°,[a] 3 d; EtOH	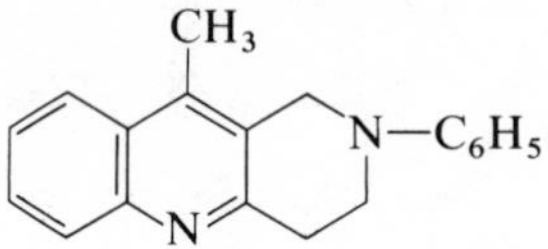 CH_3, N, $N-C_6H_5$ (38)	32
	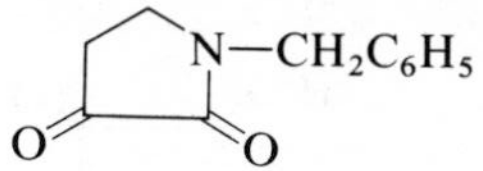 $N-CH_2C_6H_5$, O, O	H_2SO_4	Reflux,[a] 3 hr; AcOH	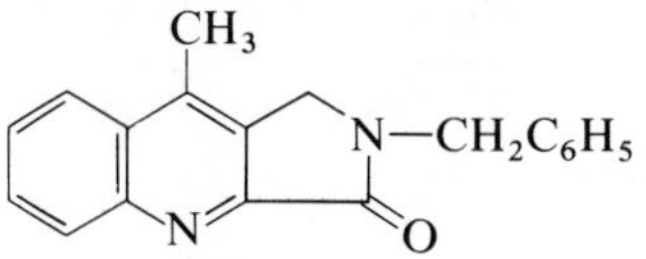 CH_3, N, $N-CH_2C_6H_5$, O (100)	88
C_{12}	$CH_3COC_{10}H_{21}$-*n*	—	160°	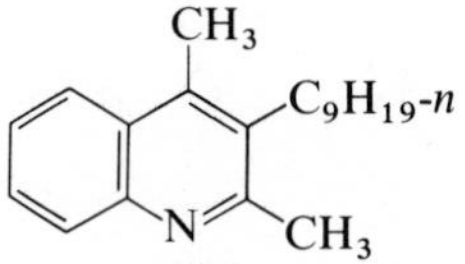 CH_3, C_9H_{19}-*n*, N, CH_3 (70)	55
	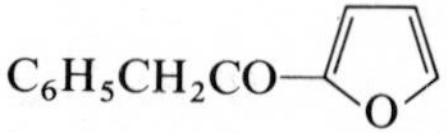 $C_6H_5CH_2CO$—(2-furyl), O	—	145°	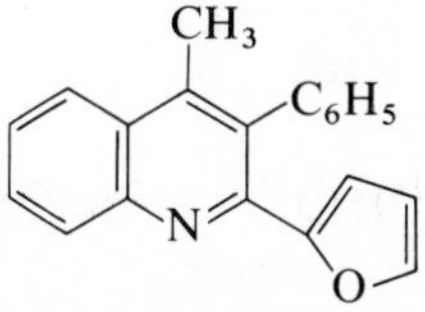 CH_3, C_6H_5, N, O (46)	152

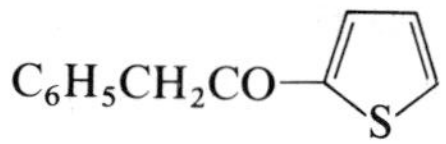 $C_6H_5CH_2CO$–(S)	—	150°	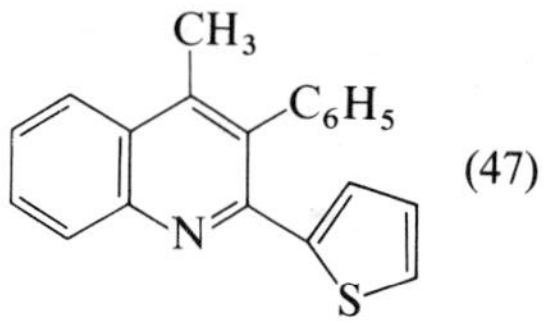 CH_3, C_6H_5, N, S (47)	152
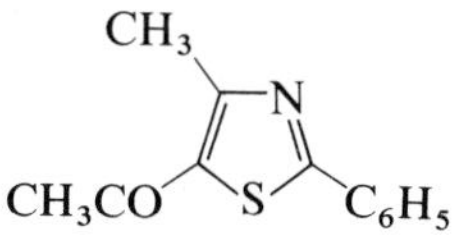 CH_3, N, CH_3CO, S, C_6H_5	—	110–130°	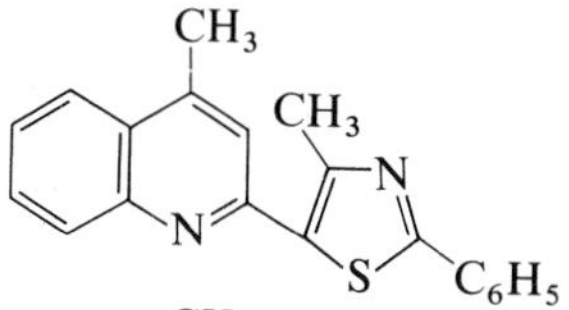 CH_3, N, CH_3, N, S, C_6H_5 (32)	153
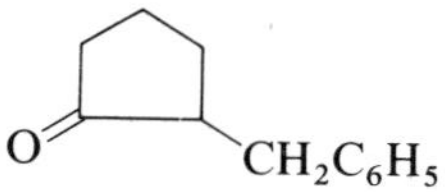 O, $CH_2C_6H_5$	—	125°	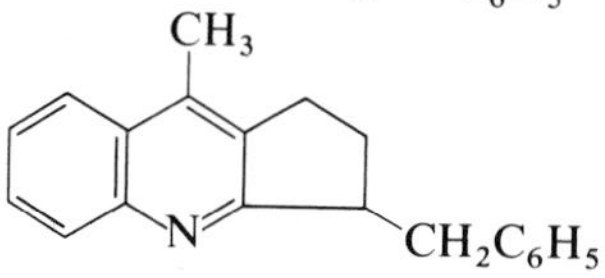 CH_3, N, $CH_2C_6H_5$ (55)	31,149
	—	180°,[a] xylene, 120°	" (75)	90
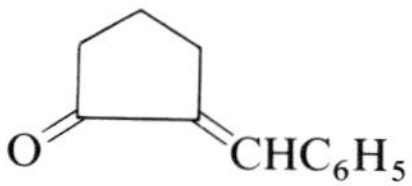 O, CHC_6H_5	—	120°	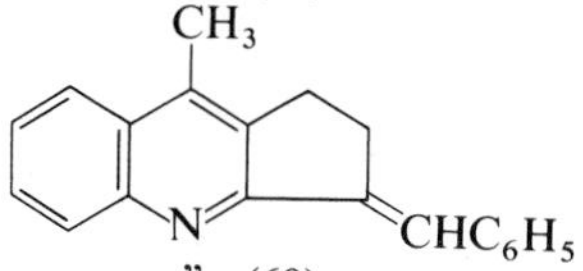 CH_3, N, CHC_6H_5 (80)	31,149
	Piperidine	150°;[a] xylene	" (60)	90
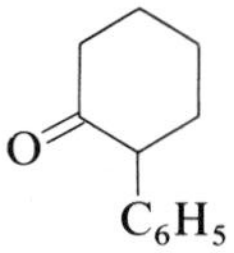 O, C_6H_5	—	125°	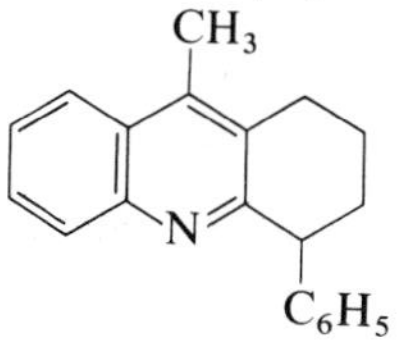 CH_3, N, C_6H_5 (67)	31,149

TABLE IV. CONDENSATIONS WITH o-AMINOACETOPHENONE HYDROCHLORIDE (Continued)

No. of C Atoms	Reactant	Catalyst	Reaction Conditions	Product(s) and Yield(s) (%)	Refs.
C_{12} (*Contd.*)	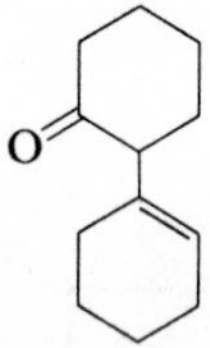 [structure]	—	130°	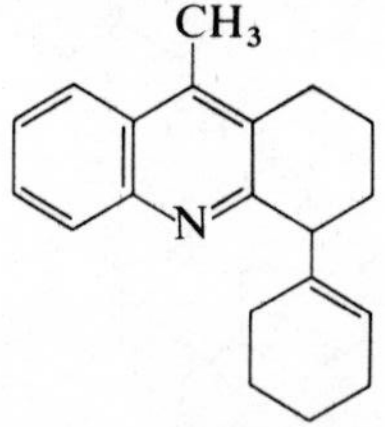 [structure] (71)	31[d],149[d]
	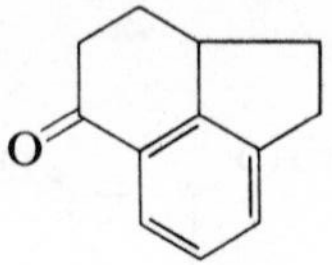 [structure]	—	140°	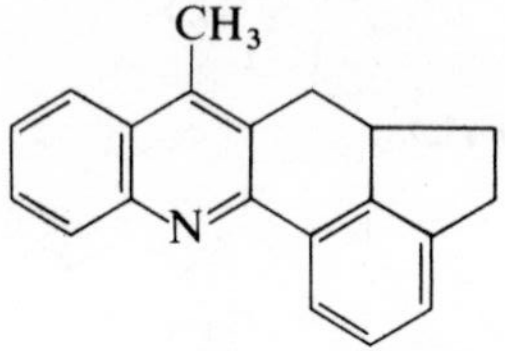 [structure] (68)	31,149
	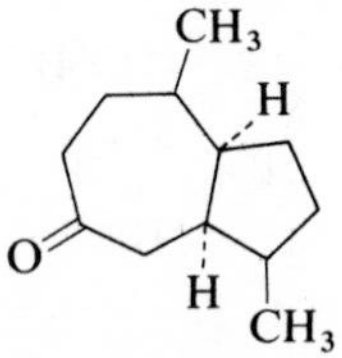 [structure] (CH_3, H, H, CH_3)	—	165°	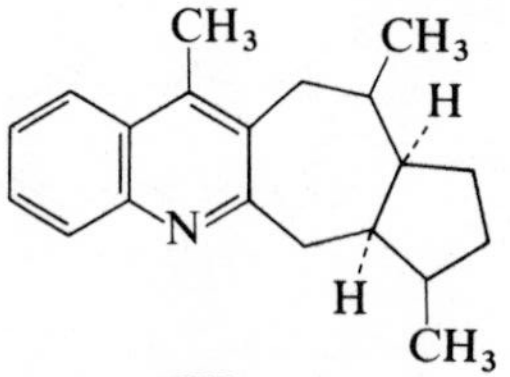 [structure] (CH_3, CH_3, H, H, CH_3) (51)	31,149
	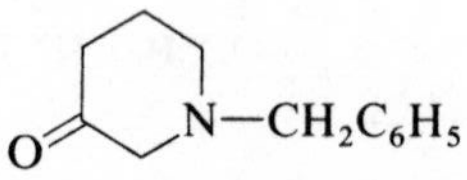 [structure] N—$CH_2C_6H_5$	—	110°	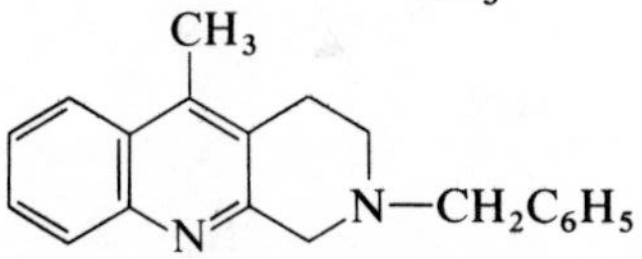 [structure] (CH_3, N—$CH_2C_6H_5$) (63)	32

N—$CH_2C_6H_5$, O	—	140°	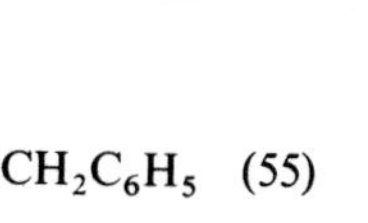 CH_3, N, N—$CH_2C_6H_5$ (55)	32
O, O	—	150°	CH_3, N, O (55)	55
N H, O	—	170°	CH_3, N, HN (37)	152
N, $COCH_2$, N	—	150°	CH_3, N, N, N (54)	152

TABLE IV. CONDENSATIONS WITH *o*-AMINOACETOPHENONE HYDROCHLORIDE (*Continued*)

No. of C Atoms	Reactant	Catalyst	Reaction Conditions	Product(s) and Yield(s) (%)	Refs.
C_{13}	$CH_3COC_{11}H_{23}$-*n*	—	165°	4-CH_3, 3-$C_{10}H_{21}$-*n*, 2-CH_3 quinoline (53)	55
	$C_6H_5COCH_2CO$–(2-furyl)	—	160°	4-CH_3-3-COC_6H_5-2-(2-furyl)quinoline + 4-CH_3-3-(2-furyl-CO)-2-C_6H_5-quinoline (46)	55
	2-$COCH_3$-3-SCH_3-naphthalene	—	150°	4-CH_3-2-(3-CH_3S-2-naphthyl)quinoline (48)	152

Ketone	Catalyst	Temperature	Product (Yield, %)	Reference
C_6H_5, O, CH_3	—	165°	CH_3, C_6H_5, N, CH_3 (84)	55
Cl, O, CH_3	—	170°	Cl, CH_3, N, CH_3 (66)	55
O	—	150°	CH_3, N (58)	31,149
O	—	150°	CH_3, N (55)	31,149
O, C_6H_5	—	140°	CH_3, N, C_6H_5 (42)	31,149

TABLE IV. CONDENSATIONS WITH *o*-AMINOACETOPHENONE HYDROCHLORIDE (*Continued*)

No. of C Atoms	Reactant	Catalyst	Reaction Conditions	Product(s) and Yield(s) (%)	Refs.
C_{13} (*Contd.*)	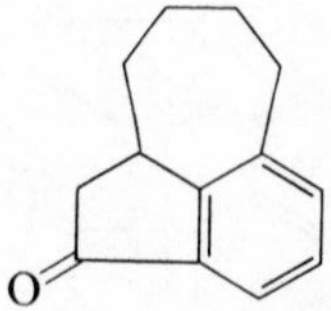 [structure]	—	135°	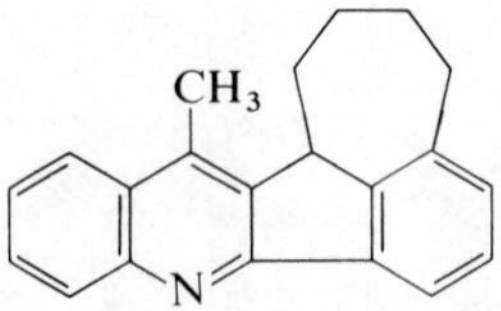 [structure: CH_3, N] (66)	31,149
	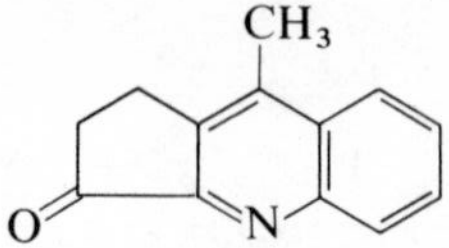 [structure: CH_3, O, N]	—	160–170°,[a] 4 hr	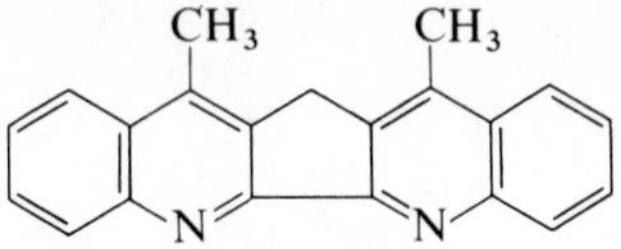 [structure: CH_3, CH_3, N, N] (80)	24
C_{14}	$C_6H_5COCH_2C_6H_5$	—	150°	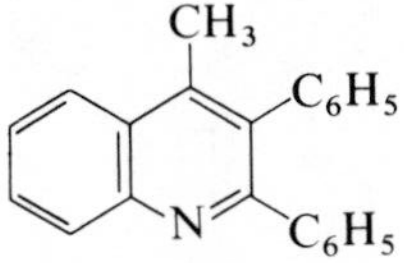 [structure: CH_3, C_6H_5, C_6H_5, N] (60)	55
	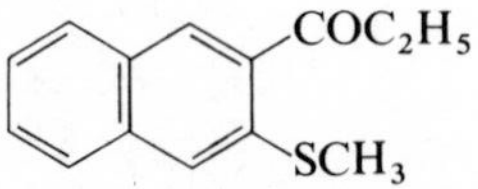 [structure: COC_2H_5, SCH_3]	—	145°	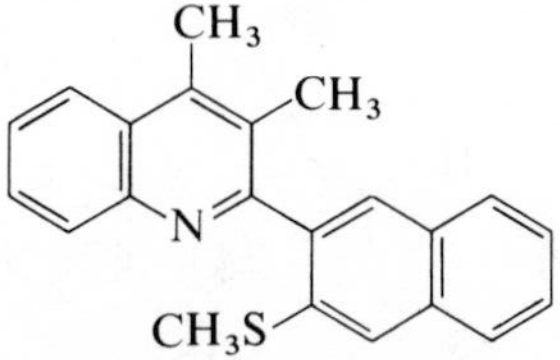 [structure: CH_3, CH_3, N, CH_3S] (85)	152

	Reactant	Catalyst	Conditions	Product(s) and Yield(s) (%)	Refs.
	[structure] OCH_3, CH_3, O	—	170°	[structure] OCH_3, CH_3, N, CH_3 (74)	55
	[structure] O, O, C_6H_5	—	140°, 1 hr	[structure] CH_3, N, O, C_6H_5 (58)	33
		KOH	Heat,[a] 6 hr; EtOH	" (5)	33
	[structure] $CO_2C_2H_5$, O, O, $CO_2C_2H_5$	—	175°	[structure] CH_3, $CO_2C_2H_5$, O, N, $CO_2C_2H_5$ (89)	15
C_{15}	$C_6H_5CH_2COCH_2C_6H_5$	—	150°	[structure] CH_3, C_6H_5, N, $CH_2C_6H_5$ (70)	55
	$C_6H_5COCH_2COC_6H_5$	—	145°	[structure] CH_3, COC_6H_5, N, C_6H_5 (80)	55

TABLE IV. CONDENSATIONS WITH *o*-AMINOACETOPHENONE HYDROCHLORIDE (*Continued*)

No. of C Atoms	Reactant	Catalyst	Reaction Conditions	Product(s) and Yield(s) (%)	Refs.
C_{15} (*Contd.*)	*o*-$CH_3COC_6H_4NHCOC_6H_5$	PPA	100°,[a] 2 hr	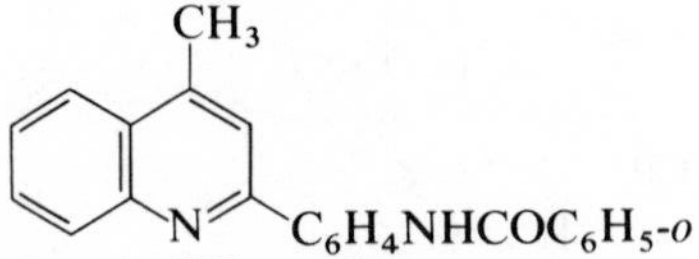 CH_3; N; $C_6H_4NHCOC_6H_5$-*o* (88)	30b
	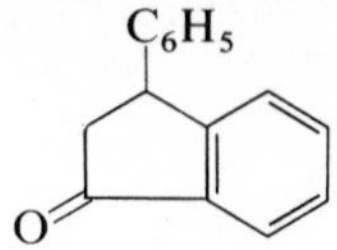 C_6H_5; O	—	140°	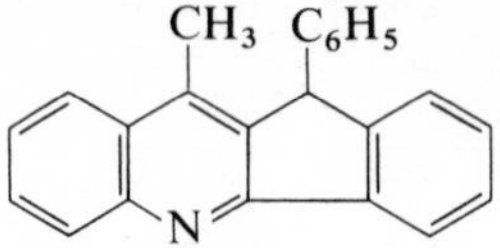 CH_3 C_6H_5; N (44)	31,149
	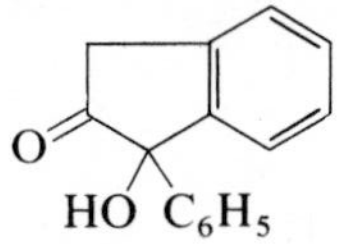 O; HO C_6H_5	—	115°	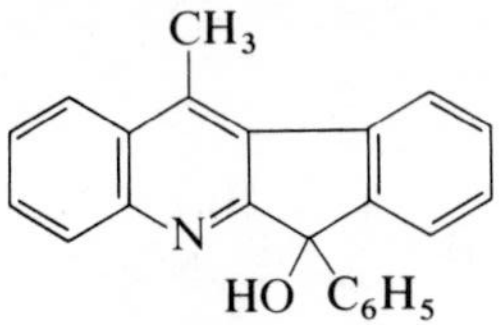 CH_3; N; HO C_6H_5 (76)	31
	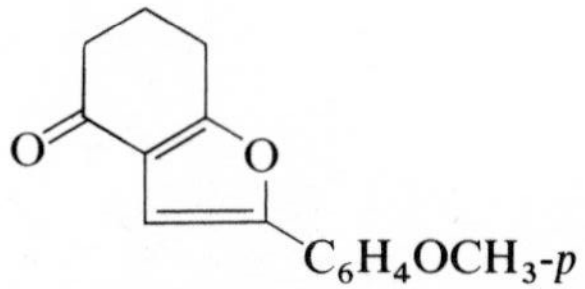 O; O; $C_6H_4OCH_3$-*p*	—	140°, 1 hr	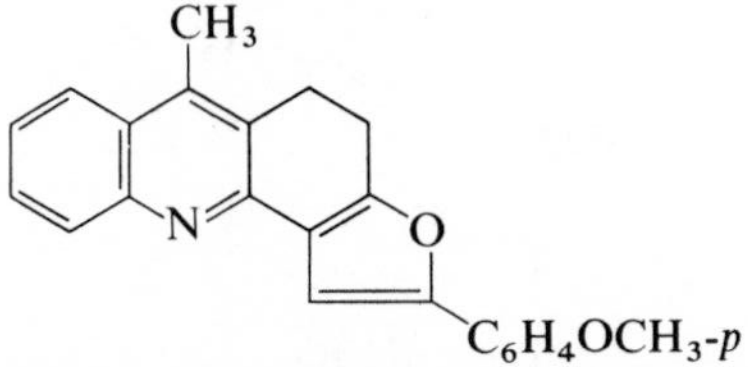 CH_3; N; O; $C_6H_4OCH_3$-*p* (50)	33

	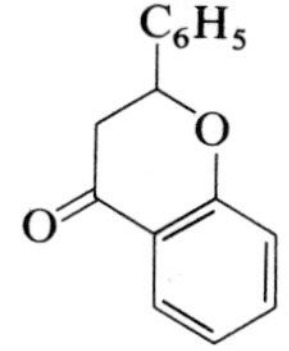 [structure; label: C_6H_5, O]	—	1. 160°, 30 min 2. 140°, 20 min	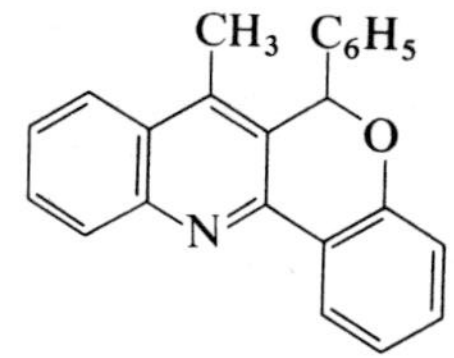 [structure; labels: CH_3, C_6H_5, O, N] (78)	32
C_{16}	$C_6H_5CH_2COCH_2COC_6H_5$	—	165°	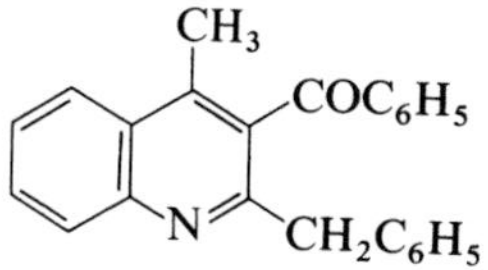 [structure; labels: CH_3, COC_6H_5, $CH_2C_6H_5$, N] (70)	55
	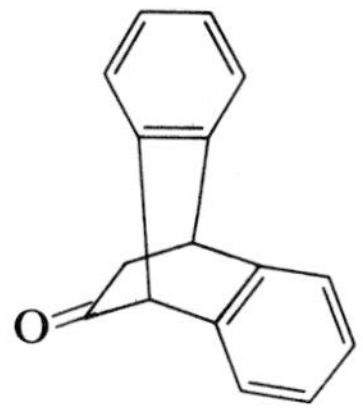 [structure; label: O]	—	165°	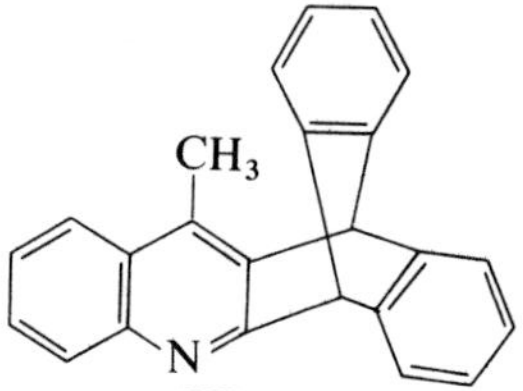 [structure; labels: CH_3, N] (3)	15
	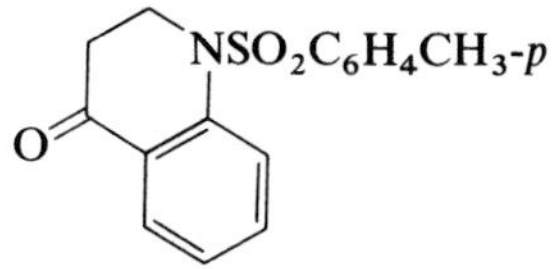 [structure; label: $NSO_2C_6H_4CH_3$-*p*, O]	—	150°	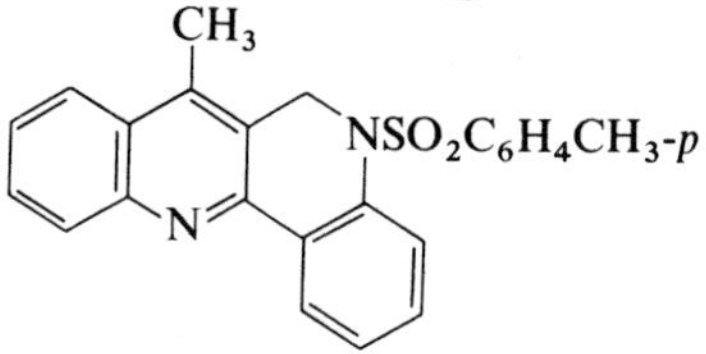 [structure; labels: CH_3, $NSO_2C_6H_4CH_3$-*p*, N] (75)	32
	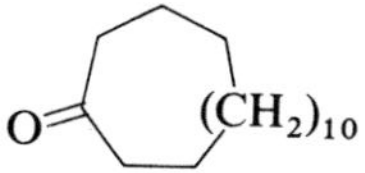 [structure; label: $(CH_2)_{10}$, O]	—	120°	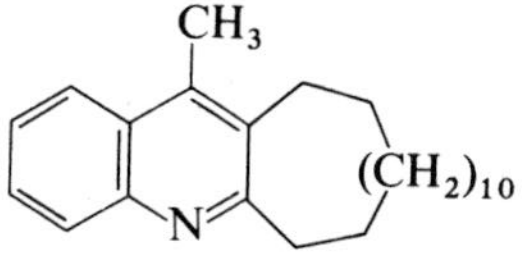 [structure; labels: CH_3, $(CH_2)_{10}$, N] (75)	31,149

TABLE IV. CONDENSATIONS WITH *o*-AMINOACETOPHENONE HYDROCHLORIDE (*Continued*)

No. of C Atoms	Reactant	Catalyst	Reaction Conditions	Product(s) and Yield(s) (%)	Refs.
C_{17}	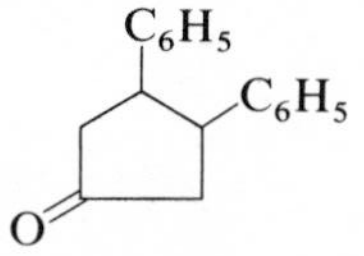 C_6H_5, C_6H_5, O	—	140°	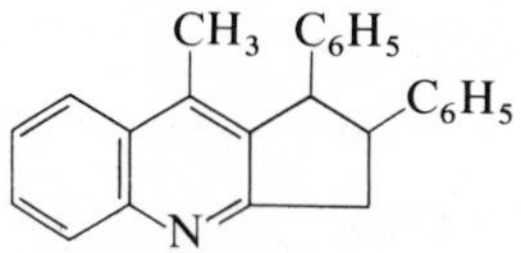 CH_3, C_6H_5, C_6H_5, N (48)	31,[e] 149[e]
	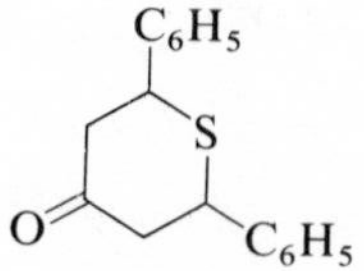 C_6H_5, S, C_6H_5, O	—	140°	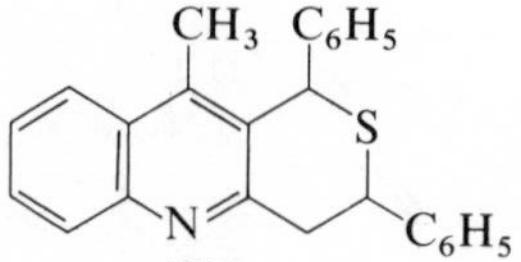 CH_3, C_6H_5, S, C_6H_5, N (76)	148
	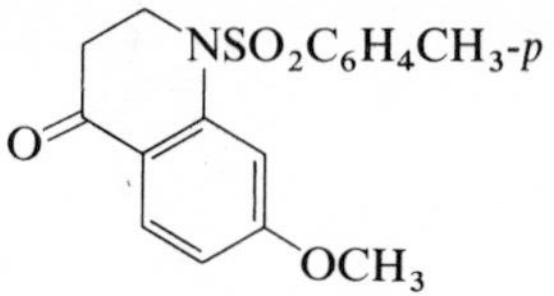 $NSO_2C_6H_4CH_3$-*p*, O, OCH_3	—	145°	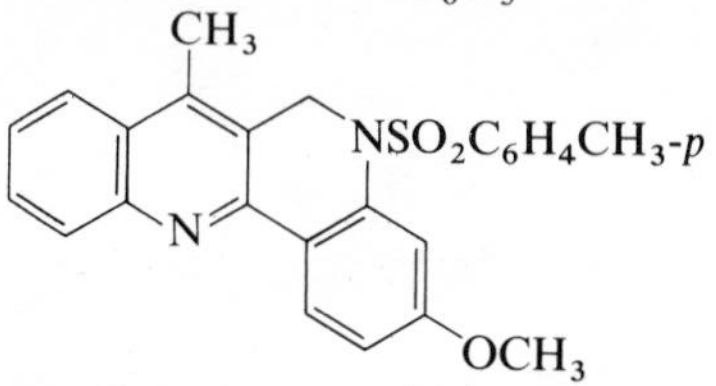 CH_3, $NSO_2C_6H_4CH_3$-*p*, N, OCH_3 (78)	32

[a] The free base rather than *o*-aminoacetophenone hydrochloride is used.

[b] The product structure is indicated by spectroscopic data in refs. 31, 58, and 149, showing the benzo ring attached to the 2, 3 position of the dihydroacridine structure.

[c] The product structure is indicated by spectroscopic data in refs. 58 and 31, showing the benzo ring attached to the 7, 8 position of the cyclohepta[*b*]quinoline structure.

[d] Reference 31 gives "cyclohexenyl"; ref. 149 gives "cyclohexylidene," identical conditions and yield.

[e] The reactant is as shown in ref. 31; ref. 149 shows reactant with $CO_2C_2H_5$ in the 5 position—same reaction, conditions and yield.

TABLE V. CONDENSATIONS WITH *o*-AMINOPHENYLGLYOXAL DIMETHYLACETALS[a]

R–, $COCH(OCH_3)_2$, NH_2 + Reactant ⟶ R–, $CH(OCH_3)_2$, R_2, N, R_1

R	Reactant	Product(s) and Yield(s) (%) R_1	R_1R_2	R_2	Refs.
H	CH_3COCH_3	CH_3		H (95)	94
	CH_3COCO_2Na	CO_2H		H (97)	60
	Cyclopentanone		$(CH_2)_3$	(95)	94
	Cyclohexanone		$(CH_2)_4$	(97)	94
	$CH_3COCH_2CO_2C_2H_5$[b]	OH		$COCH_3$ (94)	94
	$CH_2(CO_2C_2H_5)_2$[c]	OH		$CO_2C_2H_5$ (70)	94
	Cycloheptanone		$(CH_2)_5$	(95)	94
	Cyclooctanone		$(CH_2)_6$	(85)	94
	$CH_3COC_6H_5$	C_6H_5		H (91)	94
	p-$CH_3COC_6H_4Cl$	C_6H_4Cl-*p*		H (95)	94
	p-$CH_3COC_6H_4NO_2$	$C_6H_4NO_2$-*p*		H (90)	94
	m-$CH_3COC_6H_4NO_2$	$C_6H_4NO_2$-*m*		H (89)	94
	CH_3CO–$C_6H_3(NH_2)Cl$	–$C_6H_3(NH_2)Cl$		H (86)	94
H[d]	$NCCH_2CO_2C_2H_5$	OH		CN (62)	94
CH_3O	CH_3COCO_2H	CO_2H		H (100)[e]	60

[a] All reactions were run by refluxing for 15 minutes in ethanol with sodium ethoxide catalyst except where noted.

[b] The reaction was run without catalyst at 150°, 4 hours.

[c] The reaction was run without catalyst at 160°, 4 hours.

[d] The reactant was *o*-N-acetamidophenylglyoxal dimethylacetal.

[e] The product was free aldehyde obtained by hydrochloric acid hydrolysis of the acetal.

TABLE VI. CONDENSATIONS WITH *o*-AMINOBENZOPHENONE

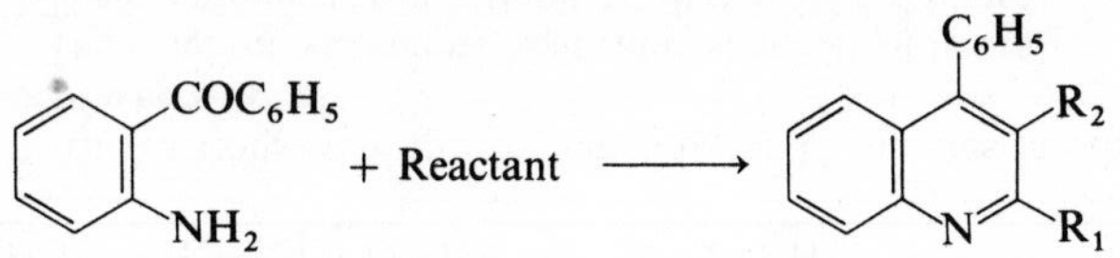

$o\text{-}H_2NC_6H_4COC_6H_5$ + Reactant ⟶ 4-C_6H_5-quinoline (R_1 at C-2, R_2 at C-3)

No. of C Atoms	Reactant	Catalyst	Reaction Conditions	Product(s) and Yield(s) (%): R_1	R_1R_2	R_2	Refs.
C_3	CH_3CO_2H	PPA	130–150°, 2 hr	$NHC_6H_4COC_6H_5$-*o*		H (40–90)	163
	CH_3COCH_3	KOH	Reflux, 6 hr; EtOH	CH_3		H (100)	22
		H_2SO_4	Reflux, 6 hr; AcOH	CH_3		H (71)	20
	$CH_3COCH_2NO_2$	HCl	24 hr; Me_2CO	CH_3		NO_2 (63)	30a
	CH_3COCH_2OH	H_2SO_4	Reflux, 4 hr; AcOH	CH_3		OH (51)	19
	$C_2H_5CO_2H$	PPA	130–150°, 2 hr	$NHC_6H_4COC_6H_5$-*o*		CH_3 (40–90)	163
C_4	$CH_3COC_2H_5$	NaOH	Reflux, 10 hr; EtOH	{CH_3 {C_2H_5		CH_3 (11)} H (71)}	20
		H_2SO_4	Reflux, 10 hr; AcOH	{CH_3 {C_2H_5		CH_3 (86)} H (trace)}	20
	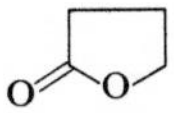 [structure]	PPA	145–150°, 2 hr		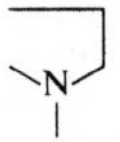 [structure] N–$C_6H_4COC_6H_5$-*o* (48)		162,163
	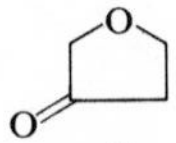 [structure]	HCl	145°		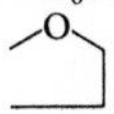 [structure] (57)		148
	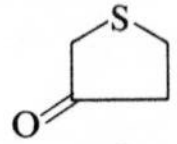 [structure]	"	130°		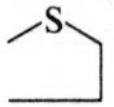 [structure] (70)		148
	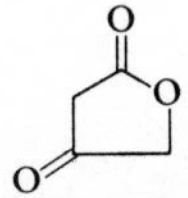 [structure]	H_2SO_4	Reflux, 1 hr; AcOH		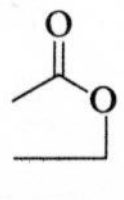 [structure] (86)		19

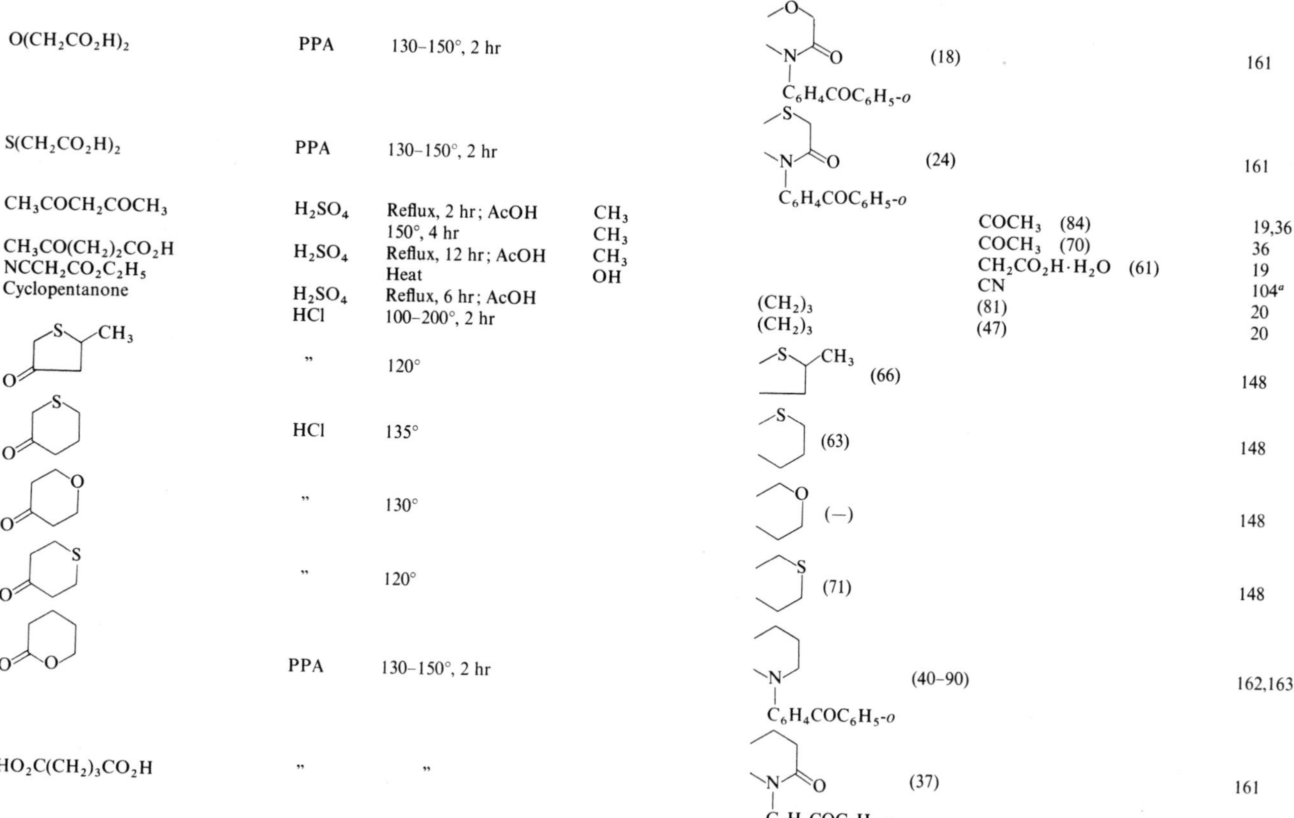

	Reactant	Catalyst	Conditions		Product		Yield (%)	Refs.
	$O(CH_2CO_2H)_2$	PPA	130–150°, 2 hr		[structure] $C_6H_4COC_6H_5$-*o*		(18)	161
	$S(CH_2CO_2H)_2$	PPA	130–150°, 2 hr		[structure] $C_6H_4COC_6H_5$-*o*		(24)	161
C_5	$CH_3COCH_2COCH_3$	H_2SO_4	Reflux, 2 hr; AcOH	CH_3		$COCH_3$	(84)	19,36
			150°, 4 hr	CH_3		$COCH_3$	(70)	36
	$CH_3CO(CH_2)_2CO_2H$	H_2SO_4	Reflux, 12 hr; AcOH	CH_3		$CH_2CO_2H\cdot H_2O$	(61)	19
	$NCCH_2CO_2C_2H_5$		Heat	OH		CN		104[a]
	Cyclopentanone	H_2SO_4	Reflux, 6 hr; AcOH		$(CH_2)_3$		(81)	20
		HCl	100–200°, 2 hr		$(CH_2)_3$		(47)	20
	[structure]	"	120°		[structure]		(66)	148
	[structure]	HCl	135°		[structure]		(63)	148
	[structure]	"	130°		[structure]		(—)	148
	[structure]	"	120°		[structure]		(71)	148
	[structure]	PPA	130–150°, 2 hr		[structure] $C_6H_4COC_6H_5$-*o*		(40–90)	162,163
	$HO_2C(CH_2)_3CO_2H$	"	"		[structure] $C_6H_4COC_6H_5$-*o*		(37)	161

TABLE VI. CONDENSATIONS WITH *o*-AMINOBENZOPHENONE (*Continued*)

COC_6H_5, NH_2 + Reactant ⟶ C_6H_5, R_2, R_1, N

No. of C Atoms	Reactant	Catalyst	Reaction Conditions	Product(s) and Yield(s) (%) R_1	R_1R_2	R_2	Refs.
C_6	$CH_3COCH_2CO_2C_2H_5$	H_2SO_4	Reflux, 2 hr; AcOH	CH_3		$CO_2C_2H_5$ (85)	19
		—	150°, 90 min	OH		$COCH_3$ (60)	23
	$HO_2C(CH_2)_4CO_2H$	PPA	130–150°, 2 hr	$NHC_6H_4COC_6H_5$-*o*		$(CH_2)_3CO_2H$ (62)	161
	O, CH_3	"	145–150°, 2 hr		CH_3 (10)		162
	Cyclohexanone	H_2SO_4	Reflux, 4 hr; AcOH		$(CH_2)_4$	(83)	20
		HCl	100–200°, 1 hr		$(CH_2)_4$	(68)	20
	O, O	—	100–120° [b]		O (80)		70
	"	H_2SO_4	Reflux, 2 hr; AcOH		"	(85)	19
	2-Acetylfuran	—	140° [b]	2-Furyl		H (38)	152
	2-Acetylthiophene	—	125° [b]	2-Thienyl		H (38)	152
	2-Acetylpyrrole	—	110–130° [b]	2-Pyrryl		H (34)	153
	CH_3CO, N, CH_3, S	—	" [b]	N, CH_3, S		H (75)	153
	CH_3CO, S, NH_2, N, CH_3	—	" [b]	S, NH_2, N, CH_3		H (45)	153

C_7	$CH_2(CO_2C_2H_5)_2$	—	130°, 2 hr	OH		$CO_2C_2H_5$	
		—	175–180°, 2 hr	OH		$CONHC_6H_4COC_6H_5$-*o* (—)	23
	$CH_3CO(CH_2)_2CO_2C_2H_5$	—	110–130° [b]	CH_3		CO_2H (95)	124
	$CO(CH_2CO_2CH_3)_2$	H_2SO_4	Reflux, 12 hr; MeOH	$CH_2CO_2CH_3$		$CH_2CO_2C_2H_5$ (75)	154
	2-Propionylfuran	—	140° [b]	2-Furyl		CO_2CH_3 (89)	19
	2-Propionylthiophene	—	130° [b]	2-Thienyl		CH_3 (65)	152
	2-Acetylpyridine	—	150° [b]	2-P yridyl		CH_3 (55)	152
						H (55)	152
	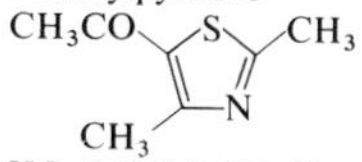 CH_3CO, S, CH_3, N, CH_3	—	110–130° [b]	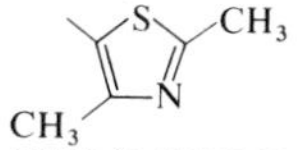 S, CH_3, N, CH_3		H (50)	153
	$HO_2C(CH_2)_5CO_2H$	PPA	130–150°, 2 hr	$NHC_6H_4COC_6H_5$-*o*		$(CH_2)_4CO_2H$ (52)	161
	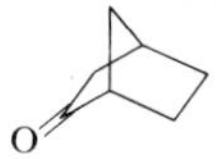 O	—	195° [b]		 (31)		15
C_8	$C_2H_5O_2CCOCH_2CO_2C_2H_5$	H_2SO_4	Reflux, 6 hr; EtOH	$CO_2C_2H_5$		$CO_2C_2H_5$ (66)	19
		—	110–130° [b]	"		" (60)	155
	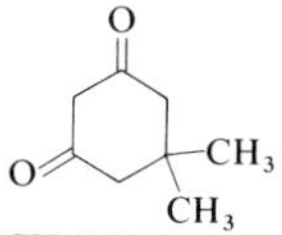 O, O, CH_3, CH_3	H_2SO_4	Reflux, 2 hr; AcOH		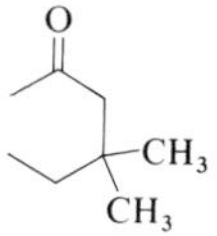 O, CH_3, CH_3 (86)		19
	$CH_3COC_6H_5$	"	Reflux, 18 hr; AcOH	C_6H_5		H (67)	20
		HCl	100–200°, 2 hr	C_6H_5		H (60)	20
		KOH	Reflux, 1 d; EtOH	C_6H_5		H (50)	23
	$C_6H_5CH_2CO_2H$	PPA	130–150°, 2 hr	$NHC_6H_4COC_6H_5$-*o*		C_6H_5 (80)	30b
	O, O	HCl	130°		O (65)		148
	O, N, S, CH_3	—	155° [b]		N, S, CH_3 (90)		55
	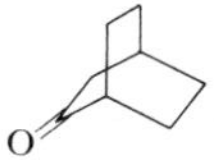 O	—	195° [b]		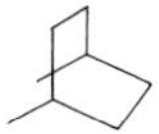 (48)		15

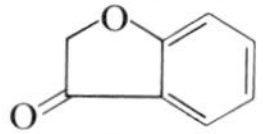

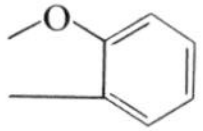

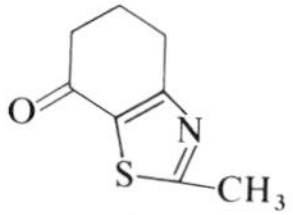

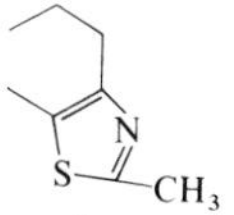

TABLE VI. Condensations with *o*-Aminobenzophenone (*Continued*)

2-Aminobenzophenone (COC_6H_5, NH_2) + Reactant ⟶ 4-phenylquinoline (C_6H_5, R_2, R_1)

No. of C Atoms	Reactant	Catalyst	Reaction Conditions	Product(s) and Yield(s) (%): R_1	R_1R_2	R_2	Refs.
C_9	$C_2H_5COC_6H_5$	H_2SO_4	Reflux, 18 hr; AcOH	C_6H_5		CH_3 (66)	20
	$CH_3COCH_2C_6H_5$	"	Reflux, 2 hr; AcOH	CH_3		C_6H_5 (64)	20
		HCl	100–200°, 1 hr	CH_3		C_6H_5 (56)	20
	$CH_3OCH_2COC_6H_5$	H_2SO_4	Reflux, 3 hr; AcOH	C_6H_5		OCH_3 (69)	19
	$C_6H_5CH_2COCO_2H$	"	Reflux, 2 hr; AcOH	H		C_6H_5 (69)	19
	$CH_3COCOC_6H_5$	"	Reflux, 4 hr; AcOH	COC_6H_5		H (57)	19
	$CO(CH_2CO_2C_2H_5)_2$	"	Reflux, 10 hr; EtOH	$CH_2CO_2C_2H_5$		$CO_2C_2H_5$ (69)	19
			110–130° [b]	"		" (95)	155
	$C_6H_5COCH_2CN$	H_2SO_4	Reflux, 4 hr; AcOH	C_6H_5		CN (72)	19
		—	110–115°,[b] 1 hr		(42)		33
	o-$HO_2CC_6H_4CH_2CO_2H$	PPA	130–160°		(63)		161
		H_2SO_4	Reflux, 2 hr; AcOH		(83)		19

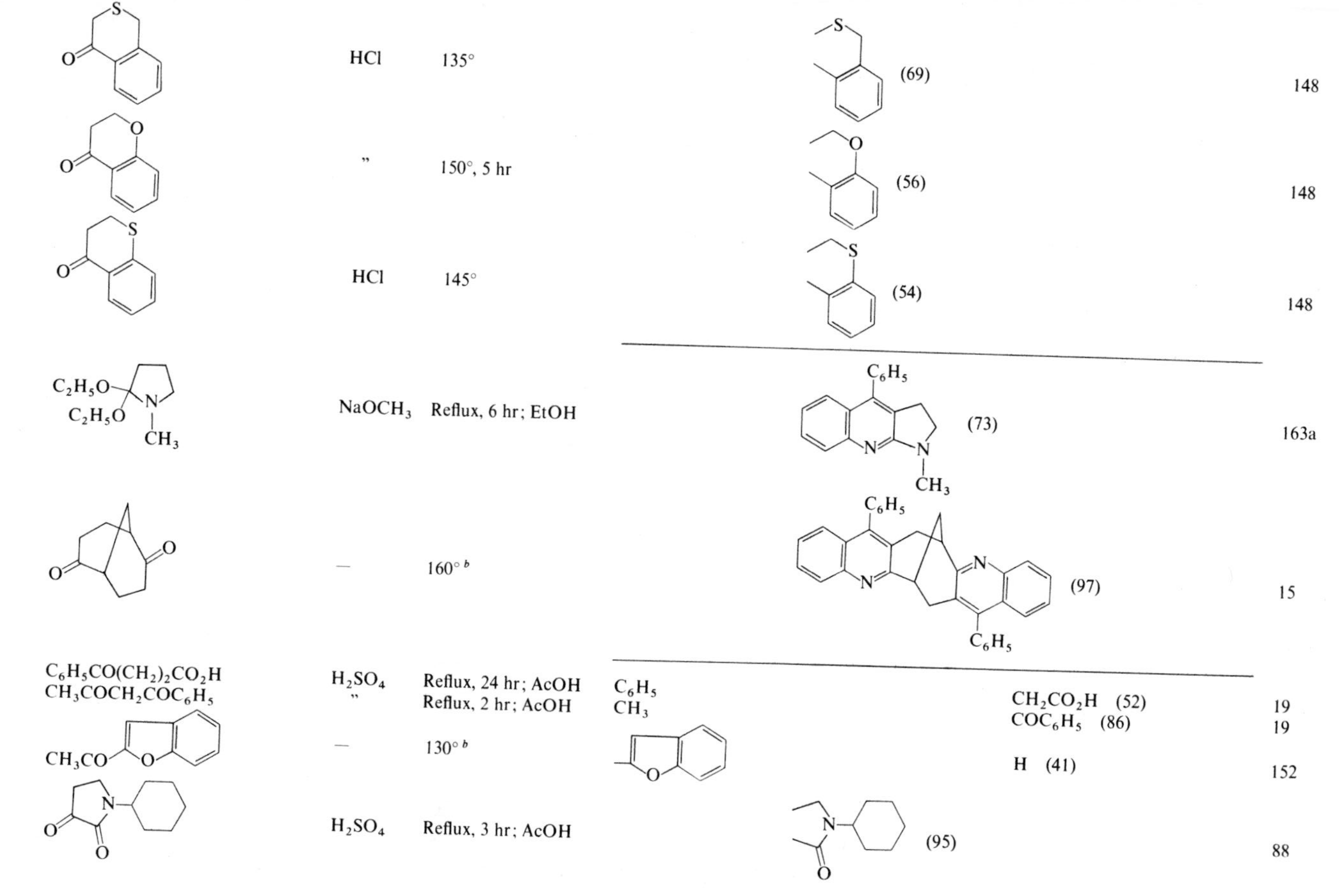

Formula	Reactant	Catalyst	Conditions	Substituent	Product (Yield, %)	Refs.
		HCl	135°		(69)	148
		"	150°, 5 hr		(56)	148
		HCl	145°		(54)	148
	C_2H_5O, C_2H_5O, CH_3	$NaOCH_3$	Reflux, 6 hr; EtOH		C_6H_5, CH_3 (73)	163a
		—	160° [b]		C_6H_5, C_6H_5 (97)	15
C_{10}	$C_6H_5CO(CH_2)_2CO_2H$	H_2SO_4	Reflux, 24 hr; AcOH	C_6H_5	CH_2CO_2H (52)	19
	$CH_3COCH_2COC_6H_5$	"	Reflux, 2 hr; AcOH	CH_3	COC_6H_5 (86)	19
	CH_3CO	—	130° [b]		H (41)	152
		H_2SO_4	Reflux, 3 hr; AcOH		(95)	88

TABLE VI. CONDENSATIONS WITH *o*-AMINOBENZOPHENONE (*Continued*)

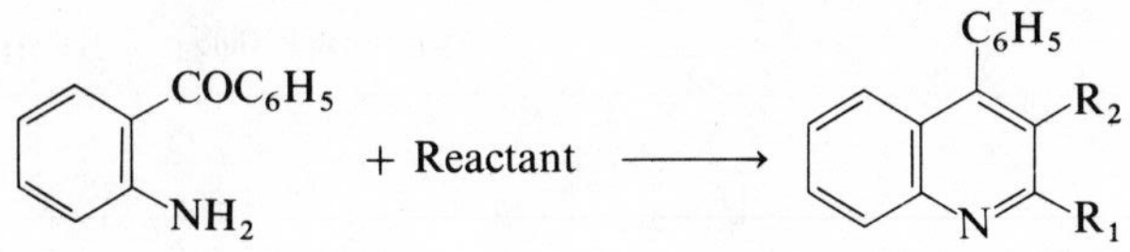

$o\text{-}H_2NC_6H_4COC_6H_5$ + Reactant ⟶ 4-C_6H_5-3-R_2-2-R_1-quinoline

No. of C Atoms	Reactant	Catalyst	Reaction Conditions	Product(s) and Yield(s) (%): R_1	R_1R_2	R_2	Refs.
C_{10} (*Contd.*)	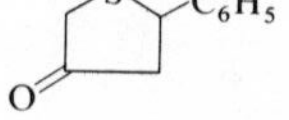 (structure: S, C_6H_5, O)	HCl	120°		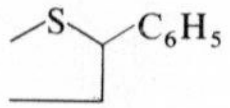 (structure: S, C_6H_5) (75)		148
	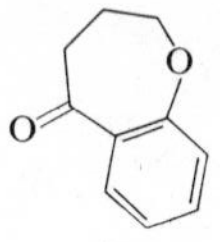 (structure: O, O, benzo ring)	"	150°		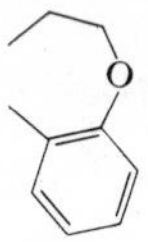 (structure: O, benzo ring) (53)		148
	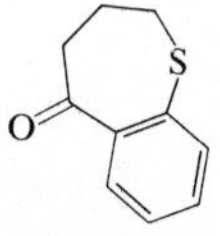 (structure: S, O, benzo ring)	HCl	130°		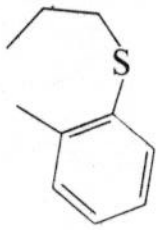 (structure: S, benzo ring) (51)		148
C_{11}	$C_6H_5CH_2COCO_2C_2H_5$	H_2SO_4	Reflux, 6 hr; AcOH	$CO_2C_2H_5$		C_6H_5 (53)	19
	$C_6H_5COCH_2CO_2C_2H_5$	"	Reflux, 12 hr; EtOH	C_6H_5		$CO_2C_2H_5$ (73)	19
		—	160°, 2.5 hr	OH		COC_6H_5 (70)	23
	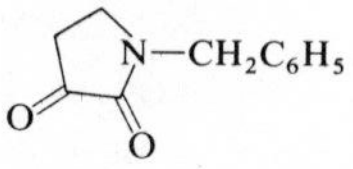 (structure: N—$CH_2C_6H_5$, O, O)	H_2SO_4	Reflux, 3 hr; AcOH		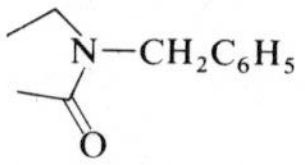 (structure: N—$CH_2C_6H_5$, O) (100)		88
	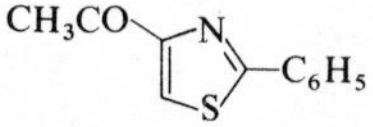 CH_3CO–(structure: N, S, —C_6H_5)	—	110–130° [b]	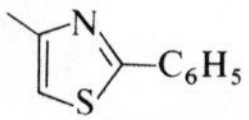 (structure: N, S, —C_6H_5)		H (78)	153

C_{12}	$C_6H_5COCH_2COCO_2C_2H_5$	”	Reflux, 6 hr; EtOH	$CO_2C_2H_5$		COC_6H_5 (65)	19
	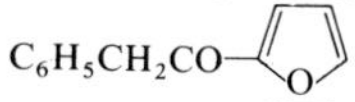 $C_6H_5CH_2CO$– (O)	—	135° [b]	2-Furyl		C_6H_5 (54)	152
	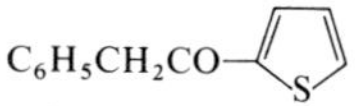 $C_6H_5CH_2CO$– (S)	—	160° [b]	2-Thienyl		C_6H_5 (62)	152
	 (N)–$COCH_2$–(N)	—	150° [b]	2-Pyridyl		2-Pyridyl (55)	152
	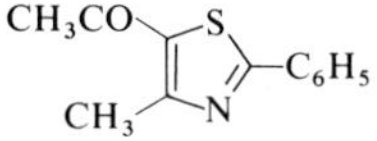 CH_3CO, S, N, C_6H_5, CH_3	—	110–130° [b]	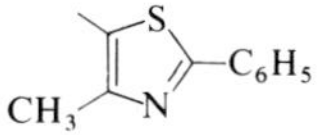 S, N, C_6H_5, CH_3		H (70)	153
	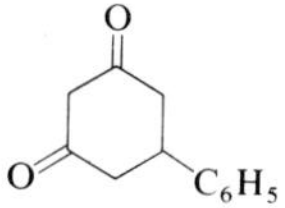 O, O, C_6H_5	—	100–120° [b]		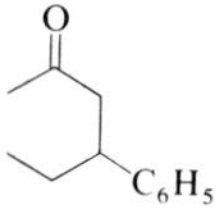 O, C_6H_5 (80)		70
	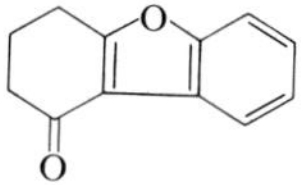 O, O	—	155° [b]		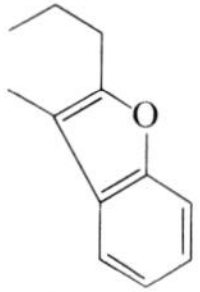 O (85)		55
	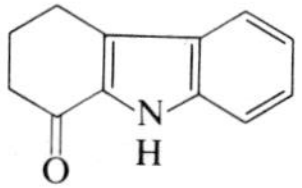 O, N, H	—	190° [b]		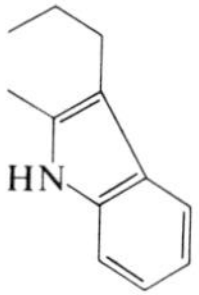 HN (71)		152
C_{13}	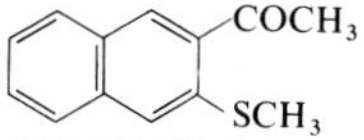 $COCH_3$, SCH_3	—	160° [b]	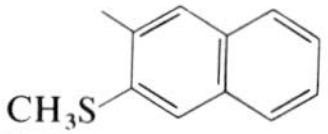 CH_3S		H (79)	152
C_{14}	$C_6H_5COCH_2C_6H_5$	H_2SO_4	Reflux, 3 hr; AcOH	C_6H_5		C_6H_5 (59)	20
		HCl	100–200°, 1 hr	C_6H_5		C_6H_5 (61)	20
		”	Reflux, 4.5 hr; C_6H_5Cl	C_6H_5		C_6H_5 (94)	39a
	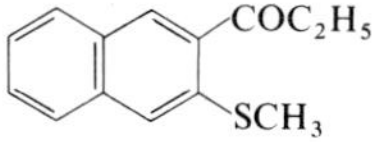 COC_2H_5, SCH_3	—	150° [b]	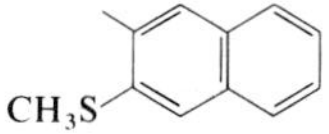 CH_3S		CH_3 (72)	152

TABLE VI. CONDENSATIONS WITH *o*-AMINOBENZOPHENONE (*Continued*)

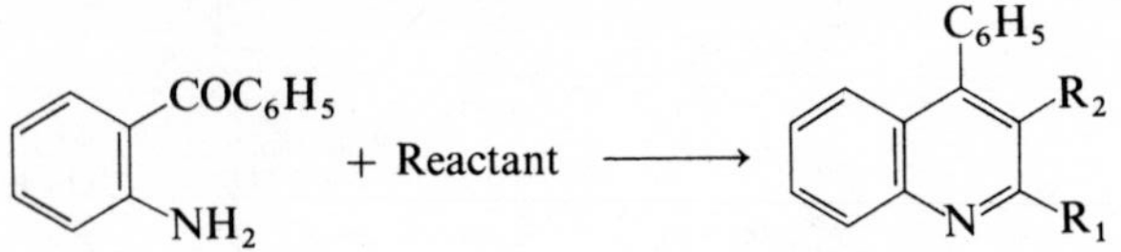

2-(COC_6H_5)-aniline (NH_2) + Reactant ⟶ 4-C_6H_5-quinoline (R_1 at C-2, R_2 at C-3)

No. of C Atoms	Reactant	Catalyst	Reaction Conditions	Product(s) and Yield(s) (%)			Refs.
				R_3	R_1R_2	R_2	
C_{14} (*Contd.*)	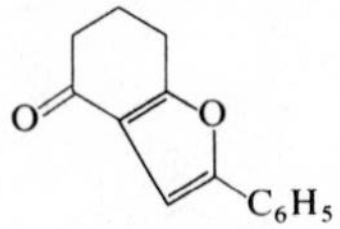 [structure: 2-C_6H_5-6,7-dihydrobenzofuran-4(5H)-one]	—	140°, 6 hr		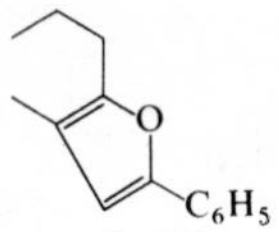 [structure: fused furan, 5-C_6H_5] (16)		33
		—	140°,[b] 1 hr		" (63)		33
	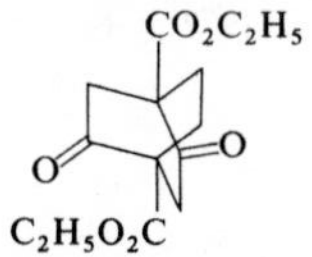 [structure: bicyclic diketone with $CO_2C_2H_5$ and $C_2H_5O_2C$]	—	195°[b]	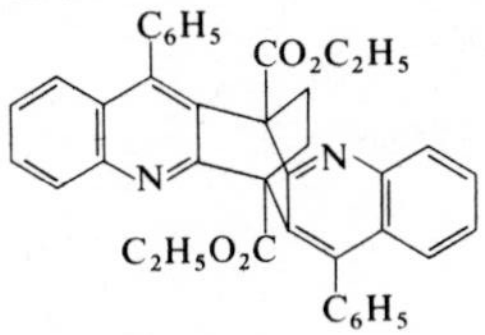 [structure: bis-quinoline with C_6H_5, $CO_2C_2H_5$, $C_2H_5O_2C$, C_6H_5] (60)			15
C_{15}	$C_6H_5CH_2COCH_2C_6H_5$	H_2SO_4	Reflux, 4 hr; AcOH	$CH_2C_6H_5$		C_6H_5 (74)	20
	$C_6H_5COCH_2COC_6H_5$	"	Reflux, 12 hr; AcOH	C_6H_5		COC_6H_5 (59)	19
	o-$CH_3CONHC_6H_4COC_6H_5$	PPA	130–135°, 2 hr	$NHC_6H_4COC_6H_5$-*o*		H (82)	30b
	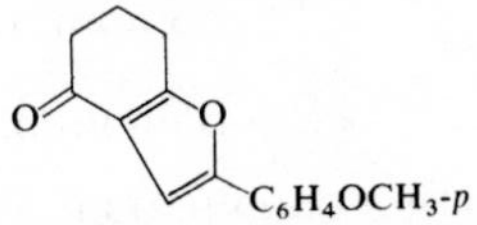 [structure: 2-($C_6H_4OCH_3$-*p*)-6,7-dihydrobenzofuran-4(5H)-one]	—	140°[b] 1 hr		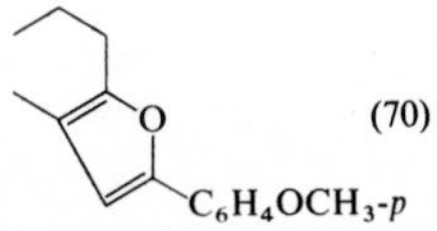 [structure: fused furan, 5-$C_6H_4OCH_3$-*p*] (70)		33

	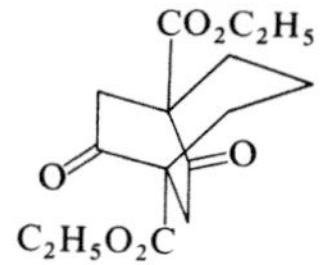 $CO_2C_2H_5$, O, O, $C_2H_5O_2C$	—	195°[b]	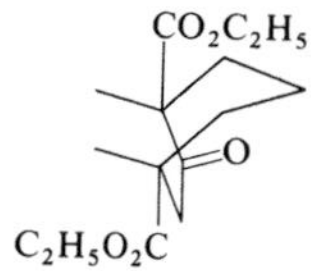 $CO_2C_2H_5$, O, $C_2H_5O_2C$ (60)			15
C_{16}	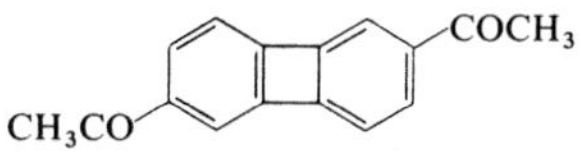 CH_3CO, $COCH_3$	”	130°, 5 hr; *m*-cresol	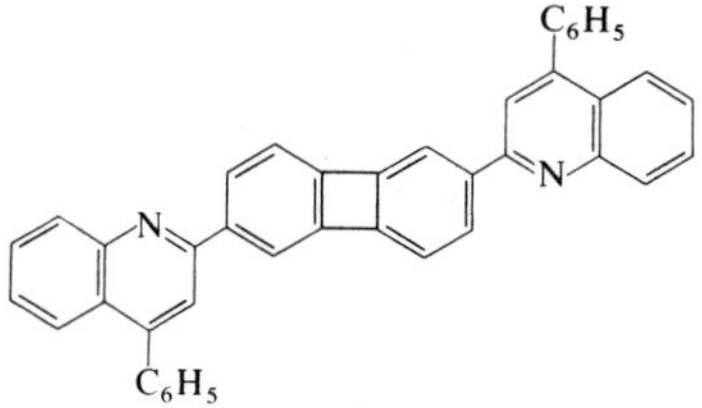 C_6H_5, N, N, C_6H_5 (95)			39d
				R_1	R_1R_2	R_2	
C_{17}	*o*-$C_6H_5COC_6H_4NHCOC_2H_5$	PPA	130–135°, 2 hr	$NHC_6H_4COC_6H_5$-*o*		CH_3 (86)	30b
	o-$C_6H_5COC_6H_4NHCOC_3H_7$-*n*	”	”	$NHC_6H_4COC_6H_5$-*o*		C_2H_5 (80)	30b
	C_6H_5, S, O, C_6H_5	HCl	150°		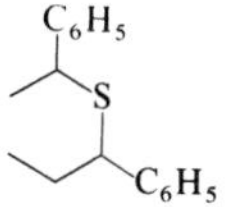 C_6H_5, S, C_6H_5 (65)		148
C_{21}	*o*-$C_6H_5COC_6H_4NHCOCH_2C_6H_5$	PPA	130–135°, 2 hr	$NHC_6H_4COC_6H_5$-*o*		C_6H_5 (80)	30b

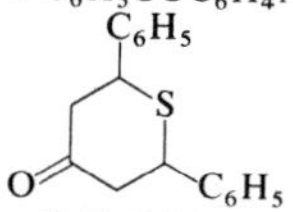

[a] This compound is in the original reference; no properties are given. It is not listed in abstracts of the article, nor in the *Beilstein* or *C.A.* formula indexes.

[b] The hydrochloride salt of *o*-aminobenzophenone rather than the free base is used.

TABLE VII. CONDENSATIONS WITH SUBSTITUTED *o*-AMINOBENZOPHENONES

A.

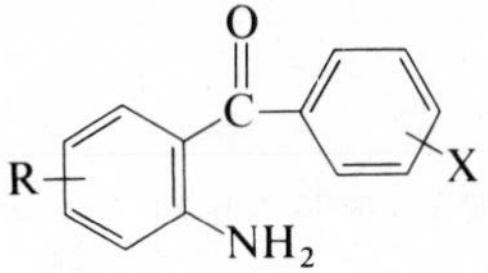

R	X	Reactant	Catalyst	Reaction Conditions	Product(s) and Yield(s) (%)	Refs.
H	p-CH_3	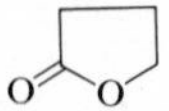 γ-butyrolactone (structure)	PPA	145–150°, 2 hr	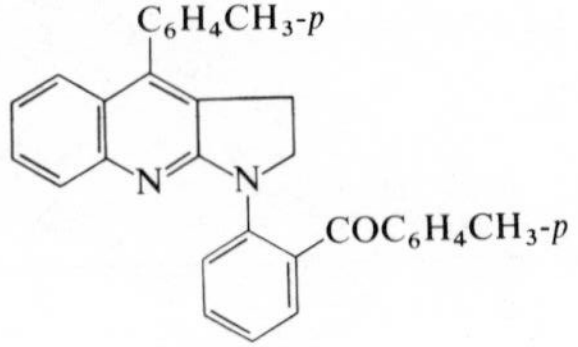 $C_6H_4CH_3$-p; $COC_6H_4CH_3$-p (structure) (57)	162
		$HO_2C(CH_2)_3CO_2H$	"	130–150°, 2 hr	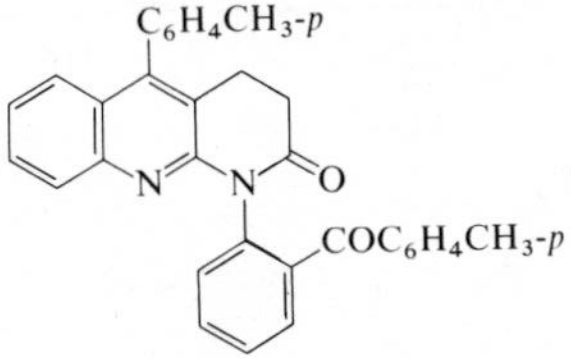 $C_6H_4CH_3$-p; $COC_6H_4CH_3$-p (structure) (44)	161
		$C_2H_5COCH_2CO_2C_2H_5$	—	110–130°[a]	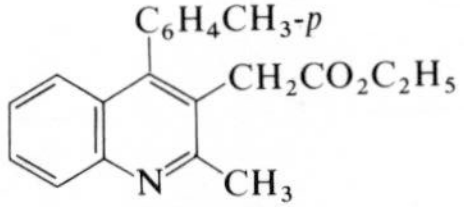 $C_6H_4CH_3$-p; $CH_2CO_2C_2H_5$; CH_3 (structure) (80)	154
		$C_2H_5O_2CCOCH_2CO_2C_2H_5$	—	110–130°[a]	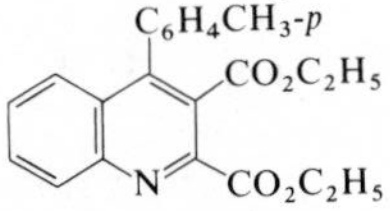 $C_6H_4CH_3$-p; $CO_2C_2H_5$; $CO_2C_2H_5$ (structure) (60)	155

		$CO(CH_2CO_2C_2H_5)_2$	—	110–130°[a]	Quinoline: 4-$C_6H_4CH_3$-*p*, 3-$CO_2C_2H_5$, 2-$CH_2CO_2C_2H_5$	(40)	155
		C_6H_4(2-CH_2CO_2H)(1-CO_2H)	PPA	130–160°	Benzopyranoquinolinone, $C_6H_4CH_3$-*p*	(76)	161
		CH_3CONH–C_6H_4–$COC_6H_4CH_3$-*p*	”	130–135°, 2 hr	Quinoline: 4-$C_6H_4CH_3$-*p*, 2-NH–C_6H_4–$COC_6H_4CH_3$-*p*	(80)	30b
		C_2H_5CONH–C_6H_4–$COC_6H_4CH_3$-*p*	”	130–135°, 2 hr	Quinoline: 4-$C_6H_4CH_3$-*p*, 3-CH_3, 2-NH–C_6H_4–$COC_6H_4CH_3$-*p*	(86)	30b
		$CH_3(CH_2)_2CONH$–C_6H_4–$COC_6H_4CH_3$-*p*	”	”	Quinoline: 4-$C_6H_4CH_3$-*p*, 3-C_2H_5, 2-NH–C_6H_4–$COC_6H_5CH_3$-*p*	(88)	30b
H	*o*-F	$CH_3COCH_2CO_2C_2H_5$	$ZnCl_2$	Reflux, 24 hr; C_6H_6	Quinoline: 4-C_6H_4F-*o*, 3-$CO_2C_2H_5$, 2-CH_3	(83)	164

TABLE VII. CONDENSATIONS WITH SUBSTITUTED *o*-AMINOBENZOPHENONES (*Continued*)

A.

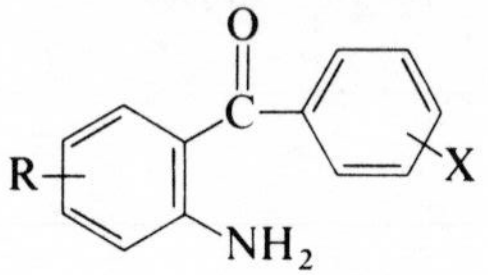

R	X	Reactant	Catalyst	Reaction Conditions	Product(s) and Yield(s) (%)	Refs.
H	2-Aza	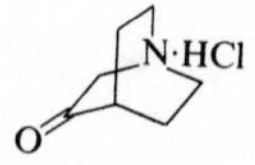 N·HCl	HCl	Reflux, 20 hr; EtOH	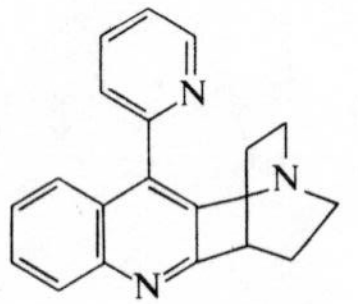 (55)	37
5-CH_3	H	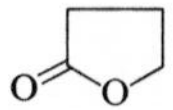	PPA	145–150°, 2 hr	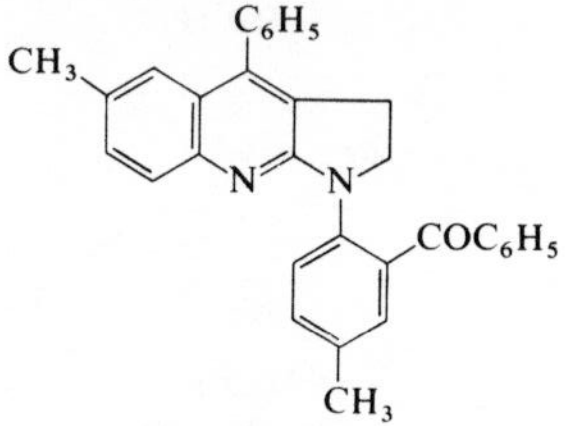 CH_3, C_6H_5, COC_6H_5, CH_3 (48)	162
		$HO_2C(CH_2)_3CO_2H$	"	130–150°, 2 hr	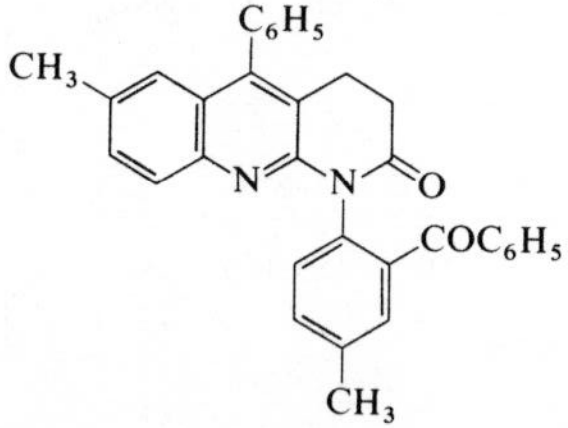 CH_3, C_6H_5, COC_6H_5, CH_3 (40)	161

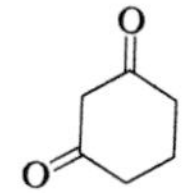 O, O (cyclohexane-1,3-dione)	—	100–120°[a]	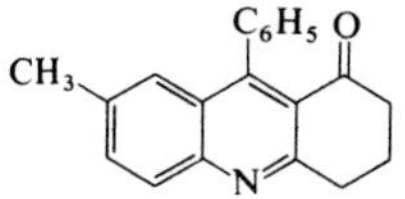 CH_3, C_6H_5, O, N (80)	70
o-$HO_2CC_6H_4CH_2CO_2H$	PPA	130–160°	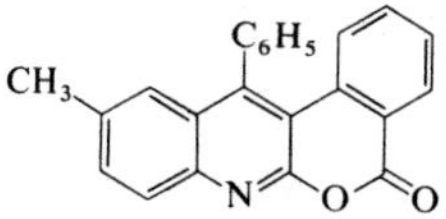 CH_3, C_6H_5, N, O, O (78)	161
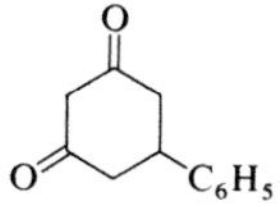 O, O, C_6H_5	—	100–120°[a]	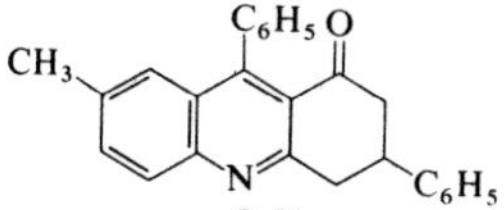 CH_3, C_6H_5, O, N, C_6H_5 (80)	70
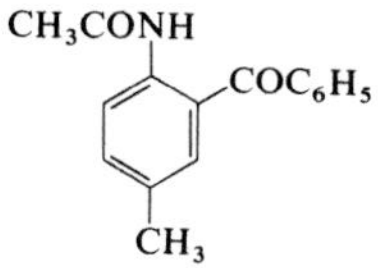 CH_3CONH, COC_6H_5, CH_3	PPA	130–135°, 2 hr	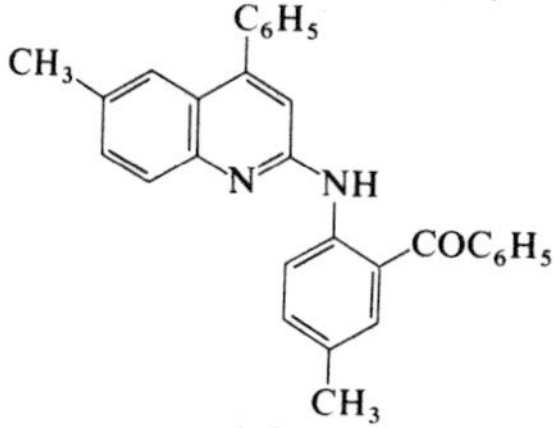 C_6H_5, CH_3, N, NH, COC_6H_5, CH_3 (75)	30b
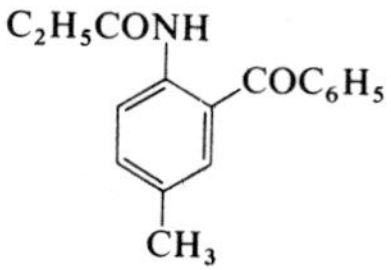 C_2H_5CONH, COC_6H_5, CH_3	"	"	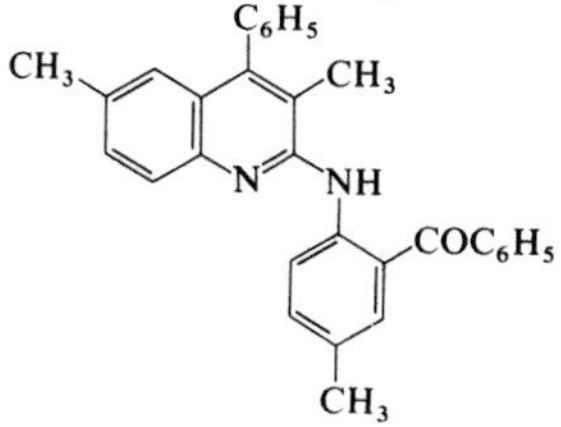 C_6H_5, CH_3, CH_3, N, NH, COC_6H_5, CH_3 (86)	30b

TABLE VII. CONDENSATIONS WITH SUBSTITUTED o-AMINOBENZOPHENONES (*Continued*)

A.

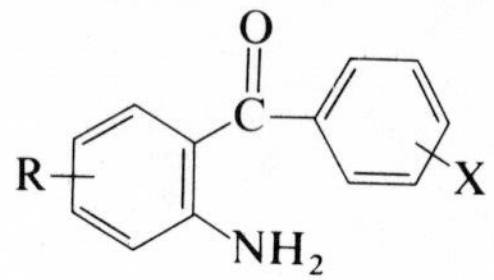

R	X	Reactant	Catalyst	Reaction Conditions	Product(s) and Yield(s) (%)	Refs.
5-CH_3 (*Contd.*)	H	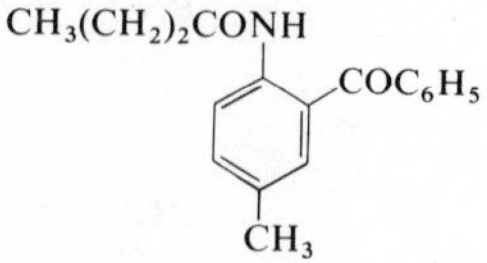 $CH_3(CH_2)_2CONH$ (2-benzoyl-4-methylphenyl)	PPA	130–135°, 2 hr	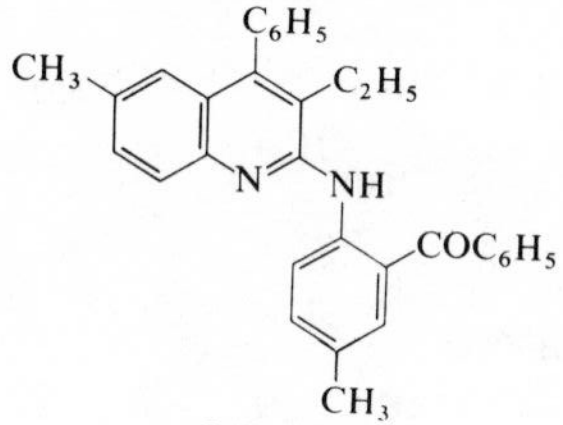 6-CH_3-4-C_6H_5-3-C_2H_5-2-[NH-(2-COC_6H_5-4-CH_3-phenyl)]quinoline (90)	30b
5-Cl	H	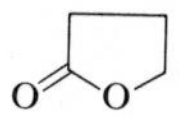 γ-butyrolactone	PPA	145–150°, 2 hr	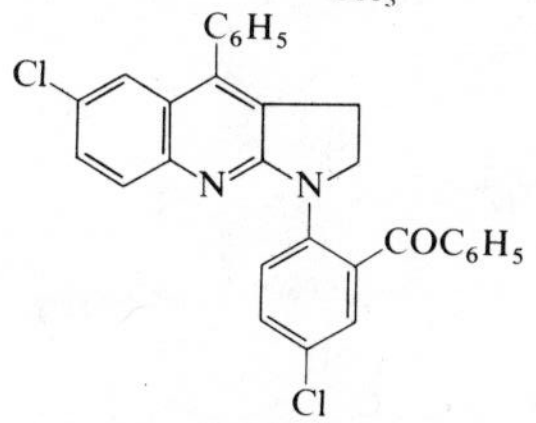 6-Cl-4-C_6H_5-pyrrolo[2,3-b]quinoline, N-(2-COC_6H_5-4-Cl-phenyl) (46)	162,163
		$O(CH_2CO_2H)_2$	"	130–150°, 2 hr 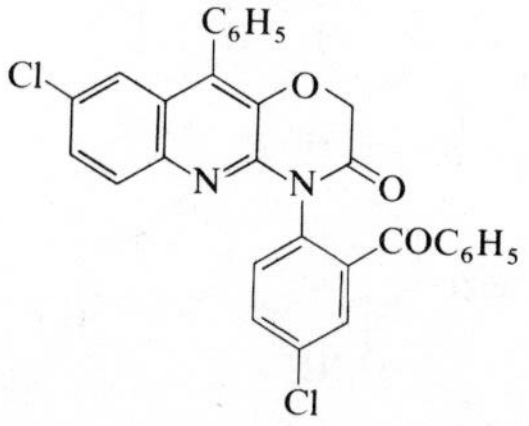	Cl, C_6H_5, oxazino-quinolinone, N-(2-COC_6H_5-4-Cl-phenyl) (32)	161

$S(CH_2CO_2H)_2$	"	"	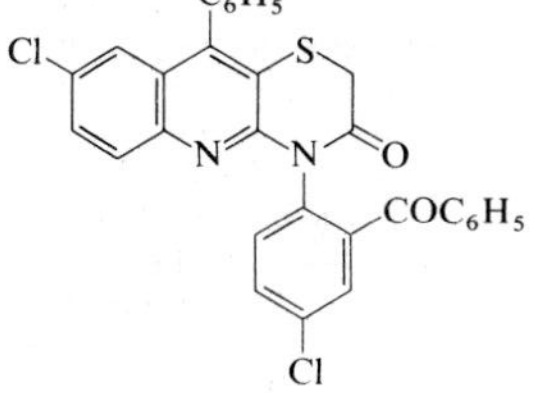 [structure: C_6H_5, Cl, S, O, N, N, COC_6H_5, Cl]	(37)	161
$HO_2C(CH_2)_3CO_2H$	"	"	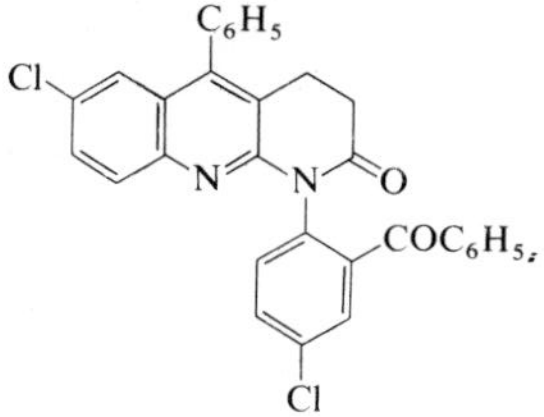 [structure: C_6H_5, Cl, O, N, N, COC_6H_5, Cl]	(46)	161
$HO(CH_2)_4CO_2H$	"	"	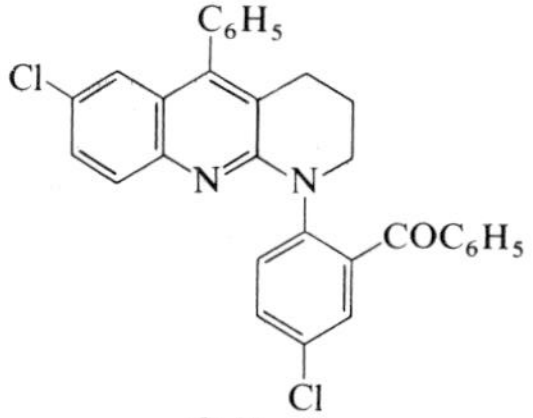 [structure: C_6H_5, Cl, N, N, COC_6H_5, Cl]	(40–90)	162,163
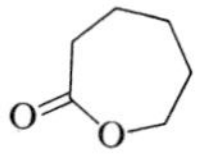 [structure: O, O]	"	—	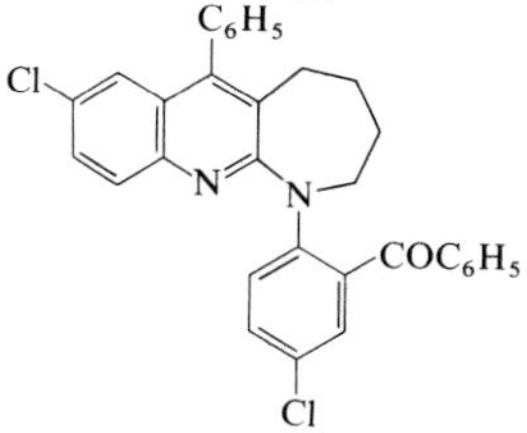 [structure: C_6H_5, Cl, N, N, COC_6H_5, Cl]	(—)	162

TABLE VII. CONDENSATIONS WITH SUBSTITUTED *o*-AMINOBENZOPHENONES (*Continued*)

A.

R	X	Reactant	Catalyst	Reaction Conditions	Product(s) and Yield(s) (%)	Refs.
5-Cl (*Contd.*)	H	1,3-cyclohexanedione	—	100–120°[a]	7-chloro-9-phenyl-3,4-dihydroacridin-1(2H)-one (80)	70
		CH_3COR_1	H_2SO_4	110–180°, 1 min	6-chloro-4-phenyl-2-R_1-quinoline	
		R_1			R_1	
		Thiazolyl-(4)			Thiazolyl-(4) (35)	165
		Furyl-(2)			Furyl-(2) (59)	165
		Thienyl-(2)			Thienyl-(2) (70)	165
		Pyrrolyl-(2)			Pyrrolyl-(2) (30)	165
		Pyridyl-(2)			Pyridyl-(2) (62)	165
		2-Methylthiazolyl-(4)			2-Methylthiazolyl-(4) (51)	165
		2,4-Dimethylthiazolyl-(5)			2,4-Dimethylthiazolyl-(5) (46)	165
		Benzimidazolyl-(2)			Benzimidazolyl-(2) (47)	165
		Benzo[*b*]furyl-(2)			Benzo[*b*]furyl-(2) (80)	165
		2-Phenylthiazolyl-(4)			2-Phenylthiazolyl-(4) (78)	165
		3-Methylbenzo[*b*]thienyl-(2)			3-Methylbenzo[*b*]thienyl-(2) (85)	165
		2-Phenyl-4-methylthiazolyl-(5)			2-Phenyl-4-methylthiazolyl-(5) (85)	165

$HO_2C(CH_2)_4CO_2H$	PPA	130–150°, 2 hr	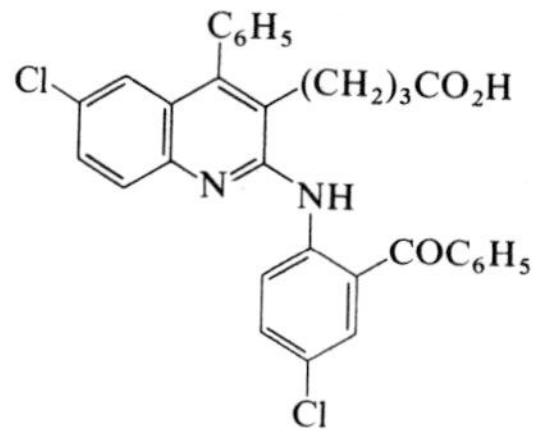 [structure: Cl, C_6H_5, $(CH_2)_3CO_2H$, N, NH, COC_6H_5, Cl] (76)	161
$HO_2C(CH_2)_5CO_2H$	"	"	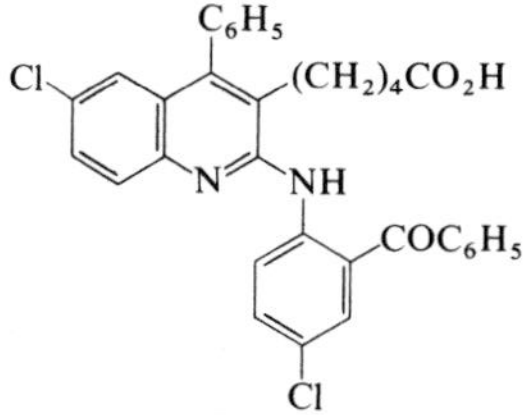 [structure: Cl, C_6H_5, $(CH_2)_4CO_2H$, N, NH, COC_6H_5, Cl] (81)	161
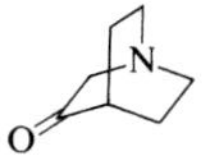 [structure: N, O]	HCl	Reflux, 4 hr; EtOH	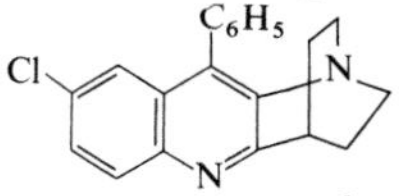 [structure: Cl, C_6H_5, N, N] (52)	37
o-$HO_2CC_6H_4CH_2CO_2H$	PPA	130–160°	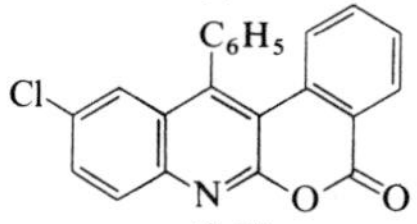 [structure: Cl, C_6H_5, N, O, O] (82)	161
$HO_2C(CH_2)_8CO_2H$	PPA	130–150°, 2 hr	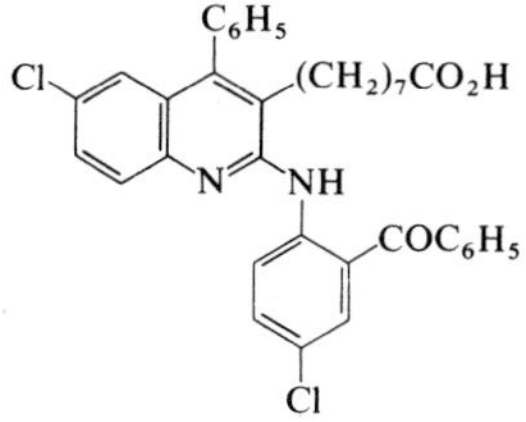 [structure: Cl, C_6H_5, $(CH_2)_7CO_2H$, N, NH, COC_6H_5, Cl] (55)	161

TABLE VII. CONDENSATIONS WITH SUBSTITUTED *o*-AMINOBENZOPHENONES (*Continued*)

A.

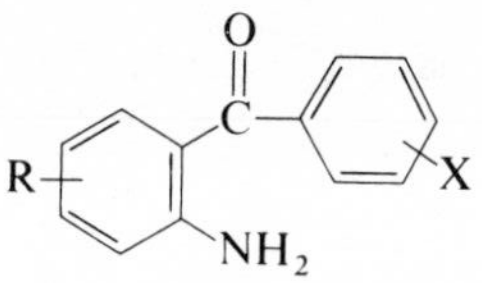

R	X	Reactant	Catalyst	Reaction Conditions	Product(s) and Yield(s) (%)	Refs.
5-Cl (*Contd.*)	H	*o*-$HO_2CC_6H_4(CH_2)_2CO_2H$	PPA	130–150°, 2 hr 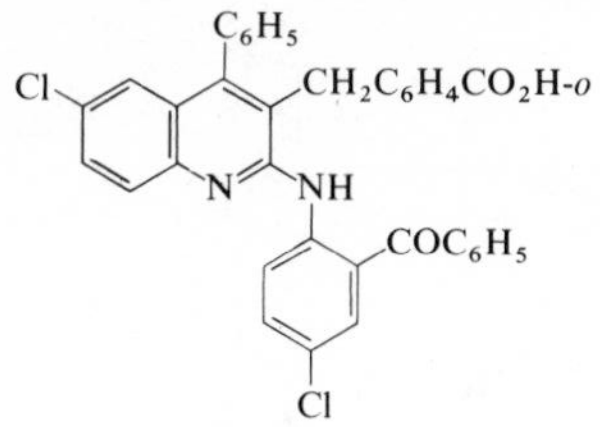	Cl, C_6H_5, $CH_2C_6H_4CO_2H$-*o*, N, NH, COC_6H_5, Cl (52)	161
		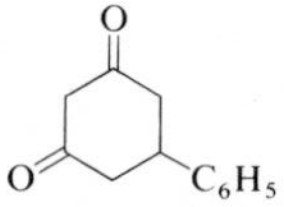 O, O, C_6H_5	"	100–120°[a] 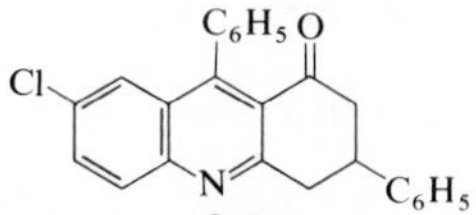	Cl, C_6H_5, O, N, C_6H_5 (80)	70
		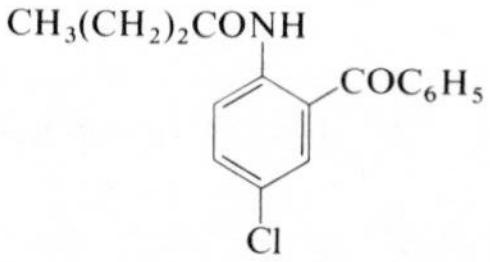 $CH_3(CH_2)_2CONH$, COC_6H_5, Cl	"	130–135°, 2 hr 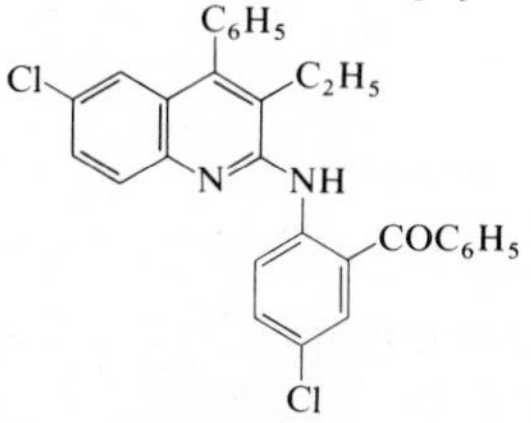	Cl, C_6H_5, C_2H_5, N, NH, COC_6H_5, Cl (84)	30b

		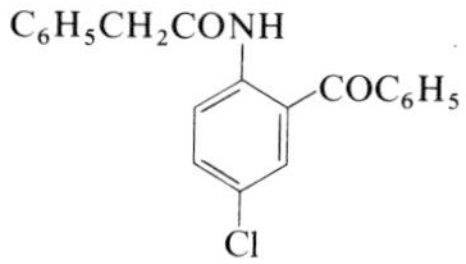 $C_6H_5CH_2CONH$, COC_6H_5, Cl	PPA	130–135°, 2 hr	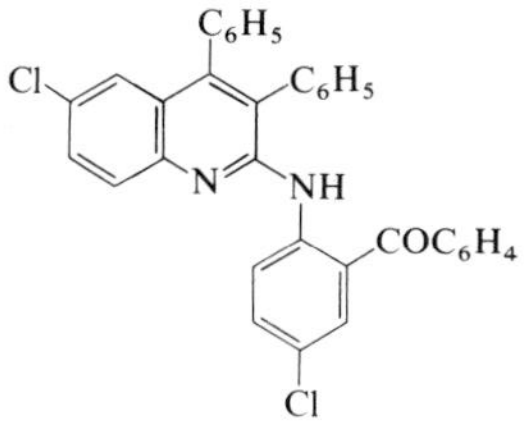 C_6H_5, C_6H_5, Cl, N, NH, COC_6H_4, Cl	(75)	30b
5-Br	H	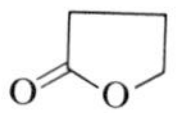 O, O	”	145–150°, 2 hr	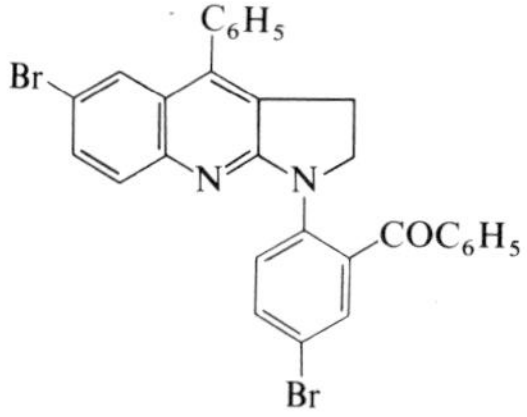 C_6H_5, Br, N, N, COC_6H_5, Br	(41)	162
		$HO_2C(CH_2)_4CO_2H$	”	130–150°, 2 hr	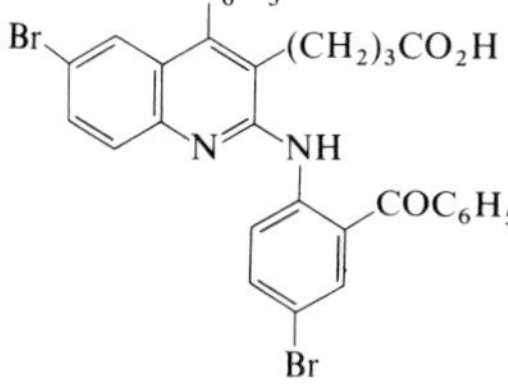 C_6H_5, Br, $(CH_2)_3CO_2H$, N, NH, COC_6H_5, Br	(69)	161
		$HO_2C(CH_2)_5CO_2H$	”	”	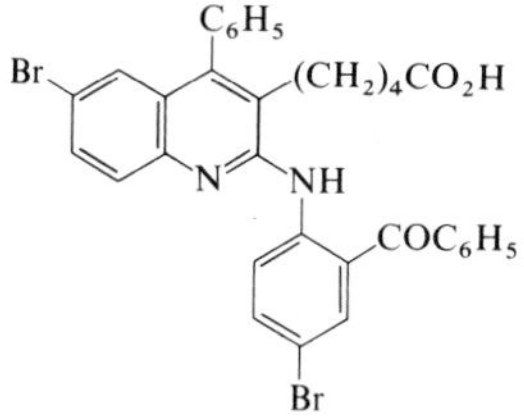 C_6H_5, Br, $(CH_2)_4CO_2H$, N, NH, COC_6H_5, Br	(65)	161

TABLE VII. CONDENSATIONS WITH SUBSTITUTED *o*-AMINOBENZOPHENONES (*Continued*)

A.

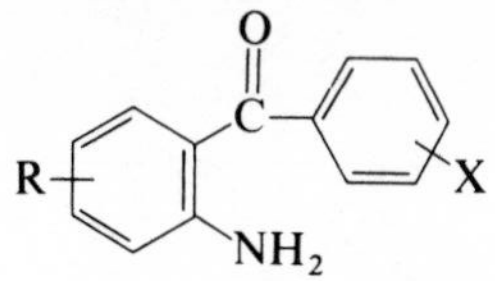

R	X	Reactant	Catalyst	Reaction Conditions	Product(s) and Yield(s) (%)	Refs.
5-Br (*Contd.*)	H	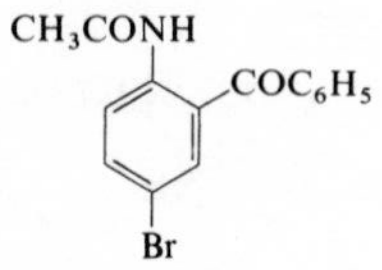 CH_3CONH, COC_6H_5, Br	PPA	130–150°, 2 hr	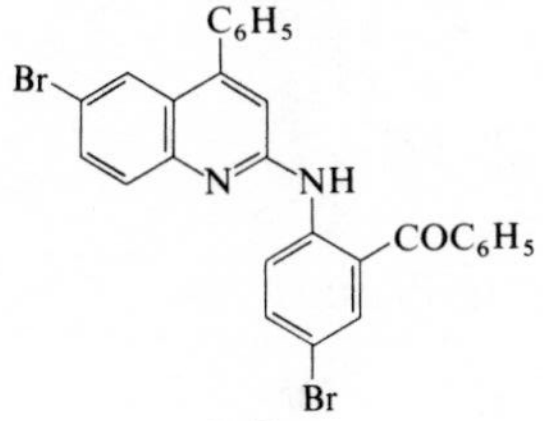 Br, C_6H_5, N, NH, COC_6H_5, Br (79)	30b
		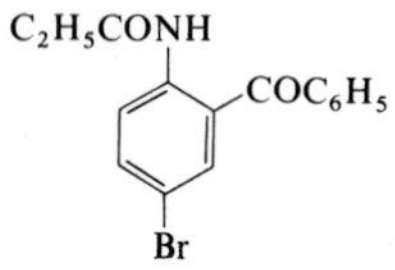 C_2H_5CONH, COC_6H_5, Br	"	"	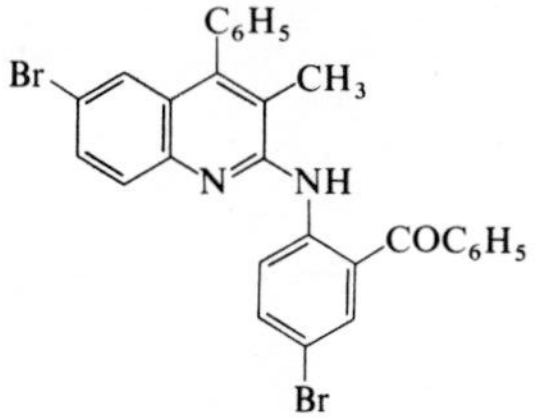 Br, C_6H_5, CH_3, N, NH, COC_6H_5, Br (86)	30b
		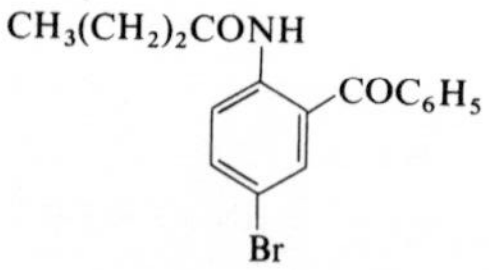 $CH_3(CH_2)_2CONH$, COC_6H_5, Br	"	"	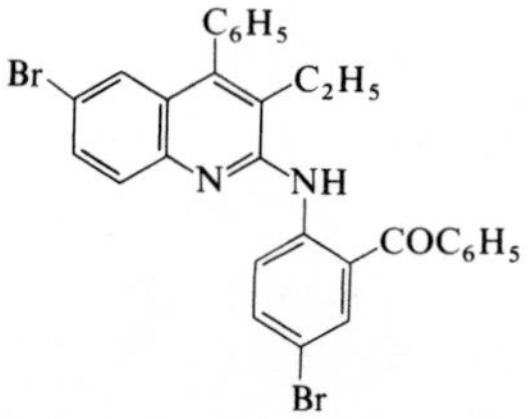 Br, C_6H_5, C_2H_5, N, NH, COC_6H_5, Br (85)	30b

5-NO_2	H	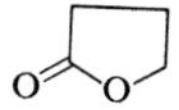 O, O	PPA	145–150°, 2 hr	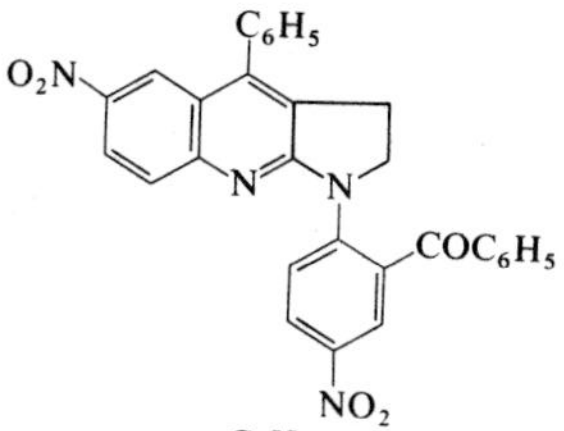 C_6H_5, O_2N, N, N, COC_6H_5, NO_2 (27)	162
		$HO_2C(CH_2)_4CO_2H$	"	130–150°, 2 hr 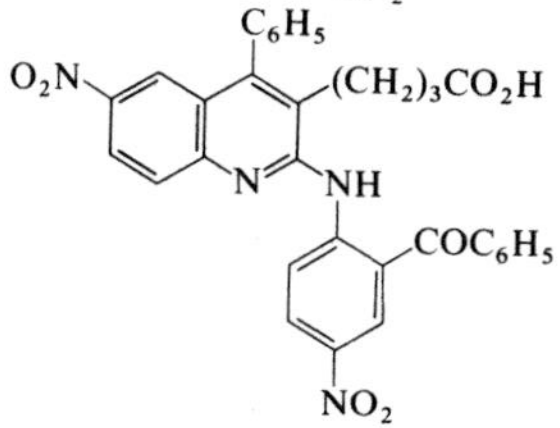	C_6H_5, O_2N, $(CH_2)_3CO_2H$, N, NH, COC_6H_5, NO_2 (23)	161
		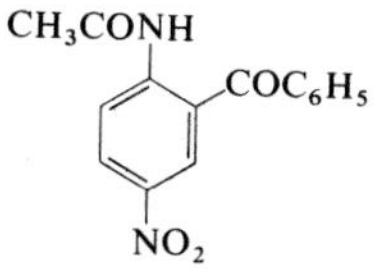 CH_3CONH, COC_6H_5, NO_2	"	130–135°, 2 hr	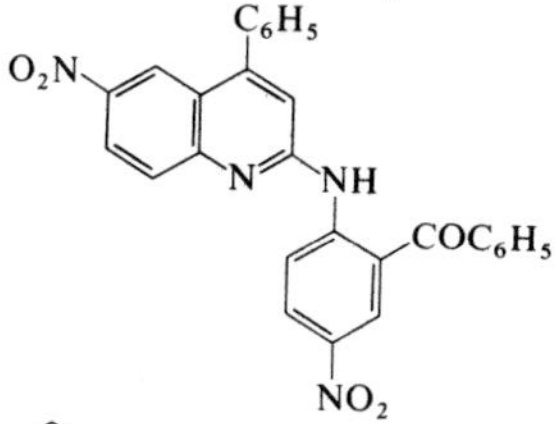 C_6H_5, O_2N, N, NH, COC_6H_5, NO_2 (59)	30b
5,6-cyclo-C_4H_4	H	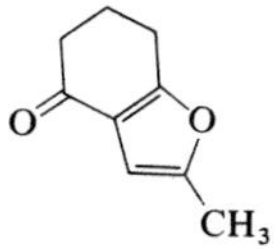 O, O, CH_3	—	160°,[a] 1 hr 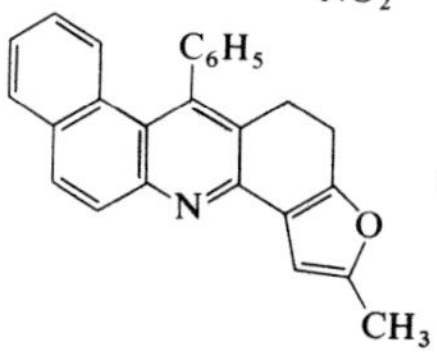	C_6H_5, N, O, CH_3 (56)	33

TABLE VII. CONDENSATIONS WITH SUBSTITUTED o-AMINOBENZOPHENONES (*Continued*)

A.

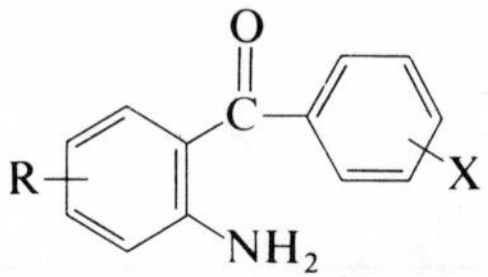

R	X	Reactant	Catalyst	Reaction Conditions	Product(s) and Yield(s) (%)	Refs.
5,6-cyclo-C_4H_4 (*Contd.*)	H	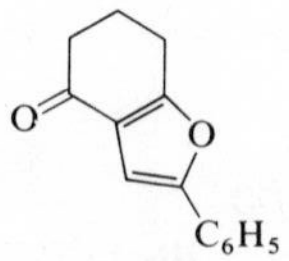 [structure] C_6H_5	—	160°,[a] 1 hr	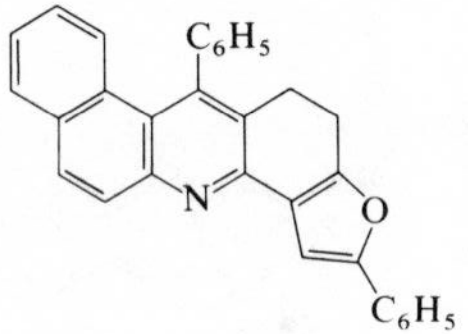 [structure] C_6H_5, C_6H_5 (67)	33
		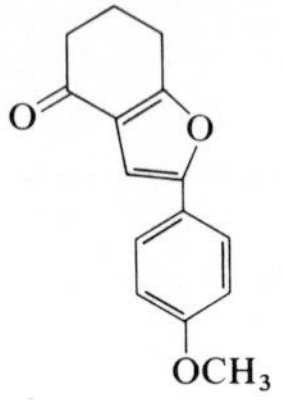 [structure] OCH_3	—	"	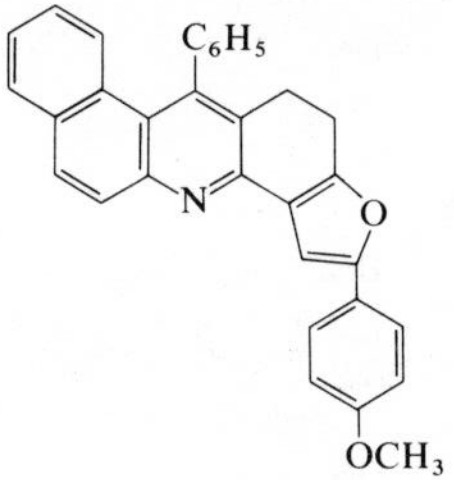 [structure] C_6H_5, OCH_3 (66)	33
4,5-Cl_2	H	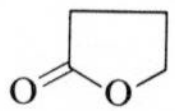 [structure]	PPA	145–150°, 2 hr	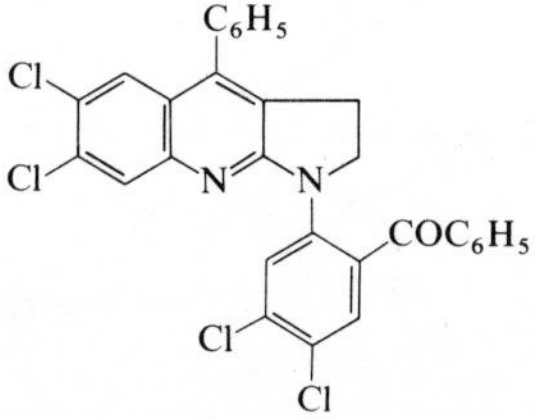 [structure] C_6H_5, Cl, Cl, COC_6H_5, Cl, Cl (37)	162

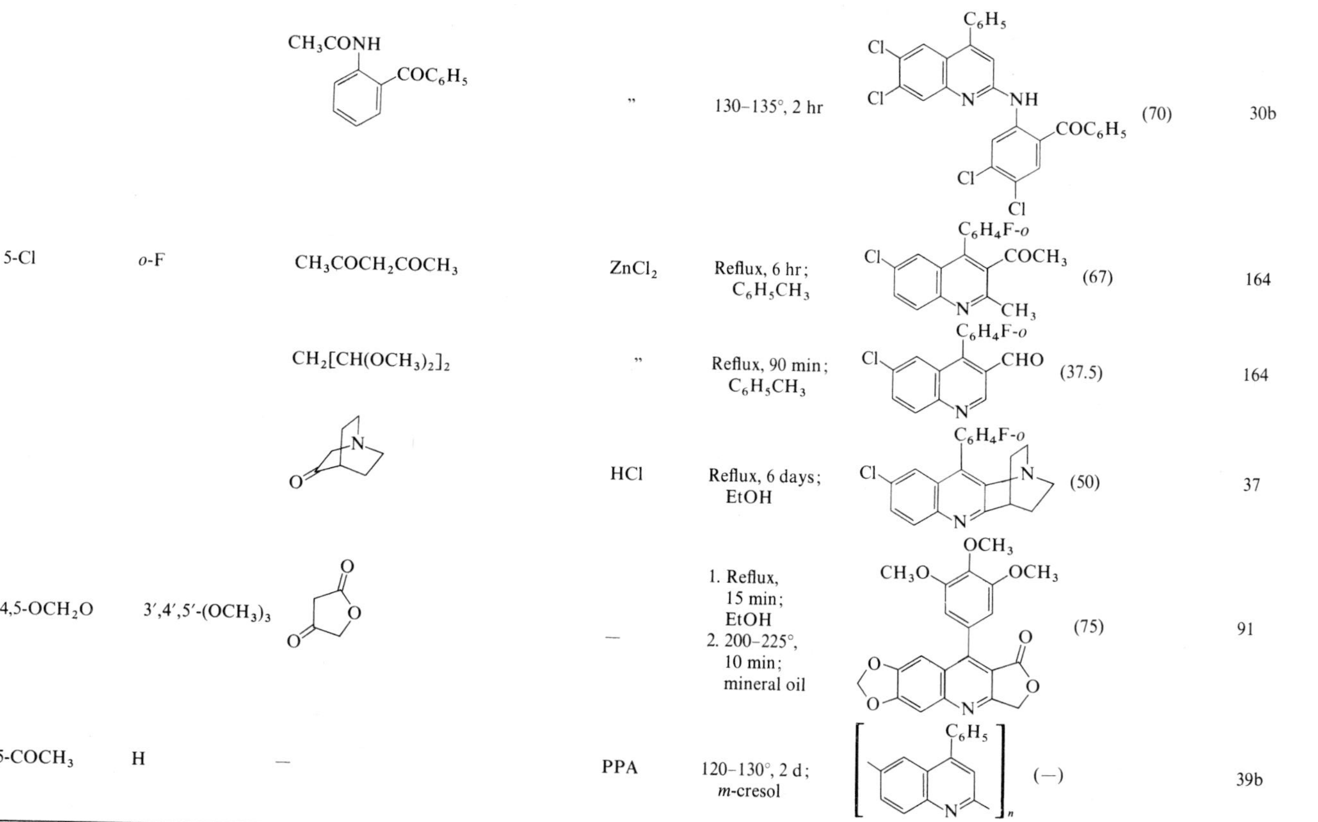

		CH_3CONH–C_6H_4–COC_6H_5	"	130–135°, 2 hr		(70)	30b
5-Cl	*o*-F	$CH_3COCH_2COCH_3$	$ZnCl_2$	Reflux, 6 hr; $C_6H_5CH_3$		(67)	164
		$CH_2[CH(OCH_3)_2]_2$	"	Reflux, 90 min; $C_6H_5CH_3$		(37.5)	164
			HCl	Reflux, 6 days; EtOH		(50)	37
4,5-OCH_2O	3′,4′,5′-$(OCH_3)_3$		—	1. Reflux, 15 min; EtOH 2. 200–225°, 10 min; mineral oil		(75)	91
5-$COCH_3$	H	—	PPA	120–130°, 2 d; *m*-cresol		(—)	39b

TABLE VII. CONDENSATIONS WITH SUBSTITUTED o-AMINOBENZOPHENONES (*Continued*)

B. 1-Aminoanthraquinone (O, NH_2, O) + $R_1CH_2—C(OCH_3)_2—R_2$ ⟶ product (R_1, R_2, N, O)

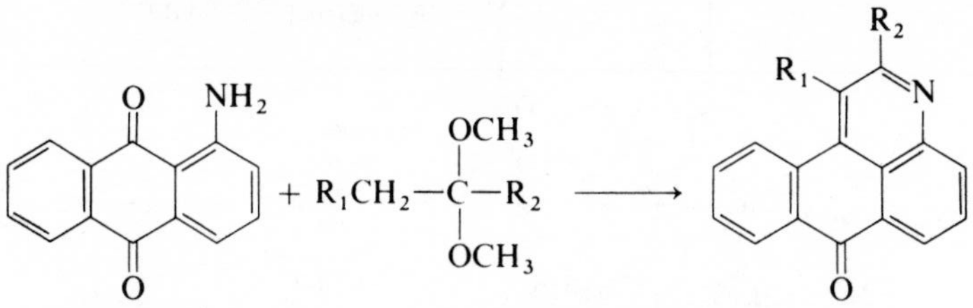

Catalyst	Reaction Conditions	Product(s) and Yield(s) (%)			Refs.
		R_1	R_1R_2	R_2	
—	Reflux, 3 hr; EtOH	H		$N(CH_3)_2$ (70)	163a
—	120°, 4 hr	H		 N (55)	163a
—	"	H		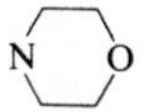 N O (50)	163a
—	"	H		 N NCH_3 (65)	163a
$NaOCH_3$	12 hr; EtOH	H		$N(C_3H_7\text{-}i)_2$ (57)	163a
"	"	CH_3		" (21)	163a
"	"	C_6H_5		 N (26)	163a
"	1. Reflux, 1 hr; EtOH 2. Air, 130°		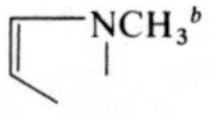 NCH_3[b] (73)		163a

C. 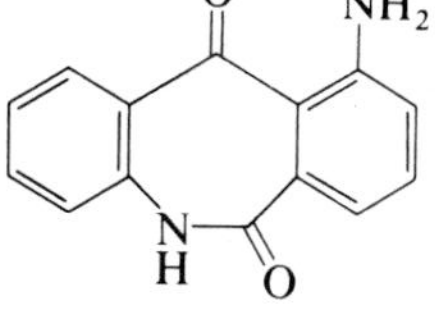 A or B + Reactant

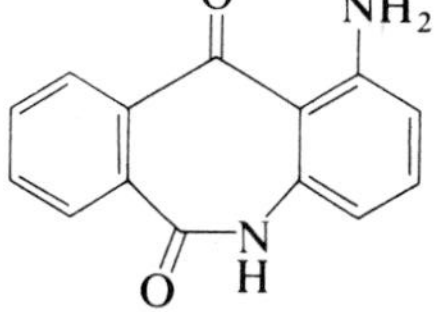

Compound	Reactant	Catalyst	Reaction Conditions	Product(s) and Yield(s) (%)	Refs.
A	$CH_3C(OCH_3)_2N(CH_3)_2$	—	Reflux, 12 hr; EtOH	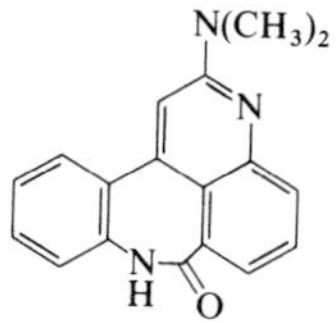 $N(CH_3)_2$ (22)	163a
B	"	—	"	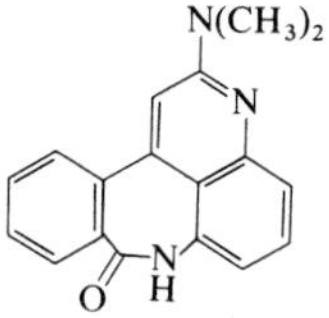 $N(CH_3)_2$ (22)	163a
A	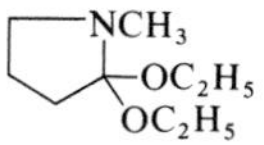 NCH_3, OC_2H_5, OC_2H_5	$NaOCH_3$	1. Reflux, 12 hr; EtOH 2. Sublime, 240°	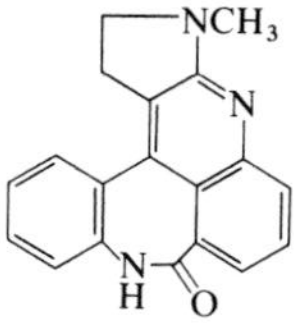 NCH_3 (19)	163a

TABLE VII. CONDENSATIONS WITH SUBSTITUTED *o*-AMINOBENZOPHENONES (*Continued*)

D. (2,4-bis(benzoyl)-1,5-diaminobenzene: C_6H_5CO, COC_6H_5, H_2N, NH_2) + $CH_3CO—Ar—COCH_3$ ⟶ $[$ pyrido-quinoline (C_6H_5, C_6H_5, N, N) $—Ar—]_n$

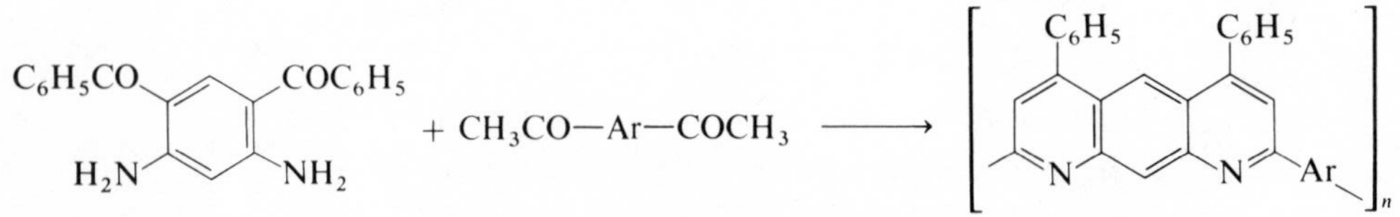

Ar	Catalyst[c]	Reaction Conditions[c]	Polymer $[\eta]$ in H_2SO_4	Refs.
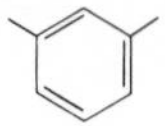 (m-phenylene)	PPA	120°, 1 hr; 170°, 12 hr; *m*-cresol	0.39	39a
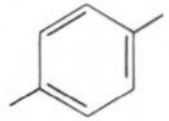 (p-phenylene)	"	"	0.15	39a
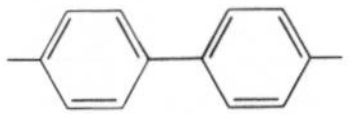 (4,4′-biphenylene)	"	30–140°, 20 hr; 84% PPA	1.35[d]	39a
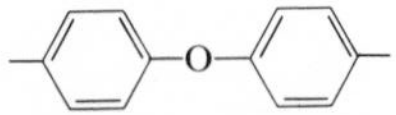 —C₆H₄—O—C₆H₄—	H_3PO_4	120°, 3.5 hr; 165°, 23 hr; sulfolane	0.77	39a
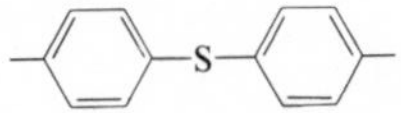 —C₆H₄—S—C₆H₄—	PPA	120°, 1 hr; 170°, 12 hr; *m*-cresol	0.13	39a
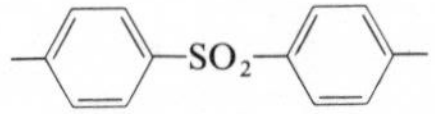 —C₆H₄—SO_2—C₆H₄—	"	"	0.12	39a

E. 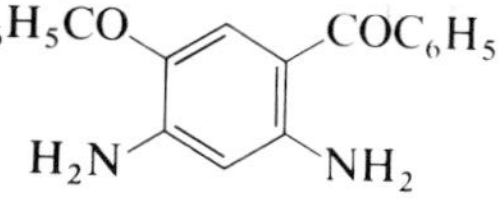 4,6-Dibenzoyl-1,3-phenylenediamine (C_6H_5CO, COC_6H_5, H_2N, NH_2) + Reactant ⟶ Polymer

Reactant	Catalyst	Reaction Conditions	Polymer	Polymer [η] in H_2SO_4	Refs.
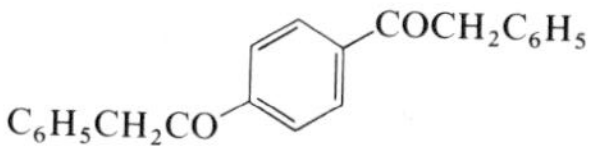 $C_6H_5CH_2CO-C_6H_4-COCH_2C_6H_5$	PPA	80–140°, 19 hr; 84% PPA	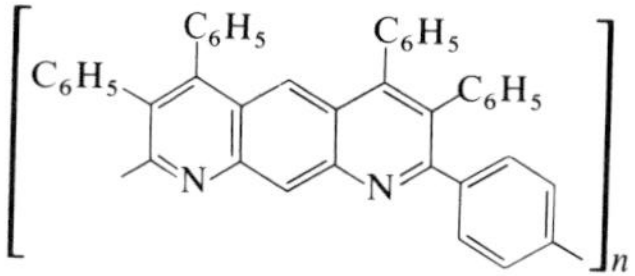 [structure: C_6H_5, C_6H_5, C_6H_5, C_6H_5, N, N]$_n$	0.30	39a
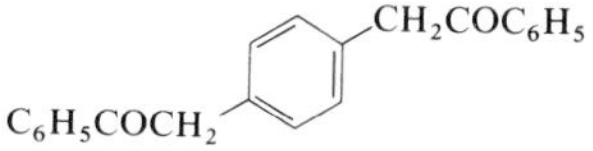 $C_6H_5COCH_2-C_6H_4-CH_2COC_6H_5$	"	60–145°, 40 hr; 84% PPA	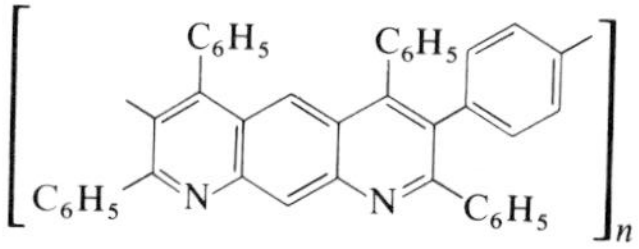 [structure: C_6H_5, C_6H_5, C_6H_5, C_6H_5, N, N]$_n$	1.00	39a

TABLE VII. CONDENSATIONS WITH SUBSTITUTED *o*-AMINOBENZOPHENONES (*Continued*)

F. 4,6-diamino-1,3-phenylene bis(phenyl ketone) (H_2N, COC_6H_5, C_6H_5CO, NH_2) + $CH_3CO—Ar—COCH_3$ ⟶ $[$ 4,6-di-C_6H_5 pyrido-quinoline (N, N) —Ar— $]_n$ [e]

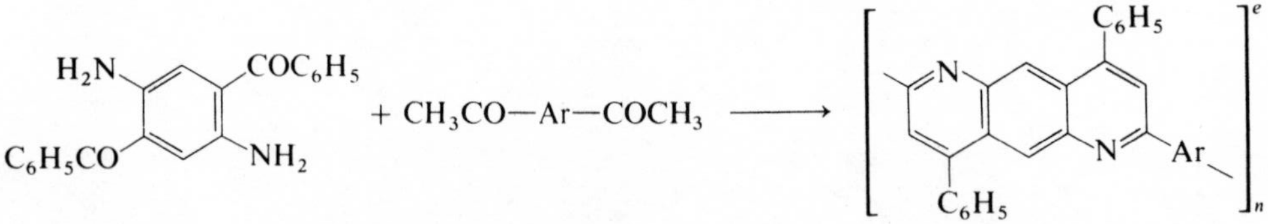

Ar	Catalyst[c]	Reaction Conditions[c]	Polymer $[\eta]$ in H_2SO_4	Refs.
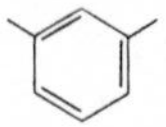 *m*-C_6H_4	PPA	120°, 1 hr; 170°, 12 hr; *m*-cresol	0.19	39a
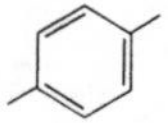 *p*-C_6H_4	"	"	0.37	39a
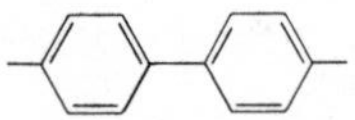 C_6H_4—C_6H_4	"	"	1.24	39a
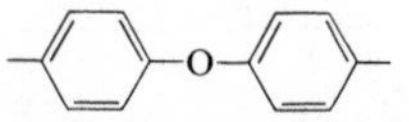 C_6H_4—O—C_6H_4	"	"	0.53	39a
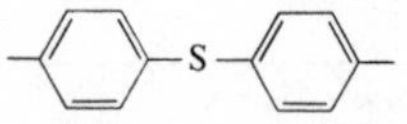 C_6H_4—S—C_6H_4	"	"	0.28	39a
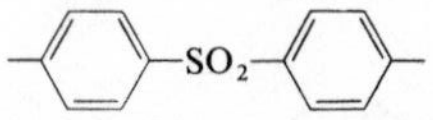 C_6H_4—SO_2—C_6H_4	"	"	0.16	39a

G. $C_6H_5CO(H_2N)C_6H_3{-}O{-}C_6H_3(NH_2)COC_6H_5$ + CH_3COR ⟶ bis[2-R-4-phenylquinolin-6-yl] ether

R	Catalyst	Reaction Conditions	Yield (%)	Refs.
C_6H_5	PPA	130°, 22 hr; *m*-cresol	(89)	39b
2-biphenylenyl	"	130°, 6 hr; *m*-cresol	(98)	39b

H. $C_6H_5CO(H_2N)C_6H_3{-}O{-}C_6H_3(NH_2)COC_6H_5$ + $RCH_2CO{-}Ar{-}COCH_2R$ ⟶ $[$quinoline(R, C_6H_5)$-O-$quinoline(C_6H_5, R)$-Ar]_n$

Ar	R	Catalyst[c]	Reaction Conditions[c]	Polymer $[\eta]$ (Solvent)	Refs.
C_6H_4	H	P_2O_5	130°, 2 d; *m*-cresol	1.0 ($CHCl_3$)	39b
$C_6H_4{-}C_6H_4$	C_6H_5	"	130°, 21 hr; *m*-cresol	3.4 (*m*-cresol)	39b
	H	"	130°, 2 d; *m*-cresol	3.1 (*m*-cresol)	39b
$C_6H_4{-}O{-}C_6H_4$	C_6H_5	"	130°, 21 hr; *m*-cresol	4.1 (*m*-cresol)	39b
	H	"	130°, 2 d; *m*-cresol	1.6 ($CHCl_3$)	39b
biphenylene	C_6H_5	"	130°, 1 d; *m*-cresol	2.0 ($CHCl_3$)	39b
	H	"	135–140°, 20 hr; *m*-cresol	4.4 (HCO_2H)	39d

TABLE VII. CONDENSATIONS WITH SUBSTITUTED *o*-AMINOBENZOPHENONES (*Continued*)

I. 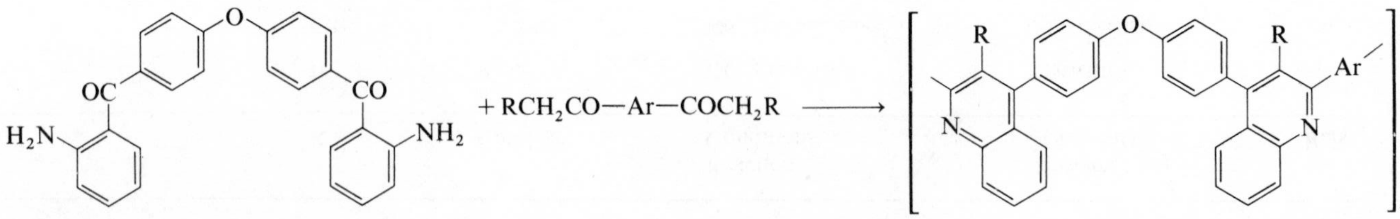 (bis(o-aminobenzoyl)diphenyl ether) + $RCH_2CO—Ar—COCH_2R$ ⟶ [polyquinoline]$_n$

Ar	R	Catalyst[c]	Reaction Conditions[c]	Polymer [η] in *m*-cresol	Refs.
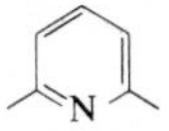 2,6-pyridinediyl	H	PPA	100–135°, 66 hr; *m*-cresol	0.08	39c
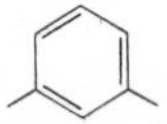 1,3-phenylene	H	"	100–130°, 61 hr; *m*-cresol	0.36	39c
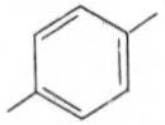 1,4-phenylene	C_6H_5	"	125–145°, 58 hr; *m*-cresol	0.26	39c
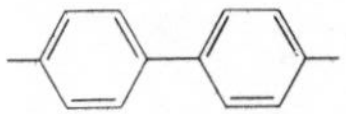 4,4′-biphenylene	H	"	110–150°, 41 hr; *m*-cresol	0.38	39c
	C_6H_5	"	110–140°, 82 hr; *m*-cresol	0.57	39c
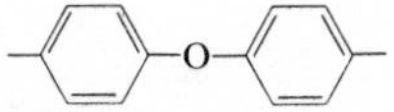 4,4′-oxydiphenylene	H	"	125–130°, 63 hr; *m*-cresol	0.62	39c

J. 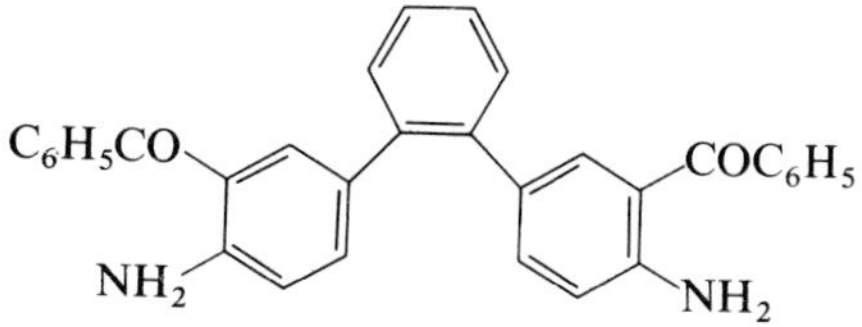 C_6H_5CO, NH_2 ... COC_6H_5, NH_2 + Reactant

Reactant	Catalyst	Reaction Conditions	Product(s) and Yield(s) or Polymer [η] (Solvent)	Refs.
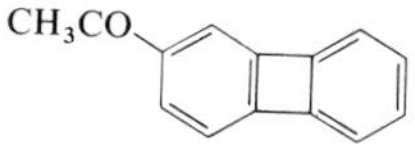 CH_3CO	PPA	130°, 11 hr; *m*-cresol 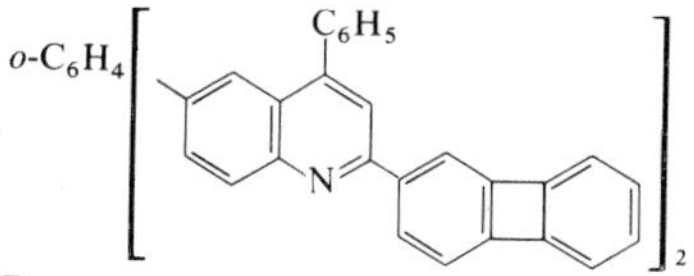	*o*-C_6H_4 [C_6H_5, N]$_2$ (79)	39d
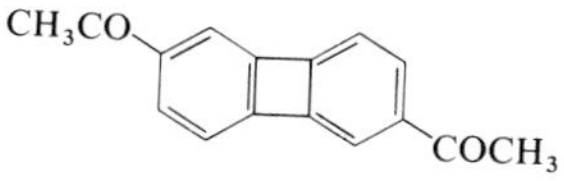 CH_3CO, $COCH_3$	P_2O_5	140°, 11hr; *m*-cresol	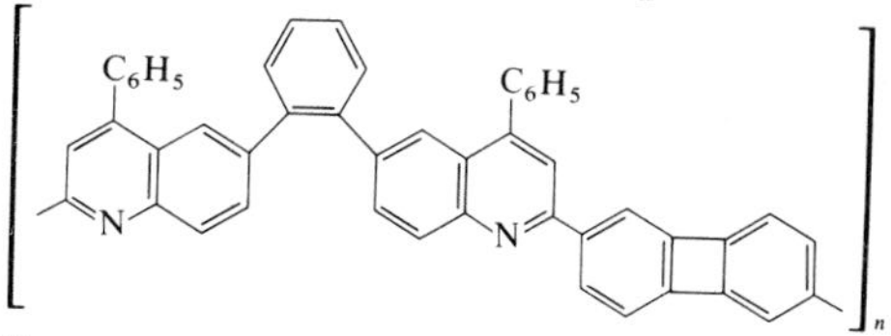 [C_6H_5, N, C_6H_5, N]$_n$ 0.04 (H_2SO_4)	39d
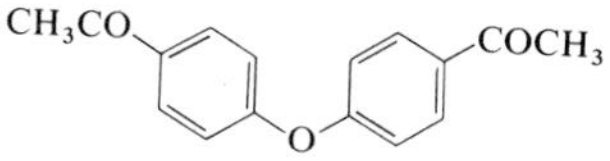 CH_3CO, O, $COCH_3$	"	140°, 20 hr; *m*-cresol	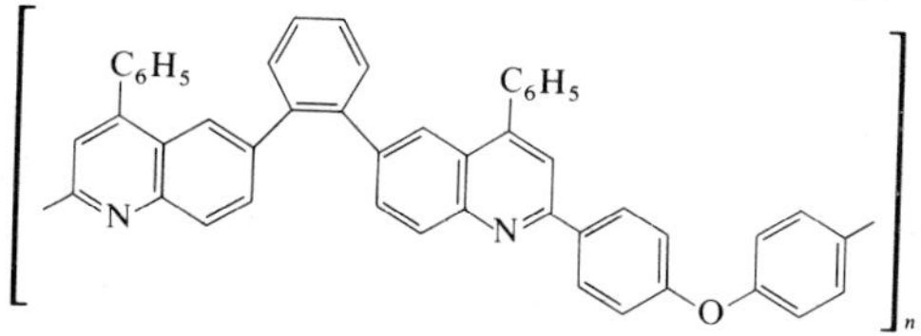 [C_6H_5, N, C_6H_5, N, O]$_n$ 0.43 ($CHCl_3$)	39d

TABLE VII. CONDENSATIONS WITH SUBSTITUTED *o*-AMINOBENZOPHENONES (*Continued*)

K. 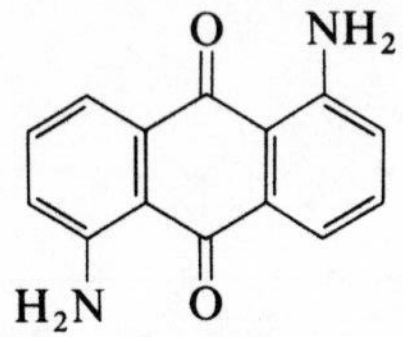 1,5-Diaminoanthraquinone + Reactant

Reactant	Catalyst	Reaction Conditions	Product(s) and Yield(s) (%) or Polymer [η] (Solvent)	Refs.
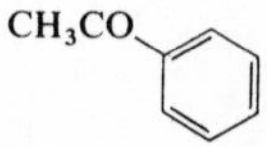 CH_3CO-phenyl	$ZnCl_2$	170°, 3.5 hr	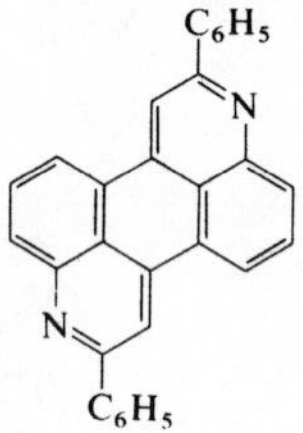 diaza-perylene with two C_6H_5 groups (—)	39e
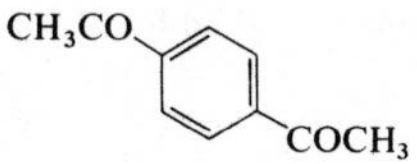 CH_3CO-p-phenylene-$COCH_3$	—	240°, 30 min	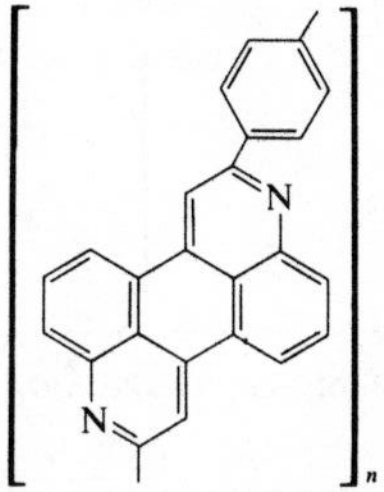 p-phenylene polymer, [ ]$_n$, 0.46 (H_2SO_4)	39e
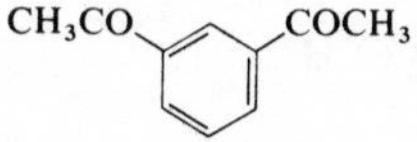 CH_3CO-m-phenylene-$COCH_3$	—	250°, sealed tube	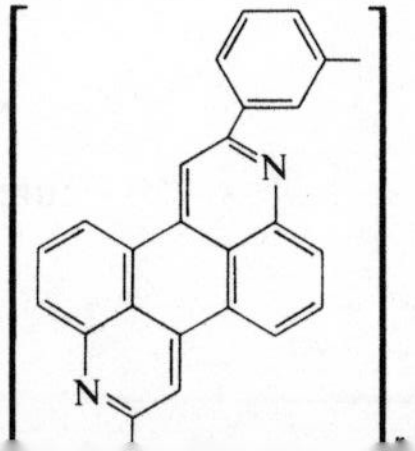 m-phenylene polymer (—)	39e

L. 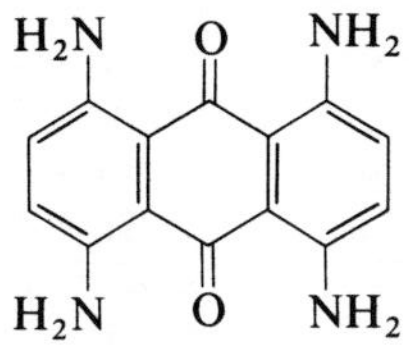 1,4,5,8-Tetraaminoanthraquinone (H_2N, O, NH_2 / H_2N, O, NH_2) + Reactant

Reactant	Catalyst	Reaction Conditions	Product	Ref.
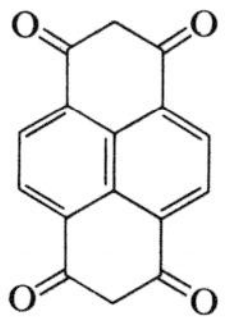 [structure]	PPA	1. 25–170°, 4.5 hr; 170°, 20 hr 2. 60–100°, 0.025 mm; CH_3SO_3H 3. 340–400°, 0.01 mm; 2–5 d	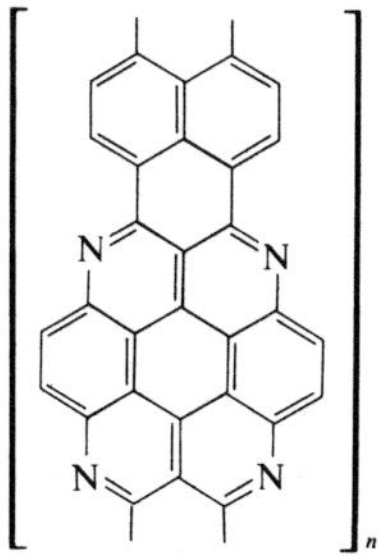 [polymer structure]$_n$	39f

[a] The hydrochloride salt of *o*-aminobenzophenone rather than the free base is used.
[b] The reactant was N-methyltetrahydropyrrolidone-2 diethylacetal.
[c] See reference for other catalysts and reaction conditions that gave lower-molecular-weight polymers.
[d] The viscosity was determined in CH_3SO_3H.
[e] The formula in the reference shows —Ar— attached to the 3-position of the right-hand ring.

TABLE VIII. CONDENSATIONS WITH AZA ANALOGS OF *o*-AMINOBENZALDEHYDE

A. 6-Aza

3-Amino-2-pyridinecarboxaldehyde (CHO, NH_2) + Reactant ⟶ 1,5-naphthyridine (R_1, R_2)

Reactant	Catalyst	Reaction Conditions	Product(s) and Yield(s) (%) R_1	R_2	Refs.
$CH_3COCH_2COCH_3$	[a]	[a]	CH_3	$COCH_3$ (—)	159
$CH_3COCH_2CO_2C_2H_5$	[a]	[a]	CH_3	$CO_2C_2H_5$ (—)	159
$CH_2(CO_2C_2H_5)_2$	[a]	[a]	OH	$CO_2C_2H_5$ (—)	159

B. 5-Aza[b]

4-Amino-3-pyridinecarboxaldehyde (CHO, NH_2) + Reactant ⟶ 1,6-naphthyridine (R_1, R_2)

Reactant	Catalyst	Reaction Conditions	Product(s) and Yield(s) (%) R_1	R_2	Refs.
CH_3CN	NaOMe	2 hr; MeOH	NH_2	H (53)	53
CH_3COCH_3	NaOH	48 hr; Me_2CO	CH_3	H (57)	18
CH_3CH_2CN	NaOMe	2 hr; MeOH	NH_2	CH_3 (55)	53
$CH_2(CN)_2$	Piperidine	15 min; EtOH	NH_2	CN (90)	53
$NCCH_2CO_2H$	"	2 hr; EtOH	NH_2	CO_2H (52)	53
$NCCH_2CONH_2$	"	1 hr; EtOH	NH_2	$CONH_2$ (87)	53

$CH_2(CONH_2)_2$	"	24 hr; EtOH	OH	$CONH_2$ (84)	18
$CH_3COC_2H_5$	"	48 hr; $CH_3COC_2H_5$	CH_3	CH_3 (44)	18
			C_2H_5	H (47)	
	NaOH	EtOH	CH_3	CH_3 (54)	52
$CH_3(CH_2)_2CN$	NaOMe	2 hr; MeOH	NH_2	C_2H_5 (25)	53
$NCCH_2CONHCH_3$	Piperidine	2 hr; EtOH	NH_2	$CONHCH_3$ (80)	53
$CH_3COCH_2COCH_3$	"	24 hr; EtOH	CH_3	$COCH_3$ (60)	18
$CH_2(CONHCH_3)_2$	"	48 hr; EtOH	OH	$CONHCH_3$ (56)	18
$NCCH_2CO_2C_2H_5$	Piperidine	1 hr; EtOH	OH	CN (92)	52
$NCCH_2CON(CH_3)_2$	"	"	OH	CN (—)	18
	NaOH	3 hr; EtOH	OH	CN (Trace)	18
			NH_2	$CON(CH_3)_2$ (63)	
$NCCH_2CONHC_2H_5$	Piperidine	2 hr; EtOH	NH_2	$CONHC_2H_5$ (65)	53
$CH_3COCH_2CO_2C_2H_5$	"	12 hr; EtOH	OH	$COCH_3$ (48)	18
			CH_3	$CO_2C_2H_5$ (46)	
	NaOH	EtOH	CH_3	$CO_2C_2H_5$ (56)	18
2-Acetylfuran	NaOH	1 hr; EtOH	2-Furyl	H (87)	18
2-Furylacetonitrile	"	3 hr; EtOH	NH_2	2-Furyl (78)	53
2-Thienylacetonitrile	"	"	NH_2	2-Thienyl (78)	53
$CH_2(CO_2C_2H_5)_2$	Piperidine	16 hr; EtOH	OH	$CO_2C_2H_5$ (97)	18
3-Acetylpyridine	NaOH	1 hr; EtOH	3-Pyridyl	H (80)	18
N-(Cyanoacetyl)morpholine	Piperidine	1 hr; EtOH	OH	CN (—)	18
	NaOH	2 hr; EtOH	OH	CN (Trace)	18
			NH_2	CON O (65)	
3-Pyridylacetonitrile	Piperidine	24 hr; EtOH	NH_2	3-Pyridyl (63)	53
$C_6H_5CH_2CHO$	"	2 hr; EtOH	H	C_6H_5 (50)	18
$CH_3COC_6H_5$	NaOH	24 hr; EtOH	C_6H_5	H (83)	18
$C_6H_5CH_2CN$	"	2 hr; EtOH	NH_2	C_6H_5 (68)	53
p-$FC_6H_4CH_2CN$	"	"	NH_2	C_6H_4F-p (67)	53
p-$ClC_6H_4CH_2CN$	"	1 hr; EtOH	NH_2	C_6H_4Cl-p (70)	53

TABLE VIII. CONDENSATIONS WITH AZA ANALOGS OF *o*-AMINOBENZALDEHYDE (*Continued*)

B. 5-Aza[b] (*Continued*)

Reactant	Catalyst	Reaction Conditions	Product(s) and Yield(s) (%) R_1	R_2	Refs.
p-$BrC_6H_4CH_2CN$	NaOH	1 hr; EtOH	NH_2	C_6H_4Br-*p* (86)	53
p-$O_2NC_6H_4CH_2CN$	Piperidine	12 hr; EtOH	NH_2	$C_6H_4NO_2$-*p* (84)	53
1-Cyclohexenylacetonitrile	NaOH	4 hr; EtOH	NH_2	1-Cyclohexenyl (59)	53
$CH_3COCH_2C_6H_5$	Piperidine	24 hr; EtOH	CH_3	C_6H_5 (61)	18
$C_2H_5COC_6H_5$	NaOH	"	C_6H_5	CH_3 (80)	18
N … $-CH_2CO_2C_2H_5$	Piperidine	EtOH	OH	3-Pyridyl (65)	52
$NCCH_2COC_6H_5$	"	1 hr; EtOH	C_6H_5	CN (71)	52
p-$CH_3C_6H_4CH_2CN$	NaOH	"	NH_2	$C_6H_4CH_3$-*p* (80)	53
$NCCH_2CONH(CH_2)_2N(C_2H_5)_2$	Piperidine	24 hr; EtOH	NH_2	$CONH(CH_2)_2N(C_2H_5)_2$ (57)	53
$NCCH_2CONH(CH_2)_2N$ O	"	"	NH_2	$CONH(CH_2)_2N$ O (75)	53
$C_6H_5CH_2CO_2C_2H_5$	"	12 hr; EtOH	OH	C_6H_5 (21)	18
p-$O_2NC_6H_4CH_2CO_2C_2H_5$	"	10 hr; EtOH	OH	$C_6H_4NO_2$-*p* (76)	18
3-Indolylacetonitrile	NaOH	4 hr; EtOH	NH_2	3-Indolyl (58)	53
$C_6H_5COCH_2CO_2C_2H_5$	Piperidine	24 hr; EtOH	C_6H_5	$CO_2C_2H_5$ (66)	18
1-Naphthylacetonitrile	NaOH	3 hr; EtOH	NH_2	1-Naphthyl (57)	53
4-Biphenylylacetonitrile	"	30 min; EtOH	NH_2	4-Biphenylyl (87)	53

C. **4-Aza**[b]

$$\text{3-amino-4-pyridinecarboxaldehyde (CHO, } NH_2\text{)} + \text{Reactant} \longrightarrow \text{1,7-naphthyridine (2-}R_1\text{, 3-}R_2\text{)}$$

Reactant	Catalyst	Reaction Conditions	Product(s) and Yield(s) (%) R₁	R₁R₂	R₂	Refs.
CH_3CH_2CHO	NaOH	30 min; EtOH	H		CH_3 (60)	166
CH_3COCH_3	"	1 hr; EtOH	CH_3		H (80)	166
$NCCH_2CONH_2$	[a]	[a]	NH_2		CN (—)	159
$CH_2(CONH_2)_2$	[a]	[a]	NH_2		$CONH_2$ (—)	159
$CH_3COCH{=}NOH$	KOH	6 hr; EtOH[c]	$CH{=}NOH$		H (56)	62
$CH_3COC_2H_5$	NaOH	1 hr; EtOH	CH_3		CH_3 (80)	166
$CH_3COCOCH_3$	"	2 hr EtOH	2-(1,7-Naphthyridyl)		H (40)	166
$NCCH_2CO_2C_2H_5$	[a]	[a]	OH		CN (—)	159
$CH_3COCH_2COCH_3$	Piperidine	8 hr[c,d]	CH_3		$COCH_3$ (59)	62,159
Cyclopentanone	NaOH	6 hr; EtOH[c]		$(CH_2)_3$	(92)	62
$CH_3COCH_2CO_2C_2H_5$	[a]	[a]	CH_3		$CO_2C_2H_5$ (—)	159
2-Acetylfuran	NaOH	15 min; EtOH	2-Furyl		H (80)	166
2-Acetylthiophene	"	1 hr; EtOH	2-Thienyl		H (50)	166
3-Acetylthiophene	"	"	3-Thienyl		H (60)	166
2-Acetylpyrrole	"	30 min; EtOH	2-Pyrryl		H (50)	166
Cyclohexanone	"	8 hr; EtOH[c]		$(CH_2)_4$ (66)		167
2-Acetylpyridine	"	30 min; EtOH	2-Pyridyl		H (80)	166
3-Acetylpyridine	"	"	3-Pyridyl		H (60)	166
4-Acetylpyridine	"	"	4-Pyridyl		H (60)	166

TABLE VIII. CONDENSATIONS WITH AZA ANALOGS OF *o*-AMINOBENZALDEHYDE (*Continued*)

C. 4-Aza[b] (*Continued*)

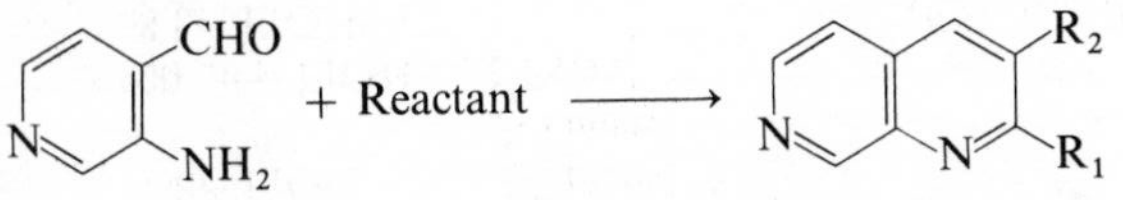

(3-amino-4-pyridinecarboxaldehyde: CHO, NH_2, N) + Reactant ⟶ (1,7-naphthyridine: N, N, R_1, R_2)

Reactant	Catalyst	Reaction Conditions	Product(s) and Yield(s) (%) R_1	R_1R_2	R_2	Refs.
$CH_2(CO_2C_2H_5)_2$	[a]	[a]	OH		$CO_2C_2H_5$ (—)	159
4-Methylcyclohexanone	NaOH	2 hr; EtOH[c]		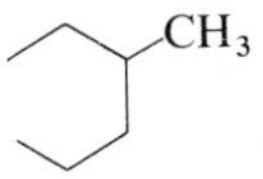 CH_3	(76)	62
$C_6H_5CH_2CHO$	NaOH	15 min; EtOH	H		C_6H_5 (60)	166
$C_6H_5COCH_3$	"	24 hr; EtOH	C_6H_5		H (80)	166
	"	8 hr; EtOH[c]	C_6H_5		H (84)	167
$C_6H_5COCH_2NO_2$	—	2 hr; EtOH[c]	C_6H_5		NO_2 (10)	62
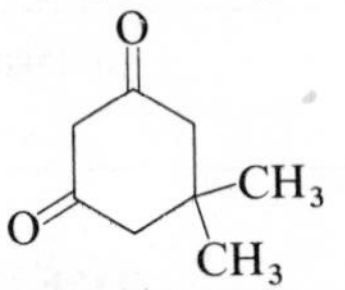 O, O, CH_3, CH_3	Piperidine	8 hr[c,d]		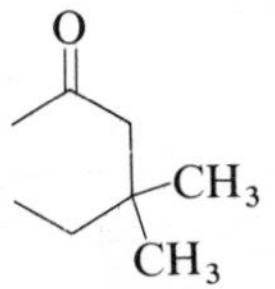 O, CH_3, CH_3	(20)	62
$C_6H_5CH_2COC_6H_5$	NaOH	30 min; EtOH	C_6H_5		C_6H_5 (60)	166

D. 3-Aza[b]

2-Aminopyridine-3-carbaldehyde (2-NH_2, 3-CHO) + Reactant ⟶ 1,8-naphthyridine with R_1 at C-2 and R_2 at C-3

Reactant	Catalyst	Reaction Conditions	Product(s) and Yield(s) (%) R_1	R_1R_2	R_2	Refs.
CH_3COCH_3	Piperidine	24 hr; Me_2CO	CH_3		H (90)	42
	NaOMe	100°, 30 min	CH_3		H (64)	167a
$CH_2(CN)_2$	Piperidine	1 hr; EtOH	NH_2		CN (96)	66
$CH_2(CONH_2)_2$	"	24 hr; EtOH	OH		$CONH_2$ (95)	168
$NCCH_2CONH_2$	"	1 hr; EtOH	NH_2		$CONH_2$ (100)	66
$CH_3COC_2H_5$	NaOH	24 hr; EtOH	CH_3		CH_3 (27)	168
	Piperidine	20 hr; $CH_3COC_2H_5$	CH_3		CH_3 (34)	169
	NaOMe	100°, 30 min	CH_3		CH_3 (32)	167a
$O_2NCH_2CO_2C_2H_5$	Piperidine	1 hr; EtOH	OH		NO_2 (74)	42
$NCCH_2CONHCH_3$	"	2 hr; EtOH	NH_2		$CONHCH_3$ (59)	168
$NCCH_2CO_2CH_3$	"	1 hr; EtOH	OH		CN (76)	168
Cyclobutanone	KOH	3 d[e]		$(CH_2)_2$	(39)	169
$CH_3COCH_2COCH_3$	Piperidine	1 hr; EtOH	CH_3		$COCH_3$ (90)	42
$CH_2(CO_2CH_3)_2$	"	24 hr; EtOH	OH		CO_2CH_3 (86)	168
$C_2H_5COC_2H_5$	NaOMe	100°, 30 min	C_2H_5		CH_3 (74)	167a
$CH_3COC_3H_7$-*n*	"	"	CH_3		C_2H_5 (45)	167a
$CH_3COC_3H_7$-*i*	"	"	C_3H_7-*i*		H (15)	167a
$NCCH_2CONHC_2H_5$	Piperidine	2 hr; EtOH	NH_2		$CONHC_2H_5$ (70)	168
$NCCH_2CON(CH_3)_2$	NaOH	10 hr; EtOH	NH_2		$CON(CH_3)_2$ (56)	168
Cyclopentanone	H_2SO_4	6 hr; AcOH		$(CH_2)_3$	(64)	169
	NaOMe	100°, 30 min		$(CH_2)_3$	(52)	167a

TABLE VIII. CONDENSATIONS WITH AZA ANALOGS OF o-AMINOBENZALDEHYDE (*Continued*)

D. 3-Aza[b] (*Continued*)

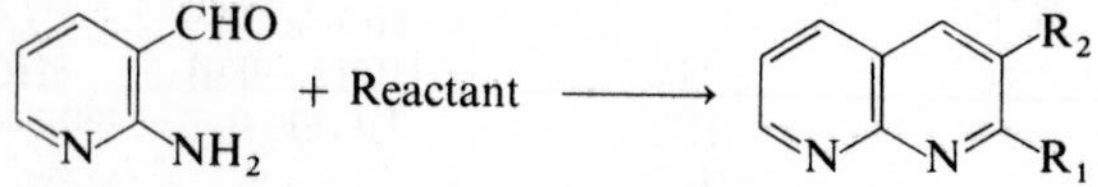

2-Aminonicotinaldehyde (CHO, NH_2) + Reactant ⟶ 1,8-naphthyridine (R_1, R_2)

Reactant	Catalyst	Reaction Conditions	Product(s) and Yield(s) (%) R_1	R_1R_2	R_2	Refs.
$CH_3COCH_2CO_2C_2H_5$	Piperidine	1 hr; EtOH	CH_3		$CO_2C_2H_5$ (81)	42
$NCCH_2CONHC_3H_7$-*n*	"	24 hr; EtOH	NH_2		$CONHC_3H_7$-*n* (80)	168
2-Furylacetonitrile	NaOH	3 hr; EtOH	NH_2		2-Furyl (84)	168
2-Thienylacetonitrile	"	1 hr; EtOH	NH_2		2-Thienyl (75)	168
$CH_3COC_4H_9$-*t*	NaOMe	100°, 30 min	C_4H_9-*t*		H (46)	167a
Cyclohexanone	H_2SO_4	8 hr; AcOH		$(CH_2)_4$	(70)	169
	NaOMe	100°, 30 min		$(CH_2)_4$	(39)	167a
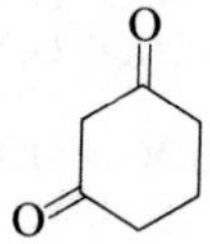 1,3-Cyclohexanedione	—	72 hr; EtOH		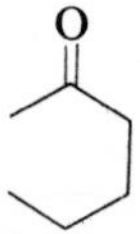 $-CO(CH_2)_3-$	(89)	17
[f]	KOH	72 hr; EtOH	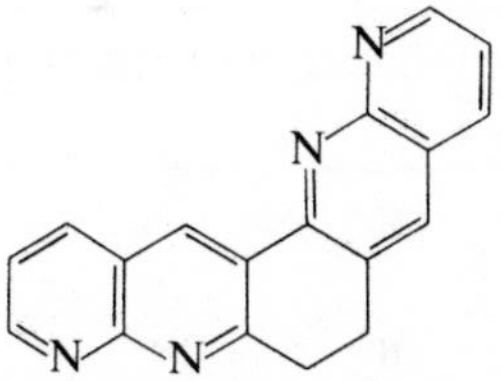 Dihydro-bis(1,8-naphthyridine) fused product (90)			17

Reactant	Catalyst	Conditions	R_1	R_2 (yield)	Ref.
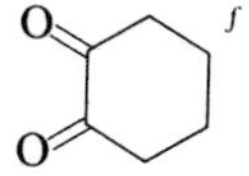 [structure][f]	”	48 hr; EtOH	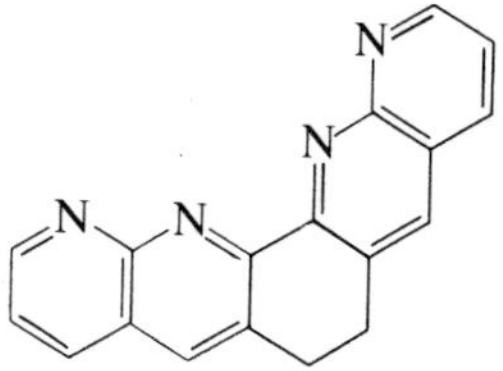 [structure]	(44)	17
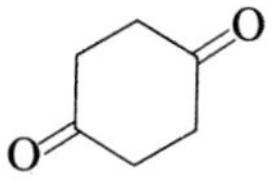 [structure]	Piperidine	1 hr; $C_6H_5CH_3$	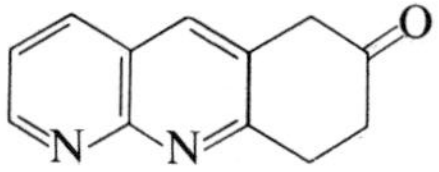 [structure]	(94)	17
$CH_2(CO_2C_2H_5)_2$	Piperidine	24 hr; EtOH	OH	$CO_2C_2H_5$ (70)	42
3-Acetylpyridine	NaOH	1 hr; EtOH	3-Pyridyl	H (61)	168
$NCCH_2CONH(CH_2)_3CH_3$	Piperidine	24 hr; EtOH	NH_2	$CONH(CH_2)_3CH_3$ (66)	168
3-Cyanomethylpyridine	”	”	NH_2	3-Pyridyl (68)	168
2-Cyanomethylpyridine	”	”	NH_2	2-Pyridyl (83)	42
$C_6H_5CH_2CHO$	”	”	H	C_6H_5 (92)	42
$CH_3COC_6H_5$	NaOH	”	C_6H_5	H (89)	168
	NaOMe	100°, 30 min	C_6H_5	H (81)	167a
$NCCH_2CO$–[structure] 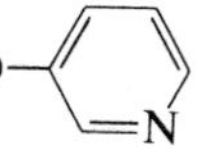	Piperidine	30 min; EtOH	3-Pyridyl	CN (80)	168
$NCCH_2SO_2C_6H_5$	NaOH	24 hr; EtOH	NH_2	$SO_2C_6H_5$ (88)	168
p-$FC_6H_4CH_2CN$	Piperidine	2 hr; EtOH	NH_2	C_6H_4F-p (91)	168
p-$O_2NC_6H_4CH_2CN$	”	6 hr; EtOH	NH_2	$C_6H_4NO_2$-p (88)	168
$C_6H_5CH_2CN$	”	5 d; EtOH	NH_2	C_6H_5 (61)	42
m-$CF_3C_6H_4CH_2CHO$	”	24 hr; EtOH	$C_6H_4CF_3$-m	H (91)	168
$C_6H_5CH_2COCH_3$	”	”	CH_3	C_6H_5 (66)	42
$C_6H_5COC_2H_5$	NaOMe	100°, 30 min	C_6H_5	CH_3 (72)	167a

TABLE VIII. CONDENSATIONS WITH AZA ANALOGS OF *o*-AMINOBENZALDEHYDE (*Continued*)

D. 3-Aza[b] (*Continued*)

2-Aminonicotinaldehyde (CHO, NH_2) + Reactant ⟶ 1,8-naphthyridine (R_1, R_2)

Reactant	Catalyst	Reaction Conditions	Product(s) and Yield(s) (%) R_1	R_2	Refs.
$NCCH_2CONH(CH_2)_2N$-morpholino	Piperidine	24 hr; EtOH	NH_2	$CONH(CH_2)_2N$-morpholino (72)	168
2-Pyridyl-$CH_2CO_2C_2H_5$	"	5 d; EtOH	OH	2-Pyridyl (76)	42
$CH_3COCH_2SO_2C_6H_5$	"	24 hr; EtOH	CH_3	$SO_2C_6H_5$ (49)	168
$C_6H_5COCH_2CN$	"	1 hr; EtOH	C_6H_5	CN (82)	66
p-$CH_3OC_6H_4CH_2CN$	NaOH	"	NH_2	$C_6H_4OCH_3$-p (96)	168
$C_6H_5CH_2CO_2C_2H_5$	Piperidine	5 d; EtOH	OH	C_6H_5 (6)	42
$C_6H_5COC_3H_7$-n	NaOMe	100°, 30 min	C_6H_5	C_2H_5 (55)	167a
$C_6H_5COC_4H_9$-n	"	"	C_6H_5	C_3H_7-n (75)	167a

Reactant	Catalyst	Reaction Conditions	Product(s) and Yield(s) (%)	Refs.
7-Oxo-5,6,7,8-tetrahydro-1,8-diazaanthracene (structure)		48 hr; $C_6H_5CH_3$	Pentacyclic bis(naphthyridine) (structure) (89)	17

E. 3-Aza-4-phenyl[b]

2-Amino-5-R_3-6-C_6H_5-pyridine-3-carboxaldehyde + Reactant ⟶ 2-R_1-3-R_2-6-R_3-7-C_6H_5-1,8-naphthyridine

Reactant	Catalyst	Reaction Conditions	Product(s) and Yield(s) (%) R_1	R_2	R_3	Refs.
CH_3COCH_3	Piperidine	24 hr; Me_2CO	CH_3	H	H (98)	42
$NCCH_2CONH_2$	"	0.1 hr; EtOH	NH_2	CN	H (85)	42
$CH_2(CONH_2)_2$	"	1 hr; EtOH	NH_2	$CONH_2$	H (100)	42
$O_2NCH_2CO_2C_2H_5$	"	"	OH	NO_2	H (47)	42
$CH_3COCH_2COCH_3$	"	"	CH_3	$COCH_3$	H (97)	42
$NCCH_2CO_2C_2H_5$	"	"	OH	CN	H (95)	42
$CH_3COCH_2CO_2C_2H_5$	"	0.1 hr; EtOH	CH_3	$CO_2C_2H_5$	H (97)	42
$CH_2(CO_2C_2H_5)_2$	"	24 hr; EtOH	OH	$CO_2C_2H_5$	H (85)	42
$C_6H_5CH_2CHO$	"	"	H	C_6H_5	H (100)	42
$C_6H_5COCH_2CHO$	"	1 hr; EtOH	H	COC_6H_5	H (56)	42
2-Pyridyl-$CH_2CO_2C_2H_5$	"	24 hr; EtOH	OH	2-Pyridyl	H (76)	42
$C_6H_5CH_2COC_6H_5$	KOH	Reflux, 3 d; EtOH	C_6H_5	C_6H_5	C_6H_5 (90)	169a

TABLE VIII. CONDENSATIONS WITH AZA ANALOGS OF o-AMINOBENZALDEHYDE (*Continued*)

F. 3-Aza-4,6-dimethyl[b]

(2-amino-4,6-dimethylpyridine-3-carbaldehyde) + Reactant ⟶ (2-R₁-3-R₂-5,7-dimethyl-1,8-naphthyridine)

Reactant	Catalyst	Reaction Conditions	Product(s) and Yield(s) (%) R_1	R_2	Refs.
$NCCH_2CO_2H$	Piperidine	1 hr; EtOH	OH	CN (88)	67
$HO_2CCH_2CONH_2$	"	24 hr; EtOH	OH	$CONH_2$ (17)	67
$NCCH_2CONH_2$	"	1 hr; EtOH	NH_2	$CONH_2$ (67)	67
$CH_2(CN)_2$	"	2 hr; EtOH	NH_2	CN (94)	67
$CH_3COC_2H_5$	NaOH	48 hr; EtOH	CH_3	CH_3 (90)	67
$NCCH_2CONHCH_3$	Piperidine	2 hr; EtOH	NH_2	$CONHCH_3$ (88)	67
$NCCH_2CONHC_2H_5$	"	24 hr; EtOH	NH_2	$CONHC_2H_5$ (84)	67
$NCCH_2CON(CH_3)_2$	NaOH	3 hr; EtOH	NH_2	$CON(CH_3)_2$ (82)	67
$NCCH_2CONHC_3H_7$-*n*	Piperidine	24 hr; EtOH	NH_2	$CONHC_3H_7$-*n* (76)	67
2-Cyanomethylthiophene	NaOH	3 hr; EtOH	NH_2	2-Thienyl (77)	67
$NCCH_2CONHC_4H_9$-*n*	Piperidine	24 hr; EtOH	NH_2	$CONHC_4H_9$-*n* (60)	67
3-Cyanomethylpyridine	"	"	NH_2	3-Pyridyl (39)	67
p-$O_2NC_6H_4CH_2CN$	"	6 hr; EtOH	NH_2	$C_6H_4NO_2$-*p* (79)	67
$C_6H_5SO_2CH_2CN$	NaOH	24 hr; EtOH	NH_2	$SO_2C_6H_5$ (98)	67
$NCCH_2CONHC_5H_{11}$-*n*	Piperidine	"	NH_2	$CONHC_5H_{11}$-*n* (64)	67
$C_6H_5CH_2CN$	NaOH	2 hr; EtOH	NH_2	C_6H_5 (71)	67
$NCCH_2CONHC_6H_{13}$-*n*	Piperidine	24 hr; EtOH	NH_2	$CONHC_6H_{13}$-*n* (55)	67
$C_6H_5COCH_2CN$	"	2 hr; EtOH	C_6H_5	CN (90)	67

G. 3-Aza-4,6-di(trifluoromethyl)[g]

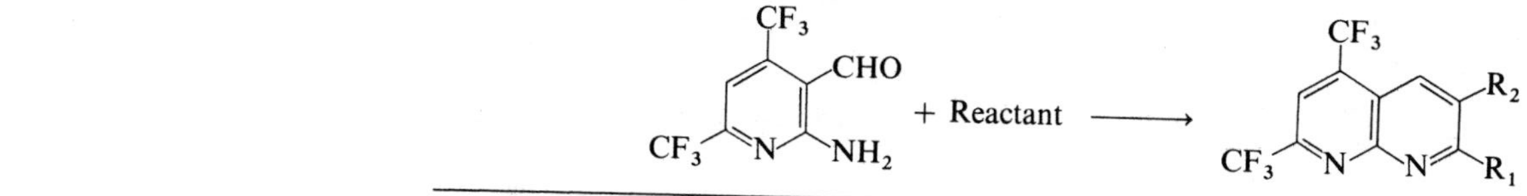

Reactant	Catalyst	Reaction Conditions	Product(s) and Yield(s) (%) R_1	R_2	Refs.
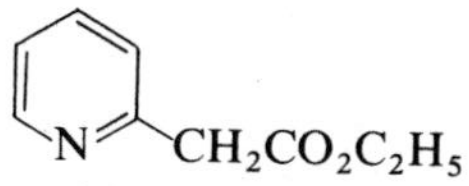	Na_2CO_3	175°; ethylene glycol[h]	OH	2-Pyridyl (31)	173
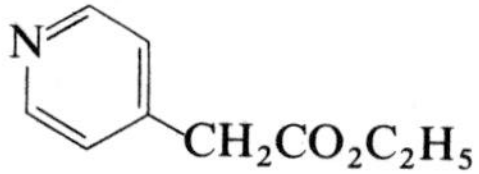	"	"[h]	OH	4-Pyridyl (30)	173

TABLE VIII. CONDENSATIONS WITH AZA ANALOGS OF *o*-AMINOBENZALDEHYDE (*Continued*)

H. 3-Aza-4-amino-5-carbaldehyde

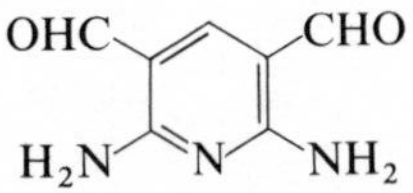

OHC, CHO, H_2N, N, NH_2 + Reactant

Reactant	Catalyst	Reaction Conditions	Product(s) and Yield(s) (%)	Refs.
$C_6H_5COCH_3$	KOH	1. Reflux, 3 hr; 1-propanol 2. HNO_3, 20°, 20 min; 80°, 30 min	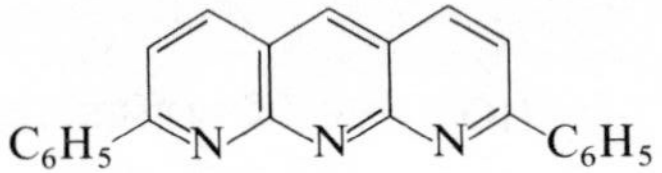 C_6H_5 N N N C_6H_5 (27)	169a
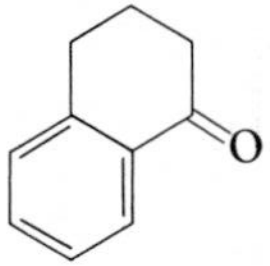	KOH	1. Reflux, 40 hr; 1-propanol 2. Reflux, 40 min; $C_6H_5NO_2$	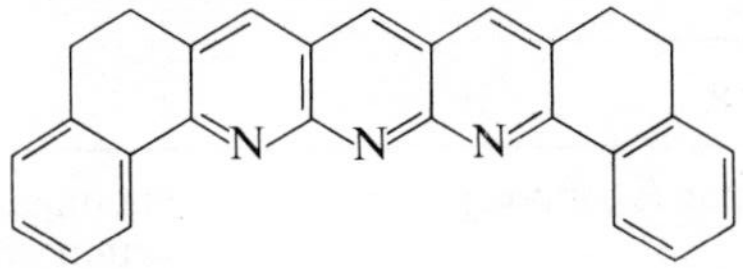 N N N (48)	169a
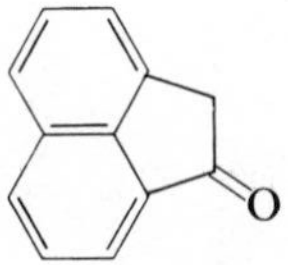	KOH	Reflux, 2 d; 1-propanol	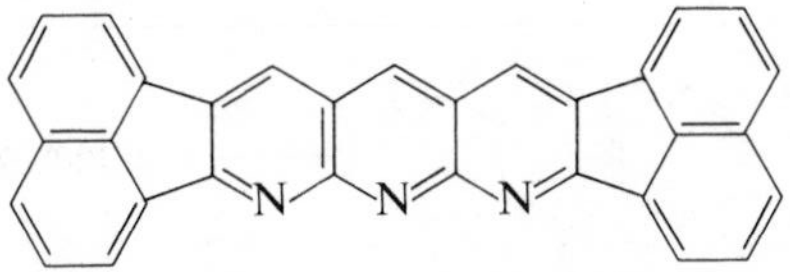 N N N (65)	169a
$C_6H_5CH_2COC_6H_5$	KOH	1. Reflux, 3 d; 1-propanol 2. Reflux, 30 min; $C_6H_5NO_2$	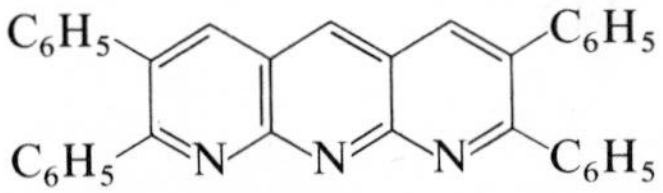 C_6H_5 C_6H_5 C_6H_5 N N N C_6H_5 (50)	169a

I. 3-Aminoquinoline-4-carbaldehydes

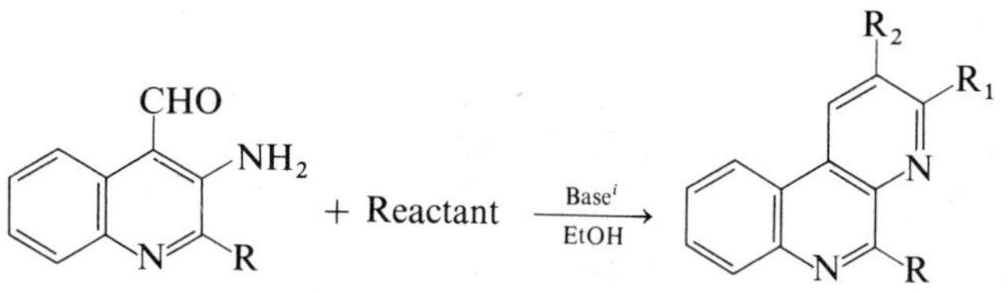

Reactant	Catalyst[i]	Product(s) and Yield(s) (%)				Refs.
		R	R_1	R_1R_2	R_2	
CH_3COCH_3	—	H	CH_3		H (—)	173a
		CH_3	CH_3		H (—)	173a
C_2H_5CHO	—	H	H		CH_3	173a
		CH_3	H		CH_3 (—)	173a
$CH_3COC_2H_5$	—	H	CH_3		CH_3 (—)	173a
		CH_3	CH_3		CH_3 (—)	173a
$CH_3COCH_2COCH_3$	—	H	CH_3		$COCH_3$ (—)	173a
		CH_3	CH_3		$COCH_3$ (—)	173a
$CH_3COCH_2CO_2C_2H_5$	—	H	CH_3		$CO_2C_2H_5$ (—)	173a
		CH_3	CH_3		$CO_2C_2H_5$ (—)	173a
$C_6H_5COCH_3$	—	H	C_6H_5		H (—)	173a
	—	CH_3	C_6H_5		H (—)	173a
2-Acetylpyridine	—	H	2-Pyridyl		H (—)	173a
	—	CH_3	"		H (—)	173a
2-Acetylthiophene	—	H	2-Thienyl		H (—)	173a
	—	CH_3	"		H (—)	173a

TABLE VIII. CONDENSATIONS WITH AZA ANALOGS OF *o*-AMINOBENZALDEHYDE (*Continued*)

I. 3-Aminoquinoline-4-carbaldehydes (*Continued*)

3-Aminoquinoline-4-carbaldehyde (CHO, NH_2, R) + Reactant $\xrightarrow[\text{EtOH}]{\text{Base}^{i}}$ product (R, R_1, R_2)

Reactant	Catalyst[i]	R	R_1	R_1R_2	R_2	Refs.
Cyclopentanone	—	H		$(CH_2)_3$ (—)		173a
	—	CH_3		$(CH_2)_3$ (—)		173a
Cyclohexanone	—	H		$(CH_2)_4$ (—)		173a
	—	CH_3		$(CH_2)_4$ (—)		173a
$CH_2(CN)_2$	—	H	NH_2		CN (—)	173a
	—	CH_3	NH_2		CN (—)	173a
$NCCH_2CONH_2$	—	H	NH_2		$CONH_2$ (—)	173a
	—	CH_3	NH_2		$CONH_2$ (—)	173a
$NCCH_2CO_2C_2H_5$	Pyridine	H	OH		CN (—)	173a
	KOH	H	NH_2		$CO_2C_2H_5$ (—)	173a
	Pyridine	CH_3	OH		CN (—)	173a
	KOH	CH_3	NH_2		$CO_2C_2H_5$ (—)	173a

J. 4-Aminoquinoline-3-carbaldehydes

$$\text{4-amino-2-R-quinoline-3-carbaldehyde} + \text{Reactant} \xrightarrow[\text{EtOH}]{\text{Base}^{i}} \text{product (R, } R_1\text{, } R_2\text{)}$$

Reactant	Catalyst[i]	R	R_1	R_1R_2	R_2	Refs.
CH_3COCH_3	—	H	CH_3		H (—)	173a
	—	CH_3	CH_3		H (—)	173a
C_2H_5CHO	—	H	H		CH_3 (—)	173a
	—	CH_3	H		CH_3 (—)	173a
$CH_3COC_2H_5$	—	H	CH_3		CH_3 (—)	173a
	—	CH_3	CH_3		CH_3 (—)	173a
$CH_3COCH_2COCH_3$	—	H	CH_3		$COCH_3$ (—)	173a
	—	CH_3	CH_3		$COCH_3$ (—)	173a
$CH_3COCH_2CO_2C_2H_5$	—	H	CH_3		$CO_2C_2H_5$ (—)	173a
	—	CH_3	CH_3		$CO_2C_2H_5$ (—)	173a
$C_6H_5COCH_3$	—	H	C_6H_5		H (—)	173a
	—	CH_3	C_6H_5		H (—)	173a
2-Acetylpyridine	—	H	2-Pyridyl		H (—)	173a
		CH_3	"		H (—)	173a
2-Acetylthiophene	—	H	2-Thienyl		H (—)	173a
		CH_3	"		H (—)	173a

TABLE VIII. CONDENSATIONS WITH AZA ANALOGS OF *o*-AMINOBENZALDEHYDE (*Continued*)

J. 4-Aminoquinoline-3-carbaldehydes (*Continued*)

$+$ Reactant $\xrightarrow[\text{EtOH}]{\text{Base}^{i}}$

Reactant	Catalyst[i]	R	R_1	R_1R_2	R_2	Refs.
Cyclopentanone	—	H		$(CH_2)_3$	(—)	173a
		CH_3		"	(—)	173a
Cyclohexanone	—	H		$(CH_2)_4$	(—)	173a
	—	CH_3		"	(—)	173a
$CH_2(CN)_2$	—	H	NH_2		CN (—)	173a
	—	CH_3	NH_2		CN (—)	173a
$NCCH_2CONH_2$	—	H	NH_2		$CONH_2$ (—)	173a
	—	CH_3	NH_2		$CONH_2$ (—)	173a
$NCCH_2CO_2C_2H_5$	Piperidine or KOH	H	NH_2		$CO_2C_2H_5$ (—)	173a
	"	CH_3	NH_2		$CO_2C_2H_5$ (—)	173a
		CH_3	OH		CN (—)	173a

K. 3,5-Diaza

Reactant	Catalyst	Reaction Conditions	Product(s) and Yield(s) (%) R_1	R_2	Refs.
2-Acetylpyridine	KOH	Reflux, 12–48 hr; EtOH	2-Pyridyl	H (82)	172
$C_6H_5COCH_3$	"	"	C_6H_5	H (84)	172
$C_6H_5COC_2H_5$	"	"	C_6H_5	CH_3 (75)	172
$CH_3COCH_2C_6H_5$	"	"	CH_3	C_6H_5 (80)	172
2-Acetylnaphthalene	"	"	2-Naphthyl	H (82)	172
$C_6H_5COCH_2C_6H_5$	"	"	C_6H_5	C_6H_5 (75)	172
	—	1. 75°, 6 hr; EtOH 2. Reflux, 15 min 3. Heat, 170°/1 mm	(20)		44
	—	1. 75°, 6 hr; EtOH 2. Reflux, 15 min 3. Reflux, 45 min, 2 *N* HCl	(88)		44
	KOH	Reflux, 12 hr; EtOH	(85)		171
	KOH	Reflux, 12 hr; EtOH	(95)		171

TABLE VIII. CONDENSATIONS WITH AZA ANALOGS OF *o*-AMINOBENZALDEHYDE (*Continued*)

L. 3,6-Diaza (3-aminopyrazine-2-carboxaldehyde)

Reactant	Catalyst	Reaction Conditions	Product(s) and Yield(s) (%)	Refs.
$CH_2(CN)_2$	N-Methylpiperidine	20°, MeOH	6-amino-pyrido[2,3-b]pyrazine-7-carbonitrile (CN, NH_2) (47)	173b
$NCCH_2CONH_2$	"	1. 20°; MeOH 2. NaOH	6-amino-pyrido[2,3-b]pyrazine-7-carboxamide ($CONH_2$, NH_2) (63)	173b

[a] The catalyst and reaction conditions were not given; consult Refs. 18, 52, and 53.

[b] All reactions were run at reflux temperature of indicated solvent for the indicated time.

[c] The *p*-toluidine azomethine derivative of 3-amino-4-pyridinecarboxaldehyde was used in the reaction.

[d] The reaction was run at 100° with no solvent.

[e] The reaction was run in EtOH at room temperature.

[f] The ratio of diketone to aminoaldehyde was 1:2.3.

[g] The aldehyde group was formed *in situ* from the corresponding benzenesulfonylhydrazide group by McFayden-Stevens reaction prior to the Friedländer condensation.

[h] 1,4-Diazabicyclo[2.2.2]octane was added to the reaction mixture.

[i] Piperidine, KOH, or NaOEt was used; the specific catalyst for each reaction was not identified unless indicated.

TABLE IX. CONDENSATIONS WITH OTHER *o*-AMINO AROMATIC AND *o*-AMINO HETEROCYCLIC CARBONYL COMPOUNDS AND THEIR DERIVATIVES

A. 2-amino-1-formylazulene (CHO, NH_2) + Reactant $\xrightarrow[\text{Reflux; EtOH}]{\text{Piperidine}}$ azuleno[2,1-b]pyridine (R_1, R_2)

Reactant	Time (hr)	Product(s) and Yield(s) (%) R_1	R_2	Refs.
$CH_2(CN)_2$	1	NH_2	CN (100)	40,41
$NCCH_2CO_2C_2H_5$	2	OH	CN (18)	40,41
$CH_3COCH_2COCH_3$[a]	6	CH_3	$COCH_3$ (50)	40,41
$NCCH_2CO_2C_2H_5$[b]	2	{OH	CN (25)}	40,41
		{NH_2	$CO_2C_2H_5$ (47)}	
$CO(CH_2CO_2C_2H_5)_2$[a]	7	$CH_2CO_2C_2H_5$	$CO_2C_2H_5$ (39)	40,41
$CH_3COCH_2CO_2C_2H_5$[a]	7	CH_3	$CO_2C_2H_5$ (35)	40,41
"[b]	5	OH	$COCH_3$ (25)	40,41
$CH_2(CO_2C_2H_5)_2$[a–c]	30	OH	$CO_2C_2H_5$ (50)	40,41

TABLE IX. CONDENSATIONS WITH OTHER *o*-AMINO AROMATIC AND *o*-AMINO HETEROCYCLIC CARBONYL COMPOUNDS AND THEIR DERIVATIVES (*Continued*)

B. 2-CHO-3-NH_2-(X-heterocycle) + Reactant $\xrightarrow[\text{EtOH–H}_2\text{O}]{\text{NaOH, 55–60°}}$ X-fused pyridine (R_1, R_2)

X	Reactant	Time (hr)	Product(s) and Yield(s) (%)		Refs.
			R_1	R_2	
O	CH_3COCH_3	16	CH_3	H (71)	174
O	CH_3COCO_2H	3	CO_2H	H (43)	174
S	CH_3COCH_3	16	CH_3	H (81)	174
S	CH_3COCO_2H	3	CO_2H	H (48)	174
S	$CH_3COCH_2COCH_3$[d]	18	CH_3	$COCH_3$ (67)	174
Se	CH_3COCH_3	16	CH_3	H (87)	174
Se	CH_3COCO_2H	3	CO_2H	H (42)	174

C.

2-Amino-3-(COR_3)-4-(R_1)-5-(R_2)thiophene + $H_2C{-}R_4$ / $O{=}C{-}R_5$ $\xrightarrow[\text{AcOH, } H_2SO_4{}^{e}]{\text{Reflux, 4–10 hr}}$ thieno[2,3-b]pyridine (R_1, R_2, R_3, R_4, R_5)

R_1	R_1R_2	R_2	R_3	R_4	R_4R_5	R_5	Yield (%)	Refs.
CH_3		H	C_6H_5	CH_3		CH_3	(45)	175
"		"	"		$-(CH_2)_3-$		(55)	175
"		"	"		$-(CH_2)_4-$		(70)	175
"		"	"		$-CH_2N(CH_3)(CH_2)_2-$		(75)	175
	$-(CH_2)_4-$		"	CH_3		CH_3	(50)	175
	"		"		$-(CH_2)_3-$		(70)	175
	"		"		$-(CH_2)_4-$		(80)	175
	"		"		$-CH_2N(CH_3)(CH_2)_2-$		(82)	175
	"		p-$CH_3OC_6H_4$		$-(CH_2)_4-$		(75)	175
	"		m-ClC_6H_4		"		(51)	175
	"		p-ClC_6H_4		"		(63)	175
	"		o-$O_2NC_6H_4$		"		(60)	175

TABLE IX. CONDENSATIONS WITH OTHER *o*-AMINO AROMATIC AND *o*-AMINO HETEROCYCLIC CARBONYL COMPOUNDS AND THEIR DERIVATIVES (*Continued*)

D. 2-C_6H_5-5-(COC_6H_5)-4-($NH_2 \cdot HClO_4$)thiazole + Reactant $\xrightarrow[15\ min]{130-150°}$ 2-C_6H_5-7-C_6H_5-6-R_2-5-R_1-thiazolo[4,5-b]pyridine

Reactant	Product(s) and Yield(s) (%)			Refs.
	R_1	R_1R_2	R_2	
$CH_3COCH_2C_6H_5$	CH_3		C_6H_5 (54)	43
$CH_3CH_2COC_6H_5$	C_6H_5		CH_3 (76)	43
$C_6H_5CH_2COC_6H_5$	C_6H_5		C_6H_5 (46)	43
Cyclopentanone		$(CH_2)_3$ (47)		43
Cyclohexanone		$(CH_2)_4$ (64)		43
1,2-Cyclohexanedione[f]		$CH_2CH_2CH_2C(=O)$ (32)		43
Cycloheptanone		$(CH_2)_5$ (49)		43
1-Tetralone		CH_2CH_2-o-C_6H_4 (48)		43
4-Chromanone		CH_2O-o-C_6H_4 (39)		43

E. (5-amino-1-R-pyrazole-4-carboxaldehyde) + Reactant ⟶ (1-R-6-R_1-5-R_2-pyrazolo[3,4-b]pyridine)

R	Reactant	Catalyst	Reaction Conditions	Product(s) and Yield(s) (%) R_1	R_1R_2	R_2	Refs.
CH_3	CH_3COCH_3	NaOEt	Reflux, 1 hr; EtOH	CH_3		H (19)	176
		H_2SO_4	Reflux, 1 d; H_2O	"		" (Trace)	176
	$CH_2(CN)_2$	"	Reflux, 1 d; EtOH	NH_2		CN (4)	176
	$CH_3COCH_2COCH_3$	"	"	CH_3		$COCH_3$ (7)	176
	Cyclopentanone	"	Reflux, 1 d; H_2O		$(CH_2)_3$	(18)	176
		NaOEt	Reflux, 1 hr; EtOH		"	(44)	176
	$CH_3COCH_2CO_2C_2H_5$	H_2SO_4	Reflux, 1 d; EtOH	CH_3		$CO_2C_2H_5$ (7)	176
	Cyclohexanone	"	Reflux, 1 d; H_2O		$(CH_2)_4$	(14)	176
	$C_6H_5COCH_3$	"	"	C_6H_5		H (5)	176
		NaOEt	Reflux, 1 hr; EtOH	"		" (69)	176
	$C_6H_5COCH_2CO_2C_2H_5$	"	"	C_6H_5		$CO_2C_2H_5$ (23)	176
		H_2SO_4	Reflux, 1 d; EtOH	"		" (8)	176
C_6H_5	CH_3COCH_3	H_2SO_4	Reflux, 1 d; H_2O	CH_3		H (15)	176
		NaOEt	Reflux, 1 hr; EtOH	"		" (11)	176
	$CH_2(CN)_2$	H_2SO_4	Reflux, 1 d; EtOH	NH_2		CN (41)	176
	$CH_3COCH_2COCH_3$	"	"	CH_3		$COCH_3$ (3)	176
	$NCCH_2CO_2C_2H_5$	NaOEt	Reflux, 1 hr; EtOH	OH		CN (69)	176
	Cyclopentanone	"	"		$(CH_2)_3$	(57)	176
		H_2SO_4	Reflux, 1 d; EtOH		"	(23)	176
	$CH_3COCH_2CO_2C_2H_5$	"	Reflux, 1 d; EtOH	CH_3		$CO_2C_2H_5$ (21)	176
	Cyclohexanone	"	Reflux, 1 d; H_2O		$(CH_2)_4$	(52)	176
	$C_6H_5COCH_3$	"	"	C_6H_5		H (7)	176
		NaOEt	Reflux, 1hr; EtOH	"		" (69)	176
	$C_6H_5COCH_2CO_2C_2H_5$	"	"	"		$CO_2C_2H_5$ (41)	176
		H_2SO_4	Reflux, 1 d; EtOH	"		" (5)	176

TABLE IX. CONDENSATIONS WITH OTHER *o*-AMINO AROMATIC AND *o*-AMINO HETEROCYCLIC CARBONYL COMPOUNDS AND THEIR DERIVATIVES (*Continued*)

F. Polycyclic Aminoaldehydes

Aminoaldehyde	Reactant	Catalyst	Reaction Conditions	Product(s) and Yield(s) (%)	Refs.
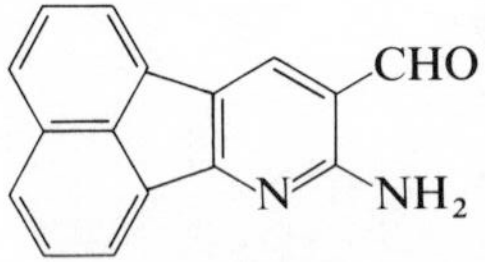 CHO, N, NH_2	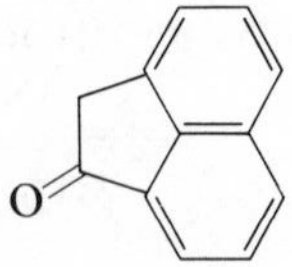 O	KOH	Reflux, 12 hr; MeOH	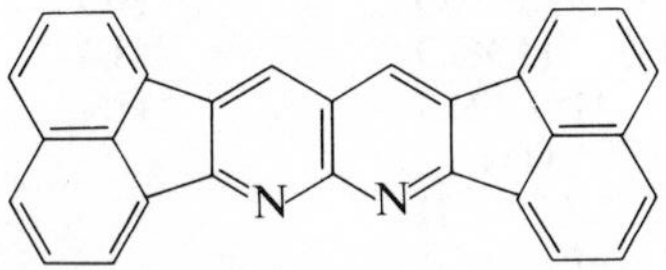 N, N (100)	171
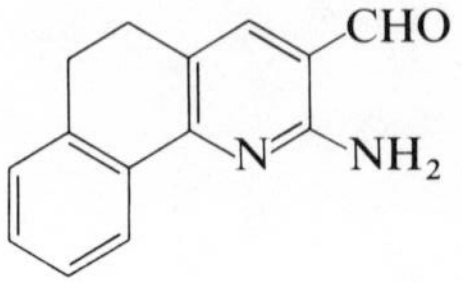 CHO, N, NH_2	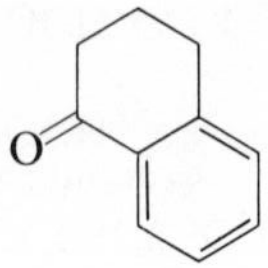 O	"	"	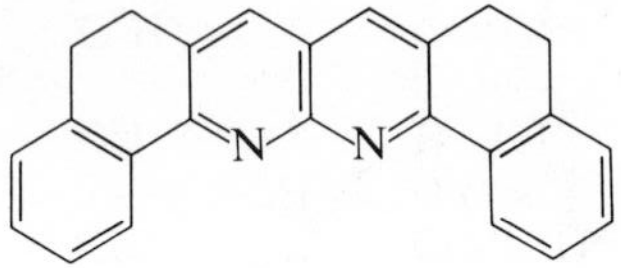 N, N (90)	171
	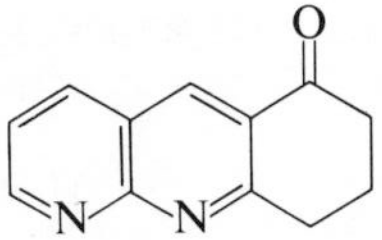 O, N, N	"	Reflux, 48 hr; EtOH	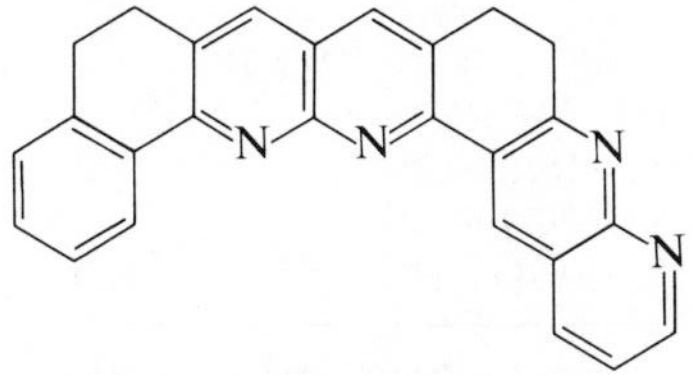 N, N, N, N (85)	171

[a] No solvent was used.
[b] The acetyl derivative of the aminoaldehyde was used.
[c] The reaction was run at 50–60°.
[d] The reaction was run in EtOH with piperidine catalyst.
[e] Similar results were obtained with no solvent and HCl catalyst at 100° for 10 min, then 200° for 1–2 hr.
[f] The reaction was run at 160° for 8 hr.

TABLE X. CONDENSATIONS WITH 3-AMINOACROLEINS[a]

$$RC(CHO)=CHNH_2 + \text{Reactant} \longrightarrow \text{pyridine (R at 5, }R_2\text{ at 3, }R_1\text{ at 2)}$$

	Reactant	Product(s) and Yield(s) (%)			Refs.
		R_1	R_1R_2	R_2	
R = H					
	$CH_3COCH_2COCH_3$[b]	CH_3		$COCH_3$ (55)	45,45a
	$CH_3COCH_2COC_2H_5$	CH_3 or C_2H_5	or	COC_2H_5 or $COCH_3$ (75)	45a
	$CH_3COCH_2CO_2C_2H_5$[c]	CH_3		$CO_2C_2H_5$ (50)	45,45a
	Cyclohexanone[b]		$(CH_2)_4$	(20)	45
	$C_2H_5COCH_2COC_2H_5$	C_2H_5		COC_2H_5 (75)	45a
	$C_2H_5COCH_2CO_2C_2H_5$	C_2H_5		$CO_2C_2H_5$ (70)	45a
	n-$C_3H_7COCH_2CO_2C_2H_5$	C_3H_7-n		$CO_2C_2H_5$ (60)	45a
	i-$C_3H_7COCH_2CO_2C_2H_5$	C_3H_7-i		$CO_2C_2H_5$ (60)	45a
	n-$C_3H_7COCH_2COC_3H_7$-n	C_3H_7-n		COC_3H_7-n (70)	45a
R = CH_3	$CH_3COCH_2COCH_3$	CH_3		$COCH_3$ (80)	45a
	$CH_3COCH_2COC_2H_5$	CH_3 or C_2H_5		COC_2H_5 or $COCH_3$ (70)	45a
	$CH_3COCH_2CO_2C_2H_5$	CH_3		$CO_2C_2H_5$ (60)	45a
	$C_2H_5COCH_2COC_2H_5$	C_2H_5		COC_2H_5 (80)	45a
	$C_2H_5COCH_2CO_2C_2H_5$	C_2H_5		$CO_2C_2H_5$ (65)	45a
	n-$C_3H_7COCH_2CO_2C_2H_5$	C_3H_7-n		$CO_2C_2H_5$ (55)	45a
	i-$C_3H_7COCH_2CO_2C_2H_5$	C_3H_7-i		$CO_2C_2H_5$ (50)	45a
	n-$C_3H_7COCH_2COC_3H_7$-n	C_3H_7-n		COC_3H_7-n (60)	45a

[a] All reactions run with $NH_4OAc + Et_3N$ catalyst at 110° for 12 hr except as noted.
[b] The catalyst was NH_4OAc only.
[c] The same result was obtained with and without Et_3N.

REFERENCES

[1] P. Friedländer, *Ber.*, **15**, 2572 (1882).

[2] O. Fischer and C. Rudolph, *ibid.*, **15**, 1500 (1882).

[3] E. Besthorn and O. Fischer, *ibid.*, **16**, 68 (1883).

[4] W. Pfitzinger, *J. Prakt. Chem.*, [2] **33**, 100 (1886).

[5] S. von Niementowski, *Ber.*, **27**, 1394 (1894).

[6] A. Combes, *Bull. Soc. Chim. Fr.*, **49**, 89 (1888).

[7] O. Doebner and W. von Miller, *Ber.*, **16**, 2464 (1883).

[8] Z. H. Skraup, *ibid.*, **13**, 2086 (1880).

[9] R. H. F. Manske and M. Kulka, *Org. Reactions*, **7**, 59 (1953).

[10] F. W. Bergstrom, *Chem. Rev.*, **35**, 77 (1944).

[11] R. C. Elderfield, in *Heterocyclic Compounds*, R. C. Elderfield, Ed., Vol. 4, Wiley, New York, 1952, Chap. 1, p. 45.

[12] L. A. Paquette, *Principles of Modern Heterocyclic Chemistry*, Benjamin, New York, 1968, p. 278.

[13] A. R. Katritzky and J. M. Lagowski, in *Heterocyclic Chemistry*, A. R. Todd, Ed., Wiley, New York, 1960, p. 39.

[14] J. A. Joule and G. F. Smith, *Heterocyclic Chemistry*, Van Nostrand-Reinhold, New York, 1972, p. 104.

[15] G. Kempter, D. Heilmann, and M. Mühlstädt, *J. Prakt. Chem.*, **314**, 543 (1972).

[16] E. A. Fehnel, J. A. Deyrup, and M. B. Davidson, *J. Org. Chem.*, **23**, 1996 (1958).

[17] T. G. Majewicz and P. Caluwe, *ibid.*, **40**, 3407 (1975).

[18] E. M. Hawes and D. K. J. Gorecki, *J. Heterocycl. Chem.*, **11**, 151 (1974).

[19] E. A. Fehnel, *ibid.*, **4**, 565 (1967).

[20] E. A. Fehnel, *J. Org. Chem.*, **31**, 2899 (1966).

[21] O. Stark, *Ber.*, **40**, 3425 (1907).

[22] R. Geigy and W. Koenigs, *ibid.*, **18**, 2400 (1885).

[23] W. Borsche and F. Sinn, *Justus Liebigs Ann. Chem.*, **538**, 283 (1939).

[24] G. Kempter and W. Stoss, *J. Prakt. Chem.*, [4] **21**, 198 (1963).

[25] W. Koenigs and E. Bischkopf, *Ber.*, **34**, 4327 (1901).

[26] W. Wislicenus and H. Elvert, *ibid.*, **42**, 1144 (1909).

[27] J. v. Braun, A. Petzold, and J. Seemann, *ibid.*, **55B**, 3779 (1922).

[28] G. R. Clemo and G. A. Swan, *J. Chem. Soc.*, **1945**, 867.

[29] G. R. Clemo and D. G. I. Felton, *ibid.*, **1952**, 1658.

[30a] A. Dornow and W. Sassenberg, *Justus Liebigs Ann. Chem.*, **602**, 14 (1957).

[30b] G. Kempter, D. Rehbaum, and J. Schirmer, *J. Prakt. Chem.*, **319**, 573 (1977).

[31] G. Kempter, P. Andratschke, D. Heilmann, H. Krausmann, and M. Mietasch, *Chem. Ber.*, **97**, 16 (1964).

[32] G. Kempter and S. Hirschberg, *ibid.*, **98**, 419 (1965).

[33] K. Takagi and T. Ueda, *Chem. Pharm. Bull.*, **20**, 380 (1972).

[34] P. Jacquignon, A. Croisy, A. Ricci, and D. Balucani, *Collect. Czech. Chem. Commun.*, **38**, 3862 (1973).

[35] G. Kempter, M. Mühlstädt, and D. Heilmann, *Angew. Chem., Int. Ed. Engl.*, **5**, 248 (1966).

[36] E. A. Fehnel and D. E. Cohn, *J. Org. Chem.*, **31**, 3852 (1966).

[37] D. L. Coffen and F. Wong, *ibid.*, **39**, 1765 (1974).

[37a] J. H. Markgraf and W. L. Scott, *Chem. Commun.*, **1967**, 296.

[38] P. Ruggli and F. Brandt, *Helv. Chim. Acta*, **27**, 274 (1944).

[39] W. Bracke, *Macromolecules*, **2**, 286 (1969).

[39a] Y. Imai, E. F. Johnson, T. Katto, M. Kurihara, and J. K. Stille, *J. Polym. Sci., Polym. Chem. Ed.*, **13**, 2233 (1975).

[39b] S. O. Norris and J. K. Stille, *Macromolecules*, **9**, 496 (1976).

[39c] J. K. Wolfe and J. K. Stille, *ibid.*, **9**, 489 (1976).

[39d] J. Garapon and J. K. Stille, *ibid.*, **10**, 627 (1977).

[39e] R. M. Mortimer, P. K. Dutt, J. Hoefnagels, and C. S. Marvel, *J. Polym. Sci., A-1*, **9**, 3337 (1971).

[39f] R. Kellman and C. S. Marvel, *J. Polym. Sci., Polym. Chem. Ed.*, **13**, 2125 (1975).

[40] T. Nozoe and K. Kikuchi, *Chem. Ind. (London)*, **1962**, 358.

[41] T. Nozoe and K. Kikuchi, *Bull. Chem. Soc. Jpn.*, **36**, 633 (1963).

[42] E. M. Hawes and D. G. Wibberley, *J. Chem. Soc., C*, **1966**, 315.

[43] V. H. Schäfer, H. Hartmann, and K. Gewald, *J. Prakt. Chem.*, **316**, 19 (1974).
[44] T. G. Majewicz and P. Caluwe, *J. Org. Chem.*, **41**, 1058 (1976).
[45] E. Breitmaier and E. Bayer, *Angew. Chem., Int. Ed. Engl.*, **8**, 765 (1969).
[45a] E. Breitmaier, S. Gassenmann, and E. Bayer, *Tetrahedron*, **26**, 5907 (1970).
[46] W. Borsche and W. Ried, *Justus Liebigs Ann. Chem.*, **554**, 269 (1943).
[47] W. Borsche and J. Barthenheier, *ibid.*, **548**, 50 (1941).
[48] W. Borsche, M. Wagner-Römmich, and J. Barthenheier, *ibid.*, **550**, 160 (1942).
[49] W. Borsche, W. Döller, and M. Wagner-Römmich, *Ber.*, **76B**, 1099 (1943).
[50] W. Borsche, *Justus Liebigs Ann. Chem.*, **532**, 127 (1937).
[51] T. K. Liao, W. H. Nyberg, and C. C. Cheng, *J. Heterocycl. Chem.*, **8**, 373 (1971).
[51a] A. Osbirk and E. B. Pederson, *Acta Chem. Scand.*, **B33**, 313 (1979).
[52] E. M. Hawes, D. K. J. Gorecki, and D. D. Johnson, *J. Med. Chem.*, **16**, 849 (1973).
[53] E. M. Hawes and D. K. J. Gorecki, *J. Heterocycl. Chem.*, **9**, 703 (1972).
[54] J. Tröger and W. Menzel, *J. Prakt. Chem.*, [2] **103**, 188 (1921).
[55] G. Kempter and G. Möbius, *ibid.*, [4] **34**, 298 (1966).
[56] J. Eliasberg and P. Friedländer, *Ber.*, **25**, 1752 (1892).
[57] C. Schöpf and G. Lehmann, *Justus Liebigs Ann. Chem.*, **497**, 7 (1932).
[58] M. Scholz, H. Limmer, and G. Kempter, *Z. Chem.*, **5**, 154 (1965).
[59] S. Danishefsky, T. A. Bryson, and J. Puthenpurayil, *J. Org. Chem.*, **40**, 796 (1975).
[60] O. Schmut and T. Kappe, *Z. Naturforsch.*, **30B**, 140 (1975).
[61] J. A. Kepler, M. C. Wani, J. N. McNaull, M. E. Wall, and S. G. Levine, *J. Org. Chem.*, **34**, 3853 (1969).
[62] H. E. Baumgarten and K. C. Cook, *ibid.*, **22**, 138 (1957).
[63] E. Besthorn and B. Geisselbrecht, *Ber.*, **53B**, 1017 (1920).
[64] S. von Niementowski and E. Sucharda, *ibid.*, **52B**, 484 (1919).
[65] J. Tröger and J. Bohnekamp, *J. Prakt. Chem.*, [2] **117**, 161 (1927).
[66] E. M. Hawes and D. G. Wibberley, *J. Chem. Soc., C*, **1967**, 1564.
[67] E. M. Hawes, D. K. J. Gorecki, and R. G. Gedir, *J. Med. Chem.*, **20**, 838 (1977).
[68] A. P. Smirnoff, *Helv. Chim. Acta*, **4**, 802 (1921).
[69] S. Yamada and I. Chibata, *Chem. Pharm. Bull.*, **3**, 21 (1955).
[70] G. Kempter and P. Klug, *Z. Chem.*, **11**, 61 (1971).
[71] S. von Neimentowski, *Ber.*, **39**, 385 (1906).
[77] O. Stark, *ibid.*, **42**, 715 (1909).
[73] M. J. Haddadin, N. C. Chelhot, and M. Pieridou, *J. Org. Chem.*, **39**, 3278 (1974).
[74a] F. G. Mann and A. J. Wilkinson, *J. Chem. Soc.*, **1957**, 3346.
[74b] J. T. Braunholtz and F. G. Mann, *ibid.*, **1958**, 3368.
[75] J. B. Hendrickson, R. Rees, and J. F. Templeton, *J. Am. Chem. Soc.*, **86**, 107 (1964).
[76] C. S. Marvel and G. S. Hiers, *Org. Syntheses, Coll. Vol. I*, 2nd Ed., 327 (1941).
[77] T. Sandmeyer, *Helv. Chim. Acta*, **2**, 234 (1919).
[78] M. Conrad and L. Limpach, *Ber.*, **20**, 944 (1887); *ibid.*, **24**, 2990 (1891).
[79] A. R. Surrey and H. F. Hammer, *J. Am. Chem. Soc.*, **68**, 113 (1946).
[80] N. J. Leonard, H. F. Herbrandson, and E. M. V. Heyningen, *ibid.*, **68**, 1279 (1946).
[81] L. Knorr, *Justus Liebigs Ann. Chem.*, **236**, 69 (1886); *ibid.*, **245**, 357 (1888).
[82] A. J. Ewins and H. King, *J. Chem. Soc.*, **103**, 104 (1913).
[83] W. M. Lauer and C. E. Kaslow, *Org. Synth., Coll. Vol. III*, 580 (1955).
[84] R. Camps, *Ber.*, **32**, 3228 (1899); *Arch. Pharm.*, **237**, 659 (1899).
[85] R. G. Gould and W. A. Jacobs, *J. Am. Chem. Soc.*, **61**, 2890 (1939).
[86] C. Tang and H. Rapoport, *ibid.*, **94**, 8615 (1972).
[87] C. Tang, C. J. Morrow, and H. Rapoport, *ibid.*, **97**, 159 (1975).
[88] R. Madhav and P. L. Southwick, *J. Heterocycl. Chem.*, **9**, 443 (1972).
[89] L. H. Zalkow, J. B. Nabors, K. French, and S. C. Bisarya, *J. Chem. Soc., C*, **1971**, 3551.
[89a] V. Ehrig, H. S. Bien, E. Klauke, and D. I. Schutze, West German Pat. 2,730,061 (1979) [*C.A.*, **90**, 152027k (1979)].
[90] W. Treibs and G. Kempter, *Chem. Ber.*, **92**, 601 (1959).
[91] E. A. Fehnel and J. E. Stuber, *J. Org. Chem.*, **24**, 1219 (1959).
[92] P. Pfeiffer and G. v. Bank, *J. Prakt. Chem.*, [2] **151**, 312 (1938).
[93] A. Rilliet and L. Kreitmann, *Helv. Chim. Acta*, **4**, 588 (1921).

[94] E. Ziegler, T. Kappe, and H. G. Foraita, *Monatsh. Chem.*, **97**, 409 (1966).
[95] T. A. Geissman and A. C. Waiss, Jr., *J. Org. Chem.*, **27**, 139 (1962).
[96] P. Friedländer and C. F. Göhring, *Ber.*, **16**, 1833 (1883).
[97] P. Friedländer and C. F. Göhring, *ibid.*, **17**, 456 (1884).
[98] J. Tröger and D. Dimitroff, *J. Prakt. Chem.*, [2] **111**, 193 (1925).
[99] W. Koenigs and F. Stockhausen, *Ber.*., **34**, 4330 (1901).
[100] E. v. Meyer, *J. Prakt. Chem.*, [2] **90**, 1 (1914).
[101] M. Conrad and H. Reinbach, *Ber.*, **34**, 1339 (1901).
[102] C. Gränacher, A. Ofner, and A. Klopfenstein, *Helv. Chim. Acta*, **8**, 883 (1925).
[103] H. Scheibler and A. Fischer, *Ber.*, **55B**, 2903 (1922).
[104] I. Guareschi, *Atti. Reale Accad. Sci. Torino, Cl. Sci. Fis. Mat. Naturali*, **28**, 719 (1893); *Chem. Zentralbl.*, **1893** II, 454.
[105] R. F. Borch, C. V. Grudzinskas, D. A. Peterson, and L. D. Weber, *J. Org. Chem.*, **37**, 1141 (1972).
[106] L. Hozer and S. v. Niementowski, *J. Prakt. Chem.*, [2] **116**, 43 (1927).
[107] G. Koller, H. Ruppersberg, and E. Strang, *Monatsh. Chem.*, **52**, 59 (1929).
[108] R. B. Woodward and E. C. Kornfeld, *J. Am. Chem. Soc.*, **70**, 2508 (1948).
[109] D. H. Hey and J. M. Williams, *J. Chem. Soc.*, **1950**, 1678.
[110] G. Bargellini and S. Berlingozzi, *Gazz. Chim. Ital.*, **53**, 3 (1923) [*C.A.*, **17**, 2287 (1923)].
[111] R. Pschorr, *Ber.*, **31**, 1289 (1898).
[112] J. Tröger and P. Köppen-Kastrop, *J. Prakt. Chem.*, [2] **104**, 335 (1922).
[113] G. Koller and E. Strang, *Monatsh. Chem.*, **50**, 48 (1928).
[114] J. Tröger and H. Meinecke, *J. Prakt. Chem.*, [2] **106**, 203 (1923).
[115] N. P. Buu-Hoï, A. Croisy, P. Jacquignon, and A. Martani, *J. Chem. Soc., C*, **1971**, 1109.
[116] T. Kametani, H. Nemoto, H. Takeda, and S. Takano, *Tetrahedron*, **26**, 5753 (1970); *Chem. Ind. (London)*, **1970**, 1323.
[117] H. Junek and B. Wolny, *Monatsh. Chem.*, **107**, 999 (1976).
[117a] A. D. Settimo, G. Primofiore, O. Livi, P. L. Ferrarini, and S. Spinelli, *J. Heterocycl. Chem.*, **16**, 169 (1979).
[118] B. Bobrański and E. Sucharda, *Rocz. Chem.*, **7**, 192 (1927) [*C.A.*, **24**, 1381 (1930)].
[119] J. Tröger and A. Ungar, *J. Prakt. Chem.*, [2] **112**, 243 (1926).
[120] G. A. Bistochi, G. D. Meo, A. Ricci, A. Croisy, and P. Jacquignon, *Heterocycles*, **9**, 247 (1978).
[121] N. P. Buu-Hoï, A. Croisy, P. Jacquignon, D.-P. Hien, A. Martani, and A. Ricci, *J. Chem. Soc., Perkin Trans. I*, **1972**, 1266.
[121a] J. Mispelter, A. Croisy, P. Jacquignon, A. Ricci, C. Rossi, and F. Schiaffela, *Tetrahedron*, **33**, 2383 (1977).
[122] T. Kametani, H. Takeda, F. Satoh, and S. Takano, *J. Heterocycl. Chem.*, **10**, 77 (1973).
[123] A. Meyer, *C.R.*, **186**, 1214 (1928).
[123a] K. Tabaković, I. Tabaković, M. Trkovnik, A. Jurić, and N. Trinajstić, *J. Heterocycl. Chem.*, **17**, 801 (1980).
[124] W. Borsche and F. Sinn, *Justus Liebigs Ann. Chem.*, **532**, 146 (1937).
[125] J. Tröger and C. Brohm, *J. Prakt. Chem.*, [2] **111**, 176 (1925).
[126] G. Stork and A. G. Schultz, *J. Am. Chem. Soc.*, **93**, 4074 (1971).
[127] A. Musierowicz, S. v. Niementowski, and Z. Tomasik, *Rocz, Chem.*, **8**, 325 (1928); *Chem. Zentralbl.*, **1928 II**, 1882.
[128] V. Gupta and I. M. Hora, *Curr. Sci.*, **45**, 756 (1976) [*C.A.*, **86**, 106333b (1977)].
[129] A. F. Bekhli and N. P. Kozyreva, *Khim. Geterotsikl. Soedin.*, **1970**, 802 [*C.A.*, **73**, 109647z (1970)].
[130] G. Koller and H. Ruppersberg, *Monatsh. Chem.*, **58**, 238 (1931).
[131] J. Quick, *Tetrahedron Lett.*, **1977**, 327.
[132] M. Shamma and L. Novak, *Tetrahedron*, **25**, 2275 (1969).
[133] M. Shamma, D. A. Smithers, and V. S. Georgiev, *ibid.*, **29**, 1949 (1973).
[134] R. Volkmann, S. Danishefsky, J. Eggler, and D. M. Solomon, *J. Am. Chem. Soc.*, **93**, 5576 (1971).
[135] E. Wenkert, K. G. Dave, R. G. Lewis, and P. W. Sprague, *ibid.*, **89**, 6741 (1967).
[136] J. Tröger and K. v. Seelen, *J, Prakt. Chem.*, [2] **105**, 208 (1923).
[137] T. Kametani, S. Takano, H. Terasawa, and H. Takeda, *Yakugaku Zasshi*, **92**, 868 (1972) [*C.A.*, **77**, 126462k (1972)].
[138] J. Tröger and G. Pahle, *J. Prakt. Chem.*, [2] **112**, 221 (1926).

[139] J. Tröger and C. Cohaus, *ibid.*, [2] **117**, 97 (1927).
[140] J. Tröger and S. Gerö, *ibid.*, [2] **113**, 293 (1926).
[140a] M. S. Khan and M. P. LaMontagne, *J. Med. Chem.*, **22**, 1005 (1979).
[141] W. Borsche and W. Ried, *Ber.*, **76B**, 1011 (1943).
[141a] K. V. Rao, *J. Heterocycl. Chem.*, **14**, 653 (1977).
[141b] K. V. Rao and H.-S. Kuo, *ibid.*, **16**, 1241 (1979).
[142a] J. W. Armit and R. Robinson, *J. Chem. Soc.*, **121**, 827 (1922).
[142b] J. W. Armit and R. Robinson, *ibid.*, **127**, 1604 (1925).
[143] S. Berlingozzi, *Gazz. Chim. Ital.*, **53**, 369 (1923) [*C.A.*, **18**, 268 (1924)].
[144] P. Ruggli and P. Hindermann, *Helv. Chim. Acta*, **20**, 272 (1937).
[145] P. Ruggli, P. Hindermann, and H. Frey, *ibid.*, **21**, 1066 (1938).
[146] S. Hibino and S. M. Weinreb, *J. Org. Chem.*, **42**, 232 (1977).
[147] A. S. Kende and P. C. Naegely, *Tetrahedron Lett.*, **1978**, 4775.
[148] G. Kempter, P. Zänker, and H. D. Zürner, *Arch. Pharm.*, **300**, 829 (1967).
[149] G. Kempter, P. Andratschke, D. Heilmann, H. Krausmann, and M. Mietasch, *Z. Chem.*, **3**, 305 (1963).
[150] G. Kempter and W. Stoss, *ibid.*, **3**, 61 (1963).
[151] G. Kempter, P. Thomas, and E. Uhlemann, East German Pat. 41936 (1965) [*C.A.*, **64**, 15856h (1966)].
[152] G. Kempter and G. Sarodnick, *Z. Chem.*, **8**, 179 (1968).
[153] G. Kempter, H. Schäfer, and G. Sarodnick, *ibid.*, **9**, 186 (1969).
[154] D. Kreysig, H. H. Stroh, and G. Kempter, *ibid.*, **9**, 187 (1969).
[155] D. Kreysig, G. Kempter, and H. H. Stroh, *ibid.*, **9**, 230 (1969).
[156] O. Fischer, *Ber.*, **19**, 1036 (1886).
[157] A. Rilliet, *Helv. Chim. Acta*, **5**, 547 (1922).
[158] A. Bischler and E. Burkart, *Ber.*, **26**, 1349 (1893).
[159] A. Decormeille, G. Queguiner, and P. Pastour, *C.R. Hebd. Seances, C*, **280**, 381 (1975) [*C.A.*, **82**, 156146f (1975)].
[160] R. Camps, *Arch. Pharm.*, **240**, 135 (1902); *Chem. Zentralbl.*, **1902 I**, 818.
[161] G. Kempter, D. Rehbaum, and J. Schirmer, *J. Prakt. Chem.*, **319**, 589 (1977).
[162] G. Kempter, D. Rehbaum, J. Schirmer, and B. Jokuff, *ibid.*, **319**, 581 (1977).
[163] G. Kempter and D. Rehbaum, East German Pat. 85075 (1971) [*C.A.*, **78**, 58382c (1973)].
[163a] F. Eiden and M. Dürr, *Arch. Pharm.*, **312**, 708 (1979).
[164] A. Walser, T. Flynn, and R. I. Fryer, *J. Heterocycl. Chem.*, **12**, 737 (1975).
[165] G. Sarodnick and G. Kempter, *Z. Chem.*, **18**, 400 (1978).
[166] A. Decormeille, F. Guignant, G. Queguiner, and P. Pastour, *J. Heterocycl. Chem.*, **13**, 387 (1976).
[167] H. E. Baumgarten and A. L. Krieger, *J. Am. Chem. Soc.*, **77**, 2438 (1955).
[167a] T. Higashino, K. Suzuki, and E. Hayashi, *Chem. Pharm. Bull.*, **23**, 2939 (1975).
[168] D. K. J. Gorecki and E. M. Hawes, *J. Med. Chem.*, **20**, 124 (1977).
[169] R. P. Thummel and D. K. Kohli, *J. Heterocycl. Chem.*, **14**, 685 (1977).
[169a] P. Caluwe and T. G. Majewicz, *J. Org. Chem.*, **42**, 3410 (1977).
[170] J. H. Markgraf, R. J. Katt, W. L. Scott, and R. N. Shefrin, *ibid.*, **34**, 4131 (1969).
[171] P. Caluwe and T. G. Majewicz, *ibid.*, **40**, 2566 (1975).
[172] G. Evens and P. Caluwe, *ibid.*, **40**, 1438 (1975).
[173] E. Eichler, C. S. Rooney, and H. W. R. Williams, *J. Heterocycl. Chem.*, **13**, 43 (1976).
[173a] A. Godard, D. Brunet, G. Queguiner, and P. Pastour, *C.R. Hebd. Seances, C*, **284**, 459 (1977) [*C.A.*, **87**, 68198x (1977)].
[173b] A. Albert and H. Mizuno, *J. Chem. Soc., Perkin Trans. I*, **1973**, 1615.
[174] S. Gronowitz, C. Westerlund, and A.-B. Hörnfeldt, *Acta Chem. Scand.*, **B29**, 233 (1975).
[175] C. Corral, R. Madroñero, and N. Ulecia, *Afinidad*, **35**, 129 (1978) [*C.A.*, **89**, 129477c (1978)].
[176] T. Higashino, Y. Iwai, and E. Hayashi, *Chem. Pharm. Bull.*, **25**, 535 (1977).

CHAPTER 3

THE DIRECTED ALDOL REACTION

TERUAKI MUKAIYAMA

The University of Tokyo, Bunkyo-ku, Tokyo, Japan

CONTENTS

ACKNOWLEDGMENTS

The author wishes to thank Drs. Toshio Sato and Tsuneo Imamoto for their help with the literature search and preparation of this manuscript. It is also a privilege to express my thanks to Miss Eriko Yamaguchi for typing the manuscript.

INTRODUCTION

The aldol reaction, usually carried out in protic solvents with base or acid as the catalyst, is one of the most versatile methods in organic synthesis.[1,2] By application of this reaction a great number of aldols and related compounds have been prepared from various carbonyl compounds. However, because of difficulty in directing the coupling, the conventional method has serious synthetic limitations. This is particularly notable when two different carbonyl compounds are used in a cross-coupling; the reaction is often accompanied by undesirable side reactions such as self-condensation and polycondensation. The synthetic limitation arises because the reaction is reversible and cannot be driven to completion if the aldol is less stable than the parent carbonyl compounds. In addition, the reverse reaction, in the presence of acid or base, generates regioisomeric enols or enolates, which in turn attack the carbonyl compounds to yield a mixture of aldols. Furthermore, the aldols are often dehydrated and the resulting unsaturated carbonyl compounds may undergo a Michael addition between enolate anions to give a complex reaction mixture.

During the last decade new methods have been developed for the directed coupling of two different carbonyl compounds (or carbonyl equivalents) to give specific carbon–carbon bond formation between the α-carbon atom of one carbonyl compound and the other carbonyl component to produce a desired crossed aldol.[3–13] These methods provide regiospecific reactions for forming carbon–carbon bonds and allow the synthesis of a wide variety of aldols by directed self- or cross-coupling. The general principle of the directed aldol reaction is shown in Eq. 1.

$$R^1R^2CHC(=X)R^3 \longrightarrow R^1R^2C{=}C(XM)R^3 \xrightarrow{R^4R^5CO}$$

X = O, NR
M = Li, Mg, Zn, B, Al, Si, C

$$\mathbf{1}\ (\text{six-membered chelate: } O\text{–}M\text{–}X;\ R^4, R^5, R^1, R^2, R^3) \xrightarrow{H_2O} \text{aldol } (O\text{–}H\cdots O;\ R^4, R^5, R^1, R^2, R^3) \qquad \text{(Eq. 1)}$$

The success of the reaction depends on effective interception of the aldol-type adduct **1** by formation of a stable six-membered chelate. This interception can usually be achieved by reaction of a carbonyl compound with a suitably reactive metal enolate or enol ether that is derived regioselectively from the other carbonyl compound. Reaction of a carbonyl compound with lithium enolates or preformed lithio derivatives of imines is a typical method used for preparation of crossed aldols, even though the reaction is carried out under strongly basic conditions. Use of magnesium, aluminum, or zinc enolates permits rather milder reaction conditions. A clean aldol reaction has been achieved using vinyloxyboranes in a process that is carried out under neutral conditions to produce various crossed aldols in excellent yields. The titanium tetrachloride–promoted coupling of silyl enol ethers, enol ethers, or enol esters with carbonyl compounds, acetals, or ketals is another particularly useful method for the preparation of aldols. The powerful electrophilic activation of the carbonyl acceptor by titanium tetrachloride provides the driving force for this reaction. The acidic reaction medium makes the method useful for compounds that have base-sensitive functional groups.

These directed aldol reactions provide efficient methods for the regiospecific formation of new carbon–carbon bonds and can be used for the preparation of key intermediates in the synthesis of important natural products.

Many so-called directed "aldol-type" reactions have been reported that involve coupling of an aldehyde or a ketone with a compound R_2CHX or RCH(X)Y (X or Y = an activating group such as CO_2R, COSR, CONHR, CN, NO_2, SOR, and SO_2R, where R = alkyl, aryl, or hydrogen). These reactions are mechanistically similar to directed aldol reactions and produce a hydroxy compound or its dehydration product. However, since these products are not aldols, ketols, or dehydration products thereof, these reactions are not classified as aldol reactions. This review includes examples where X and Y = CHO or COR only, *i.e.*, solely coupling reactions of aldehydes, ketones, or their equivalents.

STEREOCHEMISTRY

The most important stereochemical question in the directed aldol reaction concerns the formation of *threo* and/or *erythro* isomers of aldols or ketols. Consequently, extensive stereochemical studies on the geometry of enolate species, the nature of the metal, kinetic *vs.* thermodynamic control, and steric effects have been carried out.[14–27] The available data indicate that one of the stereoisomers (*threo* or *erythro*) can be formed predominantly under certain reaction conditions.*

The stereochemical outcome of the reaction is generally rationalized in terms of the geometry of the starting enolate and reaction conditions (kinetic control or thermodynamic control).[14–16,21]

Under conditions of kinetic control the stereoisomer formed is critically dependent on the geometry of the starting enolate. In general, the (*Z*)-enolate gives the *erythro* isomer and the (*E*)-enolate gives the *threo* isomer, as depicted in Eqs. 2 and 3. In both cases the carbonyl component approaches the enolate

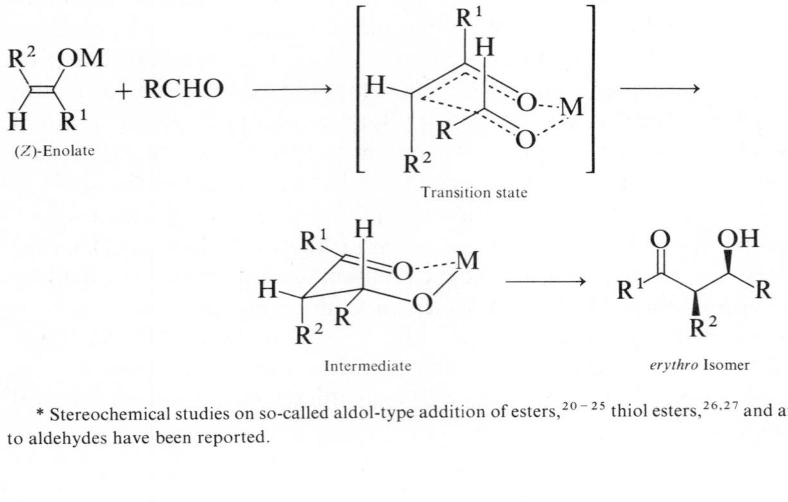

(*Z*)-Enolate + RCHO → [Transition state] → Intermediate → *erythro* Isomer (Eq. 2)

* Stereochemical studies on so-called aldol-type addition of esters,[20–25] thiol esters,[26,27] and amides[28,29] to aldehydes have been reported.

(E)-Enolate

Transition state

Intermediate

threo Isomer

(Eq. 3)

perpendicularly, and the reaction proceeds via a pericyclic process. The alkyl or aryl group of the aldehyde may occupy an equatorial position and R^1 of the enolate a pseudoaxial position in the transition state. The resulting six-membered chelate intermediates are finally hydrolyzed to the corresponding stereoisomers of β-hydroxycarbonyl compounds.

Under thermodynamic conditions (equilibrium conditions) the *threo* isomer is preferred, since the more stable chairlike conformer of the intermediate metal chelate has the maximum number of equatorial substituents, as depicted in Eq. 4.

threo Intermediate

threo Isomer

erythro Intermediate

erythro Isomer

(Eq. 4)

A typical example is provided in the reaction of (*Z*)-magnesium enolate **2** with pivalaldehyde,[14, 28–33] as shown in Eq. 5. The reaction under kinetic control (within a relatively short reaction time) affords the *erythro* aldol, whereas under more vigorous conditions the *threo* aldol is obtained exclusively. This result agrees with the postulate that the *erythro* chelate **3**, which is formed kinetically, is converted by equilibration to the thermodynamically more stable *threo* chelate **4**.

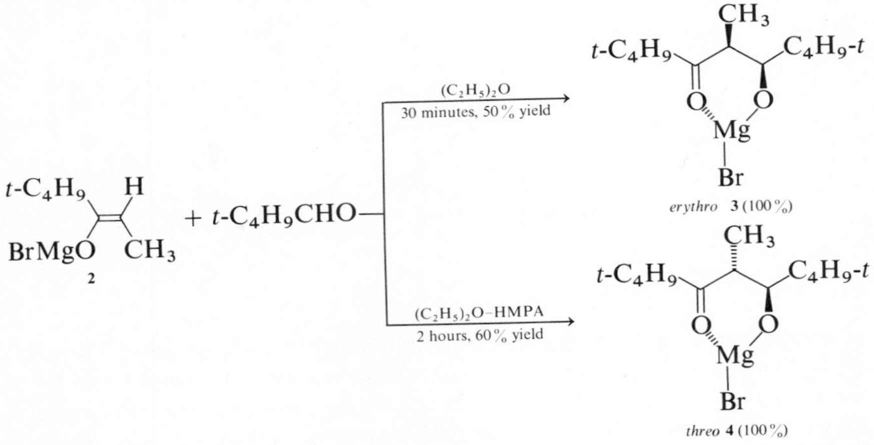

t-C_4H_9, H, BrMgO, CH_3 (**2**) + t-C_4H_9CHO — $(C_2H_5)_2O$, 30 minutes, 50% yield → *erythro* **3** (100%)

— $(C_2H_5)_2O$–HMPA, 2 hours, 60% yield → *threo* **4** (100%)

(Eq. 5)

Substituent effects on kinetic stereoselectivities have also been studied in the reaction of stereodefined magnesium enolates with various aldehydes or ketones.[14,32] The data obtained so far indicate that the size of the alkyl groups in the carbonyl acceptor is an important factor in controlling stereoselectivity. In particular, coupling with an aldehyde having a bulky alkyl group such as t-butyl or neopentyl gives high stereoselection.

Reaction of lithium enolates with aldehydes is subject to kinetic stereoselection, with (Z)-enolate giving *erythro* aldol and (E)-enolate leading to *threo* aldol (Eqs. 6 and 7).[16] However, the selectivity is limited by the substituent R; bulky groups (t-butyl, 1-adamantyl, mesityl, and trimethylsilyl)

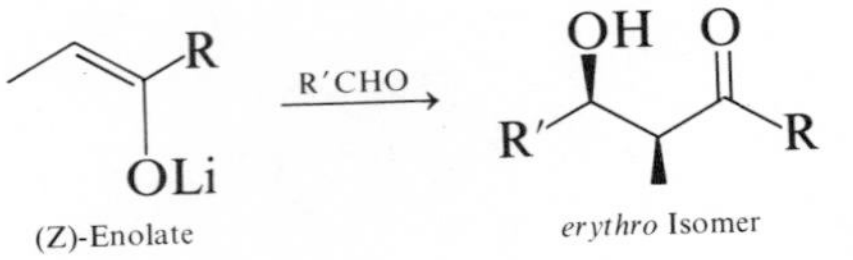

(Z)-Enolate —R′CHO→ *erythro* Isomer (Eq. 6)

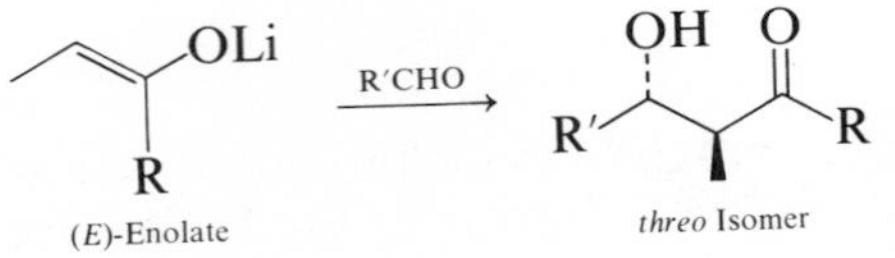

(E)-Enolate —R′CHO→ *threo* Isomer (Eq. 7)

give high stereoselection, while with smaller R groups (ethyl, isopropyl, phenyl, methoxy, t-butoxy, and diisopropylamino), stereoselectivity diminishes or disappears.[16] Kinetically controlled condensation of ethyl *t*-butyl ketone with benzaldehyde is an example of this type of reaction. Treatment of (*Z*)-enolate **5** with benzaldehyde for a very short time (5 seconds) at $-72°$ affords *erythro* aldol **6** as the sole product (Eq. 8). On the other hand,

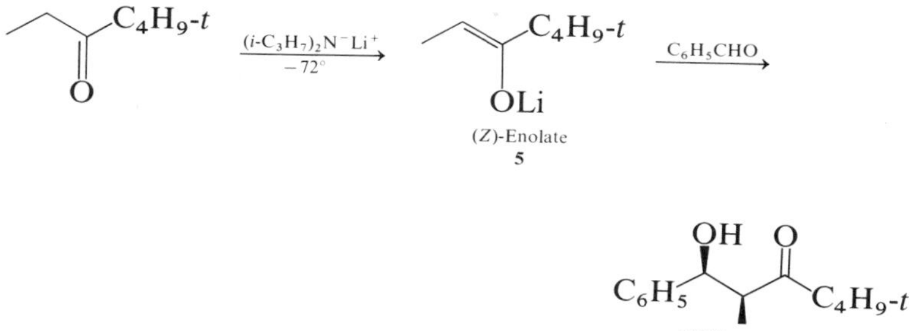

C_4H_9-*t* ketone $\xrightarrow[-72°]{(i\text{-}C_3H_7)_2N^-Li^+}$ C_4H_9-*t* / OLi

(*Z*)-Enolate
5

$\xrightarrow{C_6H_5CHO}$

OH O / C_6H_5 … C_4H_9-*t* (Eq. 8)

100% *erythro* Isomer
6

a mixture of stereoisomers of ethyl mesityl ketone enolates [92% (*E*) and 8% (*Z*)] reacts with benzaldehyde under the same conditions to afford 92% of the *threo* aldol and 8% of the *erythro* aldol (Eq. 9).[16]

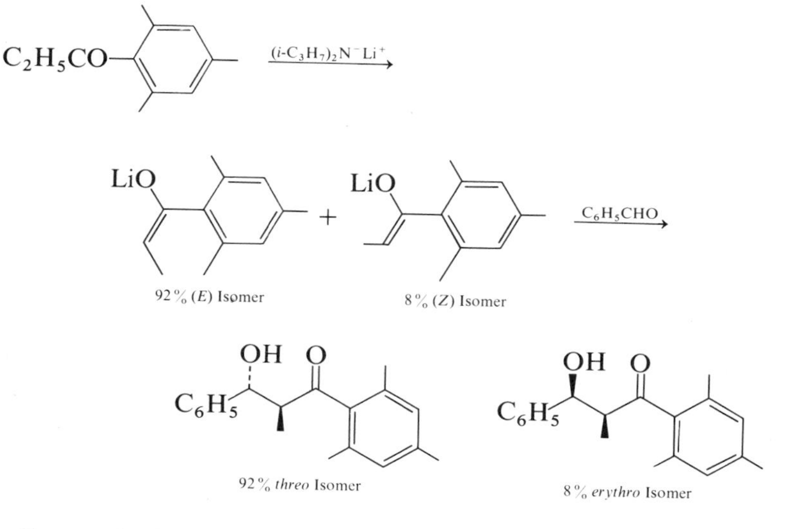

C_2H_5CO–mesityl $\xrightarrow{(i\text{-}C_3H_7)_2N^-Li^+}$

LiO … + LiO … $\xrightarrow{C_6H_5CHO}$

92% (*E*) Isomer 8% (*Z*) Isomer

OH O / C_6H_5 … OH O / C_6H_5 … (Eq. 9)

92% *threo* Isomer 8% *erythro* Isomer

Stereoselection in the reaction of vinyloxyboranes with aldehydes has recently been described by many authors.[17,18,34,36–38] The vinyloxyborane generated *in situ* from 3-pentanone and 9-borabicyclo[3.3.1]-9-nonyl trifluoromethanesulfonate (9-BBN triflate) by the action of 2,6-lutidine reacts with benzaldehyde as shown in Eq. 10. Almost complete stereoselection is

$$(C_2H_5)_2CO \xrightarrow[\text{2,6-Lutidine}]{\text{9-BBN triflate}} \text{(9-BBN)}\text{BO}C(C_2H_5)=CHCH_3 \xrightarrow{C_6H_5CHO} C_2H_5COCH(CH_3)CH(OH)C_6H_5 \quad \mathbf{7} \qquad \text{(Eq. 10)}$$

observed under mild conditions (−78°), giving *erythro* aldol **7** (96% purity).[17] This result suggests that initial generation of vinyloxyborane and subsequent reaction with benzaldehyde both proceed in a stereoselective manner. This stereochemical process is best exemplified by further study of stereodefined vinyloxyborane condensation reactions. The (*Z*) isomers **10** react with various aldehydes to yield predominantly (95%) the *erythro* aldols **11**, whereas the (*E*) isomers **8** react somewhat less stereoselectively (70–80%) to give *threo* aldols **9** as the major products.[18]

$$R^1{}_2BO(R)C{=}C(R^1)H \; \mathbf{8} \xrightarrow{R^2CHO} RCOCH(R^1)CH(OH)R^2 \; \mathbf{9}$$

$$R^1{}_2BO(R)C{=}C(H)R^1 \; \mathbf{10} \xrightarrow{R^2CHO} RCOCH(R^1)CH(OH)R^2 \; \mathbf{11}$$

It should be noted that, starting from the same ketone, the preparation of either *erythro* or *threo* aldol in a highly stereoselective manner can be achieved in some cases by proper choice of reagents.[34,35,38] For example, treatment of cyclohexyl ethyl ketone with 9-BBN triflate and N,N-diisopropylethylamine generates a (*Z*)-vinyloxyborane, which in turn reacts with an aldehyde to give the *erythro* aldol (*erythro* : *threo* = 97:3) (Eq. 11). In contrast the reaction involving dicyclopentylboryl triflate and diisopropylethylamine reverses the ratio to about 14:86, *i.e.*, predominantly to *threo*-isomer formation (Eq. 12).[38,*]

* This reversal of a large change in stereoselectivity may be explained by assuming that kinetically generated (*Z*)-enolate is readily converted into thermodynamically stable (*E*)-isomer when the sterically hindered dicyclohexylboryl triflate is used as the borylating reagent.

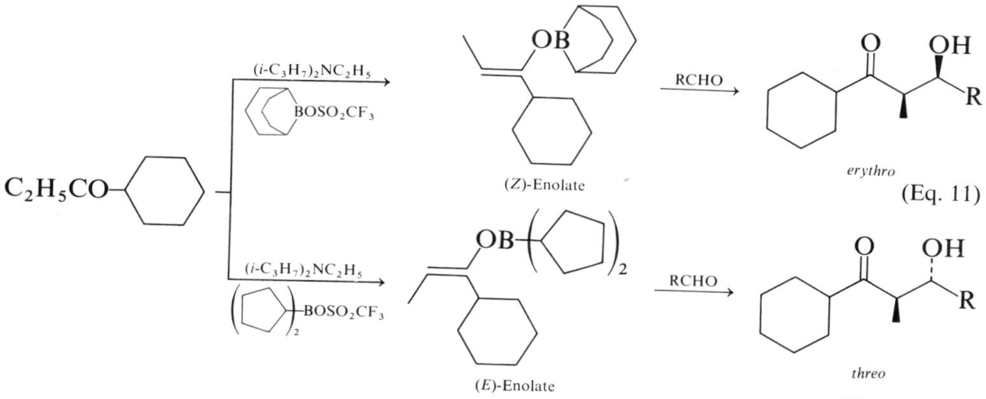

(Eq. 11)

(Eq. 12)

Condensation of aluminum enolates with aldehydes appears to be subject to kinetic stereoselection, although the available data are limited.[19] The (*Z*) and (*E*) isomers of dimethylaluminum 4,4-dimethylpent-2-en-2-olate, **12** and **13**, react with acetaldehyde in a highly stereoselective manner to yield *threo* and *erythro* aldol adducts, respectively.

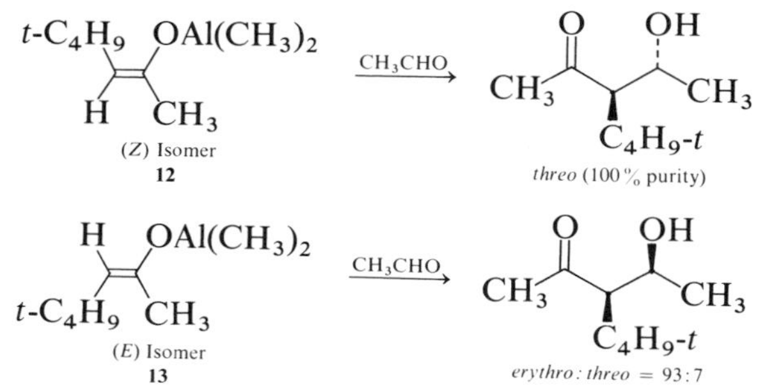

Another important stereochemical aspect of the directed aldol reaction is asymmetric induction. One approach is to utilize an optically active aldehyde having a chiral center adjacent to the carbonyl group, which influences the stereochemical course of the addition reaction. For example, the reaction of chiral aldehyde **14** with (*Z*)-enolate **15** yields a mixture of **16** and **17** in a ratio of 86:14; other diastereomers are not produced.[39] This reaction provides a useful method for the stereoselective construction of acyclic compounds containing multiple chiral centers.

The 1,2-diastereoselectivity mentioned above can be enhanced by the use of "double stereodifferentiation."[40] Reaction of the chiral ketone **18** with the chiral aldehyde **19** affords a mixture of three aldols in a ratio of 5.5:2.5:1.

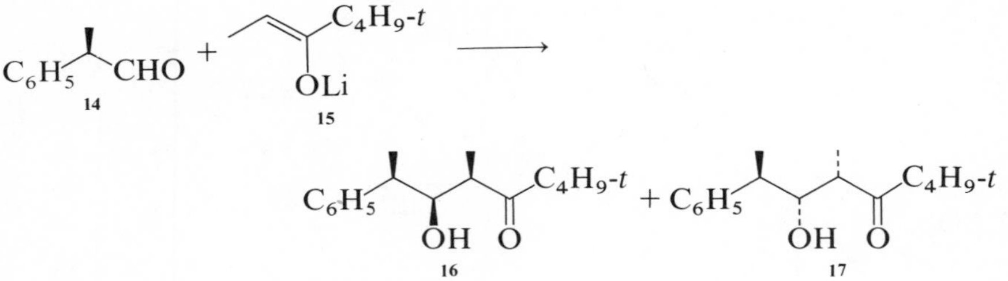

C_6H_5 CHO 14 + C_4H_9-*t* OLi 15 ⟶ C_6H_5 C_4H_9-*t* OH O 16 + C_6H_5 C_4H_9-*t* OH O 17

The two major products are the *erythro* isomers **21** and **22**, and the minor isomer is a *threo* diastereomer. A similar reaction of **18** with the enantiomeric aldehyde **20** produces only two stereoisomers (**23** and **24**), in a ratio of 13:1, with the major isomer being **23**.

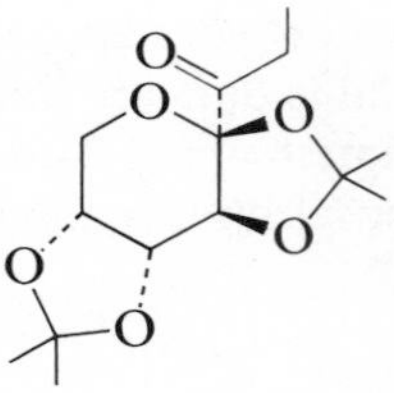

18 OHC 19 OHC 20

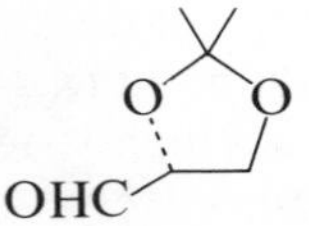

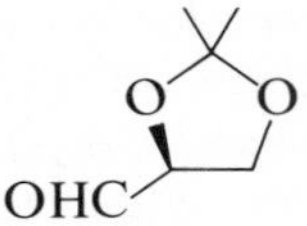

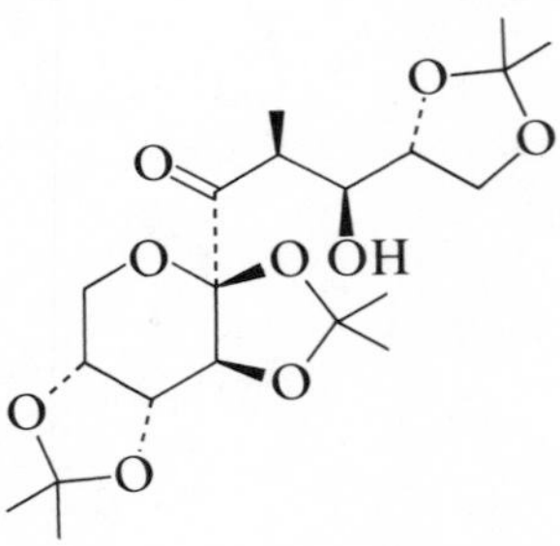

21 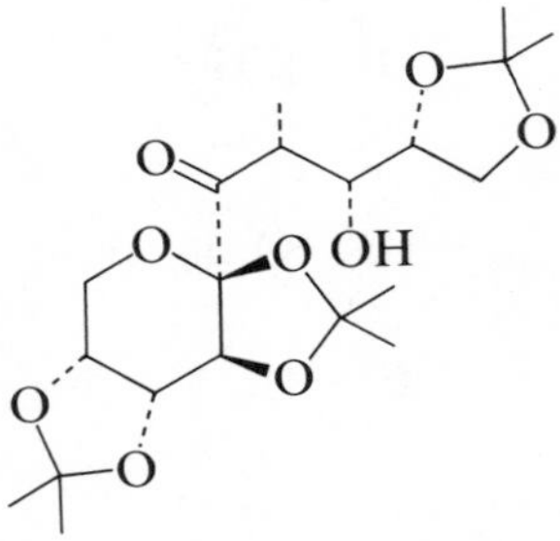 22

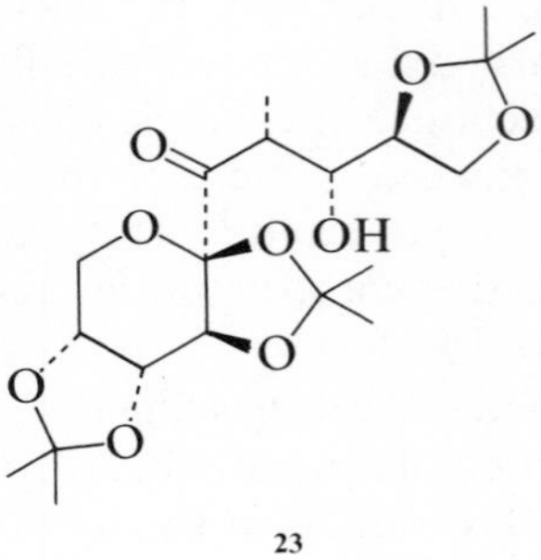

23 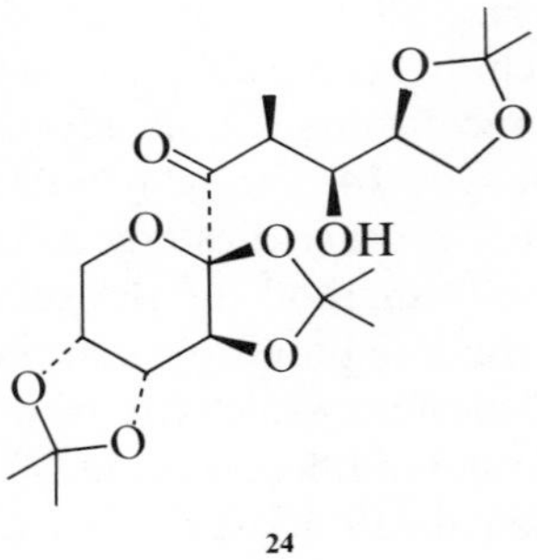 24

Another example of double stereodifferentiation is a crucial step of stereocontrolled carbon–carbon bond formation in the total synthesis of lasalocid A.[41] A chiral ketone **25** is converted, by successive treatments with lithium diisopropylamide and zinc chloride, to the corresponding zinc enolate, which in turn reacts with the other carbonyl acceptor **26** to afford four isomeric aldols. The major product has the desired stereochemistry at C_{11} and C_{12} and is identical with benzyl lasalocid A derived from lasalocid A.

A similar method has been successfully applied in the stereocontrolled total synthesis of monensin.[42]

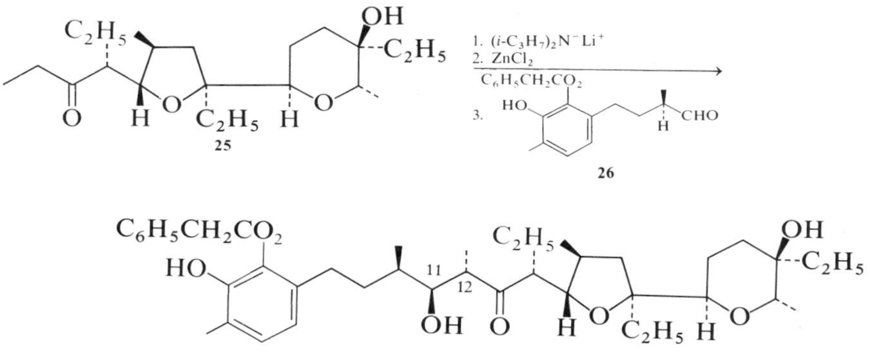

1. $(i\text{-}C_3H_7)_2N^-Li^+$
2. $ZnCl_2$
3. **26**

25

26

Benzyl lasalocid A

Almost complete asymmetric induction is achieved by the use of a chiral stereodefined lithium enolate having bulky groups.[43] (*Z*)-Enolate **27** shows very high diastereoselectivity in its reaction with a chiral aldehyde, even when both reactants are racemic. For example, **27** reacts with 2-phenylpropanal to give, after oxidation, a mixture of β-hydroxy carboxylic acids, **28** and **29**, with

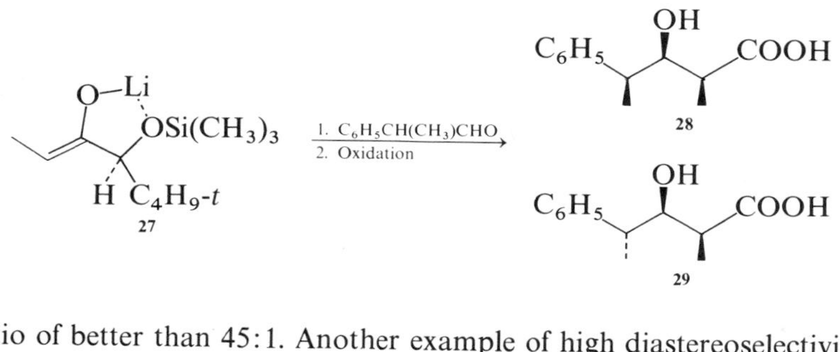

1. $C_6H_5CH(CH_3)CHO$
2. Oxidation

27

28

29

a ratio of better than 45:1. Another example of high diastereoselectivity in this "double racemic" condensation is the reaction of **27** with aldehyde **30**. Only one racemic diastereomer **31** is produced.[43]

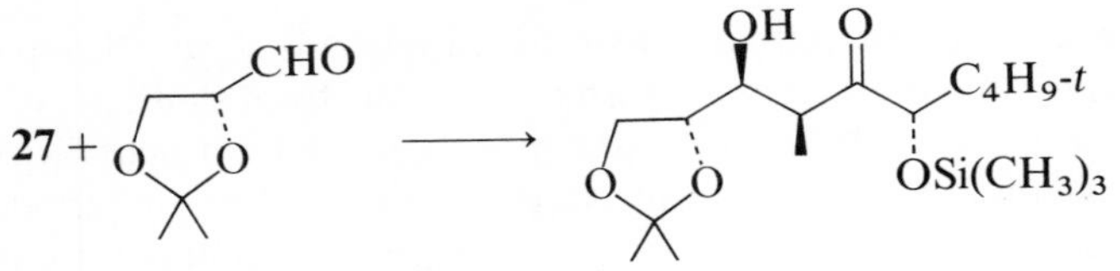

$27 +$ **30** $\longrightarrow$ **31**

As mentioned above it is well established that highly diastereoselective kinetic aldol condensations can be executed if the appropriate sterically controlling elements such as lithium and boron are corporated into the metal enolate. In these reactions the stereochemistry of the aldol product is strongly dependent on the enolate geometry. However both (*E*)- and (*Z*)-zirconium enolates have been shown to undergo selective kinetic aldol reactions to give mainly the *erythro* products.

Although detailed speculation as to the origin of the observed *erythro* selection from either enolate is premature, the following possibility has been proposed. The (*E*)-zirconium enolate reacts preferentially via a pseudo-boat transition state while the reaction of the corresponding (*Z*)-enolate proceeds preferentially via a pseudo-chair transition state.[43a]

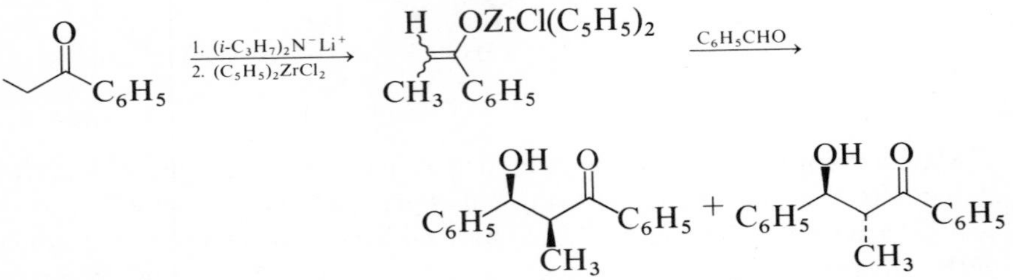

1. $(i\text{-}C_3H_7)_2N^-Li^+$
2. $(C_5H_5)_2ZrCl_2$

C_6H_5; H, $OZrCl(C_5H_5)_2$, CH_3, C_6H_5; C_6H_5CHO

OH O, C_6H_5, C_6H_5, CH_3 + OH O, C_6H_5, C_6H_5, CH_3

(90:10)

REACTION OF LITHIUM ENOLATES

The efficiently directed aldol reaction has two main requirements. The first is the regioselective formation of an enolate from a carbonyl compound. The second is effective interception by chelation of the aldol-type adduct formed from the enolate and the other carbonyl compound. The lithium enolate satisfies the above requirements, because the enolate usually equilibrates slowly and the lithium ion can effectively trap the intermediate by stable chelate formation in an aprotic solvent such as ether or tetrahydrofuran.

Regioselective or regiospecific formation of lithium enolates from unsymmetrical ketones is generally achieved by one of the following methods:[2,44,45] (1) deprotonation of ketones under kinetic or thermodynamic control, (2) reduction of α,β-unsaturated carbonyl compounds with lithium–

liquid ammonia, (3) conjugate addition of organometallic compounds to α,β-unsaturated carbonyl compounds, or (4) cleavage of enol derivatives by organolithium compounds. The lithium enolates of ketones thus formed have been employed for directed aldol reactions with aldehydes or ketones. However, this type of reaction using a lithium enolate derived from an aldehyde has not been reported.

Enolates Prepared by Deprotonation of Ketones

Several methods have been proposed for generation of kinetic lithium enolates by abstraction of the less hindered proton from unsymmetrical ketones. The most commonly used procedure is slow addition of the ketone to a solution of a slight excess of a lithium dialkylamide in tetrahydrofuran or 1,2-dimethoxyethane at low temperature (usually −78°), *i.e.*, nonequilibrating conditions. The sterically hindered lithium dialkylamides, such as lithium diisopropylamide (LDA), lithium hexamethyldisilazane (LHDS), lithium N-isopropylcyclohexylamide (LICA), lithium dicyclohexylamide (LDCA), and lithium 2,2,6,6-tetramethylpiperidide (LTMP), are commonly used for this purpose. Under these conditions ketones of the type RCH_2COCH_3, $R_2CHCOCH_3$, and $R_2CHCOCH_2R$ are kinetically deprotonated to generate the less-substituted enolate highly regioselectively. The resulting lithium enolates are highly reactive and couple rapidly with carbonyl acceptors at −78°.[15,46,47] Most examples reported utilize the terminal enolates derived from methyl alkyl ketones (RCH_2COCH_3 or $R_2CHCOCH_3$).

For example, 2-pentanone, upon deprotonation with lithium diisopropylamide at −78°, followed by addition of benzaldehyde, affords a crossed aldol **32** in 75–80% yield.

$$n\text{-}C_3H_7COCH_3 \xrightarrow[-78^\circ]{(i\text{-}C_3H_7)_2N^-Li^+} n\text{-}C_3H_7C(OLi){=}CH_2 \xrightarrow[2.\ H_2O]{1.\ C_6H_5CHO,\ -78^\circ} \underset{\mathbf{32}}{n\text{-}C_3H_7COCH_2CHOHC_6H_5}$$

Lithium diethylamide can be used for the generation of kinetic enolates in some cases, but the yields of the aldol products are moderate (30–70%).[48]

Kinetic lithium enolates of α,β-unsaturated ketones can also be used for crossed aldol formation.[47,49] The lithium enolate of 3-penten-2-one reacts with crotonaldehyde to give dienolone **33** in 70% yield.[47] Similar reactions

$$CH_3CH{=}CHCOCH_3 \xrightarrow[-78^\circ]{(i\text{-}C_3H_7)_2N^-Li^+} CH_3CH{=}CHC(OLi){=}CH_2 \xrightarrow[-78^\circ]{CH_3CH{=}CHCHO} \underset{\mathbf{33}}{CH_3CH{=}CHCOCH_2CHOHCH{=}CHCH_3}$$

are employed in the syntheses of α-bisabolalone and ocimenone (Eq. 13).[47,50]

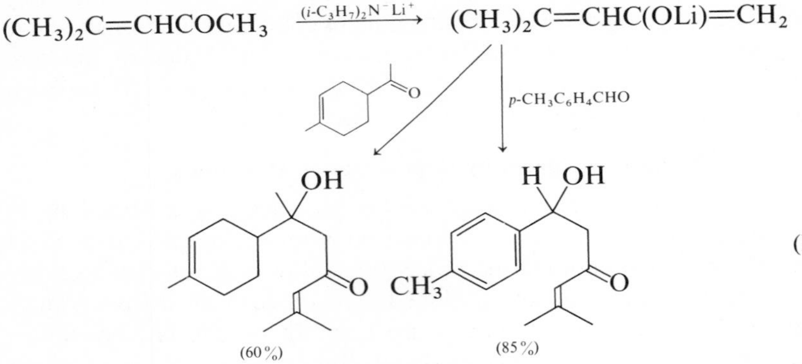

$$(CH_3)_2C{=}CHCOCH_3 \xrightarrow{(i\text{-}C_3H_7)_2N^-Li^+} (CH_3)_2C{=}CHC(OLi){=}CH_2$$

p-$CH_3C_6H_4CHO$

OH

H OH

O CH$_3$

O

(60%) (85%)

(Eq. 13)

The enol ether of a 1,3-diketone gives a kinetic enolate that subsequently reacts with carbonyl compounds to give the corresponding aldol.[51–53] This aldol can be converted into the vinylogous aldol by treatment with sodium bis(2-methoxyethoxy)aluminum hydride, followed by mild acid hydrolysis.[51] This strategy has been employed in the synthesis of polyene **34** (Eq. 14).[32]

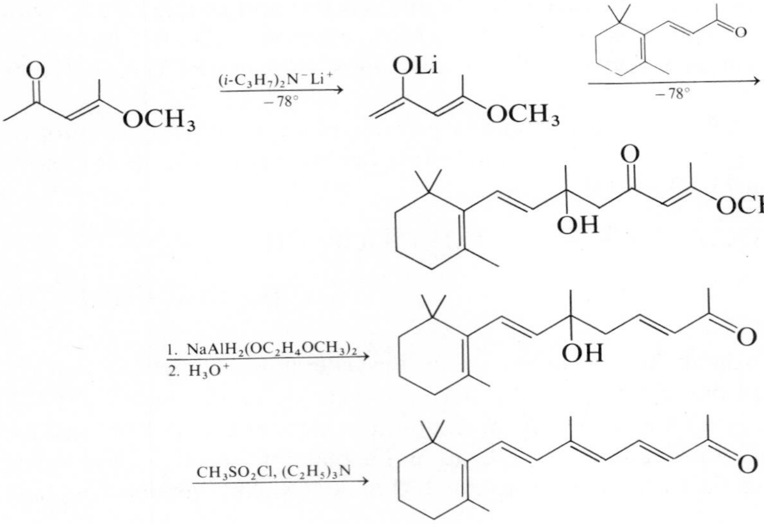

O OCH$_3$ $\xrightarrow[-78°]{(i\text{-}C_3H_7)_2N^-Li^+}$ OLi OCH$_3$ $\xrightarrow{-78°}$

O OH OCH$_3$

$\xrightarrow[2.\ H_3O^+]{1.\ NaAlH_2(OC_2H_4OCH_3)_2}$ OH O

$\xrightarrow{CH_3SO_2Cl,\ (C_2H_5)_3N}$ O (Eq. 14)

34

The overall yields of vinylogous aldols are only moderate (40–50%), but the simplicity of the sequence and its compatibility with sensitive functionality (e.g., polyenes make it generally useful.[51]

Lithium diisopropylamide generates the kinetic enolate from (S)-(+)-3-methyl-2-pentanone with less than 10% racemization. The enolate undergoes reaction with propanal to give aldol **35** in 65% yield. The newly formed asymmetric center has an (R)-configuration (15% induced).[54]

$\xrightarrow{(i\text{-}C_3H_7)_2N^-Li^+}$ OLi $\xrightarrow[-100^\circ]{C_2H_5CHO}$

HO H O, C_2H_5, H CH_3

35

Use of lithium 1,1-bis(trimethylsilyl)-3-methylbutoxide (**36**) as a sterically hindered base affords the crossed aldol product starting from a mixture of ketone and aldehyde.[55] Treatment of a mixture of 5-methyl-2-pentanone and 3-phenylpropanal with **36** gives 6-hydroxy-2-methyl-8-phenyl-4-octanone in 86% yield. This one-step procedure is fundamentally different from that usually followed for the directed aldol reaction, *i.e.*, prior formation of an enolate and subsequent addition of the other carbonyl compound to give the aldol adduct.

$$C_6H_5(CH_2)_2CHO + CH_3COCH_2C_3H_7\text{-}i \xrightarrow[\text{THF, } -40^\circ]{i\text{-}C_3H_7CH_2C(OLi)[Si(CH_3)_3]_2 \ \mathbf{36}} C_6H_5(CH_2)_2CHOHCH_2COCH_2CH(CH_3)_2$$

The enolate **38** is generated as usual by treatment of β-ketotrimethylsilane **37** with lithium diisopropylamide.[56] This enolate reacts with an aldehyde to give the aldol adduct **39**, which is desilylated in an acidic medium to afford the corresponding β-hydroxy ketone **40**. Consequently, this procedure permits differentiation of methylene groups of unsymmetrical ketones.

$$\underset{\mathbf{37}}{R^1CH(Si(CH_3)_3)COCH_2R^2} \xrightarrow[-78^\circ]{(i\text{-}C_3H_7)_2N^-Li^+} \underset{\mathbf{38}}{R^1CH(Si(CH_3)_3)C(OLi){=}CHR^2} \xrightarrow[-78^\circ]{R^3CHO}$$

$$\underset{\mathbf{39}}{R^1CH(Si(CH_3)_3)COCHR^2CHOHR^3} \xrightarrow{H_3O^+} \underset{\mathbf{40}\ (72\text{–}95\%)}{R^1CH_2COCHR^2CHOHR^3}$$

The aldol reaction of kinetic lithium enolates is a firmly established synthetic method which is valuable in natural products synthesis and is likely to become even more widely used, especially where stereocontrol is also important.

Enolates Prepared by Miscellaneous Methods

A lithium enolate can be generated regiospecifically by the reduction of an α,β-unsaturated ketone or a cyclopropyl ketone with lithium metal in liquid amonia. Ammonia is too weak an acid to protonate the enolate, which should therefore maintain its regiochemical integrity.[44] 9-Methyl-$\Delta^{4,9}$-3-octalone is reduced by lithium–liquid ammonia in the presence of aniline (better than *t*-butyl alcohol) as a proton donor to give the enolate **41**, which reacts with formaldehyde to afford the hydroxymethyl compound **42** in 60% yield.[57]

$\xrightarrow[C_6H_5NH_2]{Li-NH_3}$ LiO ... H **41** $\xrightarrow{HCHO}$ O ... H, $HOCH_2$ **42**

Successive treatment of a ketone with sodium amide and lithium bromide generates a lithium enolate that can be used for the preparation of crossed aldols (Eq. 15).[48,58]

$$R^1COCH_3 \xrightarrow[-15^\circ]{NaNH_2} R^1C(ONa){=}CH_2 \xrightarrow[-45^\circ]{LiBr}$$

$$R^1C(OLi){=}CH_2 \xrightarrow[2.\ H_2O]{1.\ R^2R^3CO} R^2R^3C(OH)CH_2COR^1 \quad \text{(Eq. 15)}$$

Lithium enolates are also generated regiospecifically by the conjugate addition of a lithium diorganocuprate to an α,β-unsaturated carbonyl compound.[59–61] The reaction has been applied to the preparation of one of the synthetic precursors of prostaglandin-$F_{2\alpha}$($PGF_{2\alpha}$) (Eq. 16).[62]

O, R^1O + R^2_2CuLi ⟶ OLi, R^1O, R^2, H $\xrightarrow[-78^\circ]{HCHO}$

$R^1 = C_6H_5C(CH_3)_2—$
$R^2 = —CH_2CH{=}CHCH(C_5H_{11}\text{-}n)(OCH_2OCH_2C_6H_5)$

O, H, CH_2OH, R^2, R^1O, H $\Longrightarrow$ $PGF_{2\alpha}$ (Eq. 16)

The intramolecular reaction utilizing 1,4 addition of lithium organocuprates to α,β-unsaturated compounds is a useful method for the construction of certain polycyclic compounds. The preparation of tetracyclic ketal **43** is a typical example.[63] The bicyclic hydroxy ketone **44** and spiro ketone **45** are also obtained by a similar procedure.[64]

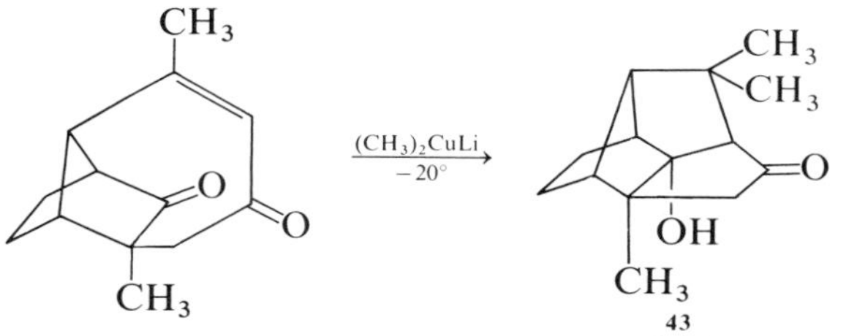

$(CH_3)_2CuLi$, −20°

43 (96%)

Both aldols **44** and **45** are products of a kinetically controlled reaction, for the indicated aldols are no longer obtained if the reaction mixture is equilibrated

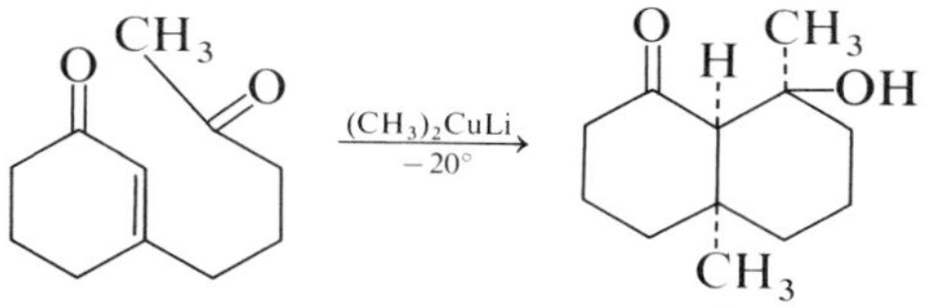

$(CH_3)_2CuLi$, −20°

44 (63%)

$(CH_3)_2CuLi$

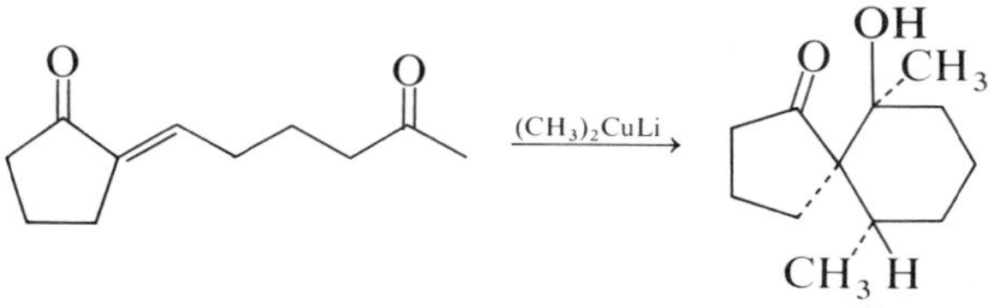

45

(24 hours at 22°). It is of interest that even at −60° the ε-formyl-α,β-enone **46** gives a considerable amount of the carbonyl addition product **47**.

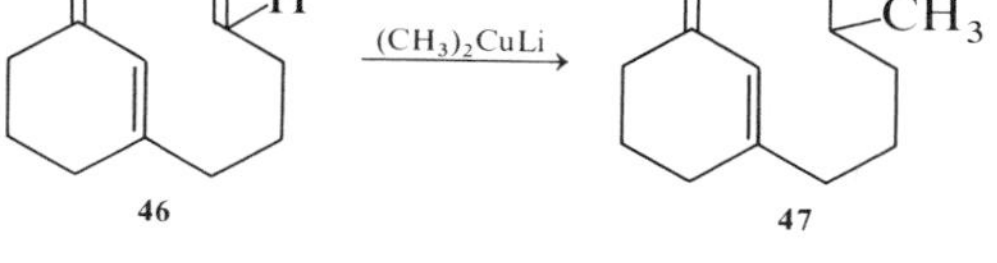

$(CH_3)_2CuLi$

46 **47**

Trimethylsilyl enol ethers can be readily converted to lithium enolates by treatment with methyllithium in tetrahydrofuran or 1,2-dimethoxyethane at room temperature.[65,66] Lithium enolates thus prepared have considerable synthetic utility, since the enolates are generated cleanly with the formation of neutral tetramethylsilane as the co-product. The utility of this procedure is demonstrated in the effective preparation of the ketone **48**.[57] Although clean

$(CH_3)SiO$ — H; 1. CH_3Li 2. HCHO → O; H; CH_2OH

48
(90%)

reactions occur, this method is of limited scope at present since the preparation of one regioisomer of a trimethylsilyl enol ether directly from an unsymmetrical ketone remains a difficult problem and requires tedious separation in some cases. Development of a convenient method for regioselective preparation of silyl enol ethers would make this aldol reaction useful.

Treatment of α,α′-dibromoketones with a large excess of lithium diorganocuprate provides alkylated enolates, *e.g.*, **49**,[67,68] which can undergo aldol reaction with aldehydes.

Br, O, Br; $(n\text{-}C_4H_9)_2CuLi$ or $t\text{-}C_4H_9O(C_4H_9\text{-}n)CuLi$ → [$n\text{-}C_4H_9$, OLi] **49**; $n\text{-}C_3H_7CHO$, Room temperature →

$n\text{-}C_4H_9$, O, H, $C_3H_7\text{-}n$

(43%)

FLUORIDE ION–CATALYZED REACTION USING TRIMETHYLSILYL ENOL ETHERS

Trimethylsilyl enol ethers can be cleaved by fluoride ion.[69] The use of catalytic amounts of tetrabutylammonium fluoride (TBAF) promotes the aldol reaction of silyl enol ethers with aldehydes (Eq. 17).[70,71] This fluoride ion–catalyzed aldol reaction is highly stereospecific; the reaction of silyl enol

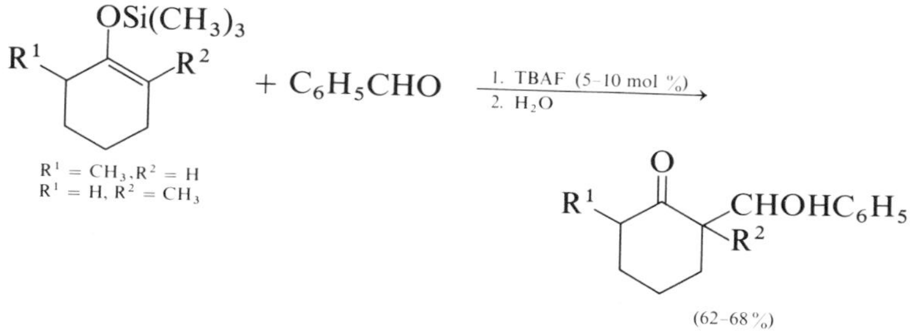

$R^1 = CH_3, R^2 = H$
$R^1 = H, R^2 = CH_3$

$+ C_6H_5CHO \xrightarrow[\text{2. } H_2O]{\text{1. TBAF (5–10 mol \%)}}$ (62–68%) (Eq. 17)

ether **50** with benzaldehyde in the presence of tetrabutylammonium fluoride at $-35°$ for 2 hours yields only the stereoisomers **51** and **52** that bear the newly introduced siloxybenzyl group in the axial position. This indicates that attack of the aldehyde on the six-membered ring occurs exclusively through axial approach via a kinetically favored chairlike transition state.[70]

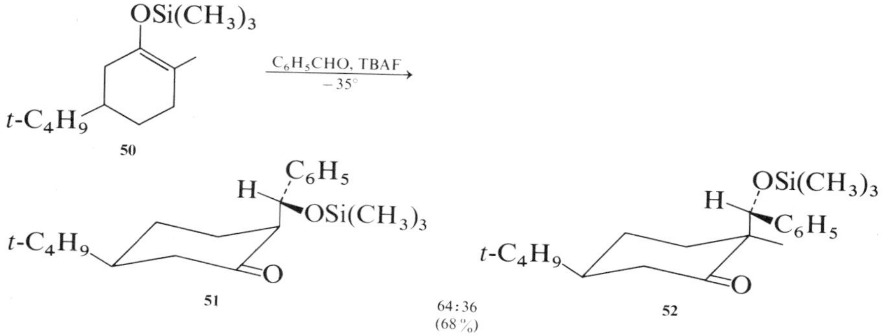

50 $\xrightarrow[-35°]{C_6H_5CHO,\ TBAF}$ **51** 64:36 (68%) **52**

The reaction pathway shown in Eq. 18 has been proposed.

$$R^1R^2C{=}CR^3[OSi(CH_3)_3] \overset{F^-}{\rightleftharpoons} R^1R^2C{=}CR^3[OSi(CH_3)_3F^-]$$

$$\overset{-(CH_3)_3SiF}{\rightleftharpoons} R^1R^2C{=}CR^3O^- \overset{R^4CHO,\ (CH_3)_3SiF}{\rightleftharpoons}$$

$$R^3COC(R^1R^2)CH(R^4)[OSi(CH_3)_3] \xrightarrow{H_2O} R^3COC(R^1R^2)CHOHR^4 \quad \text{(Eq. 18)}$$

The reaction also shows high chemical selectivity; aldehydes as electrophiles undergo this aldol reaction quite readily, while ketones do not.

Combination of this reaction with the silyl-transfer method yields a convenient crossed aldol procedure using mild reaction conditions. The coupling of isopropyl methyl ketone and benzaldehyde occurs readily in the presence of ethyl trimethylsilylacetate and tetrabutylammonium fluoride to give ketone **53** in 52% yield.

$$i\text{-}C_3H_7COCH_3 \xrightarrow[\text{2. } C_6H_5CHO,\ TBAF,\ -20^\circ]{\text{1. } (CH_3)_3SiCH_2CO_2C_2H_5,\ TBAF,\ 0^\circ} \underset{\mathbf{53}}{i\text{-}C_3H_7COCH_2CH(C_6H_5)[OSi(CH_3)_3]}$$

REACTION OF LITHIATED SCHIFF BASES, HYDRAZONES, AND OXIMES

Schiff Bases

Many directed aldol reactions using a Schiff base (azomethine compound) as the masked carbonyl compound have been reported.[3,4,72–74] A Schiff base having at least one alpha hydrogen atom is lithiated by treatment with a lithium dialkylamide to generate an organolithium compound. Subsequent reaction with a carbonyl compound gives an adduct that can be hydrolyzed with mineral acid to afford the corresponding aldol (Eq. 19).

$$R^1R^2CHC(=NR^4)R^3 \xrightarrow{R_2NLi} R^1R^2C{=}C(NLiR^4)R^3 \xrightarrow{R^5R^6CO} [\text{six-membered chelate: } R^5R^6C(O\text{-}Li\cdots N(R^4){=}C(R^3))\text{-}CR^1R^2] \xrightarrow{H_3O^+} R^5R^6C(OH)CR^1R^2COR^3 \quad \text{(Eq. 19)}$$

The standard method for the lithiation of Schiff bases and the subsequent reaction with carbonyl compounds is described below. Cyclohexylamine or *t*-butylamine is suitable as the amine component of the Schiff base, since Schiff bases having branched alkyl substituents on the nitrogen have less tendency to undergo self-addition than those with unbranched chains.[75–77] Lithium diisopropylamide is preferred over lithium diethylamide as the lithiation reagent for steric reasons.

Preparation of β-phenylcinnamaldehyde is a typical example (Eq. 20).[78] Although the reaction proceeds smoothly with aromatic ketones, this method

$$C_6H_{11}NH_2 + CH_3CHO \xrightarrow[-H_2O]{} C_6H_{11}N{=}CHCH_3 \xrightarrow{(i\text{-}C_3H_7)_2N^-Li^+}$$

$$C_6H_{11}N{=}CHCH_2Li \xrightarrow[\text{2. } H_2O]{\text{1. } (C_6H_5)_2CO} C_6H_{11}N{=}CHCH_2C(OH)(C_6H_5)_2$$

$$\xrightarrow[H_2O,\ 100^\circ]{(CO_2H)_2} \underset{(78\text{–}85\%)}{(C_6H_5)_2C{=}CHCHO} \quad \text{(Eq. 20)}$$

has limited applicability. The yield decreases with increased branching on the alpha carbon atom of the Schiff base.[3] Two α-alkyl substituents in the Schiff base cause difficulty in lithiation. For example, N-2-ethylbutylidenecyclohexylamine is not lithiated on treatment with lithium diisopropylamide under the above conditions.

The metalation of ketimines followed by reaction with benzophenone has also been reported. N-Isopropylidenecyclohexylamine is lithiated under the usual conditions to generate the corresponding organolithium compound, which is then treated with benzophenone to yield the expected adduct **54** in 64% yield.[75] Decomposition of the adduct with dilute sulfuric acid leads to the unsaturated ketone **55** in 78% yield. Treatment of the adduct with silica gel affords the corresponding aldol **56**, which cannot be obtained by the conventional aldol reaction of acetone with benzophenone using a basic catalyst.

$$\underset{\mathbf{54}}{(C_6H_5)_2C(OH)CH_2C(CH_3){=}NC_6H_{11}} \xrightarrow{H_2SO_4\text{-}H_2O} \underset{\mathbf{55}}{(C_6H_5)_2C{=}CHCOCH_3}$$

$$\underset{\mathbf{54}}{(C_6H_5)_2C(OH)CH_2C(CH_3){=}NC_6H_{11}} \xrightarrow{H_2O\text{-silica gel}} \underset{\mathbf{56}}{(C_6H_5)_2C(OH)CH_2COCH_3}$$

In the reaction of the Schiff base of the homologous diethyl ketone with benzophenone, the ketimine adduct is isolated in 62% yield (Eq. 21).[75] However, hydrolysis of the adduct yields benzophenone instead of the aldol or α,β-unsaturated carbonyl compound.

$$CH_3CHLiC(C_2H_5){=}NC_6H_{11} + (C_6H_5)_2CO \longrightarrow (C_6H_5)_2C(OH)CH(CH_3)C(C_2H_5){=}NC_6H_{11} \quad \text{(Eq. 21)}$$

The Schiff base from acetophenone and cyclohexylamine, upon reaction with lithium diisopropylamide, then benzophenone, gives the adduct in 55% yield. This is hydrolyzed with dilute mineral acid to 1,1-diphenyl-2-benzoylethylene (95%).[4]

$$LiCH_2C(C_6H_5){=}NC_6H_{11} \xrightarrow[2.\ H_2O]{1.\ (C_6H_5)_2CO} (C_6H_5)_2C(OH)CH_2C(C_6H_5){=}NC_6H_{11} \xrightarrow{H_3O^+} (C_6H_5)_2C{=}CHCOC_6H_5$$

The lithiation of the same Schiff base with 2 equivalents of butyllithium in ether followed by reaction with benzophenone leads to the bisadduct in 16% yield.[4]

$$\text{C}_6\text{H}_5\text{C}(\text{=NC}_6\text{H}_{11})\text{CH}_3 \xrightarrow{2\ n\text{-C}_4\text{H}_9\text{Li}} o\text{-LiC}_6\text{H}_4\text{C}(\text{=NC}_6\text{H}_{11})\text{CH}_2\text{Li} \xrightarrow{2\ (\text{C}_6\text{H}_5)_2\text{CO}} o\text{-}[(\text{C}_6\text{H}_5)_2\text{C(OH)}]\text{C}_6\text{H}_4\text{C}(\text{=NC}_6\text{H}_{11})\text{CH}_2\text{C}(\text{C}_6\text{H}_5)_2\text{OH}$$

The utility of the directed aldol reaction using an aldimine has been proven in the synthesis of terpenoids such as retinal and nuciferal.[80] The 13-*cis* and all-*trans* 10,14-dimethylretinals are isolated in 36% yield.[79]

$$\text{CHO} + \text{Li} \ldots \text{NC}_6\text{H}_{13}\text{-}n \longrightarrow \text{CHO} + \text{CHO}$$

The reaction of *n*-octanal with the lithiated aldimine **57** affords α,β-unsaturated aldehydes in 70% yield. This product contains more than 97% of the (*Z*) isomer.[77]

$$n\text{-C}_7\text{H}_{15}\text{CHO} + \underset{\mathbf{57}}{n\text{-C}_6\text{H}_{13}\text{CHLiCH{=}NC}_4\text{H}_9\text{-}t} \longrightarrow$$

$$\underset{(Z)\ (97\%)}{(\text{H})(n\text{-C}_7\text{H}_{15})\text{C{=}C}(\text{CHO})(\text{C}_6\text{H}_{13}\text{-}n)} + \underset{(E)\ (3\%)}{(n\text{-C}_7\text{H}_{15})(\text{H})\text{C{=}C}(\text{CHO})(\text{C}_6\text{H}_{13}\text{-}n)}$$

These reactions have been employed in the preparation of intermediates for the synthesis of natural products such as maytansine,[81–83] cembrene,[84] and dendrobine.[85]

The directed aldol reaction can be extended to azine systems. Double lithiation of acetaldazine with lithium diisopropylamide followed by reaction with benzophenone gives the corresponding bisadduct **58** in 30% yield.[4]

$$CH_3CH{=}NN{=}CHCH_3 \xrightarrow{2(i\text{-}C_3H_7)_2N^-Li^+} LiCH_2CH{=}NN{=}CHCH_2Li$$

$$\xrightarrow[2.\ H_2O]{1.\ 2(C_6H_5)_2CO} \underset{\mathbf{58}}{(C_6H_5)_2C(OH)CH_2CH{=}NN{=}CHCH_2C(OH)(C_6H_5)_2}$$

$$\xrightarrow{H_3O^+} (C_6H_5)_2C{=}CHCH{=}NN{=}CHCH{=}C(C_6H_5)_2$$

Hydrazones

A highly efficient directed aldol reaction has been performed by use of dimethylhydrazone derivatives as the masked carbonyl nucleophiles.[86–90] Dimethylhydrazones of enolizable aldehydes or ketones are lithiated quantitatively at the alpha position by *n*-butyllithium or lithium diisopropylamide in tetrahydrofuran.[86] The lithio derivatives thus obtained undergo 1,2 addition with aldehydes or ketones to produce *β*-hydroxydimethylhydrazone derivatives in high yield.[87] Subsequent oxidative hydrolysis of the resulting hydrazone adducts using sodium periodate at pH 7 yields *β*-hydroxyketones in excellent yield (Eq. 22). It should be emphasized

$$R^1C(=N\text{-}N(CH_3)_2\rightarrow Li^+)\text{=}CHR^2 \xrightarrow{R^3R^4CO} R^1C(=NN(CH_3)_3)CH(R^2)C(OH)R^3R^4 \xrightarrow[H_2O]{NaIO_4}$$

$$R^1COCH(R^2)C(OH)R^3R^4 \quad \text{(Eq. 22)}$$

that the lithiation of an unsymmetrical dimethylhydrazone proceeds in a highly regioselective manner, so that the directed aldol reaction may be realized. For example, the coupling reaction of 2-pentanone with *trans*-crotonaldehyde affords *trans,trans*-5,7-nonadien-4-one in 78% overall yield (Eq. 23).[87]

$$n\text{-}C_3H_7COCH_3 \longrightarrow n\text{-}C_3H_7C(=NN(CH_3)_2)CH_3 \longrightarrow n\text{-}C_3H_7C(=N\text{-}N(CH_3)_2\rightarrow Li^+){=}CH_2$$

$$\xrightarrow{CH_3CH{=}CHCHO} n\text{-}C_3H_7C(=NN(CH_3)_2)CH_2CH(OH)CH{=}CHCH_3 \xrightarrow[CH_3OH,\ pH\ 7]{NaIO_4}$$

$$n\text{-}C_3H_7COCH_2CH(OH)CH{=}CHCH_3 \xrightarrow[(C_2H_5)_3N]{CH_3SO_2Cl} n\text{-}C_3H_7COCH{=}CHCH{=}CHCH_3 \quad \text{(Eq. 23)}$$

This procedure is also applicable to regioselective hydroxymethylation. The protecting group is easily removed with 2 equivalents of acetic acid to form the β-hydroxyketone (Eq. 24).[87]

$$n\text{-}C_3H_7C(=NN(CH_3)_2)CH_3 \xrightarrow[3.\ 2CH_3CO_2H]{1.\ n\text{-}C_4H_9Li,\ -78^\circ;\ 2.\ (HCHO)_n} n\text{-}C_3H_7COCH_2CH_2OH \ (80\%) \quad \text{(Eq. 24)}$$

The regioselective and enantioselective aldol reaction can be carried out using chiral hydrazones.[91,92] The chiral hydrazone **59** is lithiated with n-butyllithium at -95° and treated with a carbonyl compound. The adduct is silylated with chlorotrimethylsilane to form the doubly protected ketol. Oxidative hydrolysis finally affords the chiral ketol in good chemical yield and enantiomeric excesses of 31–62% (Eq. 25).

$$R^1COCH_3 \xrightarrow{\text{(S)-1-amino-2-(methoxymethyl)pyrrolidine}} R^1C(=N\text{-}N\text{-pyrrolidine-2-}CH_2OCH_3)CH_3\ (\mathbf{59}) \xrightarrow[3.\ (CH_3)_3SiCl]{1.\ n\text{-}C_4H_9Li;\ 2.\ R^2R^3CO}$$

$$R^1C(=N\text{-}N\text{-pyrrolidine-2-}CH_2OCH_3)CH_2C^*(OSi(CH_3)_3)R^2R^3 \xrightarrow{\text{Oxidative hydrolysis}} R^1COCH_2C^*(OH)R^2R^3 \quad \text{(Eq. 25)}$$

Dianions derived from phenylhydrazones, benzoylhydrazones, and tosylhydrazones also undergo regiospecific coupling, although these methods are not widely applicable. Typically, the phenylhydrazone of acetophenone is treated with 2 equivalents of n-butyllithium in tetrahydrofuran at 0°.[93,94] The resulting dianion reacts with benzophenone to yield the corresponding adduct in 75% yield.[93]

$$C_6H_5C(CH_3){=}NNHC_6H_5 \xrightarrow{2\ n\text{-}C_4H_9Li} C_6H_5C(CH_2Li){=}NN(Li)C_6H_5$$

$$\xrightarrow{(C_6H_5)_2CO} C_6H_5C(CH_2C(OH)(C_6H_5)_2){=}NNHC_6H_5$$

The reaction of dilithiophenylhydrazones with certain aldehydes, such as benzaldehyde and acetaldehyde, produces β-hydroxyphenylhydrazones, which in turn are readily cyclized to 2-pyrazolines (Eq. 26).

$$C_6H_5C(CH_2Li){=}NN(Li)C_6H_5 + C_6H_5CHO \longrightarrow \left[C_6H_5C(CH_2CHOHC_6H_5){=}NNHC_6H_5 \right] \longrightarrow \text{1,3,5-triphenyl-2-pyrazoline } (64\%) \quad \text{(Eq. 26)}$$

Tosylhydrazone dianions, formed at $-50°$ with *n*-butyllithium in tetrahydrofuran, react with aldehydes or ketones to afford β-hydroxytosylhydrazones in good yields.[95–99] However, attempts to generate aldols from these adducts result in a retroaldol reaction under acidic conditions or no reaction under basic conditions.[97] Treatment of the adducts with excess alkyllithium at room temperature results in their smooth elimination to yield homoallylic alcohols (Eq. 27).[97] For instance, phenylacetone tosylhydrazone is converted to the dianion at $-50°$ and trapped with propanal to give 4-hydroxy-1-phenyl-2-hexanone tosylhydrazone. Elimination of this β-hydroxy tosylhydrazone with methyllithium at room temperature leads to 4-hydroxy-1-phenyl-1-hexene (*cis*:*trans* = 15:85) in 43% overall yield.

$$R^1CH_2C(=N{\sim}NHTs)CH_2R^2 \xrightarrow[-50°]{2\ n\text{-}C_4H_9Li} R^1CH_2C(=N{\sim}\bar{N}TsLi^+)CHR^2(Li) \xrightarrow[2.\ H_2O]{1.\ R^3R^4CO}$$

$$R^1CH_2C(=N{\sim}NHTs)CH(R^2)C(OH)R^3R^4 \xrightarrow{\text{Excess alkyllithium}} R^1CH{=}CHCH(R^2)C(OH)R^3R^4 \quad \text{(Eq. 27)}$$

It is assumed that a vinyl anion is an intermediate in the reaction.[97] Quenching the reaction mixture with water gives homoallylic alcohols as described above, while treatment with carbon dioxide and trifluoroacetic acid provides α-methylene-γ-lactones, as shown in Eq. 28.[98] The reaction can be carried out in one pot starting from acetone 2,4,6-trimethylphenylsulfonylhydrazone, and various α-methylene-γ-lactones (R^1, R^2 = alkyl) are isolated in good yields (40–60%). The complete sequence can be performed without isolation of intermediates. In addition, modification of this

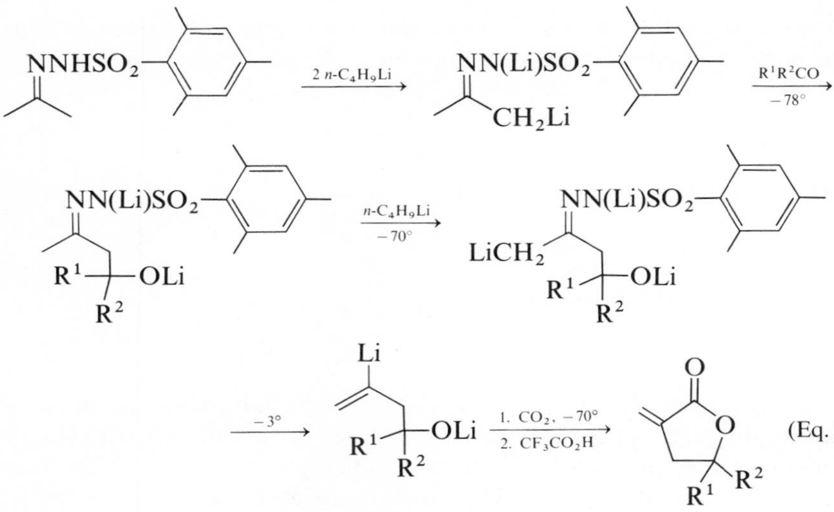

$\xrightarrow{2\ n\text{-}C_4H_9Li}$ $\xrightarrow[-78^\circ]{R^1R^2CO}$ $\xrightarrow[-70^\circ]{n\text{-}C_4H_9Li}$ $\xrightarrow{-3^\circ}$ $\xrightarrow[2.\ CF_3CO_2H]{1.\ CO_2,\ -70^\circ}$ (Eq. 28)

experimental procedure provides a convenient preparation of 3,6-dimethylenetetrahydropyran-2-one and 3,5-dimethylenetetrahydrofuran-2-one derivatives.[99]

Oximes

Dianions of substituted acetophenone oximes undergo aldol addition with aromatic ketones to give β-hydroxyoximes in fairly good yields.[100] β-Hydroxyoximes thus formed give the corresponding aldols on treatment with chromous acetate $[Cr(O_2CCH_3)_2]$. Similarly, the dianion of cyclohexanone oxime reacts with acetaldehyde to give the corresponding adduct, which when treated successively with chromous acetate and oxalic acid, is converted to 2-ethylidenecyclohexanone in 46% yield.[101]

O-Alkyloximes can also be used for the directed aldol reaction.[102] For example, treatment of the O-tetrahydropyranyl oxime of 2-heptanone with 1.1 equivalents of lithium diisopropylamide in tetrahydrofuran at -50° for 3 hours, followed by acetone, affords a single product in 95% yield (Eq. 29).

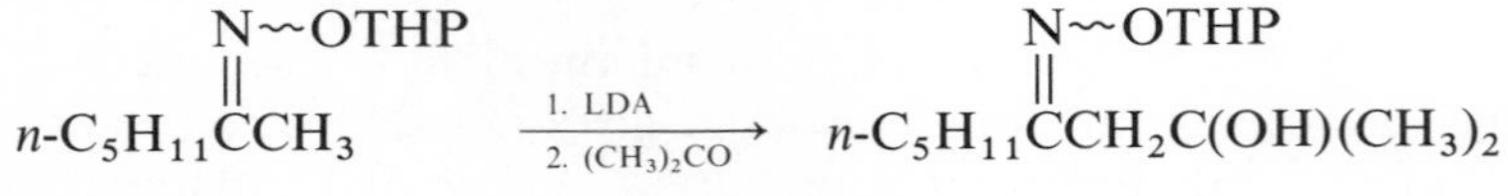

$$n\text{-}C_5H_{11}\overset{\overset{\displaystyle N\sim OTHP}{\|}}{C}CH_3 \xrightarrow[2.\ (CH_3)_2CO]{1.\ LDA} n\text{-}C_5H_{11}\overset{\overset{\displaystyle N\sim OTHP}{\|}}{C}CH_2C(OH)(CH_3)_2 \quad \text{(Eq. 29)}$$

REACTION OF MAGNESIUM, ZINC, AND ALUMINUM ENOLATES

Magnesium Enolates

The use of magnesium enolates is one of the practical methods for carrying out directed aldol reactions. These enolates are readily generated from sterically hindered ketones by the action of methylanilinomagnesium bromide, diisopropylaminomagnesium bromide, or a Grignard reagent. The enolates thus prepared react with aldehydes to give aldols in moderate yields.[103–110] However, this reaction is limited to those sterically hindered ketones that undergo little self-condensation in the presence of dialkylaminomagnesium bromide. Less sterically hindered ketones such as acetone and acetophenone cannot be used in this procedure.[107]

Magnesium enolates can also be generated by 1,4 addition of Grignard reagents to α,β-unsaturated ketones in the presence of a catalytic amount of cuprous salt.[57,111,112] Aldol reactions of these enolates with aldehydes are performed in ether at $-10°$ to $0°$ (Eq. 30).

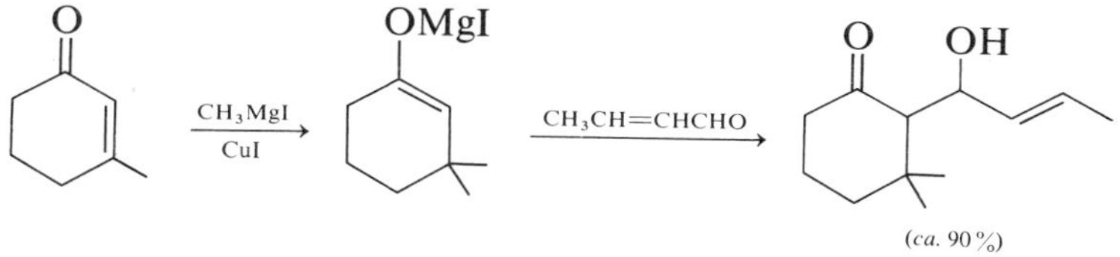
O
OMgI
O OH
CH3MgI
CuI
CH3CH=CHCHO
(ca. 90%)

(Eq. 30)

Enolates **60** generated by the reduction of sterically hindered α-haloketones such as α-bromopinacolone with magnesium metal also produce magnesium enolates that undergo the aldol reaction with aldehydes or ketones in refluxing ether.[113–116] In similar reactions using less sterically hindered

$$BrCH_2COC_4H_9\text{-}t \xrightarrow{Mg} \underset{\mathbf{60}}{CH_2{=}C(OMgBr)C_4H_9\text{-}t} \xrightarrow{R^1R^2CO} R^1R^2C(OH)CH_2COC_4H_9\text{-}t$$

α-haloketones the yields are generally low, probably because of self-condensation.[117,118] The magnesium enolate prepared from 4-bromo-3-hexanone has been used in the synthesis of isolasalocid ketone.[119] An

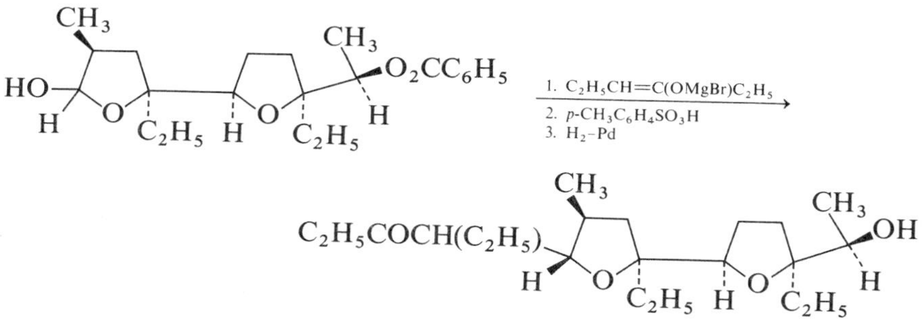
CH3
CH3
O2CC6H5
HO
H
O
C2H5
H
O
C2H5
H
1. C2H5CH=C(OMgBr)C2H5
2. p-CH3C6H4SO3H
3. H2-Pd
CH3
CH3
C2H5COCH(C2H5)
OH
H
O
C2H5
H
O
C2H5
H

alternative preparation of a magnesium enolate is the addition of anhydrous magnesium bromide to a preformed lithium enolate. This magnesium enolate undergoes directed aldol reactions in a similar fashion to zinc enolates.

Zinc Enolates

Enol acetates are cleanly cleaved by methyllithium in 1,2-dimethoxyethane at room temperature. The resulting lithium enolates are extremely reactive and subsequent reaction must be carried out at very low temperature (−78°) to avoid troublesome side reactions. Addition of an equimolar amount of zinc chloride makes the reaction more useful, since the resulting zinc enolate reacts smoothly at *ca.* 0° with an aldehyde to produce the corresponding aldol as the major product. 4-Hydroxy-3-phenyl-2-heptanone is prepared according to this procedure (Eq. 31).[120]

$$C_6H_5CH{=}C(CH_3)O_2CCH_3 \xrightarrow[\substack{3.\ n\text{-}C_3H_7CHO,\ 0\text{–}10^\circ \\ 4.\ H_2O\text{–}NH_4Cl,\ 0^\circ}]{\substack{1.\ 2CH_3Li \\ 2.\ ZnCl_2}} CH_3COCH(C_6H_5)CHOHC_3H_7\text{-}n \quad \text{(Eq. 31)}$$

A modified procedure, *i.e.*, the addition of zinc chloride to a lithium enolate prepared by using lithium diisopropylamide, has been used in the synthesis of lasalocid A[41] and ionophore A-23187.[121]

Phosphorylated enol esters are prepared in high yield by the reaction of certain α-haloketones with phosphinates, phosphonates, or phosphates.[122] These enol esters, on successive treatment with alkyllithium reagents and zinc chloride, produce regiospecifically zinc enolates, which react with carbonyl compounds to yield crossed aldols. This method has been utilized in the synthesis of (±)-$PGF_{2\alpha}$.[123]

The lithium enolate intermediates obtained by 1,4 addition of lithium dimethylcuprate to α,β-unsaturated ketones are converted to zinc enolates by

[Scheme: bromocyclopentanone bearing $C_6H_5CH_2O$ and side chain with $OCH_2OCH_2C_6H_5$, Br; $\xrightarrow[\text{Room temperature}]{(C_6H_5)_2POCH_3}$ enol phosphinate $OP(C_6H_5)_2$ (85%) $\xrightarrow{1.\ t\text{-}C_4H_9Li,\ -78^\circ;\ 2.\ ZnCl_2,\ -78^\circ;\ 3.\ HCHO,\ -78^\circ}$ 2-(hydroxymethyl)cyclopentanone, CH_2OH (80–90%) $\Longrightarrow$ $PGF_{2\alpha}$ (OH, CO_2H, HO, HO H)] adding zinc chloride. The zinc enolates thus formed react with acetaldehyde at 0° to give the aldols in high yields, whereas the direct reaction of lithium enolate **61** with carbonyl compounds does not afford the aldols.[60,61]

$$R^1R^2C{=}CR^3COR^4 \xrightarrow{(CH_3)_2CuLi}$$

$$\underset{\mathbf{61}}{R^1R^2(CH_3)CCR^3{=}C(OLi)R^4} \xrightarrow[2.\ CH_3CHO]{1.\ ZnCl_2} R^1R^2(CH_3)CC(R^3)(CHOHCH_3)COR^4$$

α-Bromoketones are also reduced by zinc powder in benzene and dimethyl sulfoxide as a mixed solvent to provide the zinc enolates, which react with aldehydes at the original site of the bromine atom. This procedure, which has its origins in the classical Reformatsky reaction, represents an extension of the directed aldol reaction. However, the zinc enolates thus formed are less reactive than those prepared from lithium enolates and zinc chloride. The reaction appears to be limited to aldehydes as carbonyl acceptors and requires a large excess of aldehyde and longer reaction time (24 hours) (Eq. 32).[124]

2-Bromo-5-methoxy-1-tetralone $\xrightarrow[2.\ \text{Excess}\ C_2H_5CHO]{1.\ Zn}$ 2-($CHOHC_2H_5$)-5-methoxy-1-tetralone (95%) (Eq. 32)

Regiospecific aldol reactions are also accomplished with 2,2,2-trichloroethyl esters of α-substituted β-keto acids, which are readily prepared from ketene dimer and 2,2,2-trichloroethanol.[125] These trichloroethyl esters, when reduced by zinc in dimethyl sulfoxide, followed by the reaction with an aldehyde, give aldols in high yields; the reaction with ketones is unsuccessful. The proposed reactive intermediate is the zinc enolate **62** of the β-keto acid. This method seems to be a modified Schöpf reaction which is carried out under almost neutral conditions.[126,127]

$$R^1COCH(R^2)COOCH_2CCl_3 \xrightarrow[25^\circ]{Zn} [\mathbf{62}] \xrightarrow[2.\ H_2O,\ -CO_2]{1.\ R^3CHO} R^1CH(OH)CH(R^2)COR^3$$

Aluminum Enolates

Organoaluminum enolates are potential reagents for regiospecific aldol reactions. The enolate has been prepared by 1,4 addition of trialkylaluminum to an α,β-unsaturated ketone in the presence of nickel acetylacetonate as a catalyst (*e.g.*, Eq. 33).[128]

$$(CH_3)_2C{=}CHCOCH_3 + (CH_3)_3Al \xrightarrow[(C_2H_5)_2O,\ -50^\circ \text{ to } -20^\circ]{Ni(acac)_2} t\text{-}C_4H_9CH{=}C(CH_3)OAl(CH_3)_2 \quad \text{(Eq. 33)}$$

An alternative preparation is the reaction of an enolizable ketone with trimethylaluminum at elevated temperature.[129] The dimeric aluminum enolate thus obtained does not react with 1,1,1-triphenylacetone, but reacts with acetone to give a crystalline aluminum complex **63** in high yield. Hydrolysis of the complex gives the corresponding β-hydroxyketone in 80% yield.[19,129,130]

$$(C_6H_5)_3CCOCH_3 + (CH_3)_3Al \xrightarrow[-CH_4]{110^\circ,\ 4\text{ hours}} [(C_6H_5)_3C(=CH_2)O\text{—}Al(CH_3)_2]_2 \ (89\%) \xrightarrow[50^\circ,\ 20\text{ min}]{2CH_3COCH_3}$$

$$[(CH_3)_2C(CH_2COC(C_6H_5)_3)O\text{—}Al(CH_3)_2]_2 \ \textbf{63} \xrightarrow{H_2O} 2(CH_3)_2C(OH)CH_2COC(C_6H_5)_3$$

High stereoselectivity is observed in the reaction of a stereodefined aluminum enolate with an aldehyde having a sterically hindered alkyl group (see Stereochemistry, p. 206).

An aluminum enolate can also be generated from an α-bromoketone by means of a dialkylaluminum chloride and zinc, as illustrated in Eq. 34.[131]

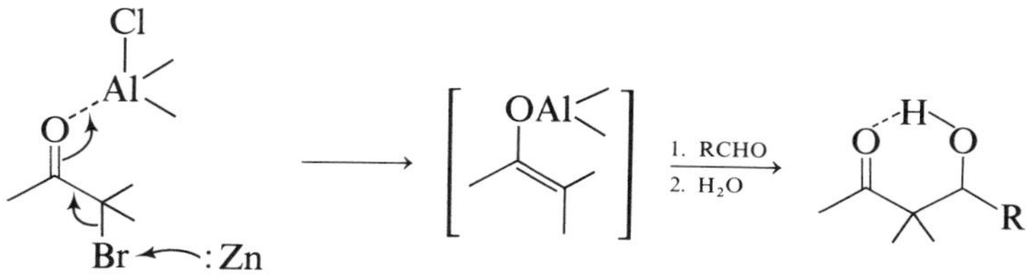

(Eq. 34)

Treatment of 2-bromo-2-methylcyclohexanone and benzaldehyde with diethylaluminum chloride–zinc in tetrahydrofuran at $-20°$ produces the expected β-hydroxy ketone (*threo*: *erythro* = 3:4) in quantitative yield.[131] The reaction of diethylaluminum 2,2,6,6-tetramethylpiperidide with an unsymmetrical ketone generates an aluminum enolate, which can be condensed with another carbonyl compound to give the crossed aldol. For instance, condensation of 2-methylcyclohexanone with benzaldehyde affords kinetically controlled products in 72% yield (Eq. 35).[132]

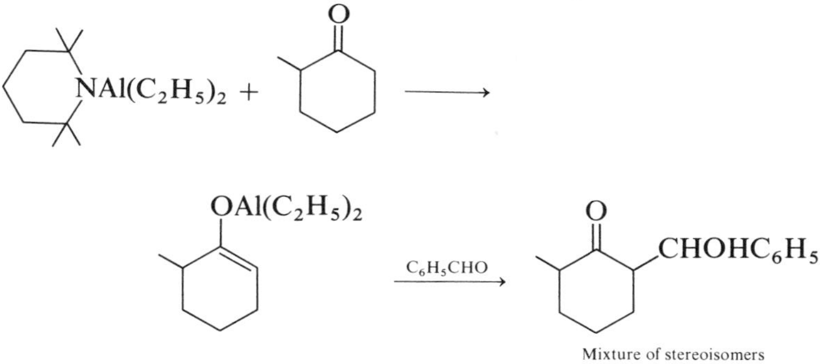

(Eq. 35)

Mixture of stereoisomers

A similar approach using the reagents dibutylaluminum phenoxide–pyridine has been applied to the preparation of *dl*-muscone.[133]

Another method of generating aluminum enolate uses $(CH_3)_2AlSC_6H_5$ or $(CH_3)_2AlSeCH_3$; these reagents add to α,β-unsaturated carbonyl compounds. The resulting aluminum enolates react with aldehydes to afford the corresponding aldols.[134] The reagents [$(CH_3)_2AlSC_6H_5$ or $(CH_3)_2AlSeCH_3$] are prepared *in situ* at 0° just before use from $(CH_3)_3Al$ and C_6H_5SH or CH_3SeH. The advantage of this method is that the α,β-enone system can be regenerated after the aldol condensation by β-elimination following oxidation of S or Se in the aldol.

REACTION OF VINYLOXYBORANES AND VINYLAMINOBORANES

Vinyloxyboranes

A highly efficient method using vinyloxyboranes has been described for the preparation of crossed aldols. Equation 36 outlines the reaction, which

$$R^1R^2C{=}CR^3(OBR^4_2) + R^5R^6CO \longrightarrow \text{[cyclic six-membered B–O chelate: } R^5R^6C–C(R^1R^2)–C(R^3){=}O \cdots B(R^4)_2–O\text{]} \xrightarrow{H_2O} R^5R^6C(OH)C(R^1R^2)COR^3 \quad \text{(Eq. 36)}$$

proceeds smoothly at room temperature to give β-hydroxycarbonyl compounds in excellent yields. For example, treatment of vinyloxyborane **64** with benzaldehyde at room temperature for 10 minutes gives crossed aldol **65** in 98% yield.[18,135,136] This method is a typical example of an aldol reaction carried out under neutral conditions.

$$\underset{\mathbf{64}}{n\text{-}C_4H_9CH{=}C(C_6H_5)OB(C_4H_9\text{-}n)_2} \xrightarrow[2.\ H_2O]{1.\ C_6H_5CHO} \underset{\mathbf{65}}{C_6H_5CHOHCH(C_4H_9\text{-}n)COC_6H_5}$$

Vinyloxyboranes can be prepared *in situ* by reaction of a diazoketone with a trialkylborane or by the 1,4 addition of an alkylborane to an α,β-unsaturated ketone.[135–141] The reaction of an enolizable ketone and a trialkylborane in the presence of diethylboryl pivalate also affords vinyloxyboranes (Eq. 37).[142,143] An alternative method is the reaction of enolizable aldehydes

$$R^1CH_2COR^2 + B(C_2H_5)_3 \xrightarrow{t\text{-}C_4H_9CO_2B(C_2H_5)_2} R^1CH{=}C(R^2)OB(C_2H_5)_2 + C_2H_6 \quad \text{(Eq. 37)}$$

with a 4-dialkylboryloxy-2-isopropyl-6-methylpyrimidine, as shown in Eq. 38.[144]

$$BR_3 + \text{2-isopropyl-6-methylpyrimidin-4(3H)-one} \longrightarrow \text{4-}(R_2BO)\text{-2-isopropyl-6-methylpyrimidine} \xrightarrow{R^1R^2CHCHO} R^1R^2C{=}CHOBR_2 \quad \text{(Eq. 38)}$$

A convenient method for the generation of vinyloxyboranes from a wide variety of enolizable ketones has been reported recently. Dialkylboryl trifluoromethanesulfonate (R_2BOTf) reacts with ketones in the presence of a tertiary amine to produce vinyloxyboranes. The vinyloxyboranes thus generated react with aldehydes to give crossed aldols in high yields (Eq. 39).[17,34–38]

$$R^1COCH_2R^2 \xrightarrow[\text{Tertiary amine}]{R_2BSO_3CF_3} R^1C(OBR_2){=}CHR^2 \xrightarrow[\text{2. } H_2O]{\text{1. } R^3CHO} R^1COCH(R^2)CHOHR^3 \quad \text{(Eq. 39)}$$

Regioselective generation of vinyloxyborane is readily controlled by a suitable combination of reagents. For example, the reaction of 4-methyl-2-pentanone with dibutylboryl trifluoromethanesulfonate and diisopropylethylamine produces the kinetically controlled vinyloxyborane **66**, which then reacts with an aldehyde to afford the β-hydroxyketone **67**.[36]

$$i\text{-}C_3H_7CH_2COCH_3 \xrightarrow[(i\text{-}C_3H_7)_2NC_2H_5]{(n\text{-}C_4H_9)_2BSO_3CF_3} \underset{\mathbf{66}}{i\text{-}C_3H_7CH_2C(OB(C_4H_9\text{-}n)_2){=}CH_2} \xrightarrow[\text{2. } H_2O]{\text{1. RCHO}} \underset{\mathbf{67}}{i\text{-}C_3H_7CH_2COCH_2CHOHR}$$

In contrast, the thermodynamically stable vinyloxyborane is generated by the reaction of the ketone and 9-borabicyclo[3.3.1]-9-nonanyl trifluoromethanesulfonate (9-BBN-triflate) in the presence of 2,6-lutidine at $-78°$ for 3 hours. Subsequent reaction with an aldehyde gives the corresponding aldol **68**.[17,37]

$$R^1CH_2COCH_3 \xrightarrow[\text{2,6-lutidine}]{\text{9-BBN}(SO_3CF_3)} R^1CH{=}C(CH_3)O\text{B(9-BBN)} \xrightarrow{R^2CHO} \underset{\mathbf{68}}{R^1CH(COCH_3)CHOHR^2}$$

γ-Ionone is synthesized without any contamination by isomeric α- and β-ionones by a crossed aldol reaction of the vinyloxyborane **69** with γ-citral.[145]

$$CH_3COCH_3 + (n\text{-}C_4H_9)_2BSO_3CF_3 \xrightarrow{(C_2H_5)_3N}$$

$$CH_2{=}C(CH_3)OB(C_4H_9\text{-}n)_2 \;\; \mathbf{69}$$

Activated alumina

Silyl enol ethers are readily converted to the corresponding boron enolates by treatment with dibutylboryl trifluoromethanesulfonate followed by removal of the resulting trimethylsilyl trifluoromethanesulfonate. The stereo- and regioselectively prepared (*Z*)-trimethylsilyl enol ethers from rearrangement of 1-trimethylsilylallylic alcohols[145a] are transformed to dibutylboron enolates by this procedure without any loss of stereo- and regiochemical integrities. The (*Z*)-boron enolates thus formed react with aldehydes in a highly stereocontrolled manner to give almost exclusively *erythro* aldols.[146]

OH, R^1, R^2, $Si(CH_3)_3$ $\xrightarrow{\text{cat. } n\text{-}C_4H_9Li}$ $OSi(CH_3)_3$, R^1, R^2 $\xrightarrow[-(CH_3)_3SiO_3SCF_3]{(n\text{-}C_4H_9)_2BO_3SCF_3}$

[$OB(C_4H_9\text{-}n)_2$, R^1, R^2] $\xrightarrow{R^3CHO}$ R^1, O, OH, R^3, R^2, CH_3

Vinylaminoboranes

Vinylaminoboranes have been found to undergo the aldol reaction with carbonyl compounds. A typical example is illustrated for the reaction of *N*-cyclohexenylcyclohexylaminodichloroborane with benzaldehyde in Eq. 40.[147] The reaction proceeds in a manner mechanistically similar to the vinyloxyborane condensation reaction, but appears to be slower and the yields of aldol products are not as high. Vinylaminoboranes such as hexenylaminodichloroborane can be isolated in pure form from the reaction of a ketimine, boron trichloride, and triethylamine in dichloromethane. However, such isolation is not always necessary for the aldol reaction since the reaction can be performed in a one-pot procedure starting from the imine. In this case the yields of aldols are at most 76%

$$\text{1-cyclohexenyl-}N(C_6H_{11})BCl_2 \xrightarrow[\text{Room temperature, 2 hours}]{C_6H_5CHO,\ N(C_2H_5)_3} [\text{cyclic N=}C\text{, }BCl_2\text{, O, }C_6H_5] \xrightarrow{H_2O} \text{2-}(CHOHC_6H_5)\text{cyclohexanone} \quad (\text{Eq. 40})$$

threo:*erythro* = 1:2
(71%)

An enantioselective aldol reaction has been reported using a chiral vinylaminodichloroborane. The aldol **70** is obtained in 34% yield and 41% enantiomeric excess.[148]

$$C_6H_5C(CH_3){=}N\text{-bornyl (H)} + BCl_3 \xrightarrow{(C_2H_5)_3N}$$

$$C_6H_5C(=CH_2)N(BCl_2)\text{-bornyl (H)} \xrightarrow[2.\ H_2O]{1.\ C_6H_5CHO,\ (C_2H_5)_3N} C_6H_5COCH_2\overset{*}{C}HOHC_6H_5$$

70

Some vinyloxyaminochloroboranes, which are prepared from ketones, diethylaminodichloroboranes, and triethylamine, are stable and isolable compounds that undergo crossed aldol reactions with aldehydes in the presence of triethylamine (Eq. 41).[149]

$$\text{cyclohexanone} + (C_2H_5)_2NBCl_2 + (C_2H_5)_3N \xrightarrow{4\text{–}8^\circ}$$

$$\text{1-cyclohexenyl-}OB(Cl)N(C_2H_5)_2 \xrightarrow[\text{Room temperature}]{p\text{-}O_2NC_6H_4CHO,\ (C_2H_5)_3N} \text{2-}(CHOHC_6H_4NO_2\text{-}p)\text{cyclohexanone} \quad (\text{Eq. 41})$$

(56%)

(74%)
(*threo*:*erythro* = 1:2.4)

LEWIS ACID–PROMOTED ALDOL REACTIONS

Acid catalysts are used less than base catalysts for conventional aldol reactions because the resulting aldol is often dehydrated to the α,β-unsaturated carbonyl compound and the yield of the aldol is generally low. In contrast, the reaction of enol ethers with acetals or ketals in the presence of Lewis acids such as boron trifluoride etherate or zinc chloride in an aprotic solvent yields aldol-type adducts.[150,151] However, when stoichiometric amounts of carbonyl compounds and enol ethers are employed for the reaction, yields of 1:1 adducts are generally low because undesirable polymerization of enol ethers occurs under these conditions.

The use of stoichiometric amounts of titanium tetrachloride, trimethylsilyl enol ether, and a carbonyl compound is a great advance in the directed crossed aldol reaction.[152] Trimethylsilyl enol ethers are highly versatile enol derivatives that can be prepared regioselectively from various ketones under conditions of either kinetic or thermodynamic control.[65,153–164] Furthermore, regiospecific syntheses of trimethylsilyl enol ethers can also be accomplished by (1) trapping the enolate anions formed from α,β-unsaturated carbonyl compounds[165–168] or α-haloketones,[169] and (2) thermal rearrangement of trimethylsilyl-β-ketoesters[170] or siloxyvinyl cyclopropanes.[171,172]

Reaction of Trimethylsilyl Enol Ethers with Aldehydes and Ketones

The reaction of trimethylsilyl enol ethers with carbonyl compounds is promoted by suitable Lewis acids such as titanium tetrachloride, stannic chloride, aluminum chloride, boron trifluoride etherate, or zinc chloride. Reaction of 1-trimethylsiloxycyclohexene with benzaldehyde in the presence of various Lewis acids has been examined (Eq. 42).[173]

$OSi(CH_3)_3$ (71) $+ C_6H_5CHO \xrightarrow[CH_2Cl_2]{MX_n} \xrightarrow{H_2O}$ $CHOHC_6H_5$ (72)

$+$ $CH[OSi(CH_3)_3]C_6H_5$ (73) $+$ CHC_6H_5 (74) (Eq. 42)

These results with silyl enol ether **71** can be summarized as follows:

Lewis Acid	Conditions: Temperature	Conditions: Time (hours)	Yield of Products (%): **72**	**73**	**74**
$TiCl_4$	RT[a]	2	82	Trace	2
$TiCl_4$	−78°	1	92	0	0
$SnCl_4$	RT	1	33	Trace	28
$SnCl_4$	−78°	1	83	Trace	Trace
$FeCl_3$	RT	1	0	0	12
$AlCl_3$	RT	1	55	Trace	Trace
BCl_3	RT	1	26	0	24
$Et_2O \cdot BF_3$	−78°	1	80	12	0
$ZnCl_2$	RT	10	69	8	3
$ZnCl_2$	−78°	12	Trace	0	0
$(n\text{-}C_4H_9)_3SnCl$	RT	24	0	0	0
$MgCl_2$	RT	24	0	0	0
$CdCl_2$	RT	24	0	0	0
LiCl	RT	24	0	0	0

[a] The reaction was performed at room temperature.

Titanium tetrachloride appears to be superior to other Lewis acids with regard to yield. This titanium tetrachloride–promoted aldol reaction is assumed to proceed by the pathway shown in Eq. 43. Trimethylsilyl enol ethers attack carbonyl compounds activated by titanium tetrachloride to

$$R^1R^2C{=}CR^3[OSi(CH_3)_3] + R^4R^5CO \xrightarrow[-(CH_3)_3SiCl]{TiCl_4} \underset{\mathbf{75}}{R^3C(=O\cdots TiCl_3)CR^1R^2CR^4R^5O\text{-}} \xrightarrow{H_2O} R^3C(=O\cdots H)CR^1R^2CR^4R^5O\text{-} \quad \text{(Eq. 43)}$$

form trimethylsilyl chloride and titanium salts of the adducts. The undesirable retro aldol reaction of the adducts is inhibited by the formation of stable titanium chelates **75**, hydrolysis of which yields the desired aldols.[173] The reaction proceeds with retention of the regiochemical integrity of the starting silyl enol ethers to afford the corresponding crossed aldols.[173–177] Enol ethers derived from ketones or aldehydes react smoothly with aldehydes at −78°, whereas more vigorous reaction conditions (0° to room temperature) are required for ketones.

The reaction usually gives a mixture of stereoisomers (*threo* and *erythro*) of the aldol in comparable amounts; for example, the reaction of 1-trimethylsiloxycyclopentene with β-phenylpropionaldehyde in the presence of titanium tetrachloride at −78° yields a mixture of aldols (*threo*:*erythro* = 1:1) in 68% yield.

$$\text{1-}(OSi(CH_3)_3)\text{cyclopentene} + C_6H_5(CH_2)_2CHO \xrightarrow[\text{2. } H_2O]{\text{1. } TiCl_4,\ -78°} \text{2-}[CHOH(CH_2)_2C_6H_5]\text{cyclopentanone}$$

threo:*erythro* = 1:1

The trimethylsilyl enol ether **76** reacts with 1,3,5-trioxane or paraldehyde at −78° in the presence of titanium tetrachloride to give the aldol product in good yields.[173]

$$(RCHO)_3 + \underset{\mathbf{76}}{C_6H_5CH{=}C(CH_3)[OSi(CH_3)_3]} \xrightarrow[\text{2. } H_2O]{\text{1. } TiCl_4} RCHOHCH(C_6H_5)COCH_3$$

R = H (64%)
R = CH_3 (92%)

Chemoselectivity is observed with acceptors having two different kinds of carbonyl functions such as aldehyde and ketone or ester in the same molecule. Treatment of phenylglyoxal with enol ether **77** at −78° affords the α-hydroxy-γ-diketone **78**.[173]

$$C_6H_5COCHO + \underset{\mathbf{77}}{C_6H_5CH{=}C(CH_3)[OSi(CH_3)_3]} \xrightarrow[\text{2. } H_2O]{\text{1. } TiCl_4} \underset{\mathbf{78}}{C_6H_5COCHOHCH(C_6H_5)COCH_3}$$

The reaction of ketoesters other than β-ketoesters with the enol ether **77** gives hydroxyketoesters **79** as sole products.[176,177]

$$CH_3CO(CH_2)_nCO_2R + \underset{\mathbf{77}}{C_6H_5CH{=}C(CH_3)[OSi(CH_3)_3]} \xrightarrow[\text{2. } H_2O]{\text{1. } TiCl_4} \underset{\mathbf{79}}{CH_3COCH(C_6H_5)C(OH)(CH_3)(CH_2)_nCO_2R}$$

n = 0, 2, 3
R = CH_3, C_2H_5

The reaction of (+)-1-methylbutanal with 2-trimethylsiloxy-2-butene proceeds quite smoothly at $-78°$ without racemization of the aldehyde to give a diastereomeric mixture of aldol products, which in turn is dehydrated on refluxing in benzene in the presence of a catalytic amount of *p*-toluenesulfonic acid to give optically pure (+)-manicone, an alarm pheromone.[178]

$$(+)\text{-}C_2H_5\overset{*}{C}H(CH_3)CHO + CH_3CH{=}C(C_2H_5)OSi(CH_3)_3 \xrightarrow[2.\ H_2O]{1.\ TiCl_4,\ -78°}$$

$$C_2H_5\overset{*}{C}H(CH_3)CHOHCH(CH_3)COC_2H_5 \xrightarrow{-H_2O}$$

threo : *erythro* = 45 : 1
(92%)

$$\underset{(83\%)}{C_2H_5(CH_3)\overset{*}{C}H{-}CH{=}C(CH_3)COC_2H_5}$$

Asymmetric induction is observed in the similar reaction of silyl enol ethers with (−)-menthyl pyruvate (Eq. 44).[179]

$$R^1COCO_2\text{-Menthyl} + R^2R^3C{=}CR^4OSiR_3 \xrightarrow[2.\ H_2O]{1.\ TiCl_4}$$

$$R^4COC(R^2)(R^3)\overset{*}{C}(OH)(R^1)CO_2\text{-Menthyl} \quad \text{(Eq. 44)}$$

(up to 68% e.e.)

Reaction of Trimethylsilyl Enol Ethers with Acetals and Ketals

The advantage of using acetals or ketals instead of aldehydes or ketones is that they act only as electrophiles and probably coordinate with Lewis acids more strongly than the parent carbonyl compounds. Trimethylsilyl enol ethers react readily with acetals or ketals at $-78°$ in the presence of titanium tetrachloride to afford β-alkoxy carbonyl compounds in high yields (Eq. 45).[180]

$$R^1R^2C(OR)_2 + R^3R^4C{=}CR^5OSi(CH_3)_3 \xrightarrow[2.\ H_2O]{1.\ TiCl_4}$$

$$R^1R^2C(OR)C(R^3)(R^4)COR^5 \quad \text{(Eq. 45)}$$

Titanium tetrachloride–promoted reactions of silyl enol ethers with α-bromoacetals or ketals provide β-alkoxy-γ-bromoketones, **80**, which are easily converted into furans by refluxing in toluene. However, in one case (R^1, R^2, R^4 = H, R^3 = C_6H_5) a γ-bromo-α,β-unsaturated ketone becomes the major product (Eq. 46).[181] The reaction of 1-trimethylsiloxy-1-cyclopentene with α-bromoacetals affords β-alkoxy-γ-bromoketones, which on successive treatments with *p*-toluenesulfonic acid and triethylamine are converted into cyclopentenone derivatives **81** in good yields (Eq. 47).[182]

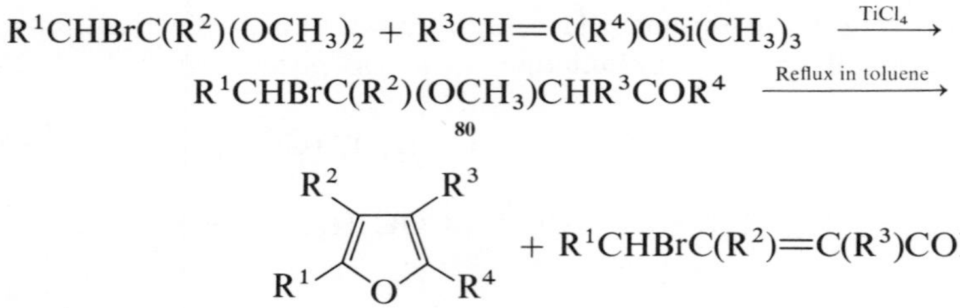

$$R^1CHBrC(R^2)(OCH_3)_2 + R^3CH{=}C(R^4)OSi(CH_3)_3 \xrightarrow{TiCl_4} R^1CHBrC(R^2)(OCH_3)CHR^3COR^4 \ (\mathbf{80}) \xrightarrow{\text{Reflux in toluene}}$$

$$\text{2-}R^1\text{-3-}R^2\text{-4-}R^3\text{-5-}R^4\text{-furan} + R^1CHBrC(R^2){=}C(R^3)COR^4 \qquad \text{(Eq. 46)}$$

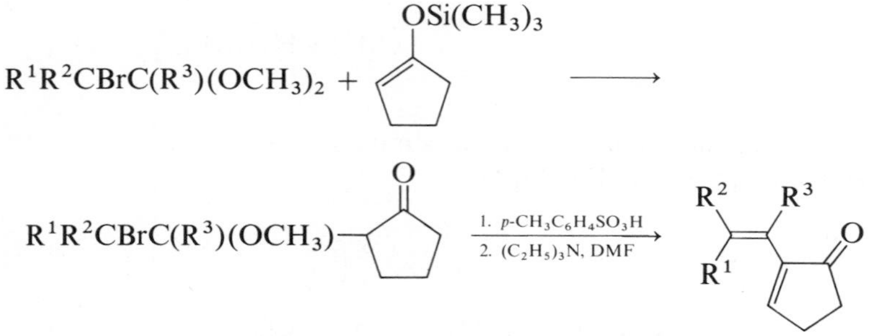

$$R^1R^2CBrC(R^3)(OCH_3)_2 + \text{1-}OSi(CH_3)_3\text{-cyclopentene} \longrightarrow$$

$$\text{2-}[R^1R^2CBrC(R^3)(OCH_3)]\text{cyclopentanone} \xrightarrow[\text{2. } (C_2H_5)_3N,\ DMF]{\text{1. } p\text{-}CH_3C_6H_4SO_3H} \text{2-}[R^1R^2C{=}C(R^3)]\text{cyclopent-2-enone}\ (\mathbf{81}) \qquad \text{(Eq. 47)}$$

A related aldol reaction of 1-trimethylsiloxycyclohexene with 1-O-acetyl-2,3,5-tri-O-benzoyl-*β*-D-ribofuranose in the presence of stannic chloride at 25° affords 2-(2,3,5-tri-O-benzoyl-D-ribofuranosyl) cyclohexanone in high yield.[183]

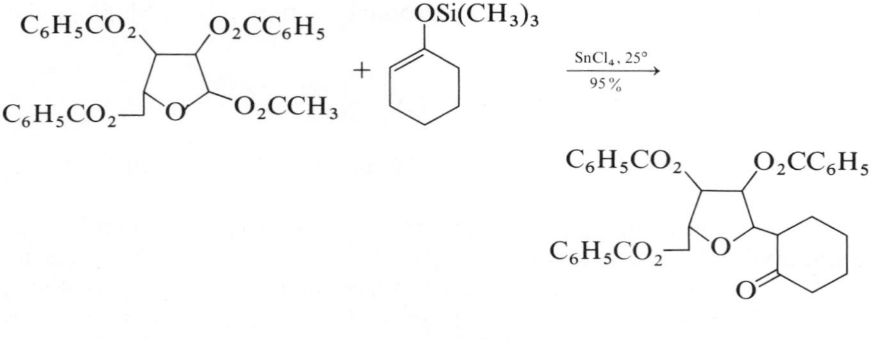

$$\text{1-O-acetyl-2,3,5-tri-O-benzoyl-ribofuranose } (C_6H_5CO_2, O_2CC_6H_5, O_2CCH_3) + \text{1-}OSi(CH_3)_3\text{-cyclohexene} \xrightarrow[95\%]{SnCl_4,\ 25^\circ} \text{2-(2,3,5-tri-O-benzoyl-ribofuranosyl)cyclohexanone}$$

The aldol reaction of trimethylsilyl enol ethers of trimethylsilyl ketones with acetals in the presence of boron trifluoride etherate at low temperature provides alkoxyacylsilanes, which are readily converted into (*E*)-α,β-unsaturated aldehydes in high yields by treatment with catalytic amounts of tetrabutylammonium hydroxide (Eq. 48). However, ketals react sluggishly under similar conditions.[184]

$$R^1CH(OR^2)_2 + R^3CH{=}C(OSi(CH_3)_3)Si(CH_3)_3 \xrightarrow[-78^\circ \text{ to } -30^\circ]{BF_3\cdot O(C_2H_5)_2}$$

$$R^1CH(OR^2)CH(R^3)COSi(CH_3)_3 \xrightarrow{(n\text{-}C_4H_9)_4NOH} (E)\text{-}R^1CH{=}C(R^3)CHO \qquad \text{(Eq. 48)}$$

1,2-Bis(trimethylsiloxy)cyclobutene **82** is used as a strained silyl enol ether in aldol-type reactions with aldehydes, acetals, and ketals in the presence of titanium tetrachloride or boron trifluoride etherate. The adducts (**83** and **84**) are easily converted into cyclopentanedione derivatives in high yields via pinacol rearrangement.[185]

$$R^1CHO + \underset{\mathbf{82}}{1,2\text{-}[(CH_3)_3SiO]_2\text{-cyclobutene}} \xrightarrow[-78^\circ]{TiCl_4 \text{ or } BF_3\cdot O(C_2H_5)_2}$$

$$R^1CH[OSi(CH_3)]\text{-}C_4H_4O(OSi(CH_3)_3)\ (\mathbf{83}) \xrightarrow{CF_3CO_2H} \text{2-}R^1\text{-3-hydroxy-2-cyclopentenone}$$

$$R^1R^2C(OR)_2 + 1,2\text{-}[(CH_3)_3SiO]_2\text{-cyclobutene} \xrightarrow{BF_3\cdot O(C_2H_5)_2}$$

$$R^1R^2C(OR)\text{-}C_4H_4O(OSi(CH_3)_3)\ (\mathbf{84}) \xrightarrow{CF_3CO_2H} \text{2-}R^1\text{-2-}R^2\text{-1,3-cyclopentanedione}$$

Dienoxylsilanes **85**, prepared from crotonaldehyde or dimethylacrolein and chlorotrimethylsilane, react with acetals at the C4 carbon atom in the presence of titanium tetrachloride and titanium tetraisopropoxide to give vinylogous δ-alkoxy-α,β-unsaturated aldehyde condensation products **86**.[186–188]

$$R^1CH(OR)_2 + CH_2{=}C(R^2)CH{=}CHOSi(CH_3)_3 \xrightarrow{TiCl_4-Ti(OC_3H_7\text{-}i)_4}$$

85

$$R^1CH(OC_3H_7\text{-}i)CH_2C(R^2){=}CHCHO \xrightarrow[\text{Molecular sieves}]{\text{Tertiary amine}} R^1CH{=}CHC(R^2){=}CHCHO$$

86

This method has been used successfully in the syntheses of natural products such as vitamin A,[189] variotin,[190,191] and hypacrone.[192]

Intramolecular cyclization is a useful method for the construction of ring systems. An example using titanium tetrachloride as the condensation reagent has been reported (Eq. 49).[193]

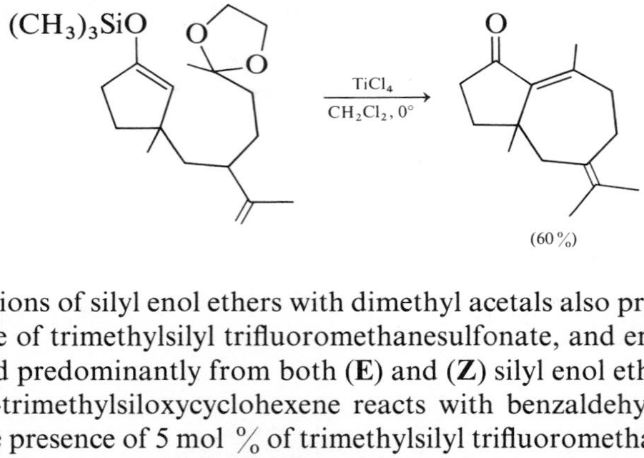

$(CH_3)_3SiO$ … $\xrightarrow[CH_2Cl_2,\ 0°]{TiCl_4}$ … (60%) (Eq. 49)

The reactions of silyl enol ethers with dimethyl acetals also proceed by the catalytic use of trimethylsilyl trifluoromethanesulfonate, and erythro aldols are obtained predominantly from both **(E)** and **(Z)** silyl enol ethers.[193a] For example, 1-trimethylsiloxycyclohexene reacts with benzaldehyde dimethyl acetal in the presence of 5 mol % of trimethylsilyl trifluoromethanesulfonate at −78°C to give the corresponding β-methoxyketone (erythro: threo = 93 : 7).

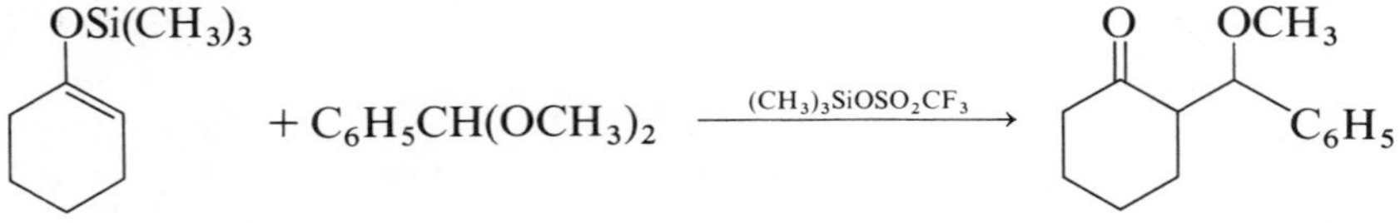

$OSi(CH_3)_3$ + $C_6H_5CH(OCH_3)_2$ $\xrightarrow{(CH_3)_3SiOSO_2CF_3}$ O, OCH_3, C_6H_5

erythro : threo = 93 : 7

Reaction of Alkyl Enol Ethers with Aldehydes, Acetals, and Ketals

The reaction of alkyl enol ethers is much slower than that of trimethylsilyl enol ethers. The silicon–oxygen bond of silyl enol ethers is easily cleaved leading to rapid formation of the aldol-type adduct **87**. On the other hand, the stable carbocation **88** may be formed in the case of the reaction of alkyl enol ethers, because the carbon–oxygen bond is hard to cleave. In this case the carbocation **88** often reacts further with enol ether, leading to polymerization.

$OSi(CH_3)_3$

87

OR

88

Alkyl enol ethers react with 2 molar equivalents of aldehyde in the presence of a catalytic amount of boron trifluoride etherate to give 1,3-dioxane derivatives, which on treatment with mineral acid are converted to α,β-unsaturated carbonyl compounds in varying yields (20–80%) (Eq. 50).[194–198]

$$2R^1CHO + R^2CH{=}C(R^3)OR \xrightarrow[\text{Reflux in }(C_2H_5)_2O]{BF_3\cdot O(C_2H_5)_2}$$

$$\text{(1,3-dioxane: } R^1, R^1, R^2, R^3, OR) \xrightarrow{H_3O^+} R^1CH{=}C(R^2)COR^3 \quad \text{(Eq. 50)}$$

The combined use of titanium tetrachloride and titanium tetraisopropoxide gives more favorable results. Such a reaction proceeds under mild conditions to give a 1:1 adduct, which on treatment with an alcohol or water affords alkoxyacetals **89** or α,β-unsaturated carbonyl compounds **90**, respectively.[199]

$$R^1CHO + R^2CH{=}C(R^3)OR \xrightarrow[CH_2Cl_2,\ -78^\circ \text{ to } 0^\circ]{TiCl_4-Ti(OC_3H_7\text{-}i)_4}$$

$\xrightarrow{R^4OH}$ $R^1CH(OC_3H_7\text{-}i)CH(R^2)C(R^3)(OR^4)(OR)$ **89**

$\xrightarrow{H_2O}$ $R^1CH{=}C(R^2)COR^3$ **90**

The reaction of enol ethers with aldehydes and ketones in the presence of boron trifluoride etherate gives α,β-unsaturated aldehydes or β-alkoxyketones in poor yields.[196,200]

The reaction of acetals with alkyl enol ethers catalyzed by boron trifluoride etherate or zinc chloride has been investigated since 1939, and its scope and limitations are well described.[150,151,201,202]

The reaction of α,β-unsaturated acetals with enol ethers has been applied to the synthesis of naturally occurring polyenes.[203–209] Alkoxydienes have been used similarly as the enol ether component for the reaction with α,β-unsaturated acetals in the presence of zinc chloride (Eq. 51).[210–213] However, lower yields are obtained than in the titanium tetrachloride–promoted reaction of trimethylsilyl enol ethers.

$$(CH_3)_2C{=}CCH(OR)_2 + CH_2{=}C(CH_3)CH{=}CHOR \xrightarrow{ZnCl_2}$$

$$(CH_3)_2C{=}CHCH(OR)CH_2C(CH_3){=}CHCH(OR)_2 \quad \text{(Eq. 51)}$$

Reaction of Enol Esters with Aldehydes, Acetals, and Ketals

Enol esters are effective nucleophiles for certain Lewis acid–promoted aldol reactions, although their reactivity is much lower than that of silyl enol ethers. Isopropenyl acetate reacts with acetals in the presence of titanium tetrachloride in methylene chloride to give the β-alkoxyketones in good yields (Eq. 52).[214] However, the reaction with benzaldehyde gives a mixture of aldol-type products such as β-acetoxy, β-hydroxy, β-chloro, and α,β-unsaturated methyl ketones.[214]

$$RCH(OCH_3)_2 + CH_2{=}C(CH_3)O_2CCH_3 \xrightarrow[CH_2Cl_2]{TiCl_4}$$

$$RCH(OCH_3)CH_2COCH_3 \quad \text{(Eq. 52)}$$

Diketene, an enol ester activated by strain, reacts with acetals and ketals at low temperature in the presence of titanium tetrachloride to give δ-alkoxy-β-ketoesters in good yields (Eq. 53).[215] Aldehydes are good acceptors for the

$$R^1R^2C(OR)_2 + \text{diketene} \xrightarrow[2.\ R^3OH]{1.\ TiCl_4,\ -78\ to\ -20^\circ}$$

$$R^1R^2C(OR)CH_2COCH_2CO_2R^3 \quad \text{(Eq. 53)}$$

reaction of diketene, and the products are δ-hydroxy-β-ketoesters. Thus the functional group $-CH_2COCH_2CO_2R$ is introduced into a carbonyl compound under acidic conditions.[216,217] This reaction has been used in syntheses of (±)-pestalotin and (±)-epipestalotin, as shown in Eq. 54.

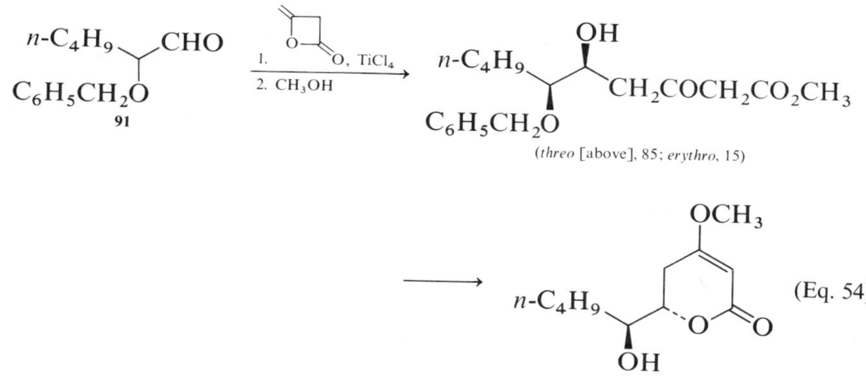
n-C4H9
CHO
C6H5CH2O
1. O, TiCl4
2. CH3OH
OH
n-C4H9
CH2COCH2CO2CH3
C6H5CH2O
OCH3
n-C4H9
O
O
OH

91

(*threo* [above], 85; *erythro*, 15)

(Eq. 54)

[(±)-pestalotin (above), 85; (±)-epipestalotin, 15]

Predominant formation of the *threo* isomer [(±)pestalotin] is rationalized by assuming a fixed conformation of aldehyde **91** by coordination of titanium tetrachloride to the two oxygen atoms as illustrated in formula **92**. Nucleophilic attack of diketene on the complex **92** then occurs from the less hindered side to give the *threo* adduct **93** preferentially.

δ-Hydroxy-β-ketoesters have also been prepared by condensation of aldehydes or ketones with the 1,3-dianion of acetoacetic esters under basic conditions.[218,219]

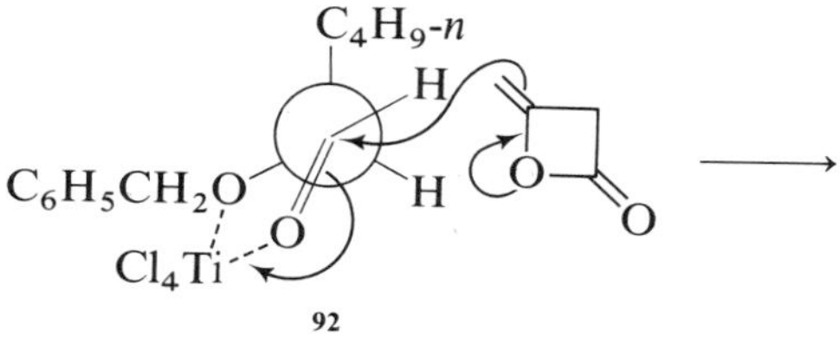
C4H9-n
H
C6H5CH2O
H
O
Cl4Ti
O
O

92

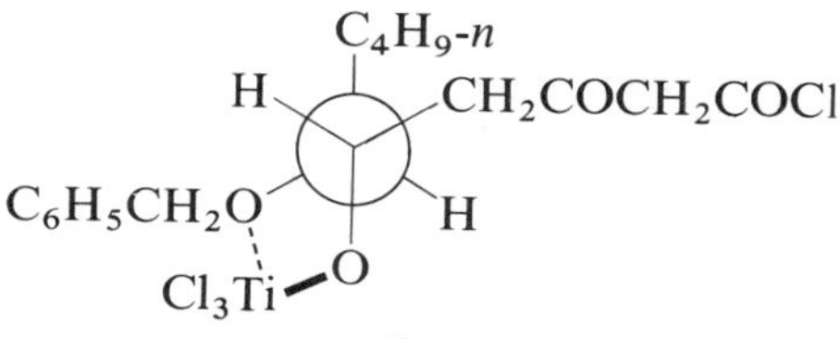
C4H9-n
H
CH2COCH2COCl
C6H5CH2O
H
Cl3Ti
O

93

The reaction of diketene with acetaldehyde diethyl acetal in the presence of boron trifluoride etherate is reported to give ethyl 2-(1-ethoxyethyl)acetoacetate as the major product.[220]

$$CH_3CH(OC_2H_5)_2 + \text{(diketene)} \xrightarrow{BF_3 \cdot O(C_2H_5)_2} CH_3COCH(CH(CH_3)OC_2H_5)CO_2C_2H_5$$

Cyclic enol esters such as α-angelicalactone and 3,4-dihydro-6-methyl-2H-pyran-2-one react with aldehydes in methylene chloride at 0° in the presence of Lewis acids to give β-acetyl-γ-lactones and γ-acetyl-δ-lactones in high yields.[221,222] In this reaction boron trifluoride etherate is superior to titanium tetrachloride and stannic chloride (Eq. 55). Reaction of 2-methyl-α-angelica-lactone with formaldehyde provides predominantly *cis*-β-acetyl-α-methyl-γ-butyrolactone (92% yield, *cis*: *trans* = 85: 15), which has been converted to the antibiotic botryodiplodin (Eq. 56).[223]

$$RCHO + \text{(cyclic enol lactone, } (CH_2)_n\text{)} \xrightarrow[CH_2Cl_2,\ 0^\circ]{BF_3 \cdot O(C_2H_5)_2} \text{(lactone with } CH_3CO, (CH_2)_n, R\text{)} \quad \text{(Eq. 55)}$$

$n = 1, 2$

Mixture of *cis* and *trans* isomers

$$RCHO + \text{(2-methyl-α-angelicalactone)} \xrightarrow[CH_2Cl_2,\ 0^\circ]{BF_3 \cdot O(C_2H_5)_2} \text{(}CH_3CO\text{-substituted methyl lactone)} \longrightarrow \text{(}CH_3CO\text{-substituted methyl tetrahydrofuranol, OH)} \quad \text{(Eq. 56)}$$

Reaction of β-Ketotrimethylsilanes with Aldehydes

Directed aldol reaction of the kinetic enolate of the α-trimethylsilyl ketone **94** with aldehydes under basic conditions is described in the section Reaction of Lithium Enolates (p. 214). Reversed regioselectivity in its reaction with aldehydes is observed on treatment of silane **94** with Lewis acids, such as boron trifluoride etherate, titanium tetrachloride, or stannic chloride, in methylene chloride at low temperature.[56] Only one regioisomer of the aldol is obtained in good yield. This demonstrates the regiocontrol that is sometimes attainable by changing the reagent (*i.e.*, lithium diisopropyl-amide *vs.* Lewis acid).

$$R^1CHO + \underset{\mathbf{94}}{R^2CH[Si(CH_3)_3]COCH_2R^3} \xrightarrow[CH_2Cl_2,\ -78^\circ \text{ to } -50^\circ]{BF_3 \cdot O(C_2H_5)_2} R^1CHOHCHR^2COCH_2R^3$$

EXPERIMENTAL CONDITIONS

The experimental conditions for the directed aldol reaction are significantly different from those of the conventional aldol reaction that uses base or acid as the catalyst. The directed aldol reaction generally requires highly reactive intermediates, such as an organolithium compound or a vinyloxyborane, which means that the reaction must be carried out in an inert atmosphere (nitrogen or argon). The best yields are obtained when all reactants and solvents are carefully purified immediately before use. The most commonly used solvents are tetrahydrofuran, diethyl ether, 1,2-dimethoxyethane, dimethyl sulfoxide, benzene, and methylene chloride. The ether solvents, such as diethyl ether, tetrahydrofuran, and 1,2-dimethoxyethane, should be freshly distilled before use from lithium aluminum hydride or benzophenone ketyl. Methylene chloride appears preferable for the titanium tetrachloride–promoted reactions, presumably because titanium tetrachloride is soluble in methylene chloride.

Workup under mild conditions is necessary to avoid undesirable further transformation of aldols. Special care should be taken if products are to be distilled. Because many aldols and ketols readily dissociate to reactants when heated, isolation of these substances by distillation at low temperature under reduced pressure in a nitrogen atmosphere is recommended. Complete removal of acidic or basic substances prior to distillation is necessary for efficient recovery of liquid aldols or ketols because these impurities catalyze dissociation or dehydration.

EXPERIMENTAL PROCEDURES

The procedures given here have been chosen to illustrate the uses of a variety of different enolate species in directed aldol reactions.

1-Hydroxy-4,4-dimethyl-1-phenyl-3-pentanone (Reaction of a Methyl Ketone with an Aldehyde using Lithium Diisopropylamide).[15] To a cold (−30°) solution of lithium diisopropylamide (from 25 mmol of methyllithium and 25 mmol of diisopropylamine) and a few milligrams of 2,2′-bipyridyl (an indicator for excess lithium diisopropylamide)[154] in 15 mL of diethyl ether was added 2.5 g (25 mmol) of pinacolone under a nitrogen atmosphere. The resulting orange solution was cooled to −60°, and 2.65 g (25 mmol) of benzaldehyde was added. The solution was stirred at −50 to −60° for 5 minutes and then partitioned between diethyl ether and cold (0°), aqueous 1 M hydrochloric acid. The organic layer was washed successively with aqueous sodium bicarbonate, water, and aqueous sodium chloride, and then dried and concentrated. Short-path distillation of the residual liquid

(5.67 g) separated 4.1 g (80%) of the ketol, 1-hydroxy-4,4-dimethyl-1-phenyl-3-pentanone, as a colorless liquid, bp 86° (0.07 mm), n_D^{25} 1.5077. The ketol was crystallized from a hexane–pentane mixture as white prisms; mp 22–23°; infrared (carbon tetrachloride) cm^{-1}: 3530 (assoc OH) and 1695 (C=O); ultraviolet (95% ethanol): series of weak maxima (ε 46–120) in the region 240–270 nm with a maximum at 291 nm (ε 85); proton magnetic resonance (carbon tetrachloride) δ: 7.0–7.4 (multiplet, 5H, aryl CH), 4.9–5.2 (multiplet, 1H, benzylic CHO), 3.70 (doublet, J = 3.0 Hz, 1H, exchanged with D_2O), 2.5–2.9 (multiplet, 2H, CH_2CO), and 1.05 [singlet, 9H, $(CH_3)_3C$]; mass spectrum m/e (relative intensity): 131 (100), 106 (28), 105 (28), 103 (25), 77 (41), 57 (53), 51 (20), and 43 (18).

6-Hydroxy-2-methyl-4-undecanone (Reaction of a Methyl Ketone with an Aldehyde using Lithium *t*-Butoxide).[55] To a solution of *t*-butanol (0.44 g, 6 mmol) in 10 mL of dry tetrahydrofuran was added *n*-butyllithium (6 mmol) under an argon atmosphere at room temperature. The solution was cooled to −40°, and a mixture of *n*-hexanal (0.50 g. 5 mmol) and 4-methyl-2-pentanone (0.60 g, 6 mmol) in 10 mL of tetrahydrofuran was added. After the mixture was stirred for 3 hours at the same temperature, it was treated with saturated aqueous ammonium chloride and then extracted with ether three times. The combined extracts were dried over sodium sulfate and the solvent was removed with a rotary evaporator. The residual oil was purified by column chromatography on silica gel (*n*-hexane–ether = 3:1), followed by short-path distillation to give 0.65 g (65%) of the product, bp 80° (0.1 mm); infrared (neat) cm^{-1}: 3420 and 1705; proton magnetic resonance (carbon tetrachloride) δ: 0.8–1.8 (multiplet, 18H), 2.3 [doublet, J = 2 Hz, 2H, $(CH_3)_2CHCH_2CO$], 2.5 [doublet, J = 6 Hz, 2H, $CH(OH)CH_2CO$], 3.2 (broad, 1H, OH), and 3.9 (multiplet, 1H, CHOH).

3-Hydroxy-1,3-diphenyl-1-propanone (Reaction of Trimethylsilyl Enol Ether with an Aldehyde using Methyllithium and Magnesium Bromide).[15] A cold (−40°) solution of the lithium enolate, prepared from methyllithium (5.0 mmol) and α-trimethylsiloxystyrene (0.73 g, 4.1 mmol) in dry tetrahydrofuran (10 mL), was treated with a solution of magnesium bromide (5.0 mmol) and benzaldehyde (0.53 g, 5.0 mmol) in tetrahydrofuran (10 mL). The resulting mixture was stirred at −35° to −50° for 10 minutes and then partitioned between ether and cold (0°) aqueous hydrochloric acid. The organic layer was washed successively with aqueous sodium bicarbonate, water, and brine, then dried. After concentration of the solution, the residual liquid (0.94 g) was crystallized from pentane to give the ketol (0.77 g, 81%), mp 48–50°; infrared (in carbon tetrachloride) cm^{-1}: 3540 (assoc OH) and 1675 (C=O); ultraviolet (95% ethanol), nm max (ε): 241 (13,200); proton magnetic resonance (carbon tetrachloride) δ: 7.0–8.1 (multiplet, 10 H, aryl CH); 5.0–5.4 (triplet of doublet, 1 H, benzylic CHO), 3.43 (doublet, $J = 2.5$ Hz, 1 H, OH, exchanged with D_2O), and 3.16 (doublet, $J = 6.0$ Hz, 2 H, $COCH_2$); mass spectrum m/e (relative intensity: 120 (22), 106 (43), 105, (100), 77 (83), and 51 (31).

***threo*-4-Hydroxy-3-phenyl-2-heptanone (Reaction of an Enol Acetate with an Aldehyde using Methyllithium and Zinc Chloride).**[120] Full details for the preparation of *threo*-4-hydroxy-3-phenyl-2-heptanone (64–69%) from *trans*-2-acetoxy-1-phenylpropene by the use of methyllithium and zinc chloride are described in *Organic Syntheses.*[120]

2-(1-Hydroxyethyl)-3,3-dimethylcyclohexanone (Reaction of a Magnesium Enolate with an Aldehyde).[111] Methylmagnesium iodide, prepared from magnesium (1.44 g, 60 mmol) and methyl iodide (8.52 g, 60 mmol) in diethyl ether (40 mL), was treated with finely powdered cuprous iodide (200 mg) at $-5°$. After the mixture had been stirred for 5 minutes at $-5°$, 3-methyl-2-cyclohexenone (5.5 g, 50 mmol) in ether (20 mL) was added, and stirring was continued for an additional 30 minutes at $-5°$. A solution of acetaldehyde (2.2 g, 50 mmol) in absolute ether (10 mL) was added with efficient stirring at $-15°$ to $-10°$. The reaction mixture was stirred at 0° for 30 minutes and at 25° for 30 minutes, and then poured into ice-cold 2 *N* hydrochloric acid (25 mL). The product was extracted with ether, washed (brine), and dried over anhydrous magnesium sulfate. Distillation gave 2-(1-hydroxyethyl)-3,3-dimethylcyclohexanone (6.34 g, 75%), bp 69–73° (0.01 mm); infrared (in chloroform) cm^{-1}: 3400 and 1685; proton magnetic resonance (in chloroform-d) δ: 1.05 (singlet, 3 H), 1.12 (singlet, 3 H), 1.3 (doublet, $J = 7$ Hz, 3 H), 2.0–2.5 (multiplet, 3 H), 3.5 (singlet, 1 H), and 4.10 (multiplet, 1 H); mass spectrum m/e (relative intensity): 170 (M^+, 3), 152 (9), 137 (9), 126 (23), 111 (100), 95 (7), 83 (95), 69 (32), 55 (62), 43 (36), 41 (42), 39 (20), and 29 (32).

β-Phenylcinnamaldehyde (Reaction of a Schiff Base with a Ketone).[78] Full details for the preparation of β-phenylcinnamaldehyde from ethylidenecyclohexylamine and benzophenone in 69–78% yield are given in *Organic Syntheses.*[78]

β-Cyclocitrylidene Acetaldehyde (Reaction of a Schiff Base with an Aldehyde).[75] To a solution of lithium diisopropylamide (0.1 mol) in ether (160 mL) was added N-ethylidenecyclohexylamine (12.5 g, 0.1 mol) at 0°. The resulting solution was cooled to $-70°$, whereupon some metalated Schiff base precipitated. Freshly distilled β-cyclocitral [bp 97–98° (16 mm), n_D^{20} 1.4978] (15.2 g, 0.1 mol) was added dropwise to the mixture at $-70°$ under nitrogen. The mixture was stirred for 12 hours at room temperature.

Acetic acid (25%) was added at 0°. After the mixture was stirred for 3 hours at room temperature, the red ether layer was neutralized with aqueous sodium bicarbonate and dried over anhydrous sodium sulfate. Distillation gave β-cyclocitrylidene acetaldehyde (10.0 g, 56%), bp 76–91° (0.5 mm). The product was further purified by recrystallization (twice from petroleum ether at −70°), followed by distillation, bp 87.5–88.5° (0.8 mm), n_D^{20} 1.5378.

6-Hydroxy-7-nonen-4-one and 5,7-Nonadien-4-one (Reaction of an N,N-Dimethylhydrazone with an Aldehyde).[86,87] The N,N-dimethylhydrazone of 2-pentanone (1.28 g, 10 mmol) was metalated with *n*-butyllithium (10 mmol) in dry tetrahydrofuran (30 mL) at −78° for 20 minutes.

To a suspension of the lithiated N,N-dimethylhydrazone in tetrahydrofuran was added *trans*-crotonaldehyde (0.70 g, 10 mmol) at −78°. After the temperature was kept at −78° for 3 hours then at 0° for 12 hours, neutralization with acetic acid at −40° and extractive workup (methylene chloride–water) afforded the N,N-dimethylhydrazone of 6-hydroxy-7-nonen-4-one (1.98 g, 100%) as a bright-yellow viscous oil.

The hydroxy N,N-dimethylhydrazone (0.99 g, 5 mmol) was oxidatively hydrolyzed (2.2 equivalents of aqueous sodium periodate in methanol and 1 *N* pH 7 phosphate buffer solution, at room temperature for 5 hours)[86] to give 0.80 g (98% spectroscopic yield) of 6-hydroxy-7-nonen-4-one, which was distilled without any residue to give a yellow oil, bp 70° (0.05 mm); infrared (neat) cm^{-1}: 3425 (OH), 1710 (C=O), and 1675 (sh, C=C); proton magnetic resonance (chloroform-*d*) δ: 0.91 (triplet, 3H), 1.20–2.00 (multiplet, 2H), 1.70 (doublet, 3H), 2.20–2.75 (triplet and ABX, 4H), 3.40 (broad singlet, 1H) 4.5 (ABX, 1H), and 5.60 (multiplet, 2H); mass spectrum *m/e* (relative intensity): 156 (M^+), 71 (100).

To a mixture of 6-hydroxy-7-nonen-4-one (0.5 g, 32 mmol) and triethylamine (1.4 mL, 10 mmol) in methylene chloride (15 mL) was added methanesulfonyl chloride (0.27 mL, 3.5 mmol) under reflux. After 1 hour, the reaction mixture was washed successively with 1.4 *N* hydrochloric acid, aqueous sodium bicarbonate, and water. The organic layer was dried over anhydrous sodium sulfate and concentrated. Distillation gave 5,7-nonadien-4-one as a yellow oil (0.38 g, 85%), bp 50° (0.01 mm); infrared (neat) cm^{-1}: 3030 (=CH), 1670 (C=O), and 1598 (C=C); proton magnetic resonance (chloroform-*d*) δ: 0.95 (triplet, 3H), 1.72 (septet, 2H), 1.88 (doublet, 3H), 2.55 (triplet, 2H), 5.90–6.50 (multiplet, 3H), and 6.95–7.48 (multiplet, 1H); mass spectrum *m/e* (relative intensity): 138 (M^+) and 95 (100).

3-[(Hydroxy)phenylmethyl]-2-octanone (Reaction of an α,β-Unsaturated Ketone with an Aldehyde via Vinyloxyborane).[137] A solution of methyl vinyl ketone (150 mg, 2.1 mmol) and tri-*n*-butylborane (450 mg, 2.4 mmol) in absolute tetrahydrofuran containing a small amount of oxygen (as an initiator of the 1,4-addition reaction of tri-*n*-butylborane with methyl vinyl ketone)[224] was stirred overnight at room temperature under nitrogen to give (2-octen-2-yloxy)di-*n*-butylborane. To the solution was added benzaldehyde (160 mg, 1.5 mmol) in absolute tetrahydrofuran. After 30 minutes the solvent was removed, and the residue was treated with a mixture of methanol (20 mL) and 30% aqueous hydrogen peroxide (1 mL) for 1 hour. The mixture was concentrated under reduced pressure to remove methanol and was then extracted with ether. The extract was washed with 5% sodium bicarbonate and water, dried over anhydrous sodium sulfate, and concentrated. The residual oil was purified by preparative thin-layer chromatography on silica gel with chloroform as the developing solvent to give 3-(hydroxyphenylmethyl)-2-octanone [*threo* (240 mg, 68%) and *erythro* (79 mg, 23%)]. *Threo* isomer: infrared (neat) cm^{-1}: 3420 and 1700; proton magnetic resonance (in carbon tetrachloride) δ: 0.6–1.8 (multiplet, 11 H), 2.06 (single, 3 H), 2.73 (multiplet, 1 H), 3.32 (singlet, 1 H), 4.56 (doublet, 1 H), and 7.23 (singlet, 5 H). *Erythro* isomer: infrared (neat) cm^{-1}: 3420 and 1695; proton magnetic resonance (carbon tetrachloride) δ: 0.6–1.8 (multiplet, 11 H), 1.83 (singlet, 3 H), 2.70 (multiplet, 1 H), 3.15 (singlet, 1 H), 4.70 (doublet, 1 H), and 7.23 (singlet, 5 H).

2-[(Hydroxy)phenylmethyl]-1-phenyl-1-hexanone (Reaction of an α-Diazo-Ketone with an Aldehyde via Vinyloxyborane).[136] To a stirred solution of diazoacetophenone (154 mg, 1.05 mmol) in 5 mL of dry tetrahydrofuran was added tri-*n*-butylborane (216 mg, 1.19 mmol) at room temperature under argon. Nitrogen gas was evolved immediately. After 45 minutes, the stirred mixture was treated with a solution of benzaldehyde (92 mg, 0.87 mmol) in 5 mL tetrahydrofuran for 10 minutes at room temperature. After removal of tetrahydrofuran the oily residue was treated with 30% hydrogen peroxide (2 mL) in methanol (10 mL) at room temperature. The solution was allowed to stand overnight and then water was added. The mixture was concentrated *in vacuo* to remove methanol and the residue was extracted with ether. The ether was dried over anhydrous sodium sulfate, and the solvent evaporated to give an oil which was purified by preparative thin-layer chromatography (silica gel) using methylene chloride to give 2-(hydroxyphenylmethyl)-1-phenyl-1-hexanone (240 mg, 98%) as an oil. Infrared (neat) cm^{-1}: 3440 and 1660; proton magnetic resonance (carbon tetrachloride) δ: 0.3–2.0 (multiplet, 9 H), 3.4–3.9 (multiplet, 2 H), 4.6–4.9 (multiplet, 1 H), 6.8–7.4 (multiplet, 8 H), and 7.5–8.0 (multiplet, 2 H).

6-Hydroxy-2-methyl-8-phenyl-4-octanone (Reaction of a Methyl Ketone with an Aldehyde via Vinyloxyborane).[17]

A. Dibutylboryl Trifluoromethanesulfonate (Triflate). Trifluoromethanesulfonic acid (12.51 g, 83.3 mmol) was added to tributylborane (15.16 g, 83.3

mmol) at room temperature under argon. After being stirred for 3 hours, the reaction mixture was distilled under reduced pressure under argon. Bp 37° (0.12 mm), 19.15 g (84%). Infrared (in carbon tetrachloride) cm^{-1}: 1405, 1380, 1320, 1200, and 1150.

B. 6-Hydroxy-2-methyl-8-phenyl-4-octanone.[36] To an ethereal solution (1.5 mL) of dibutylboryl triflate (0.301 g, 1.1 mmol) and ethyldiisopropylamine (0.142 g, 1.1 mmol) was added dropwis an ether solution (1.5 mL) of 2-methyl-4-pentanone (0.10 g, 1.0 mmol) at −78° under argon. After the mixture was stirred for 30 minutes, an ether solution (1.5 mL) of 3-phenylpropanal (0.134 g, 1.0 mmol) was added at that temperature. The reaction mixture was allowed to stand for 1 hour, and was then added to pH 7.0 phosphate buffer at room temperature and extracted with ether. After removal of ether, the residue was treated with 30% hydrogen peroxide (1 mL) in methanol (3 mL) for 2 hours and water was added. This mixture was concentrated under reduced pressure to remove most of the methanol and the residue was extracted with ether. The combined extracts were washed with a 5% solution of sodium bicarbonate and then saturated brine, dried over anhydrous sodium sulfate, and the solvent was removed. The crude oil was purified by preparative thin-layer chromatography on silica gel using chloroform as developing solvent to give 6-hydroxy-2-methyl-8-phenyl-4-octanone (0.192 g, 82%). Infrared (neat) cm^{-1}: 1695; proton magnetic resonance (carbon tetrachloride) δ: 0.86 (doublet, 6 H), 1.70–2.00 (multiplet, 3 H), 2.06–2.92 (multiplet, 6 H), 3.05 (broad singlet, 1 H), 4.00 (multiplet, 1 H), 7.11 (singlet, 5 H).

2-[(Hydroxy)phenylmethyl]cyclohexanone (Reaction of a Trimethylsilyl Enol Ether with an Aldehyde).[173] A methylene chloride (10 mL) solution of 1-trimethylsiloxycyclohexene (0.426 g, 2.5 mmol) was added dropwise to a solution of benzaldehyde (0.292 g, 2.75 mmol) and titanium tetrachloride (0.55 g, 2.75 mmol) in dry methylene chloride (20 mL) under an argon atmosphere at −78°, and the mixture was stirred for 1 hour. After hydrolysis at that temperature, the resulting organic layer was extracted with ether, and the extract was washed with water and dried over anhydrous sodium sulfate. The solution was concentrated under reduced pressure, and the residue was purified by column chromatography on silica gel with elution with methylene chloride to afford *erythro*-2-(hydroxyphenylmethyl)cyclohexanone (115 mg, 23%), mp 103.5–104.5° (recrystallized from 2-propanol); infrared (KBr) cm^{-1}: 3530 (OH) and 1700 (C=O); proton magnetic resonance (chloroform-*d*) δ: 7.27 (singlet, 5 H), 5.40 (doublet, $J = 2.5$ Hz, 1 H), 3.05 (singlet, 1 H, exchanged with D_2O), and 1.1–2.7 (multiplet, 9 H). From the last fraction 346 mg (69%) of *threo* isomer was obtained, mp 75° (recrystallized from *n*-hexane–ether); infrared (KBr) cm^{-1}: 3495 (OH) and

1695 (C=O); proton magnetic resonance (carbon tetrachloride) δ: 7.29 (singlet, 5 H), 4.83 (doublet, $J = 9.0$ Hz, 1 H), 3.77 (singlet, 1 H, exchanged with D_2O), and 1.1–2.9 (multiplet, 9 H).

5-Isopropoxy-7-phenyl-2,6-heptadienal (Reaction of a Dienoxysilane with Cinnamaldehyde Dimethylacetal).[187] To a mixture of titanium tetraisopropoxide (2.0 mmol), cinnamaldehyde dimethyl acetal (356 mg, 2.0 mmol), and methylene chloride (25 mL) was added titanium tetrachloride (2.3 mmol) in 1 mL of methylene chloride at $-40°$ under an argon atmosphere. After a minute of stirring 1-trimethylsiloxybutadiene (355 mg, 2.5 mmol) in methylene chloride (4 mL) was added to the mixture. The mixture was kept at $-40°$ for 30 minutes, then quenched with 20% aqueous potassium carbonate, followed by extraction with ether. The combined extracts were dried over anhydrous sodium sulfate and the solvent was removed. The residual oil was purified by preparative thin-layer chromatography on silica gel [hexane–ethyl acetate (4:1)] to afford 5-isopropoxy-7-phenyl-2,6-heptadienal (441 mg, 90%); infrared (neat) cm^{-1}: 1690, 1640, and 1600; proton magnetic resonance (carbon tetrachloride) δ: 9.50 (doublet, $J = 8$ Hz, 1 H), 7.30 (singlet, 5 H), 5.8–7.2 (multiplet, 4 H), 3.4–4.3 (multiplet, 2 H), 2.4–2.8 (multiplet, 2 H), and 1.15 (doublet, $J = 8$ Hz, 6 H).

1,1,3-Triethoxy-4-methyl-6-(2,6,6-trimethylcyclohexen-1-yl)-4-hexene (Reaction of an Enol Ether with an α,β-Unsaturated Acetal).[204] To a mixture of 1,1-diethoxy-2-methyl-4-(2,6,6-trimethylcyclohexen-1-yl)-2-butene (280 g, 1 mol) and 10% zinc chloride solution in ethyl acetate (10 mL) was added ethyl vinyl ether (76 g, 1.06 mol) at 40–45°. The rate of the addition was controlled to keep the reaction temperature at 40–45°. After 1 hour the reaction mixture was diluted with ether, washed with aqueous sodium hydroxide, and dried over anhydrous potassium carbonate. Distillation gave 320 g (91%) of product as a colorless oil, bp 127–129° (0.01 mm), n_D^{24} 1.4705.

Methyl 5-Hydroxy-3-oxo-5-phenylpentanoate (Reaction of Diketene with an Aldehyde).[216] To a vigorously stirred solution of benzaldehyde (0.165 g, 1.25 mmol) and diketene (0.5 mL of 5 *M* solution, 2.5 mmol) in methylene chloride (6.5 mL) was added at once a 3 *M* solution of titanium tetrachloride (0.5 mL, 1.5 mmol) in methylene chloride at $-78°$ under argon. The mixture was stirred for 2 minutes, dry methanol (2 mL) was added, and after stirring for 30 minutes at -20 to $-10°$, the mixture was poured into ice-cold aqueous potassium carbonate (10 mL, 4.4 mmol). The resulting insoluble pale-yellow precipitate was removed by filtration and the filtrate was extracted with ether. The organic layer was washed with saturated aqueous sodium bicarbonate to remove the accompanying 4-hydroxy-2-pyrone derivative, then with saturated brine, and the solvent was evaporated. Column chromatography on silica gel with elution with methylene chloride gave methyl 5-hydroxy-3-oxo-5-phenylpentanoate (0.253 g, 91% yield); infrared (neat) cm^{-1}: 3460, 1740, 1715, 1330, 1260, and 1150; proton magnetic resonance (carbon tetrachloride) δ: 1.4 (singlet, 9 H), 2.75 (multiplet, 2 H), 3.7 (broad singlet, 1 H), 5.0 (multiplet, 1 H), and 7.2 (singlet, 5 H). The combined aqueous alkaline solution was evaporated under reduced pressure. The residue was acidified with 2 *N* hydrochloric acid, and was extracted thoroughly with ethyl acetate. The extract was washed with water and saturated brine. After removal of the solvent, the resulting crystalline product was washed with a small amount of ether to give 5,6-dihydro-4-hydroxy-6-phenyl-2-pyrone (0.018 g, 7% yield), mp 127–128° (dec); infrared (KBr) cm^{-1}: 1745, 1720, 1300, and 1020; proton magnetic resonance (in chloroform-*d*) δ: 1.05 (doublet, $J = 7$ Hz, 3 H), 1.1 (doublet, $J = 7$ Hz, 3 H), 1.95 (multiplet, 1 H), 2.25–2.85 (multiplet, 2 H), 3.3 (doublet, $J = 18$ Hz, 1 H), 3.7 (doublet, $J = 18$ Hz, 1 H), and 4.45 (multiplet, 1 H).

ADDED IN PROOF

Recent advances in the directed aldol reaction are included in this Appendix, which covers the literature from March 1980 to September 1981. Extensive work on the aldol and related reactions has flourished and a large number of valuable results have been published during this period. One of the notable developments is kinetic stereoselection with lithium enolates. Heathcock and co-workers have revealed the main factors controlling the stereochemistry of the aldol additions.[230,251–255] A recent review of stereoselective aldol condensations emphasizes the importance of lithium enolates.[256]

A current example of an enantioselective aldol reaction is that of a chiral lithium enolate **95** with the chiral aldehyde **96**, yielding two stereoisomeric aldols in a significant stereoselection ratio of 1 : 8.[193a]

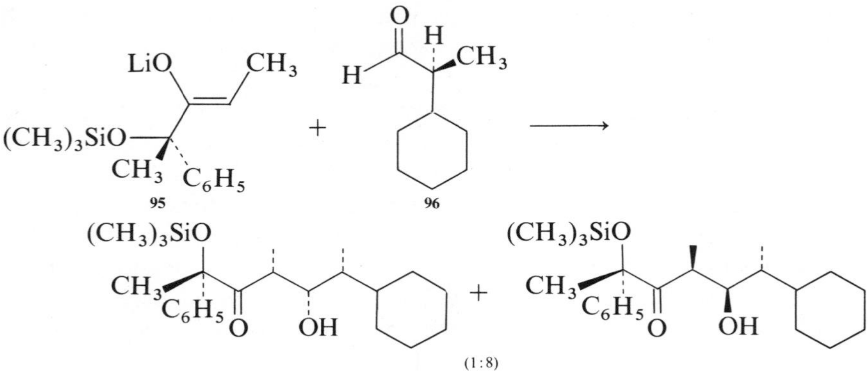

LiO, CH_3, $(CH_3)_3SiO$, CH_3, C_6H_5 **95** + O, H, H, CH_3 **96** →

$(CH_3)_3SiO$, CH_3, C_6H_5, O, OH + $(CH_3)_3SiO$, CH_3, C_6H_5, O, OH

(1 : 8)

Higher enantioselectivity is achieved with boron enolates. Indeed, among the various metal enolates, boron enolates result in the highest stereoselectivity in the formation of β-hydroxy carbonyl compounds. Consequently, new methods for the generation of boron enolates have been devised. Phenyl-substituted boron enolates can be generated by the acylation of boron-stabilized carbanions with methyl benzoate, as is illustrated in the following scheme; the carbanions are prepared from 1-alkynes and 9-BBN.[257] The boron enolate **97** thus generated adds to an aldehyde to afford predominantly a *threo* aldol **98**.

$$\mathrm{RC{\equiv}CH} \xrightarrow{\text{2 eq. 9-BBN}} \mathrm{RCH_2CH}\left[-\mathrm{B}(\text{9-BBN})\right]_2 \xrightarrow{\mathrm{CH_3Li}} \mathrm{RCH_2CH(Li)B}(\text{9-BBN})$$

$$\xrightarrow{\mathrm{C_6H_5CO_2CH_3}} \left[\mathrm{RCH_2CH(COC_6H_5)B}(\text{9-BBN})\right] \longrightarrow \underset{\mathbf{97}}{\mathrm{RCH_2CH{=}C(C_6H_5)OB}(\text{9-BBN})}$$

$$\xrightarrow{\mathrm{R'CHO}} \underset{\mathbf{98}}{\mathrm{RCH_2CH(CHOHR')COC_6H_5}}$$

Reaction of a silyl enol ether with dibutylboryl trifluoromethanesulfonate generates a boron enolate quantitatively.[258,259] Thus treatment of a (*Z*)-trimethylsilyl enol ether **99** with dibutylboryl trifluoromethanesulfonate, followed by removal of the resulting trimethylsilyl trifluoromethanesulfonate, affords a (*Z*)-boron enolate **100** in high regio- and stereoselectivity. Its subsequent reaction with an aldehyde yields the corresponding *erythro* aldol **101** with high specificity.[258]

$$\underset{\mathbf{99}}{\mathrm{RC(OSi(CH_3)_3){=}CHR'}} + (n\text{-}\mathrm{C_4H_9})_2\mathrm{BOSO_2CF_3} \longrightarrow \underset{\mathbf{100}}{\mathrm{RC(OB(C_4H_9\text{-}}n)_2){=}\mathrm{CHR'}} + \mathrm{(CH_3)_3SiOSO_2CF_3}$$

$$\xrightarrow{\mathrm{R''CHO}} \underset{\mathbf{101}}{\mathrm{RCOCH(R')CH(OH)R''}}$$

Evans and co-workers have studied enantioselective aldol reactions of boron enolates, employing the procedure exploited by Mukaiyama.[260–262] Highly selective asymmetric induction can be achieved in aldol additions using chiral boron enolates. For example, the reaction of a boron enolate of chiral ketone **102** with isobutyraldehyde affords *erythro* and *threo* aldols in a ratio of 91:9. Of the two possible *erythro* diastereomers, isomer **103** is produced exclusively; the other isomer (**104**) is not detected.[262]

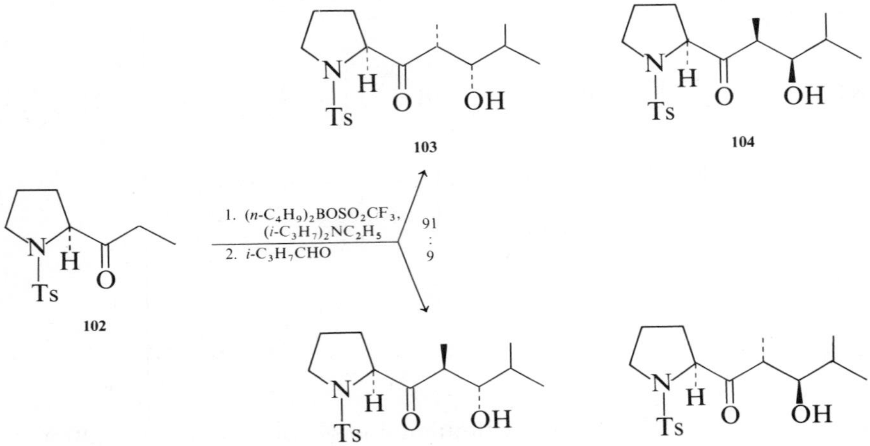

103 **104**

102

1. $(n\text{-}C_4H_9)_2BOSO_2CF_3$, $(i\text{-}C_3H_7)_2NC_2H_5$
2. $i\text{-}C_3H_7CHO$

91 : 9

Another example is the total synthesis of 6-deoxyerythronolide B, where the aldol strategy using chiral boron enolates is successfully utilized.[265] The following scheme shows the important steps in the synthetic sequence.

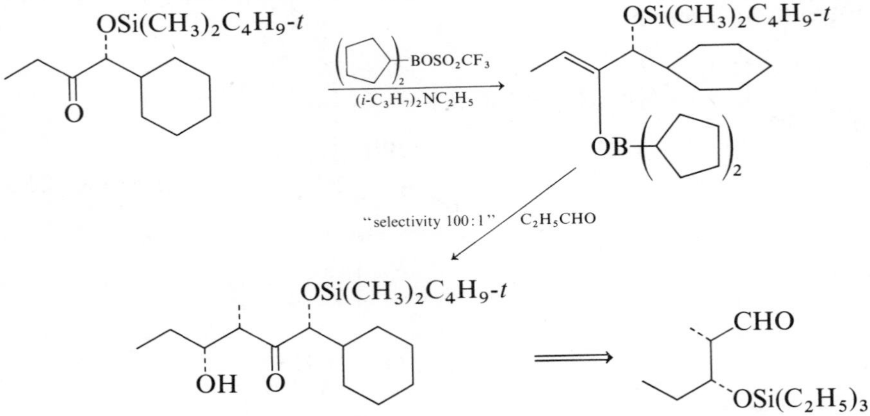

$OSi(CH_3)_2C_4H_9\text{-}t$

$(C_5H_9)_2BOSO_2CF_3$, $(i\text{-}C_3H_7)_2NC_2H_5$

$OSi(CH_3)_2C_4H_9\text{-}t$ $OB(C_5H_9)_2$

"selectivity 100:1" C_2H_5CHO

$OSi(CH_3)_2C_4H_9\text{-}t$ OH O ⟹ CHO $OSi(C_2H_5)_3$

105

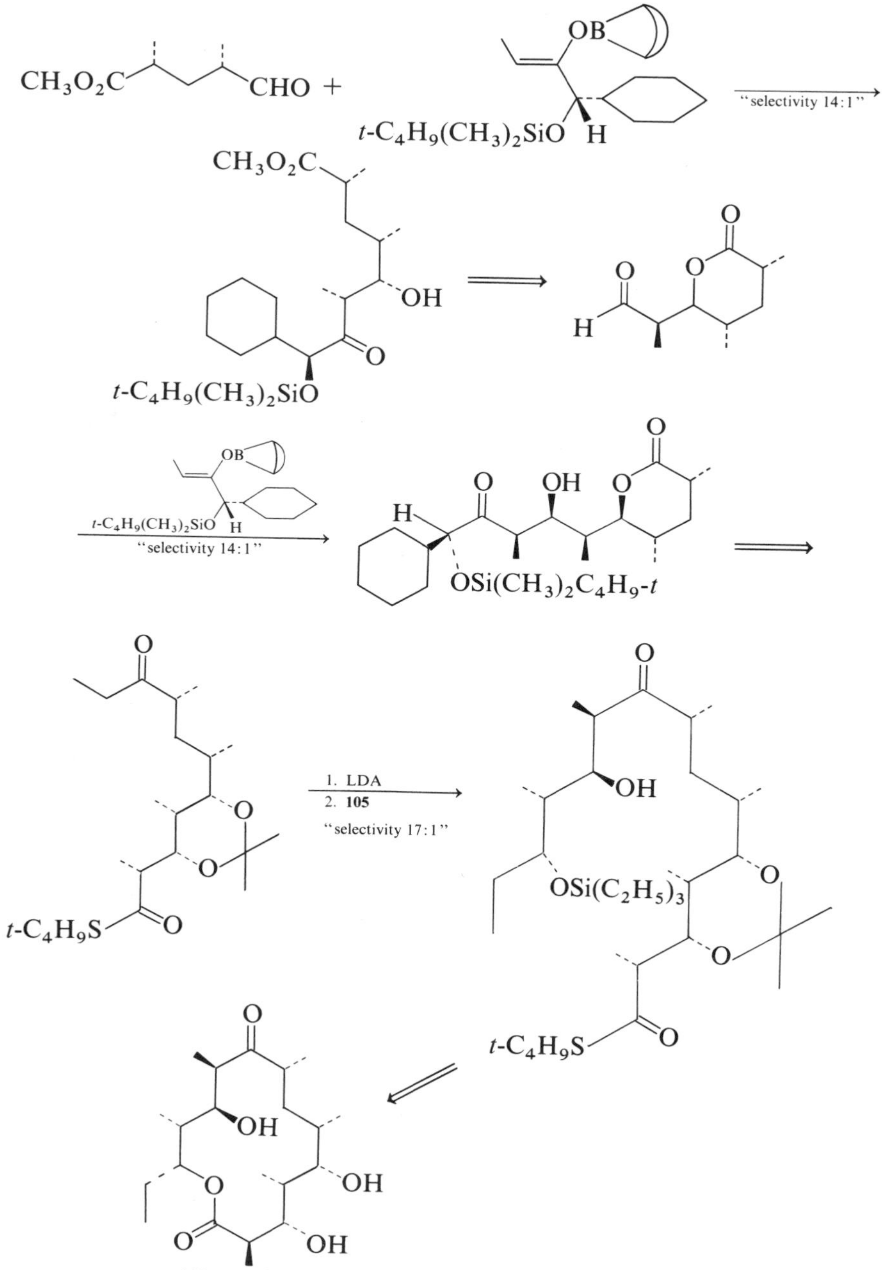

CH_3O_2C … CHO + t-$C_4H_9(CH_3)_2SiO$ H → "selectivity 14:1"

CH_3O_2C … OH … O … t-$C_4H_9(CH_3)_2SiO$ ⟹ O, H, O, O

t-$C_4H_9(CH_3)_2SiO$ H, "selectivity 14:1" → H, O, OH, O, O, $OSi(CH_3)_2C_4H_9$-t ⟹

O, O, O, t-C_4H_9S, O → 1. LDA 2. **105** "selectivity 17:1" → O, OH, $OSi(C_2H_5)_3$, O, O, t-C_4H_9S, O

⟸ O, OH, O, OH, O, OH

6-Deoxyerythronolide B

In this total synthesis, all of the crucial carbon–carbon bond-forming reactions involved in the construction of the carbon framework are aldol reactions. More important, the overall stereoselection of these four reactions reaches 85%.[265]

An interesting aldol reaction that proceeds via an extended transition state occurs when a mixture of an (*E*)- or (*Z*)-silyl enol ether, aldehyde, and a catalytic amount of tris(diethylamino)-sulfonium (TAS) difluorotrimethylsiliconate in tetrahydrofuran is allowed to stand at low temperature (usually at $-78°$).[193a,266,267] The corresponding β-trimethylsiloxy ketone is obtained in good yield. This reaction gives predominantly *erythro* aldols, regardless of the geometry of the starting silyl enol ethers. The reaction may proceed by the following pathway.[267]

$$\underset{\mathbf{106}}{\mathrm{R\overset{OSi(CH_3)_3}{\overset{|}{C}}{=}CHR'}} + \underset{\mathbf{107}}{[(C_2H_5)_2N]_3S^+Si(CH_3)_3F_2^-} \rightleftharpoons \underset{\mathbf{108}}{\mathrm{R\overset{O^-S^+[N(C_2H_5)_2]_3}{\overset{|}{C}}{=}CHR'}} + \underset{\mathbf{109}}{2\,(CH_3)_3SiF}$$

$$\mathbf{108} + \mathrm{R''CHO} \rightleftharpoons \underset{\mathbf{110}}{\mathrm{R\overset{O}{\overset{\|}{C}}CHR'\overset{O^-S^+[N(C_2H_5)_2]_3}{\overset{|}{C}}HR''}}$$

$$\mathbf{109} + \mathbf{110} \longrightarrow \mathrm{R\overset{O}{\overset{\|}{C}}CHR'\overset{OSi(CH_3)_3}{\overset{|}{C}}HR} + [(C_2H_5)_2N]_3S^+F^-$$

The stereoselectivity leading to *erythro* aldols is best interpreted in terms of an extended transition state in which electrostatic repulsion of oxygens is minimized. In the reaction of (*E*)-enolates, the transition state **A** is sterically favored over the diastereomeric transition state **B**, since the latter suffers a gauche interaction (R/R″). By the same reasoning, the *erythro* transition state **C** derived from a (*Z*)-enolate is more stable than the *threo* transition state **D**. Thus both (*Z*)- and (*E*)-enolates afford predominantly the *erythro* aldol. This characteristic feature of stereoselectivity sharply contrasts with the stereoselection observed in reactions of lithium and boron enolates that proceed through six-membered pericyclic transition states.[267]

$\delta-$O R″ H R H O R′

A

$\delta-$O H H R R″ $O^{\delta-}$ R′

B

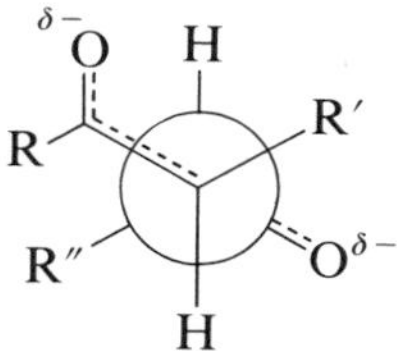

C

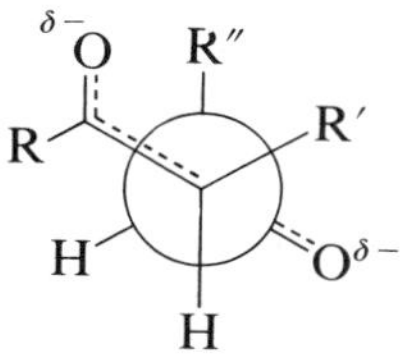

D

Silicon and tin enolates have also been used effectively. For example, in another *erythro*-selective aldol reaction, silyl enol ether **111** and acetal **112** were treated with a catalytic amount (1–5 mole %) of trimethylsilyl trifluoromethanesulfonate.[193a] Aldol product **113** is formed. This reaction

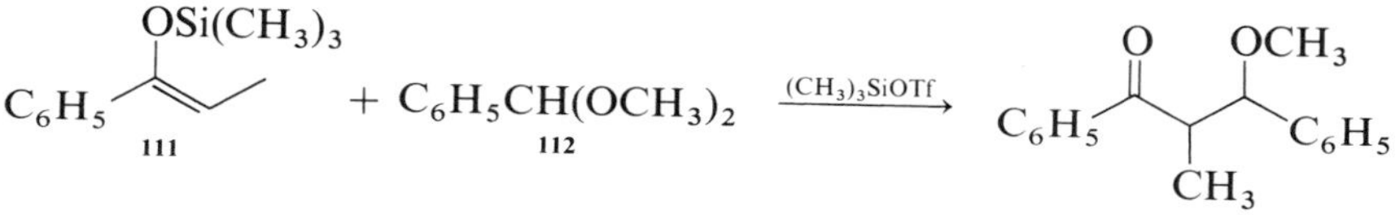

111 112 113

(*erythro*: *threo* = 84:16)

proceeds in the catalytic sense by the use of the trimethylsilyl moiety both as protecting group of the enolate and as initiator of the reaction.

One convenient method for the generation of tin enolates is the reduction of haloketones with tin metal or stannous fluoride. These enolates readily afford the aldol products by subsequent treatment with aldehydes.[268,269]

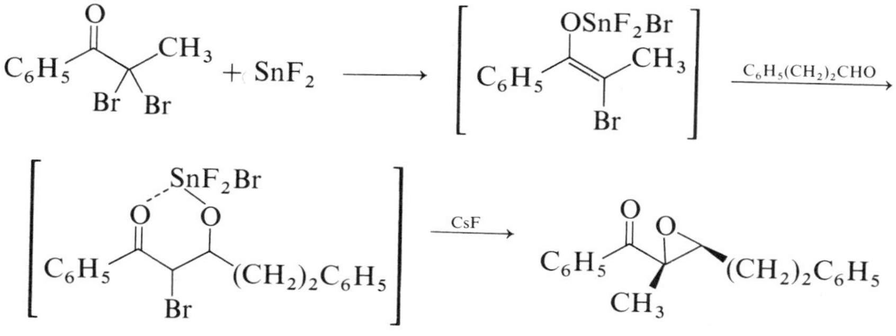

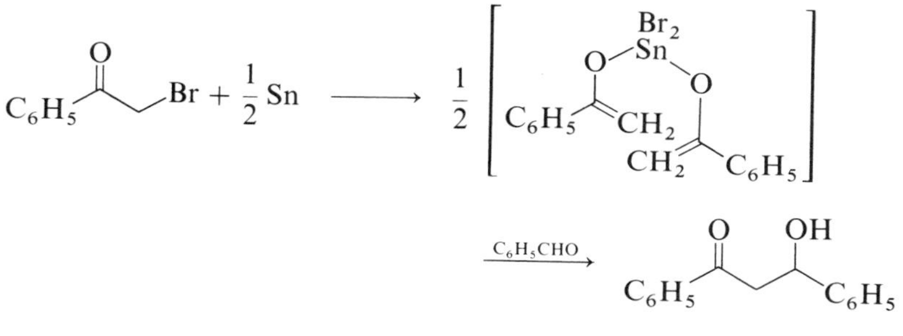

Trialkyltin enolates are easily prepared by the metal exchange reaction with lithium enolates, and the stereoselectivity of the aldol reactions has been studied. The reaction of trimethyl or triethyltin enolates **114** of propiophenone (or cyclohexanone) with benzaldehyde at −78°, under kinetic control, gives predominantly the *threo* aldol diastereomer **115**. In these reactions, the tin enolates were isolated before use. The NMR spectrum of **114** showed a 9:1 ratio of C-derivative to O-derivative.[270]

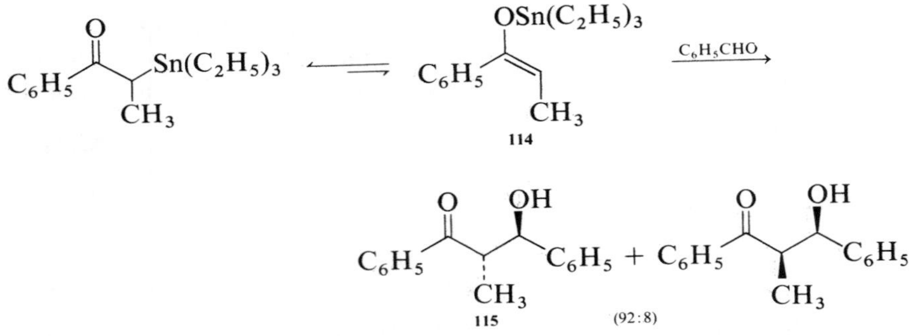
O
Sn(C2H5)3
C6H5
CH3
OSn(C2H5)3
C6H5CHO
114
OH
115
(92:8)

In contrast, the reaction of aldehydes with triphenyltin enolates generated *in situ* under kinetic control yields *erythro* aldols **116** in moderate selectivity.[271] This result does not correspond with other observations.[270]

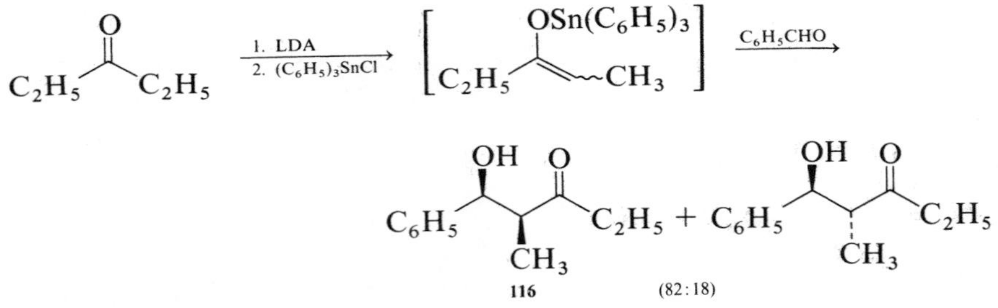
C2H5
1. LDA
2. (C6H5)3SnCl
OSn(C6H5)3
C6H5CHO
116
(82:18)

As with boron enolates, the selectivity of these reactions appears to depend on the auxiliary substituent groups. However, further examination of the mechanistic aspects of these reactions is required before any clear interpretation of the results can be made.

Another *erythro*-selective aldol reaction that is independent of starting enolate geometry has been realized by using zirconium enolates.[272,273] The zirconium enolate **117** is prepared *in situ* by the treatment of the corresponding lithium enolate with bis(cyclopentadienyl)zirconium dichloride. It undergoes a facile aldol reaction with benzaldehyde to give predominantly the *erythro* aldol **118**. This method has been used to develop a highly enantioselective aldol-type condensation starting from optically active amide enolates.[274] A divalent tin enolate formed from stannous trifluoromethanesulfonate and a ketone in the presence of N-ethylpiperidine reacts with aldehydes to give *erythro* aldol products with high stereoselectivity.[279] This tin enolate also gives aldols from ketones.[280]

TABULAR SURVEY

The following five tables contain examples of the reactions covered in the text. The literature survey covers articles appearing up to September 1981. The general arrangement is explained by the titles of the tables. Within each table the substances are arranged by increasing number of carbon atoms, and then by increasing number of hydrogen atoms, first for the carbonyl acceptor, then for the nucleophilic aldehyde or ketone equivalent. Carbon and hydrogen atoms of OR or NR in the following functional groups are not counted: C(OR) in an acetal or a ketal, C=COR in an enol derivative, and C=NR in an imine, a hydrazone, or an oxime. Carbon atoms and hydrogen atoms of the trimethylsilyl group of β-keto trimethylsilanes are also not included.

Column 3 in the tables indicates the reagent for generating the enolate or enolate equivalent, or the Lewis acid in Lewis acid–promoted reactions. The structures (or partial structures) of products are shown in column 4. A dash in parentheses means that the yield is not given in the literature. If there is more than one reference, the yield is from the first reference listed. In certain cases the ratio of *threo* and *erythro* isomers of an aldol or (E) and (Z) isomers of an α,β-unsaturated aldehyde is given in parentheses.

Abbreviations used in the tables are as follows; LDA: lithium diisopropylamide; LDCA: lithium dicyclohexylamide; LTMP: lithium tetramethylpiperidide; Ac: acetyl; 9-BBN(OSO_2CF_3): 9-borabicyclo[3.3.1]nonan-9-yl trifluoromethanesulfonate; and THP: tetrahydropyranyl.

TABLE I. Reaction of Aldehyde Equivalents with Aldehydes or Acetals

No. of C Atoms	Aldehyde or Acetal	Aldehyde Equivalent	Reagent	Product(s)[a] and Yield(s) (%)	Refs.
C_2	$BrCH_2CH(OCH_3)_2$	$C_6H_5CH{=}CHOSi(CH_3)_3$	$TiCl_4$	C_6H_5, O (32)	181
	CH_3CHO	$CH_3CH{=}CHOC_2H_5$	$BF_3 \cdot O(C_2H_5)_2$	$RCH{=}C(CH_3)CHO$ (51)[b]	197
		$NCCH_2CH{=}CHOC_2H_5$	"	$RCH{=}C(CH_2CN)CHO$ (31)[b]	197
		$C_2H_5CH{=}CHOC_2H_5$	"	$RCH{=}C(C_2H_5)CHO$ (55)[b]	197
	CH_3CH, O, O	$CH_2{=}CHOC_2H_5$	$BF_3 \cdot O(C_2H_5)_2$	OC_2H_5, O, O, R (57)	227
	$CH_3CH(OCH_3)_2$	$CH_2{=}CHOCH_3$	$BF_3 \cdot O(C_2H_5)_2$	$RCH(OR')CH_2CH(OR')_2$ (79)	225
		$C_2H_5CH{=}CHOCH_3$	"	$RCH(OR')CH(C_2H_5)CH(OR')_2$ (74)	225
	$CH_3CH(OC_2H_5)_2$	$CH_2{=}CHOCH_3$	$BF_3 \cdot O(C_2H_5)_2$	$RCH(OR')CH_2CH(OR')_2$ (67)	226
		$CH_3CH{=}CHOC_2H_5$	"	$RCH(OR')CH(CH_3)CH(OR')_2$ (58)	226
		$C_2H_5CH{=}CHOC_2H_5$	"	$RCH(OR')CH(C_2H_5)CH(OR')_2$ (48)	225
		$C_2H_5(C_4H_9\text{-}n)C{=}CHOC_2H_5$	"	$RCH(OR')C(C_2H_5)(C_4H_9\text{-}n)CH(OR')_2$ (24)	225
		$C_6H_5CH_2CH{=}C[OSi(CH_3)_3]Si(CH_3)_3$	"	$RCH{=}CH(CH_2C_6H_5)CHO$ (64)[b] [100% (*E*)]	184
C_3	C_2H_5CHO	$CH_2{=}CHOC_2H_5$	$TiCl_4$, $Ti(OC_3H_7\text{-}i)_4$	$RCH(OC_3H_7\text{-}i)CH_2CH(OR')_2$ (61)	199
		$CH_3CH{=}CHOC_2H_5$	$BF_3 \cdot O(C_2H_5)_2$	$RCH{=}C(CH_3)CHO$ (92)[b]	197
		$C_2H_5CH{=}NN(CH_3)_2$	LDA	$RCHOHCH(CH_3)CH{=}NN(CH_3)_2$ (90)	87
		$NCCH_2CH{=}CHOC_2H_5$	$BF_3 \cdot O(C_2H_5)_2$	$RCH{=}C(CH_2CN)CHO$ (25)[b]	197
		$C_2H_5CH{=}CHOC_2H_5$	"	$RCH{=}C(C_2H_5)CHO$ (68)[b]	197
	$C_2H_5CH(OC_2H_5)_2$	$CH_2{=}CHOC_2H_5$	$BF_3 \cdot O(C_2H_5)_2$	$RCH(OR')CH_2CH(OR')_2$ (56)	226
C_4	$NC(CH_2)_2CHO$	$CH_3CH{=}CHOC_2H_5$	$BF_3 \cdot O(C_2H_5)_2$	$RCH{=}C(CH_3)CHO$ (29)[b]	197
		$C_2H_5CH{=}CHOC_2H_5$	"	$RCH{=}C(C_2H_5)CHO$ (33)[b]	197
	$CH_3CH{=}CHCHO$	$CH_2{=}CHOC_2H_5$	BF_3	$RCH(OR')CH_2CHO$ (46)	196
		$CH_2{=}CHOC_4H_9\text{-}n$	"	$RCH(OR')CH_2CHO$ (39)	196
		$C_2H_5CH{=}CHOC_2H_5$	"	$RCH(OR')CH(C_2H_5)CHO$ (61)	196
		$(CH_3)_2C{=}CHOCH_3$	"	$RCH(OR')C(CH_3)_2CHO$ (33)	196
	$CH_3CH{=}CHCH(OCH_3)_2$	$CH_2{=}CHCH{=}CHOSi(CH_3)_3$	$TiCl_4$	$RCH(OR')CH_2CH{=}CHCHO$ (86)	186,187
		"	$TiCl_4$, $Ti(OC_3H_7\text{-}i)_4$	" (80)	186,187

	$CH_3CH{=}CHCH(OC_2H_5)_2$	$CH_2{=}CHOC_2H_5$	$ZnCl_2$	$RCH(OR')CH_2CH(OR')_2$ (79)	228
	$n\text{-}C_3H_7CHO$	$CH_2{=}CHOC_2H_5$	$BF_3 \cdot O(C_2H_5)_2$	$RCH{=}CHCHO$ (65)[b]	195
		"	BF_3	$RCH(OR')CH_2CHO$ (16)	196
		$CH_3CH{=}NC_6H_{11}$	LDA	$RCH{=}CHCHO$ (65)[b]	75
		$CH_3CH{=}CHOC_2H_5$	$BF_3 \cdot O(C_2H_5)_2$	$RCH{=}CH(CH_3)CHO$ (68)[b]	197
		$C_2H_5CH{=}CHOCH_3$	"	$RCH{=}CH(C_2H_5)CHO$ (78)[b]	195
		$C_2H_5CH{=}CHOC_2H_5$	"	" (55)[b]	197
		"	BF_3	$RCH(OR')CH(C_2H_5)CHO$ (51)	196
	$n\text{-}C_3H_7CH(OCH_3)_2$	$C_6H_5CH_2CH{=}CHOSi(CH_3)_2$	$TiCl_4$	$RCH{=}C(CH_2C_6H_5)CHO$ (78)	173
		$CH_3CH{=}C[OSi(CH_3)_3]Si(CH_3)_3$	$BF_3 \cdot O(C_2H_5)_2$	$RCH{=}C(CH_3)CHO$ (79)[b]	184
		$C_2H_5CH{=}C[OSi(CH_3)_3]Si(CH_3)_3$	"	$RCH{=}C(C_2H_5)CHO$ (69)[b]	184
		$C_6H_5CH_2CH{=}C[OSi(CH_3)_3]Si(CH_3)_3$	"	$RCH{=}C(CH_2C_6H_5)CHO$ (69)[b]	184
	$n\text{-}C_3H_7CH(OC_2H_5)_2$	$(CH_3)_2C{=}CHOSi(CH_3)_3$	$(CH_3)_3SiO_3SCF_3$	$RCH(OCH_3)C(CH_3)_2CHO$ (89)	193a
		$CH_2{=}CHOC_2H_5$	$BF_3 \cdot O(C_2H_5)_2$	$RCH(OR')CH_2CH(OR')_2$ (72) $RCH(OR')CH_2CH(OR')CH_2CH(OR')_2$ (16)	225
	$i\text{-}C_3H_7CHO$	$C_2H_5CH{=}NN(CH_3)_2$	LDA	$RCHOHCH(CH_3)CH{=}NN(CH_3)_2$ (80)	87
		$CH_3CH{=}NC_4H_9\text{-}t$	LDA	R–CH=CH–CHO (43)	85
C_5	$(CH_3)_2C{=}CHCH(OC_2H_5)_2$	$CH_2{=}C(CH_3)CH{=}CHOC_2H_5$	$ZnCl_2$	$RCH(OR')CH_2C(CH_3){=}CHCH(OR')_2$ (44) $RCH(OR')[CH_2C(CH_3){=}CHCH(OR')]_2OR'$ (30) $RCH(OR')[CH_2C(CH_3){=}CHCH(OR')]_3OR'$ (15)	210
	5-nitro-2-furyl-$CH(OCH_3)_2$ (O_2N–furan–$CH(OCH_3)_2$)	$CH_2{=}CHOCH_3$	$BF_3 \cdot O(C_2H_5)_2$	$RCH{=}CHCHO$ (95)[b]	198
C_6	$CH_3CH{=}CHCH{=}CHCHO$	$C_2H_5CH{=}CHOC_2H_5$	BF_3	$RCH(OR')CH(C_2H_5)CHO$ (49)	196
	$CH_3(CH{=}CH)_2CH(OC_2H_5)_2$	$CH_2{=}CHOC_2H_5$	$ZnCl_2$	$RCH{=}CHCHO$ (71)[b]	228
	$n\text{-}C_3H_7CH{=}CHCH(OCH_3)_2$	$CH_2{=}CHCH{=}CHOSi(CH_3)_3$	$TiCl_4$, $Ti(OC_3H_7\text{-}i)_4$	$RCH(OR')CH_2CH{=}CHCHO$ (84)	187
	$n\text{-}C_5H_{11}CHO$	$CH_2{=}C(CH_3)CH{=}CHOSi(CH_3)_3$	"	$RCH(OR')CH_2C(CH_3){=}CHCHO$ (74)	187
	$n\text{-}C_5H_{11}CH(OCH_3)_2$	$CH_2{=}CHOCH_3$	BF_3	$RCH{=}CHCHO$ (45)[b]	195
		$CH_2{=}CHCH{=}CHOSi(CH_3)_3$	$TiCl_4$, $Ti(OC_3H_7\text{-}i)_4$	$RCH(OC_3H_7\text{-}i)CH_2CH{=}CHCHO$ (75)	187
	$n\text{-}C_3H_7CH{=}CHCHO$	$C_2H_5CH{=}CHOC_2H_5$	BF_3	$RCH(OR')CH(C_2H_5)CHO$ (67)	196
	$(C_2H_5)_2CHCH(OC_2H_5)_2$	$C_2H_5CH{=}CHOC_2H_5$	$BF_3 \cdot O(C_2H_5)_2$	$RCH(OR')CH(C_2H_5)CH(OR')_2$ (54)	225
C_7	$p\text{-}O_2NC_6H_4CHO$	$ClCH{=}CHOC_2H_5$	$TiCl_4$, $Ti(OC_3H_7\text{-}i)_4$	$RCH(OC_3H_7\text{-}i)CHClCH(OC_2H_5)(OC_3H_7\text{-}i)$ (81)	199
		$(CH_3)_2CHCH{=}NC_6H_{11}$	$BCl_2\text{–}N(C_2H_5)_2$, $N(C_2H_5)_3$	$RCH(OAc)C(CH_3)_2CHO$ (33)[b]	146
	C_6H_5CHO	$ClCH{=}CHOC_2H_5$	$TiCl_4$, $Ti(OC_3H_7\text{-}i)_4$	$RCH(OC_3H_7\text{-}i)CHClCH(OC_2H_5)_2$ (100)	199
		$CH_2{=}CHOCH_3$	BF_3	$RCH{=}CHCHO$ (60)[b]	194
		$CH_3CH{=}CHOC_2H_5$	$BF_3 \cdot O(C_2H_5)_2$	$RCH{=}C(CH_3)CHO$ (53)[b]	197
		$CH_3CH{=}NC_6H_{11}$	LDA	$RCHOHCH_2CH{=}NC_6H_{11}$ (94)	3

TABLE I. REACTION OF ALDEHYDE EQUIVALENTS WITH ALDEHYDES OR ACETALS (*Continued*)

No. of C Atoms	Aldehyde or Acetal	Aldehyde Equivalent	Reagent	Product(s)[a] and Yield(s) (%)	Refs.
C_7 (*Contd.*)	C_6H_5CHO	$CH_3CH{=}CHOSi(CH_3)_3$ (*trans*)	CH_3Li	$RCH(OH)CH(CH_3)CHO$ (75) (*threo*:*erythro* = 35:65)	230
		" (*cis*)	"	(−) (*threo*:*erythro* = 50:50)	251
		$C_2H_5CH{=}CHOC_2H_5$	$TiCl_4$, $Ti(OC_3H_7\text{-}i)_4$	$RCH(OC_3H_7\text{-}i)CH(C_2H_5)CH(OC_2H_5)_2$ (73)	199
		"	$BF_3 \cdot O(C_2H_5)_2$	$RCH{=}C(C_2H_5)CHO$ (64)[b]	197
		$C_2H_5CH{=}CHOB(Cl)N(C_2H_5)_2$	$(C_2H_5)_3N$	$RCH{=}C(C_2H_5)CHO$ (17)	148
		$C_2H_5CH{=}CHOC_2H_5$	BF_3	$RCH(OR')CH(C_2H_5)CHO$ (78)	196
	O_2N–(furan-2,5-diyl)–$CH{=}CHCHO$	$CH_2{=}CHOCH_3$	$BF_3 \cdot O(C_2H_5)_2$	$RCH{=}CHCHO$ (93)	198
	O_2N–(furan-2,5-diyl)–$CH{=}CHCH(OCH_3)_2$	$CH_2{=}CHOCH_3$	$BF_3 \cdot O(C_2H_5)_2$	$RCH{=}CHCH(OR')_2$ (93)	198
	C_6H_5CH (1,3-dioxolane ring)	$CH_2{=}CHOC_2H_5$	BF_3	R-substituted seven-membered ring (1,3-dioxepane) bearing OC_2H_5 (36)	227
		"	$BF_3 \cdot O(C_2H_5)_2$	$RCH{=}CHCH$ (1,3-dioxolane ring) (47)	229
	$C_6H_5CH(OCH_3)_2$	$CH_2{=}CHCH{=}CHOSi(CH_3)_3$	$TiCl_4$, $Ti(OC_3H_7\text{-}i)_4$	$RCH(OC_3H_7\text{-}i)CH_2CH{=}CHCHO$ (83)	187
	$C_6H_5CH(OC_2H_5)_2$	$CH_2{=}CHOC_2H_5$	$BF_3 \cdot O(C_2H_5)_2$	$RCH(OR')CH_2CH(OR')_2$ (72)	231
		$CH_3CH{=}C[OSi(CH_3)_3]Si(CH_3)_3$	"	$RCH{=}C(CH_3)CHO$ (86)[b] [100% (*E*)]	184
		$C_6H_5CH_2CH{=}C[OSi(CH_3)_3]Si(CH_3)_3$	"	$RCH{=}C(CH_2C_6H_5)CHO$ (86)[b] [(*E*)-(*Z*) = 90:8]	184
C_8	$C_6H_5CH_2CHO$	$(CH_3)_2C{=}CHOSi(CH_3)_3$	$TiCl_4$	$RCHOHC(CH_3)_2CHO$ (86)	173
	$CH_3(CH{=}CH)_3CH(OC_2H_5)_2$	$CH_2{=}CHOC_2H_5$	$ZnCl_2$	$RCH{=}CHCHO$ (83)[b]	228
	$n\text{-}C_7H_{15}CHO$	$n\text{-}C_7H_{15}CH{=}NC_4H_9\text{-}t$	LDA	$RCH{=}C(C_6H_{13}\text{-}n)CHO$ (70)[b]	77
	$n\text{-}C_4H_9CH(C_2H_5)CH(OC_2H_5)_2$	$CH_2{=}CHOC_2H_5$	$BF_3 \cdot O(C_2H_5)_2$	$RCH(OR')CH_2CH(OR')_2$ (67)	225
		$C_2H_5CH{=}CHOC_2H_5$	"	$RCH(OR')CH(C_2H_5)CH(OR')_2$ (40)	225

	THPOCH(CH_3)CHO	$C_2H_5CH{=}NC_6H_{11}$	LDA	R–CH=C(CH_3)–CHO (70)[b]	82
C_9	O_2N–(2-furyl)–$(CH{=}CH)_2CHO$	$CH_2{=}CHOCH_3$	$BF_3 \cdot O(C_2H_5)_2$	$RCH{=}CHCHO$ (80)	198
	O_2N–(2-furyl)–$(CH{=}CH)_2CH(OCH_3)_2$	$CH_2{=}CHOCH_3$	$BF_3 \cdot O(C_2H_5)_2$	$RCH{=}CHCH(OR')_2$ (90)	198
	$C_6H_5CH{=}CHCH$(1,3-dioxolan-2-yl)	$CH_2{=}CHCH{=}CHOSi(CH_3)_3$	$TiCl_4$	$RCH(OCH_2CH_2OH)CH_2CH{=}CHCHO$ (60)	186,187
	$C_6H_5CH{=}CHCH(OCH_3)_2$	$CH_2{=}CHCH{=}CHOSi(CH_3)_3$	$TiCl_4$, $Ti(OC_3H_7\text{-}i)_4$	$RCH(OC_3H_7\text{-}i)CH_2CH{=}CHCHO$ (90)	187
		"	$TiCl_4$	$RCH(OR')CH_2CH{=}CHCHO$ (88)	186,187
		$CH_2{=}C(CH_3)CH{=}CHOSi(CH_3)_3$	"	$RCH(OR')CH_2C(CH_3){=}CHCHO$ (85)	187
		$C_2H_5CH{=}CHCH{=}CHOSi(CH_3)_3$	"	$RCH(OR')CH(C_2H_5)CH{=}CHCHO$ (69)	187
	$C_6H_5CH{=}CHCH(OC_2H_5)_2$	$CH_3CH{=}C[OSi(CH_3)_3]Si(CH_3)_3$	$BF_3 \cdot O(C_2H_5)_2$	R–CH=C(CH_3)–CHO (77)[b]	184
		$C_6H_5CH_2CH{=}C[OSi(CH_3)_3]Si(CH_3)_3$	"	$RCH{=}C(CH_2C_6H_5)CHO$ (80)[b] [(E)-(Z) = 84:16]	184
	$C_6H_5(CH_2)_2CHO$	$ClCH{=}CHOC_2H_5$	$TiCl_4$, $Ti(OC_3H_7\text{-}i)_4$	$RCH(OC_3H_7\text{-}i)CHClCH(OC_2H_5)_2$ (96)	199
		$CH_2{=}CHOC_2H_5$	"	$RCH(OC_3H_7\text{-}i)CH_2CH(OC_2H_5)_2$ (100)	199
		$C_2H_5CH{=}CHOC_2H_5$	"	$RCH(OC_3H_7\text{-}i)CH(C_2H_5)CH(OC_2H_5)_2$ (56)	199
		$(CH_3)_2C{=}CHOSi(CH_3)_3$	$TiCl_4$	$RCHOHC(CH_3)_2CHO$ (95)	173
	$C_6H_5(CH_2)_2CH(OCH_3)_2$	$CH_2{=}CHCH{=}CHOSi(CH_3)_3$	$TiCl_4$, $Ti(OC_3H_7\text{-}i)_4$	$RCH(OC_3H_7\text{-}i)CH_2CH{=}CHCHO$ (80)	187
		$(CH_3)_2{=}CHOSi(CH_3)_3$	$TiCl_4$	$RCH(OR')C(CH_3)_2CHO$ (100)	180
		$CH_2{=}C(CH_3)CH{=}CHOSi(CH_3)_3$	$TiCl_4$, $Ti(OC_3H_7\text{-}i)_4$	$RCH(OC_3H_7\text{-}i)CH_2C(CH_3){=}CHCHO$ (76)	187
	$C_6H_5(CH_2)_2CH(OC_2H_5)_2$	$CH_2{=}CHCH{=}CHOSi(CH_3)_3$	$TiCl_4$, $Ti(OC_3H_7\text{-}i)_4$	$RCH(OC_3H_7\text{-}i)CH_2CH{=}CHCHO$ (79)	187
	$n\text{-}C_8H_{17}CHO$	$n\text{-}C_3H_7CH{=}NC_4H_9\text{-}t$	LDA	$RCH{=}C(C_2H_5)CHO$ (—)[b]	77
C_{10}	$CH_3(C_6H_5)C{=}CHCH(OC_2H_5)_2$	$CH_2{=}CHOC_2H_5$	$ZnCl_2$	$RCH(OR')CH_2CH(OR')_2$ (50)	212
	$[(C_2H_5O)_2CH\text{–}CH{=}C(CH_3)\text{–}CH_2\text{–}C{\equiv}]_2$	$CH_2{=}CHOC_2H_5$	$ZnCl_2$, $BF_3 \cdot O(C_2H_5)_2$	$[(RO)_2CH\text{–}CH_2\text{–}CH(OR')\text{–}C(CH_3){=}CH\text{–}CH_2\text{–}C{\equiv}]_2$ (—)	206

TABLE I. REACTION OF ALDEHYDE EQUIVALENTS WITH ALDEHYDES OF ACETALS (*Continued*)

No. of C Atoms	Aldehyde or Acetal	Aldehyde Equivalent	Reagent	Product(s)[a] and Yield(s) (%)	Refs.
C_{10} (*Contd.*)	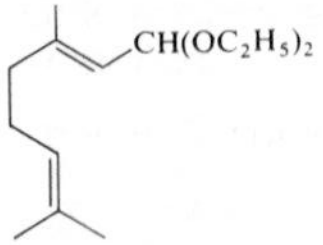 $CH(OC_2H_5)_2$	$CH_2{=}C(CH_3)CH{=}CHOC_2H_5$	$ZnCl_2$	$RCH(OR')CH_2C(CH_3){=}CHCH(OR')_2$ (45), $RCH(OR')\text{-}[CH_2C(CH_3){=}CHCH(OR')]_2\text{-}OR'$ (18)	232
	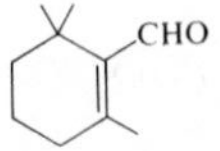 CHO	$CH_3CH{=}NC_6H_{11}$	LDA	$RCH{=}CHCHO$ (50)[b]	75
	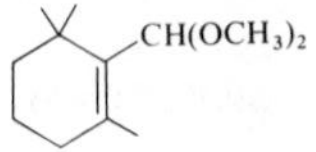 $CH(OCH_3)_2$	$CH_2{=}CHCH{=}CHOSi(CH_3)_3$	$TiCl_4$, $Ti(OC_3H_7\text{-}i)_4$	$RCH(OR')CH_2{=}CHCHO$ (80)	187
	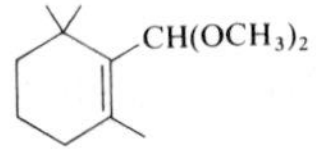 $CH(OCH_3)_2$	$CH_2{=}C(CH_3)CH{=}CHOSi(CH_3)_3$	$TiCl_4$, $Ti(OC_3H_7\text{-}i)_4$	$RCH(OR')CH_2C(CH_3){=}CHCHO$ (80)	187
	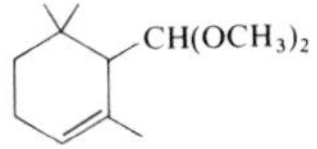 $CH(OCH_3)_2$	$CH_2{=}C(CH_3)CH{=}CHOSi(CH_3)_3$	$TiCl_4$, $Ti(OC_3H_7\text{-}i)_4$	$RCH(OR')CH_2C(CH_3){=}CHCHO$ (81)	187
C_{11}	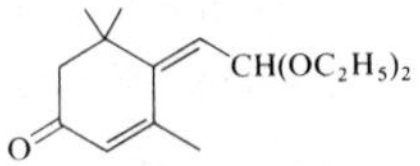 O, $CH(OC_2H_5)_2$	$CH_3CH{=}CHOC_2H_5$	$ZnCl_2$	$RCH{=}C(CH_3)CHO$ (—)[b]	233
	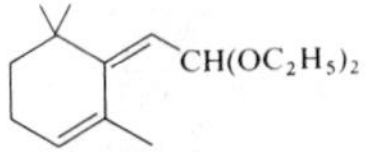 $CH(OC_2H_5)_2$	$CH_3CH{=}CHOC_2H_5$	$ZnCl_2$	$RCH(OR')CH(CH_3)CH(OR')_2$ (—)	205
	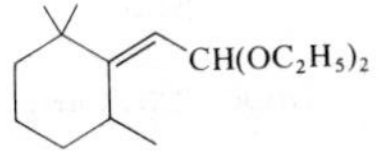 $CH(OC_2H_5)_2$	$CH_3CH{=}CHOC_2H_5$	$ZnCl_2$	$RCH(OR')CH(CH_3)CH(OR')_2$ (—)	205

	$n\text{-}C_4H_9CH(OAc)CH{=}C(CH_3)CH(OCH_3)_2$	$CH_2{=}CHCH{=}CHOSi(CH_3)_3$	$TiCl_4$, $Ti(OC_3H_7\text{-}i)_4$	$RCH(OR')CH_2CH{=}CHCHO$ (85)	190
C_{12}	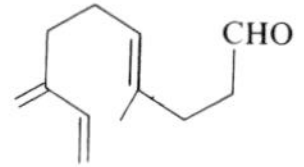 CHO	$C_2H_5CH{=}NC_4H_9\text{-}t$	LDA	$RCH{=}C(CH_3)CHO$ (60)[b]	76
	$p\text{-}CH_3C_6H_4CH(CH_3)(CH_2)_2CHO$	$C_2H_5CH{=}NC_4C_9\text{-}t$	LDA	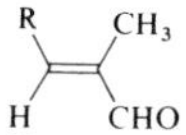 R, CH_3, H, CHO (83)[b]	80
C_{14}	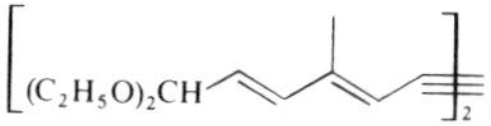 $[(C_2H_5O)_2CH]_2$	$CH_3CH{=}CHOC_2H_5$	$ZnCl_2$	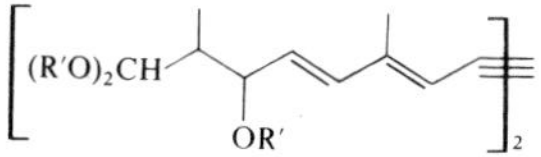 $[(R'O)_2CH, OR']_2$ (—)	206
	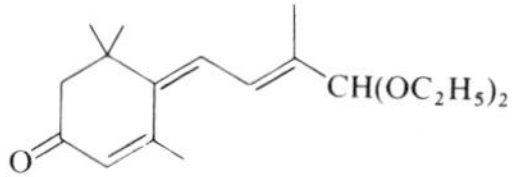 $CH(OC_2H_5)_2$	$CH_2{=}CHOC_2H_5$	$ZnCl_2$	$RCH{=}CHCHO$ (—)[b]	233
	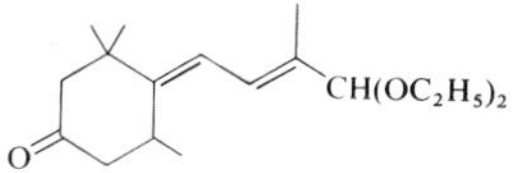 $CH(OC_2H_5)_2$	$CH_2{=}CHOC_2H_5$	$ZnCl_2$	$RCH(OR')CH_2CH(OR')_2$ (—)	207
	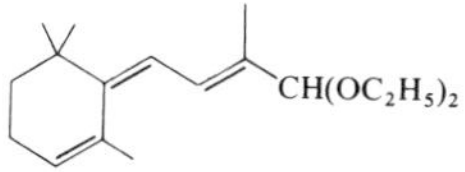 $CH(OC_2H_5)_2$	$CH_2{=}CHOC_2H_5$	$ZnCl_2$	$RCH(OR')CH_2CH(OR')_2$ (—)	205
	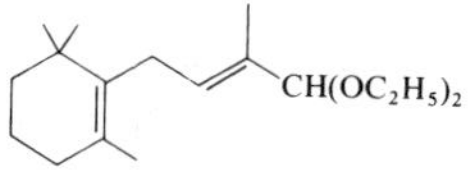 $CH(OC_2H_5)_2$	$CH_2{=}CHOC_2H_5$	$ZnCl_2$	$RCH(OR')CH_2CH(OR')_2$ (91)	204,205
	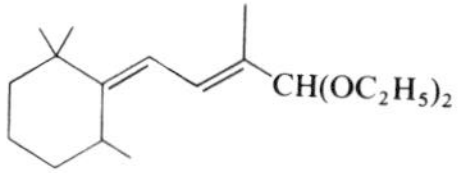 $CH(OC_2H_5)_2$	$CH_2 = CHOC_2H_5$	$ZnCl_2$	$RCH(OR')CH_2CH(OR')_2$ (—)	205
C_{15}	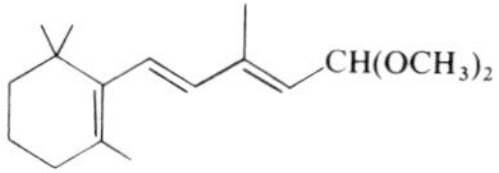 $CH(OCH_3)_2$	$CH_2{=}C(CH_3)CH{=}CHOSi(CH_3)_3$	$TiCl_4$, $Ti(OC_3H_7\text{-}i)_4$	$RCH(OR')CH_2C(CH_3){=}CHCHO$ (80)	189

TABLE I. REACTION OF ALDEHYDE EQUIVALENTS WITH ALDEHYDES OR ACETALS (*Continued*)

No. of C Atoms	Aldehyde or Acetal	Aldehyde Equivalent	Reagent	Product(s)[a] and Yield(s) (%)	Refs.
C_{15} (*Contd.*)	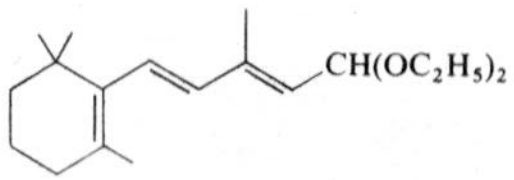 $CH(OC_2H_5)_2$	$CH_2{=}C(CH_3)CH{=}CHOC_2H_5$	$ZnCl_2$	$RCH(OR')CH_2C(CH_3){=}CHCH(OR')_2$ (62)	213
C_{16}	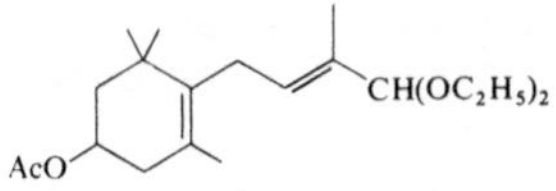 AcO, $CH(OC_2H_5)_2$	$CH_2{=}CHOC_2H_5$	$ZnCl_2$	$RCH{=}CHCHO$ (69)[b]	208
	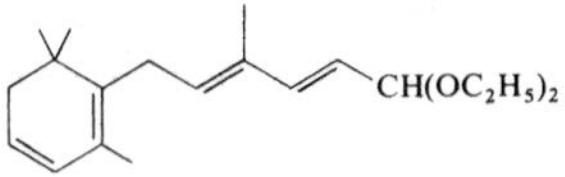 $CH(OC_2H_5)_2$	$CH_3CH{=}CHOC_2H_5$	$ZnCl_2$	$RCH(OR')CH(CH_3)CH(OR')_2$ (—)	205
	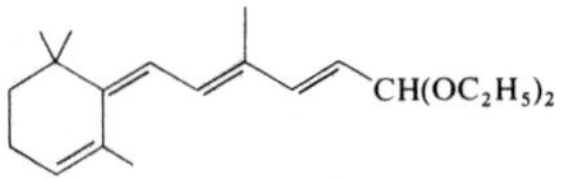 $CH(OC_2H_5)_2$	$CH_3CH{=}CHOC_2H_5$	$ZnCl_2$	$RCH(OR')CH(CH_3)CH(OR')_2$ (—)	205
	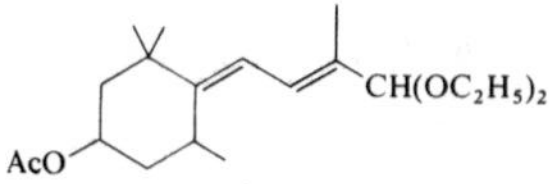 AcO, $CH(OC_2H_5)_2$	$CH_2{=}CHOC_2H_5$	$ZnCl_2$	$RCH{=}CHCHO$ (98)[b]	233
	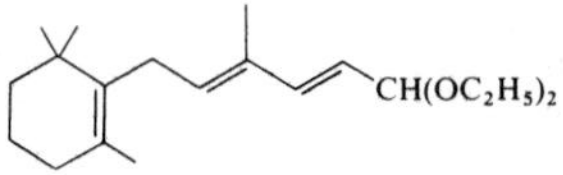 $CH(OC_2H_5)_2$	$CH_3CH{=}CHOC_2H_5$	$ZnCl_2$	$RCH(OR')CH(CH_3)CH(OR')_2$ (—)	204

	Aldehyde or acetal	Enol ether or imine	Catalyst	Product (Yield, %)[a]	Ref.
	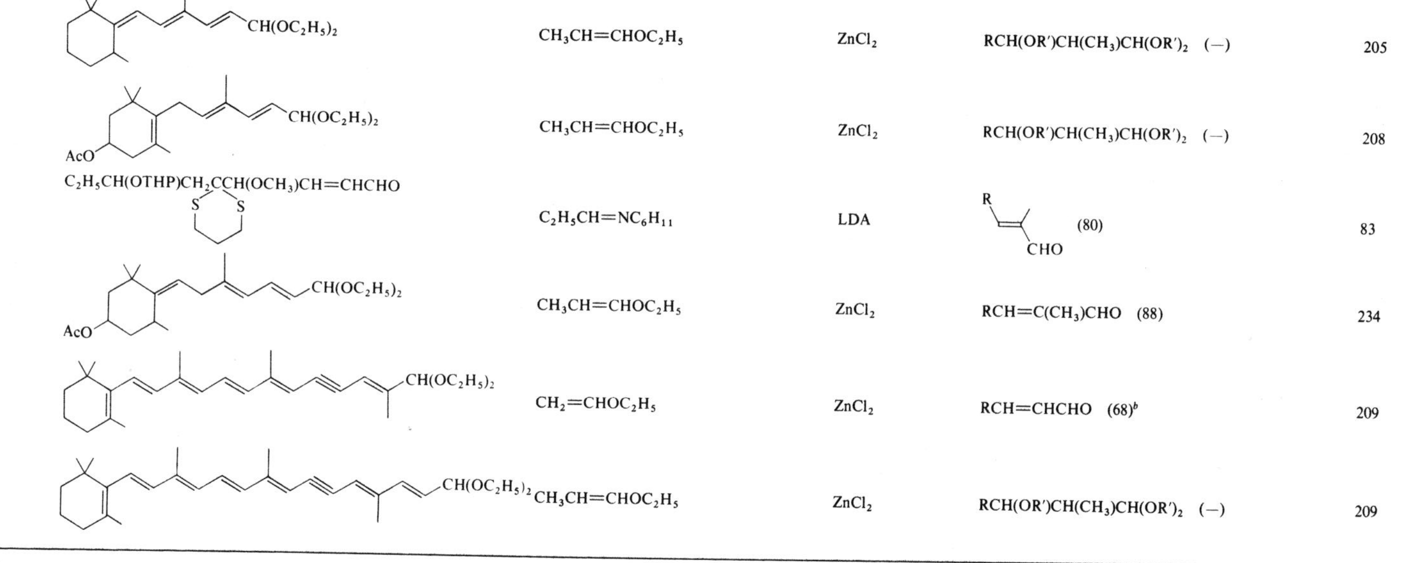	$CH_3CH{=}CHOC_2H_5$	$ZnCl_2$	$RCH(OR')CH(CH_3)CH(OR')_2$ (—)	205
C_{18}	AcO, $CH(OC_2H_5)_2$	$CH_3CH{=}CHOC_2H_5$	$ZnCl_2$	$RCH(OR')CH(CH_3)CH(OR')_2$ (—)	208
	$C_2H_5CH(OTHP)CH_2CCH(OCH_3)CH{=}CHCHO$ (S, S)	$C_2H_5CH{=}NC_6H_{11}$	LDA	R, CHO (80)	83
C_{19}	AcO, $CH(OC_2H_5)_2$	$CH_3CH{=}CHOC_2H_5$	$ZnCl_2$	$RCH{=}C(CH_3)CHO$ (88)	234
C_{25}	$CH(OC_2H_5)_2$	$CH_2{=}CHOC_2H_5$	$ZnCl_2$	$RCH{=}CHCHO$ (68)[b]	209
C_{27}	$CH(OC_2H_5)_2$	$CH_3CH{=}CHOC_2H_5$	$ZnCl_2$	$RCH(OR')CH(CH_3)CH(OR')_2$ (—)	209

[a] R and R′ in the product are the groups R and R′ in the aldehyde RCHO or in the acetal $RCH(OR')_2$.
[b] The yield given is after hydrolysis.

TABLE II. Reaction of Ketones or Their Equivalents with Aldehydes or Acetals

No. of C Atoms	Aldehyde or Acetal	Ketone or Equivalent	Reagent	Product(s)[a] and Yield(s) (%)	Refs.
C_1	HCHO	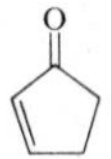 O	$(CH_3)_2AlSC_2H_5$	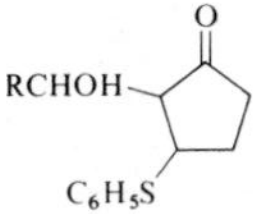 RCHOH, O, C_6H_5S (56)	134
		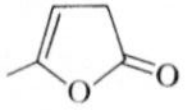 O, O	$BF_3 \cdot O(C_2H_5)_2$	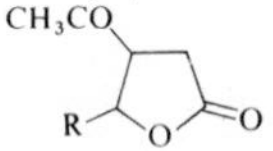 CH_3CO, R, O, O (93)	222
		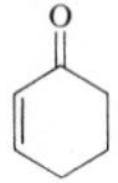 O	CH_3MgBr, $CuI \cdot Bu_3P$	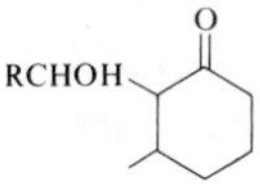 RCHOH, O (70)	57
		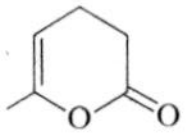 O, O	$BF_3 \cdot O(C_2H_5)_2$	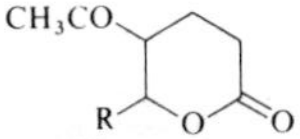 CH_3CO, R, O, O (90)	222
		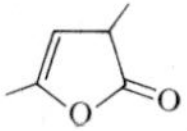 O, O	"	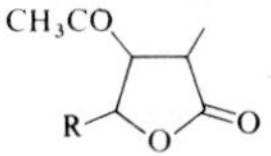 CH_3CO, R, O, O (92)	223
		$C_6H_5CH{=}C[OSi(CH_3)_3]CH_3$	$TiCl_4$	$RCHOHCH(C_6H_5)COCH_3$ (64)	173
		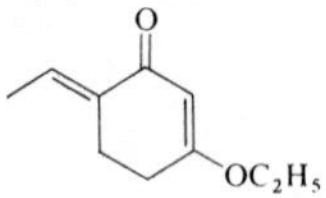 O, OC_2H_5	LDCA	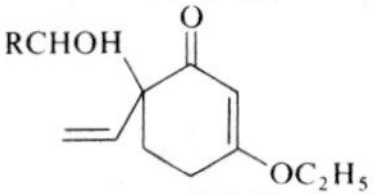 RCHOH, O, OC_2H_5 (40)	52
		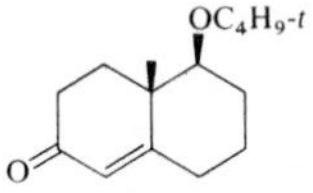 OC_4H_9-*t*, O	Li–NH_3	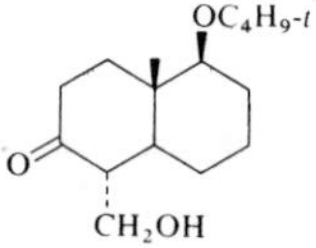 OC_4H_9-*t*, O, CH_2OH (—)	57

		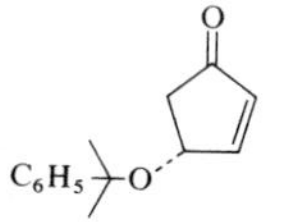 C_6H_5–O (cyclopentenone)	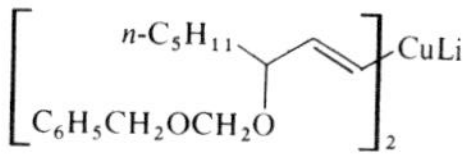 $[n\text{-}C_5H_{11}CH(OCH_2OCH_2C_6H_5)CH{=}CH]_2CuLi$	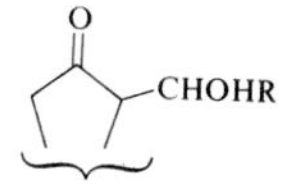 CHOHR (50–60)	62
		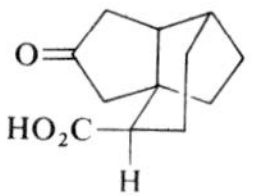 HO_2C, H	LDA	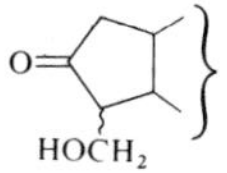 $HOCH_2$ (62)	235
		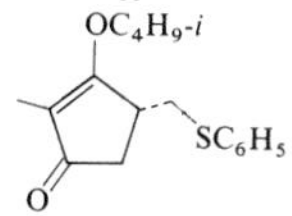 OC_4H_9-i, SC_6H_5	LDA	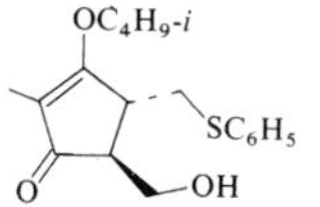 OC_4H_9-i, SC_6H_5, OH (82)	277
		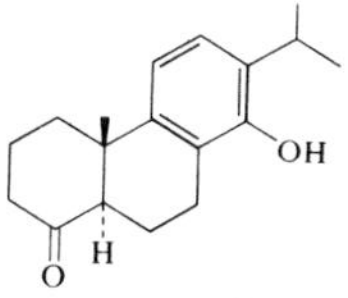 OH, H	LDA	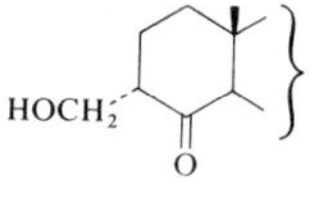 $HOCH_2$ (73)	236
		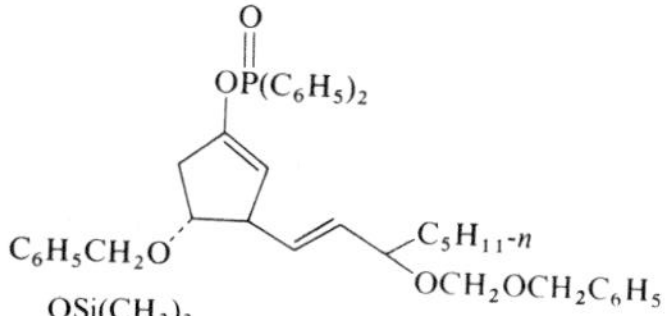 $OP(C_6H_5)_2$, $C_6H_5CH_2O$, C_5H_{11}-n, $OCH_2OCH_2C_6H_5$	t-C_4H_9Li, $ZnCl_2$	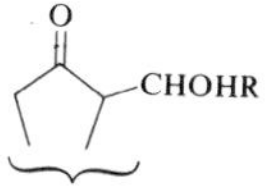 CHOHR (80–90)	123
C_2	CCl_3CHO	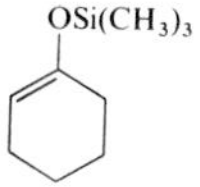 $OSi(CH_3)_3$ (cyclohexenyl)	$TiCl_4$	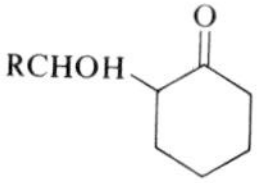 RCHOH (94)	177
		$C_2H_5COC(CH_3)_3$	LDA	$RCHOHCH(CH_3)COC(CH_3)_3$ (83) (*erythro*)	230
		$CH_2{=}C(C_6H_5)OSi(CH_3)_3$	"	$RCHOHCH_2COC_6H_5$ (92)	177
	Cl_2CHCHO	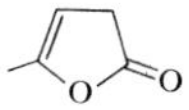 (methylfuranone)	$BF_3 \cdot O(C_2H_5)_2$	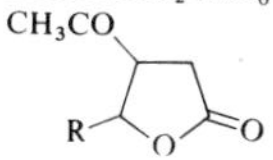 CH_3CO, R (60)	222

TABLE II. REACTION OF KETONES OR THEIR EQUIVALENTS WITH ALDEHYDES OR ACETALS (*Continued*)

No. of C Atoms	Aldehyde or Acetal	Ketone or Equivalent	Reagent	Product(s)[a] and Yield(s) (%)	Refs.
C_2 (*Contd.*)	$ClCH_2CHO$	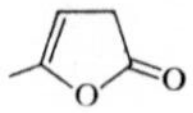	$BF_3 \cdot O(C_2H_5)_2$	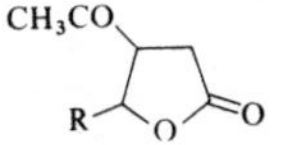 CH_3CO, R (90)	222
		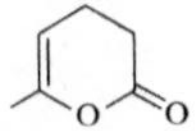	"	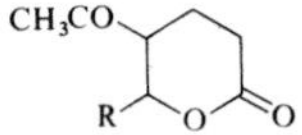 CH_3CO, R (65)	222
	$ClCH_2CH(OCH_3)_2$	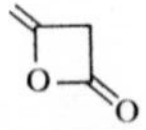	$TiCl_4$	$RCH(OR')CH_2COCH_2CO_2CH_3$ (58)	215
	$BrCH_2CH(OCH_3)_2$	$CH_2{=}C[OSi(CH_3)_3]C_6H_5$	$TiCl_4$	$RCH(OR')CH_2COC_6H_5$ (—)	181
		$C_6H_5CH{=}C[OSi(CH_3)_3]CH_3$	"	$RCH(OR')CH(C_6H_5)COCH_3$ (—)	181
	CH_3CHO	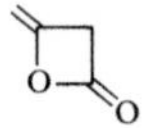	$TiCl_4$	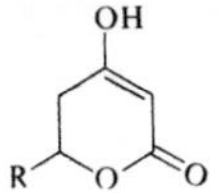 OH, R (78), $RCHOHCH_2COCH_2CO_2CH_3$ (3)	216
		$CH_2{=}CHCOCH_3$	$(CH_3)_2AlSC_6H_5$	$RCHOHCH(CH_2SC_6H_5)COCH_3$ (60)	134
		"	$(CH_3)_2AlSeCH_3$	$RCHOHCH(CH_2SeCH_3)COCH_3$ (55)	134
		"	$(n\text{-}C_4H_9)_2CuLi$, $ZnCl_2$	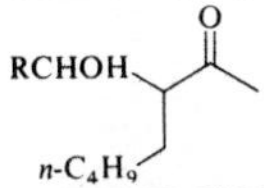 RCHOH, $n\text{-}C_4H_9$ (92)	61
		$CH_3COC_2H_5$	$(C_2H_5)_2NLi$	$RCHOHCH_2COC_2H_5$ (11)	48
		$CH_3COC_3H_7\text{-}n$	LDA	$RCHOHCH_2COC_3H_7\text{-}n$ (28), $RCHOHCH(C_2H_5)COCH_3$ (3)	46
		$CH_3COC_3H_7\text{-}n$	$(C_2H_5)_2NLi$	$RR'C(OH)CH_2COC_3H_7\text{-}n$ (34)	48
		$CH_3COC_3H_7\text{-}i$	"	$RR'C(OH)CH_2COC_3H_7\text{-}i$ (42)	48
		(*S*)-$CH_3COCH(CH_3)C_2H_5$	LDA	$RCHOHCH_2COCH(CH_3)C_2H_5$ (65)	54
		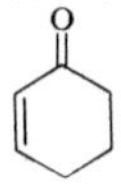	$(CH_3)_2CuLi$, $ZnCl_2$	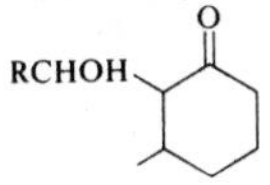 RCHOH (97)	60,61

"	$(CH_3)_2AlSC_6H_5$	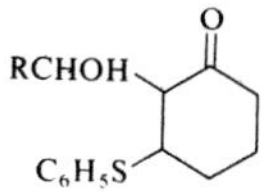 RCHOH, C_6H_5S, O (94)	134
"	$(CH_3)_2AlSeCH_3$	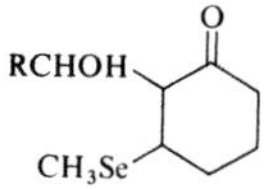 RCHOH, CH_3Se, O (77)	134
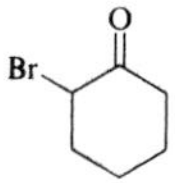 Br, O	Zn	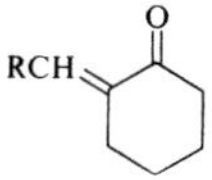 RCH=, O (57)	124
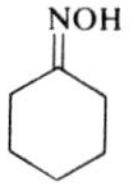 NOH	n-C_4H_9Li	" (45)[b]	101
$(CH_3)_2C{=}CHCOCH_3$	$(CH_3)_2CuLi$, $ZnCl_2$	$RCHOHCH(C_4H_9\text{-}t)COCH_3$ (96) (*threo*)	60,61
i-$C_4H_9COCH_3$	$(C_2H_5)_2NLi$	$RR'C(OH)CH_2COC_4H_9\text{-}i$ (44)	48
t-$C_4H_9COCHBrCH_3$	Mg	$RCHOHCH(CH_3)COC_4H_9\text{-}t$ (56) (*erythro*)	14
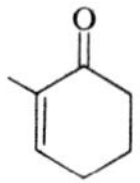 O	$(CH_3)_2CuLi$, $ZnCl_2$	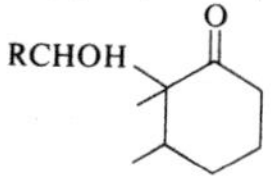 RCHOH, O (32)	60,61
"	$(CH_3)_2AlSC_6H_5$	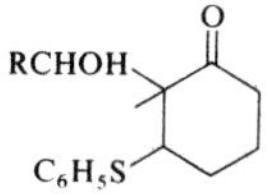 RCHOH, C_6H_5S, O (75)	134
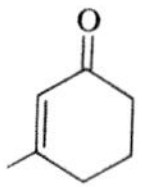 O	$(CH_3)_2AlSC_6H_5$	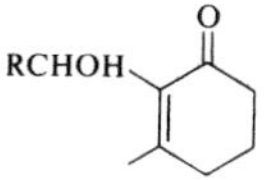 RCHOH, O (50)	134
"	$(CH_3)_2CuLi$, $ZnCl_2$	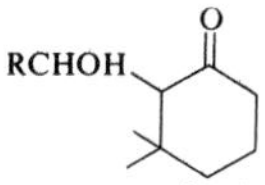 RCHOH, O (98) (*threo*)	60,61
"	CH_3MgI, CuI	" (75)	111

TABLE II. Reaction of Ketones or Their Equivalents with Aldehydes or Acetals (*Continued*)

No. of C Atoms	Aldehyde or Acetal	Ketone or Equivalent	Reagent	Product(s)[a] and Yield(s) (%)	Refs.
C_2 (*Contd.*)	CH_3CHO	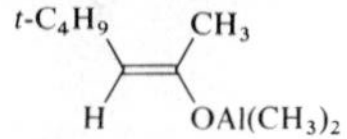 $t\text{-}C_4H_9$, CH_3, H, $OAl(CH_3)_2$	—	$RCHOHCH(C_4H_9\text{-}t)COCH_3$ (62) (*erythro*:*threo* = 93:7)	19
		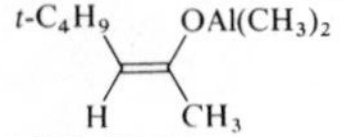 $t\text{-}C_4H_9$, $OAl(CH_3)_2$, H, CH_3	—	$RCHOHCH(C_4H_9\text{-}t)COCH_3$ (66) (100% *threo*)	19
		$i\text{-}C_3H_7COC_3H_7\text{-}i$	$CH_3(C_6H_5)NMgBr$	$RCHOHC(CH_3)_2COC_3H_7\text{-}i$ (73)	107
		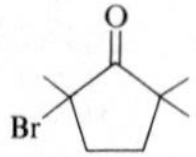 O, Br	Mg	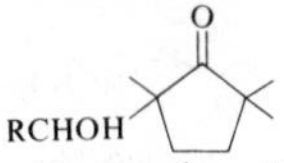 O, RCHOH (65) (*threo*:*erythro* = 95:5)	14
		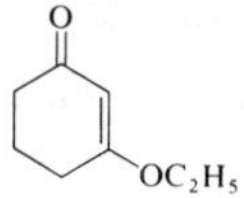 O, OC_2H_5	LDA	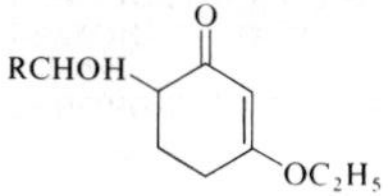 RCHOH, O, OC_2H_5 (92)	52
		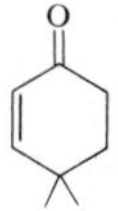 O	$(CH_3)_2CuLi$, $ZnCl_2$	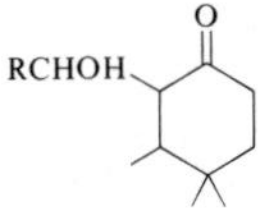 RCHOH, O (87)	60,61
		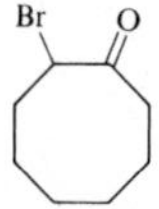 Br, O	Zn	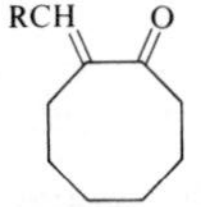 RCH, O (57)	124
		$CH_2{=}CHCOC_6H_5$	$(C_2H_5)_3B$	$RCH[OB(C_2H_5)_2]CH(C_3H_7\text{-}n)COC_6H_5$ (55)	140
		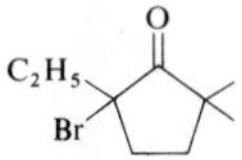 O, C_2H_5, Br	Mg	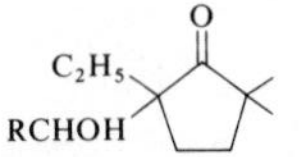 O, C_2H_5, RCHOH (52) (*threo*:*erythro* = 90:10)	14

Reactant	Reagent	Product (Yield, %)	Refs.
$C_6H_5CH{=}C[OSi(CH_3)_3]CH_3$	$TiCl_4$	$RCHOHCH(C_6H_5)COCH_3$ (68)	173
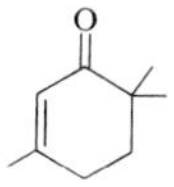 [structure]	$(CH_3)_2CuLi, ZnCl_2$	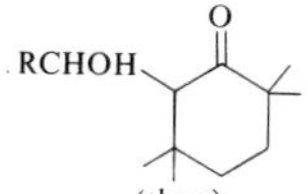 RCHOH [structure] (82) (*threo*)	60,61
$CH_3COCH{=}CHC_6H_5$	"	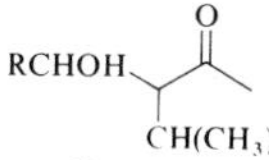 RCHOH [structure] $CH(CH_3)(C_6H_5)$ (83)	61
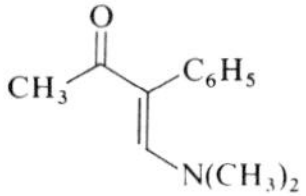 CH_3 [structure] C_6H_5 $N(CH_3)_2$	LTMP	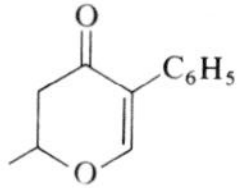 [structure] C_6H_5 (67)[b]	237
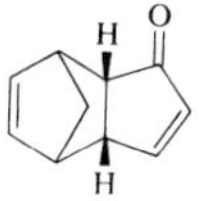 [structure]	n-C_4H_9MgBr, CuCl (catalyst)	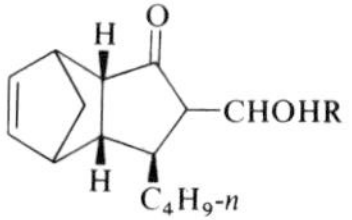 [structure] CHOHR C_4H_9-n (—)	112
$CH_3COCH(C_4H_9\text{-}n)CO_2CH_2CCl_3$	Zn	$RCHOHCH(C_4H_9\text{-}n)COCH_3$ (90)	125
$C_6H_5CH_2COCH{=}CHCH_3$	$(CH_3)_2CuLi, ZnCl_2$	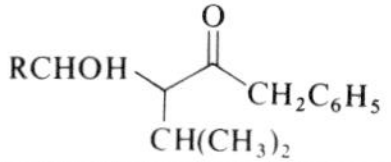 RCHOH [structure] $CH_2C_6H_5$ $CH(CH_3)_2$ (91)	60,61
n-$C_4H_9COC_6H_5$	$CH_3(C_6H_5)NMgBr$	$RCHOHCH(C_3H_7\text{-}n)COC_6H_5$ (50)	107
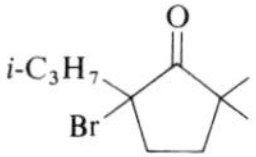 i-C_3H_7 Br [structure]	Mg	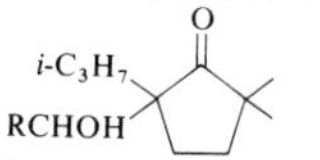 i-C_3H_7 RCHOH [structure] (59) (*threo*: *erythro* = 50:50)	14
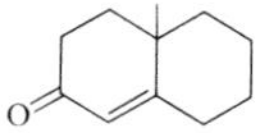 [structure]	$(CH_3)_2CuLi, ZnCl_2$	 [structure] RCHOH (*threo*) (76)	60,61
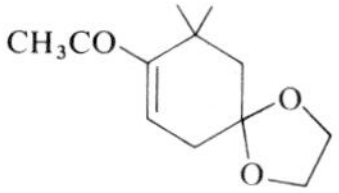 CH_3CO [structure]	LDA	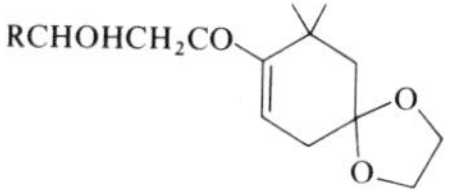 $RCHOHCH_2CO$ [structure] (—)	49

TABLE II. Reaction of Ketones or Their Equivalents with Aldehydes of Acetals (*Continued*)

No. of C Atoms	Aldehyde or Acetal	Ketone or Equivalent	Reagent	Product(s)[a] and Yield(s) (%)	Refs.
C_2 (*Contd.*)	CH_3CHO	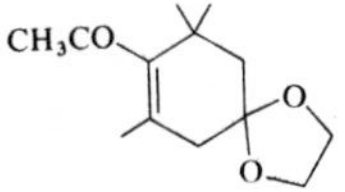 CH_3CO O O	LDA	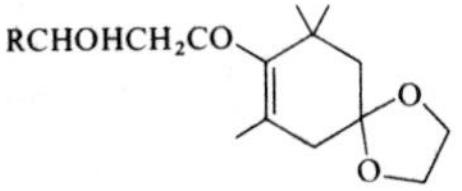 $RCHOHCH_2CO$ O O (−)	49
		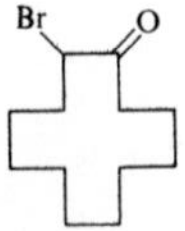 Br O	Zn	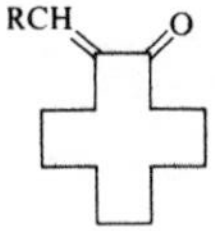 RCH O (70)	124
		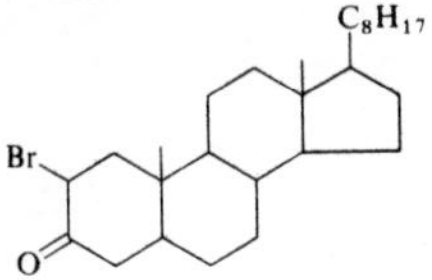 C_8H_{17} Br O	"	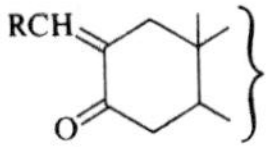 RCH O } (90)	124
	$CH_3CH(OC_2H_5)_2$	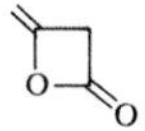 O O	$TiCl_4$	$RCH(OR')CH_2COCH_2CO_2C_2H_5$ (67)	215
C_3	$CH_2{=}CHCHO$	$(CH_3)_2C{=}NOH$	$n\text{-}C_4H_9Li$	$RCH{=}CHCOCH_3$ (15)[b]	101
	$CH_3CHBrCH(OC_2H_5)_2$	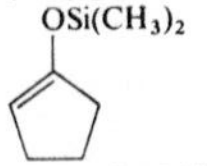 $OSi(CH_3)_2$	$TiCl_4$	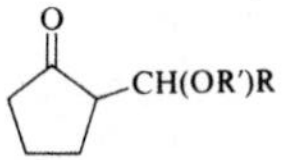 O CH(OR')R (−)	182
	C_2H_5CHO	$(CH_3)_2C{=}NNHSO_2C_6H_4CH_3\text{-}p$	$n\text{-}C_4H_9Li$	$RCHOHCH_2C(CH_3){=}NNHSO_2C_6H_4CH_3\text{-}p$ (78)	97
		$CH_3COCH{=}CH_2$	$(n\text{-}C_4H_9)_2CuLi, ZnCl_2$	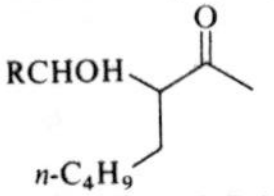 RCHOH O $n\text{-}C_4H_9$ (65)	61
		$C_2H_5C(CH_3){=}NNHSO_2C_6H_4CH_3\text{-}p$	$n\text{-}C_4H_9Li$	$RCHOHCH_2C(C_2H_5){=}NNHSO_2C_6H_4CH_3\text{-}p$ (78)	97

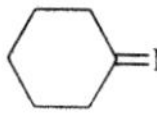 $=NNHSO_2C_6H_4CH_3$-*p*	*n*-C_4H_9Li	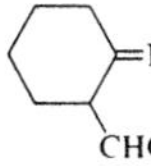 $=NNHSO_2C_6H_4CH_3$-*p*, CHOHR (74)	97
$CH_3COCH{=}C(CH_3)_2$	$(n\text{-}C_4H_9)_2CuLi$, $ZnCl_2$	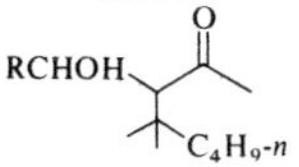 RCHOH, O, C_4H_9-*n* (93)	61
"	$(CH_3)_2CuLi$, $ZnCl_2$	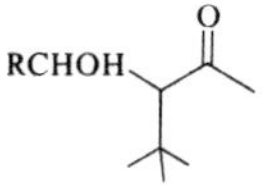 RCHOH, O (50)	61
(*S*)-$CH_3COCH(CH_3)C_2H_5$	LDA	$RCHOHCH_2COCH(CH_3)C_2H_5$ (65)	54
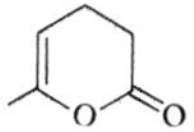 O, O	$BF_3 \cdot O(C_2H_5)_2$	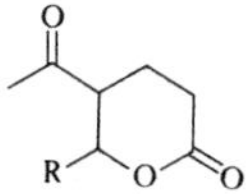 O, R, O, O (78)	222
$C_2H_5COC_4H_9$-*t*	LDA	$RCHOHCH(CH_3)COC_4H_9$-*t* (81) (*erythro*)	230
$C_6H_5COCHN_2$	$(n\text{-}C_4H_9)_3B$, $LiOC_6H_5$	$RCHOHCH(C_4H_9\text{-}n)COC_6H_5$ (87) (*erythro*)	18
O, Br	Mg	O, RCHOH (69) (*threo*:*erythro* = 89:11)	14
t-$C_4H_9COCHBrCH_3$	Mg	$RCHOHCH(CH_3)COC_4H_9$-*t* (71)	14
$C_2H_5COC_6H_5$	$(C_2H_5)_3B$, *t*-$C_4H_9CO_2B(C_2H_5)_2$	$RCH[OB(C_2H_5)_2]CH(CH_3)COC_6H_5$ (91)	142
$C_6H_5CH_2C(CH_3){=}NNHSO_2C_6H_4CH_3$-*p*	*n*-C_4H_9Li	$RCHOHCH_2C(CH_2C_6H_5){=}NNHSO_2C_6H_4CH_3$-*p* (77)	97
n-$C_5H_{11}COC_5H_{11}$-*n*	$C_6H_5N(CH_3)MgBr$	$RCHOHCH(C_4H_9\text{-}n)COC_5H_{11}$-*n* (15)	107
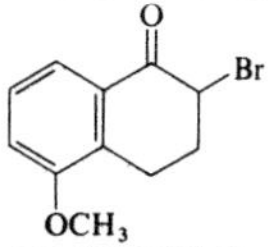 O, Br, OCH_3	Zn	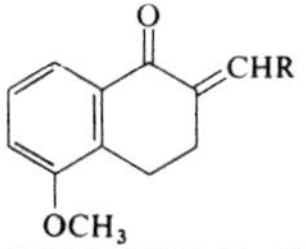 O, CHR, OCH_3 (95)	124
$CH_3COCH(C_4H_9\text{-}n)CO_2CH_2CCl_3$	Zn	$RCHOHCH(C_4H_9\text{-}n)COCH_3$ (89)	125

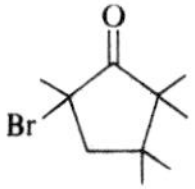

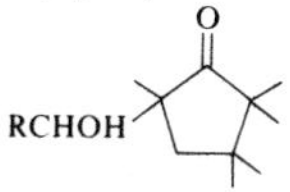

TABLE II. REACTION OF KETONES OR THEIR EQUIVALENTS WITH ALDEHYDES OR ACETALS (*Continued*)

No. of C Atoms	Aldehyde or Acetal	Ketone or Equivalent	Reagent	Product(s)[a] and Yield(s) (%)	Refs.
C_3 (*Contd.*)	C_2H_5CHO	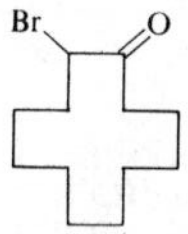 Br, O (structure)	Zn	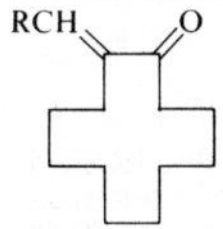 RCH, O (structure) (85)	124
		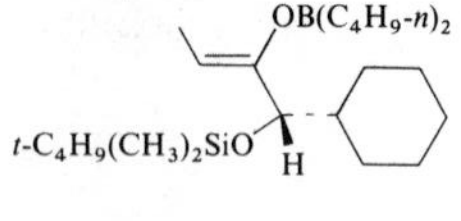 $OB(C_4H_9\text{-}n)_2$; $t\text{-}C_4H_9(CH_3)_2SiO$, H (structure)		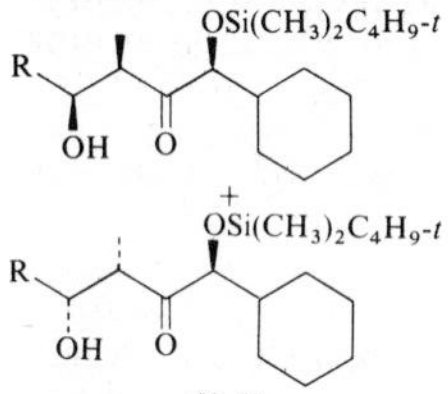 R, OH, O, $OSi(CH_3)_2C_4H_9\text{-}t$ (structure) + R, OH, O, $OSi(CH_3)_2C_4H_9\text{-}t$ (structure) (50:1)	(—)	264
		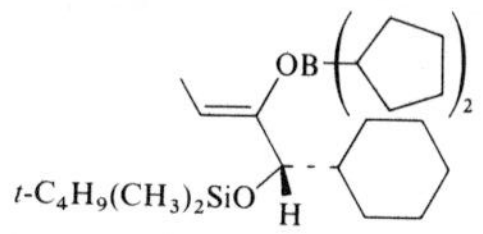 OB(cyclopentyl)$_2$; $t\text{-}C_4H_9(CH_3)_2SiO$, H (structure)		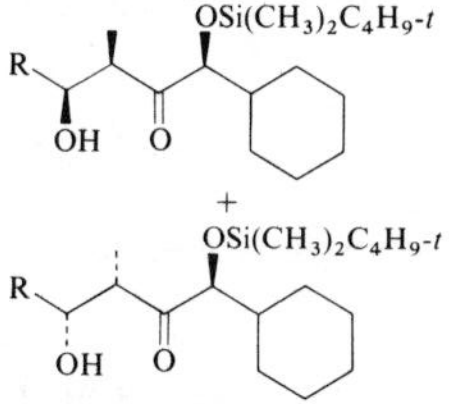 R, OH, O, $OSi(CH_3)_2C_4H_9\text{-}t$ (structure) + R, OH, O, $OSi(CH_3)_2C_4H_9\text{-}t$ (structure) (100:1) (—)	264
		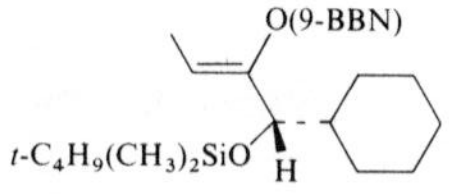 O(9-BBN); $t\text{-}C_4H_9(CH_3)_2SiO$, H (structure)		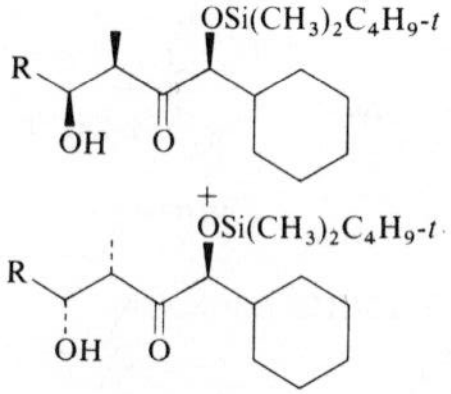 R, OH, O, $OSi(CH_3)_2C_4H_9\text{-}t$ (structure) + R, OH, O, $OSi(CH_3)_2C_4H_9\text{-}t$ (structure) (17:1) (—)	264

		O(9-BBN) $C_6H_5C{=}CHC_6H_{13}$-*n*		R–CH(OH)–CH(C_6H_{13}-*n*)–CO–C_6H_5 + R–CH(OH)–CH(C_6H_{13}-*n*)–CO–C_6H_5 (40:60) (80)	257
C_4	$NC(CH_2)_2CH(OCH_3)_2$	$CH_2{=}C[OSi(CH_3)_3]C_6H_5$	$TiCl_4$	$RCH(OR')CH_2COC_6H_5$ (60)	180
	$CH_3CH{=}CHCHO$	$CH_2{=}C(CH_3)OC_2H_5$	BF_3	$RCH(OR')CH_2COCH_3$ (30)	196
		$CH_3COCH{=}CHCH_3$	LDA	$RCHOHCH_2COCH{=}CHCH_3$ (70)	47
		$CH_3(C_3H_7\text{-}n)C{=}NN(CH_3)_2$	*n*-C_4H_9Li	$RCHOHCH_2COC_3H_7$-*n* (91)[b]	87
		O, OC_2H_5 (cyclohexenone)	LDA	RCHOH, O, OC_2H_5 (92) (*threo:erythro* = 90:10)	53
		O, OC_2H_5 (dimethylcyclohexenone)	"	RCHOH, O, OC_2H_5 (98) (*threo:erythro* = 93:7)	53
		O (methylcyclohexenone)	CH_3MgI, CuI	RCHOH, O (90)	111
	$CH_2{=}C(CH_3)CHO$	O (cyclohexenone)	$(CH_3)_2AlSC_6H_5$	RCHOH, O, C_6H_5S (97)	134
		$(C_2H_5)_2CO$	$(n\text{-}C_4H_9)_2BOSO_2CF_3$, $(i\text{-}C_3H_7)_2NC_2H_5$	$RCHOHCH(CH_3)COC_2H_5$ (68) (*erythro:threo* >97:3)	262
	$CH_3CH{=}CHCH(OC_2H_5)_2$	$CH_2{=}CHOC_2H_5$	$ZnCl_2$	$RCH{=}CHCHO$ (66)[b]	228
	$C_2H_5CHBrCH(OCH_3)_2$	$OSi(CH_3)_3$ (cyclopentene)	$TiCl_4$	$RCH(OCH_3)$, O (cyclopentanone) (—)	182
		$CH_2{=}C[OSi(CH_3)_3]C_6H_5$	"	$RCH(OR')CH_2COC_6H_5$ (—)	181
		$C_6H_5CH{=}C[OSi(CH_3)_3]CH_3$	"	$RCH(OR')CH(C_6H_5)COCH_3$ (—)	181
	$(CH_3)_2CBrCH(OCH_3)_2$	$OSi(CH_3)_3$ (cyclopentene)	$TiCl_4$	$RCH(OCH_3)$, O (cyclopentanone) (—)	182

TABLE II. REACTION OF KETONES OR THEIR EQUIVALENTS WITH ALDEHYDES OR ACETALS (*Continued*)

No. of C Atoms	Aldehyde or Acetal	Ketone or Equivalent	Reagent	Product(s)[a] and Yield(s) (%)	Refs.
C_4 (*Contd.*)	n-C_3H_7CHO	CH_3COCH_3	$NaNH_2$, LiBr	$RCHOHCH_2COCH_3$ (50)	58
		$CH_2{=}CHCOCH_3$	$(n\text{-}C_4H_9)_3B$	$RCHOHCH(C_5H_{11}\text{-}n)COCH_3$ (71)	137
		"	$(n\text{-}C_4H_9)_2BSC_6H_5$	$RCHOHCH(CH_2SC_6H_5)COCH_3$ (83)	137
		$C_2H_5COCH_3$	$(C_2H_5)_2NLi$	$RCHOHCH_2COC_2H_5$ (40)	48
		$(C_2H_5)_2CO$	$(n\text{-}C_4H_9)_2BOSO_2CF_3$ $(i\text{-}C_3H_7)_2NC_2H_5$	$RCHOHCH(CH_3)COC_2H_5$ (65) (*erythro* : *threo* >97:3)	262
		$CH_3COC_3H_7$-n	LDA	$RCHOHCH_2COC_3H_7$-n (65)	46,47
		"	$NaNH_2$, LiBr	" (77)	58
		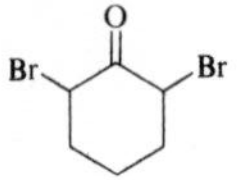	n-$C_4H_9(t\text{-}C_4H_9O)CuLi$	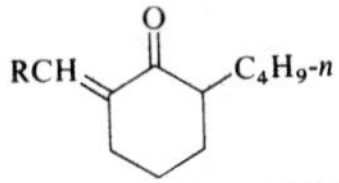 (43)	67
		i-$C_3H_7COC_3H_7$-i	$C_6H_5N(CH_3)MgBr$	$RCHOHC(CH_3)_2COC_3H_7$-i (81)	107
		$N_2CHCOC_6H_5$	$(n\text{-}C_4H_9)_3B$	$RCHOHCH(C_4H_9\text{-}n)COC_6H_5$ (88)	136
		$C_6H_5CH{=}C(OAc)CH_3$	CH_3Li, $ZnCl_2$	$RCHOHCH(C_6H_5)COCH_3$ (81)	15,120
		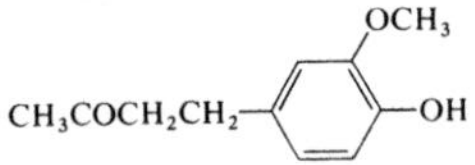	$[(CH_3)_3Si]_2NLi$	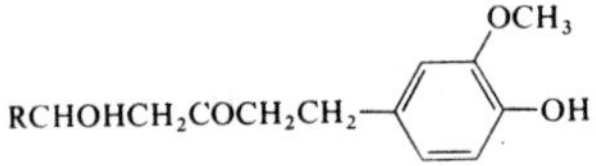 (—)	238
	n-$C_3H_7CH(OCH_3)_2$	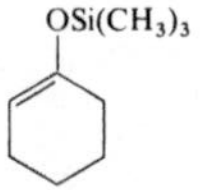	$(CH_3)_3SiO_3SCF_3$	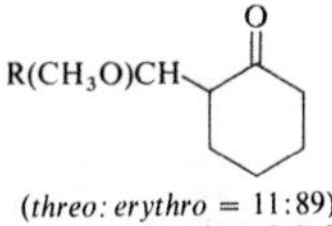 (91) (*threo* : *erythro* = 11:89)	193a
		$CH_2{=}C[OSi(CH_3)_3]C_6H_5$	$(CH_3)_3SiO_3SCF_3$	$RCH(OCH_3)CH_2COC_6H_5$ (75)	193a
	i-C_3H_7CHO	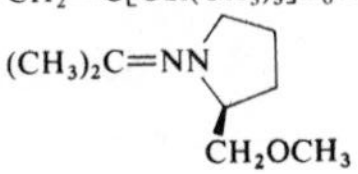	n-C_4H_9Li	$RCHOHCH_2COCH_3$ (72)[b]	91

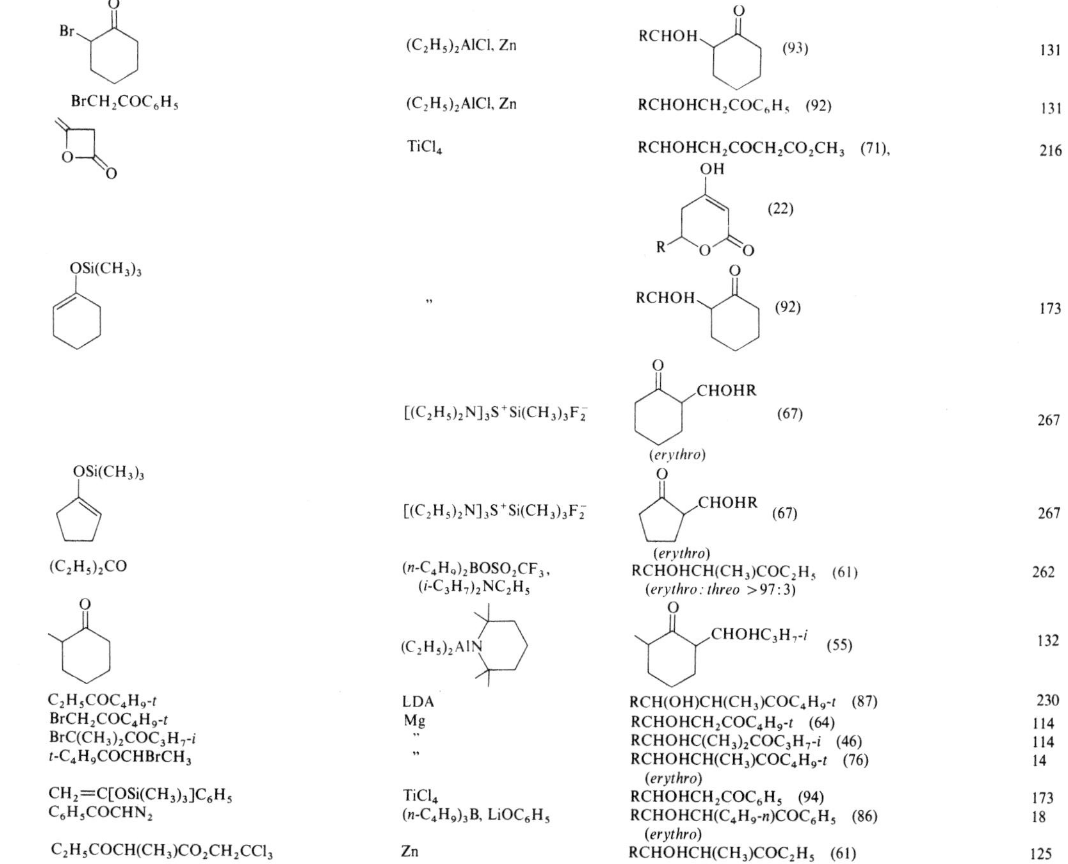

	$(C_2H_5)_2AlCl$, Zn	RCHOH– (93)	131
$BrCH_2COC_6H_5$	$(C_2H_5)_2AlCl$, Zn	$RCHOHCH_2COC_6H_5$ (92)	131
	$TiCl_4$	$RCHOHCH_2COCH_2CO_2CH_3$ (71), (22)	216
	"	RCHOH– (92)	173
	$[(C_2H_5)_2N]_3S^+Si(CH_3)_3F_2^-$	CHOHR (67) (*erythro*)	267
	$[(C_2H_5)_2N]_3S^+Si(CH_3)_3F_2^-$	CHOHR (67) (*erythro*)	267
$(C_2H_5)_2CO$	$(n\text{-}C_4H_9)_2BOSO_2CF_3$, $(i\text{-}C_3H_7)_2NC_2H_5$	$RCHOHCH(CH_3)COC_2H_5$ (61) (*erythro* : *threo* >97:3)	262
	$(C_2H_5)_2AlN$	$CHOHC_3H_7\text{-}i$ (55)	132
$C_2H_5COC_4H_9\text{-}t$	LDA	$RCH(OH)CH(CH_3)COC_4H_9\text{-}t$ (87)	230
$BrCH_2COC_4H_9\text{-}t$	Mg	$RCHOHCH_2COC_4H_9\text{-}t$ (64)	114
$BrC(CH_3)_2COC_3H_7\text{-}i$	"	$RCHOHC(CH_3)_2COC_3H_7\text{-}i$ (46)	114
$t\text{-}C_4H_9COCHBrCH_3$	"	$RCHOHCH(CH_3)COC_4H_9\text{-}t$ (76) (*erythro*)	14
$CH_2{=}C[OSi(CH_3)_3]C_6H_5$	$TiCl_4$	$RCHOHCH_2COC_6H_5$ (94)	173
$C_6H_5COCHN_2$	$(n\text{-}C_4H_9)_3B$, $LiOC_6H_5$	$RCHOHCH(C_4H_9\text{-}n)COC_6H_5$ (86) (*erythro*)	18
$C_2H_5COCH(CH_3)CO_2CH_2CCl_3$	Zn	$RCHOHCH(CH_3)COC_2H_5$ (61)	125

TABLE II. Reaction of Ketones or Their Equivalents with Aldehydes or Acetals (*Continued*)

No. of C Atoms	Aldehyde or Acetal	Ketone or Equivalent	Reagent	Product(s)[a] and Yield(s) (%)	Refs.
C_4 (*Contd.*)	i-C_3H_7CHO	$C_2H_5COC_6H_5$	$C_6H_5N(CH_3)MgBr$	$RCHOHCH(CH_3)COC_6H_5$ (88)	107
		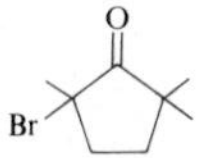 Br (cyclopentanone structure)	Mg	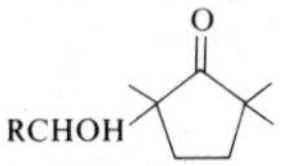 RCHOH (cyclopentanone structure) (62) (*threo*)	14
		$C_6H_5CH{=}C[OSi(CH_3)_3]CH_3$	$TiCl_4$	$RCHOHCH(C_6H_5)COCH_3$ (93)	173
		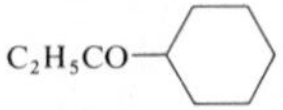 C_2H_5CO–(cyclohexyl)	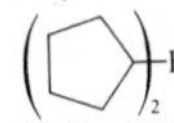 (cyclopentyl)$_2BOSO_2CF_3$, $[(CH_3)_2CH]_2NC_2H_5$	 $RCHOHCH(CH_3)CO$–(cyclohexyl) (86) (*erythro*: *threo* = 12:88)	
		"	9-BBN(OSO_2CF_3), $[(CH_3)_2CH]_2NC_2H_5$	(*erythro*: *threo* > 97:3) (87)	38
		$CH_3COCH[OSi(CH_3)_3]C_4H_9$-$t$	LDA	$RCHOHCH_2COCH[OSi(CH_3)_3]C_4H_9$-$t$ (63) (*erythro* : *threo* = 50:50)	255
		$C_2H_5COCH[OSi(CH_3)_3]C_4H_9$-$t$	LDA	$RCHOHCH(CH_3)COCH[OSi(CH_3)_3]$ (93) (*erythro*)	43, 255
		$BrCH(C_3H_7$-$i)COC_4H_9$-i	Mg	$RCHOHCH(C_3H_7$-$i)COC_4H_9$-i (56)	114
		$C_2H_5COC(CH_3)_2OSi(CH_3)_3$	LDA	$RCHOHCH(CH_3)COC(CH_3)_2OSi(CH_3)_3$ (86) (100% *erythro*)	39
		$CH_3CO(CH_2)_2C_6H_5$	i-$C_4H_9C[Si(CH_3)_3]_2OLi$	$RCHOHCH_2CO(CH_2)_2C_6H_5$ (71)	55
	i-$C_3H_7CH(OCH_3)_2$	$CH_2{=}C(CH_3)OAc$	$TiCl_4$	$RCH(OR')CH_2COCH_3$ (92)	214
	i-C_3H_7CHO	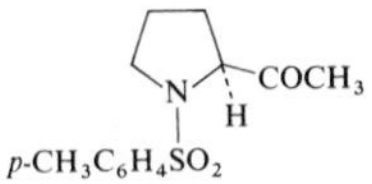 N–p-$CH_3C_6H_4SO_2$ pyrrolidine, COCH$_3$, H (structure)	$(n$-$C_4H_9)_2BOSO_2CF_3$, $(i$-$C_3H_7)_2NC_2H_5$	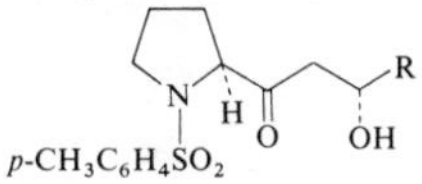 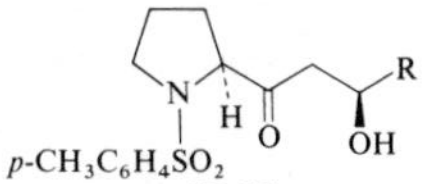 p-$CH_3C_6H_4SO_2$ (structure: N, H, O, R, OH) + p-$CH_3C_6H_4SO_2$ (structure: N, H, O, R, OH) (77) (72:28)	260
		"	LDA	" (54:46) (75)	260

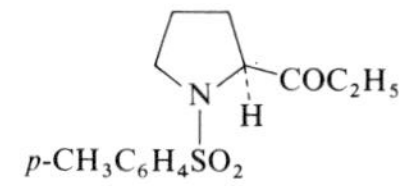

"	(pyrrolidine) COC_2H_5, H, N–p-$CH_3C_6H_4SO_2$	$(n$-$C_4H_9)_2BOSO_2CF_3$, $(i$-$C_3H_7)_2NC_2H_5$	(pyrrolidine) $COCH(CH_3)CH(OH)R$, H, N–p-$CH_3C_6H_4SO_2$ (*erythro* : *threo* = 91:9)	262
	"	(cyclopentyl)$_2$$BOSO_2CF_3$, $(i$-$C_3H_7)_2NC_2H_5$	" (*erythro* : *threo* = 87:13)	262
	$OSi(CH_3)_3$ (mesityl enol ether)	$[(C_2H_5)_2N]_3S^+Si(CH_3)_3F_2^-$	$RCHOHCH(CH_3)CO$–(mesityl) (25) (*eryhtro* : *threo* = 95:5)	267
	$OSi(CH_3)_3$ (mesityl enol ether)	$[(C_2H_5)_2N]_3S^+Si(CH_3)_3F_2^-$	$RCHOHCH(CH_3)CO$–(mesityl) (17) (*erythro* : *threo* = 95:5)	267
	O(9-BBN), C_6H_{11}, t-$C_4H_9(CH_3)_2SiO$ H		R, OH, O, $OSi(CH_3)_2C_4H_9$-t, C_6H_{11} + R, OH, O, $OSi(CH_3)_2C_4H_9$-t, C_6H_{11} (—) (> 100:1)	264
	$OB(C_4H_9$-$n)_2$, C_6H_{11}, t-$C_4H_9(CH_3)_2SiO$ H		R, OH, O, $OSi(CH_3)_2C_4H_9$-t, C_6H_{11} + R, OH, O, $OSi(CH_3)_2C_4H_9$-t, C_6H_{11} (—)	264
i-$C_3H_7CH(OCH_3)_2$	(diketene) O, O	$TiCl_4$	$RCH(OR')CH_2COCH_2CO_2CH_3$ (77)	215

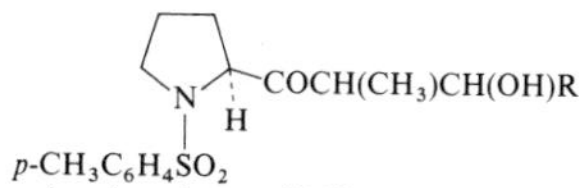

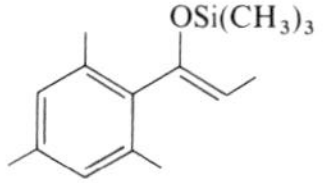

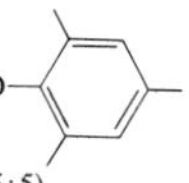

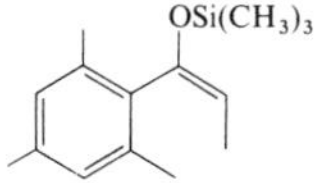

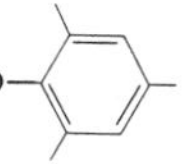

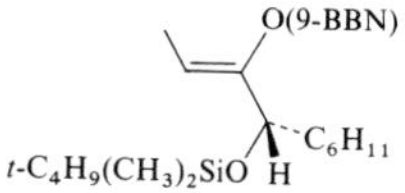

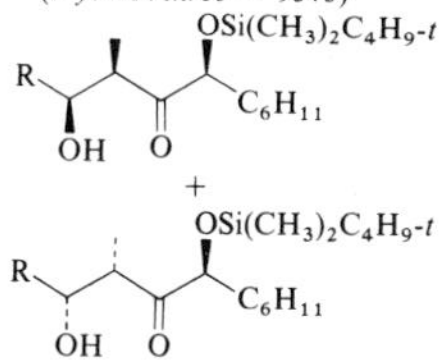

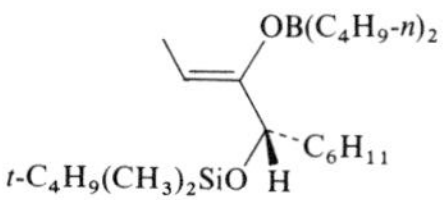

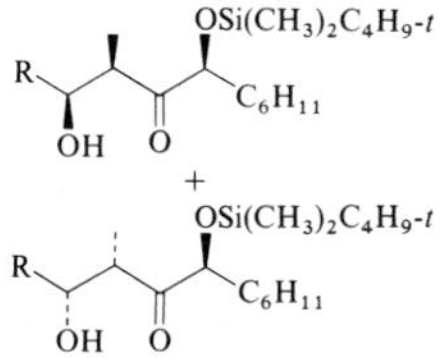

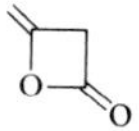

TABLE II. Reaction of Ketones or Their Equivalents with Aldehydes or Acetals (*Continued*)

No. of C Atoms	Aldehyde or Acetal	Ketone or Equivalent	Reagent	Product(s)[a] and Yield(s) (%)	Refs.
C_4 (*Contd.*)	$i\text{-}C_3H_7CH(OCH_3)_2$	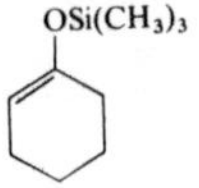 $OSi(CH_3)_3$	$(CH_3)_3SiO_3SCF_3$	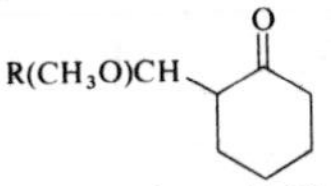 $R(CH_3O)CH$, O (95) (*threo*:*erythro* = 14:86)	193a
C_5	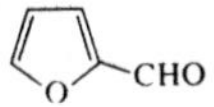 CHO, O	$CH_3COCH(C_4H_9\text{-}n)CO_2CH_2CCl_3$	Zn	$RCHOHCH(C_4H_9\text{-}n)COCH_3$ (80)	125
		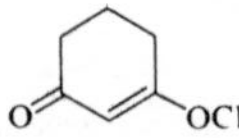 O, OCH_3	LDA	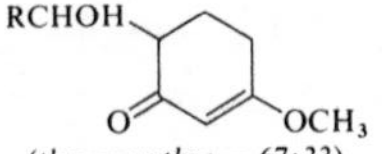 RCHOH, O, OCH_3 (78) (*threo*:*erythro* = 67:33)	53
		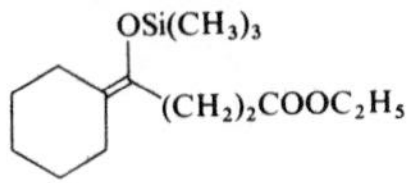 $OSi(CH_3)_3$, $(CH_2)_2COOC_2H_5$	$(n\text{-}C_4H_9)_4\overset{+}{N}\overset{-}{F}$	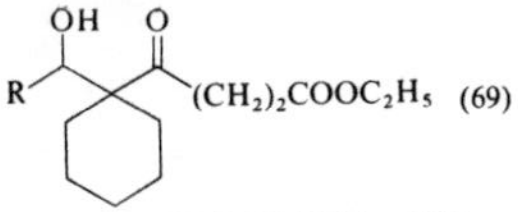 OH, O, R, $(CH_2)_2COOC_2H_5$ (69)	71
	$(CH_3)_2C{=}CHCH(OC_2H_5)_2$	$CH_2{=}C(OC_2H_5)CH_3$	$ZnCl_2$	$RCH(OR')CH_2CCH_3(OR')_2$ (60)	232
	$n\text{-}C_4H_9CHO$	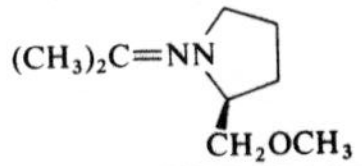 $(CH_3)_2C{=}NN$, CH_2OCH_3	$n\text{-}C_4H_9Li$	$RCHOHCH_2COCH_3$ (77)[b]	91
		$CH_3CH{=}C[OSi(CH_3)_3]C_2H_5$	$n\text{-}C_4H_9Li$, Cp_2ZrCl_2	$RCHOHCH(CH_3)COC_2H_5$ (85) (*eryhtro*:*threo* = 75:25)	273
		"	$TiCl_4$	$RCHOHCH(CH_3)COC_2H_5$ (80) (*erythro*:*threo* = 55:45)	273
	$CH_3CH(C_2H_5)CHO$	$CH_3CH{=}C[OSi(CH_3)_3]C_2H_5$	"	$RCHOHCH(CH_3)COC_2H_5$ (92) (*threo*:*erythro* = 98:2)	178
		$CH_2{=}C[OSi(CH_3)_3]C_3H_7\text{-}n$	"	$RCHOHCH_2COC_3H_7\text{-}n$ (—)	178
	$i\text{-}C_3H_7CH_2CHO$	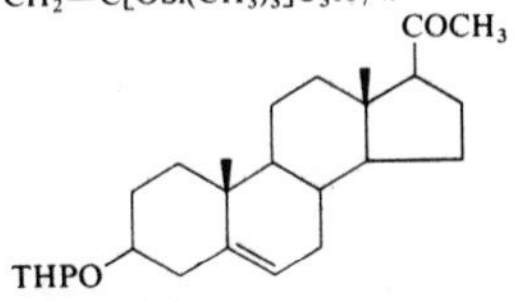 $COCH_3$, THPO	LDA	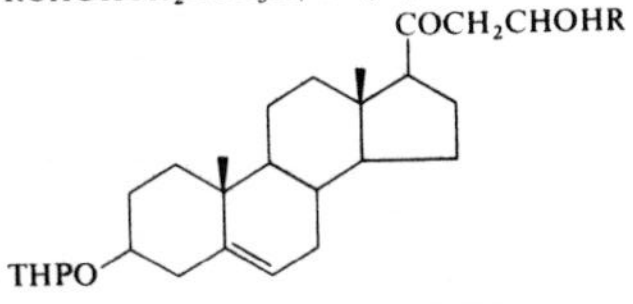 $COCH_2CHOHR$, THPO (82)	239

		O, Br (ring structure)	Mg	RCHOH (ring structure, O) (48) (*threo*:*erythro* = 87:13)	14
	$t\text{-}C_4H_9CHO$	$CH_3(C_3H_7\text{-}i)C{=}NN$ (ring structure, CH_2OCH_3)	$n\text{-}C_4H_9Li$	$RCHOHCH_2COC_3H_7\text{-}i$ (42)[b]	91
		$CH_2{=}C(OAc)C_4H_9\text{-}t$	$CH_3Li, ZnCl_2$	$RCHOHCH_2COC_4H_9\text{-}t$ (82)	15,120
		$CH_3COC_4H_9\text{-}t$	LDA	" (82)	15
		O, Br (ring structure)	Mg	RCHOH (ring structure, O) (50) (*threo*–*erythro* = 94:6)	14
		$t\text{-}C_4H_9COCHBrC_3H_7\text{-}n$	"	$RCHOHCH(C_3H_7\text{-}n)COC_4H_9\text{-}t$ (69) (*erythro*)	14
		$t\text{-}C_4H_9COCHBrCH_3$	"	$RCHOHCH(CH_3)COC_4H_9\text{-}t$ (50) (*erythro*)	14
		$t\text{-}C_4H_9COCHBrC_2H_5$	"	$RCHOHCH(C_2H_5)COC_4H_9\text{-}t$ (52) (*erythro*)	14
		$C_2H_5COCH[OSi(CH_3)_3]C_4H_9\text{-}t$	LDA	$RCHOHCH(CH_3)COCH[OSi(CH_3)_3]C_4H_9\text{-}t$ (47) (*erythro*)	43
		$t\text{-}C_4H_9COCHBrC_3H_7\text{-}i$	Mg	$RCHOHCH(C_3H_7\text{-}i)COC_4H_9\text{-}t$ (46) (*erythro*: *threo* = 26:74)	14
		$t\text{-}C_4H_9COCHBrC_4H_9\text{-}i$	"	$RCHOHCH(C_4H_9\text{-}i)COC_4H_9\text{-}t$ (72) (*erythro*)	14
		$t\text{-}C_4H_9COCHBrC_4H_9\text{-}t$	"	$RCHOHCH(C_4H_9\text{-}t)COC_4H_9\text{-}t$ (74) (*threo*)	14
		$t\text{-}C_4H_9COCHBrCH_2C_4H_9\text{-}t$	"	$RCHOHCH(CH_2C_4H_9\text{-}t)COC_4H_9\text{-}t$ (49) (*erythro*: *threo* = 58:42)	14
	CH_3, CH_3, H, CHO (alkene structure)	$(C_2H_5)_2CO$	$(n\text{-}C_4H_9)_2BOSO_2CF_3$, $(i\text{-}C_3H_7)_2NC_2H_5$	$RCHOHCH(CH_3)COC_2H_5$ (65) (*erythro*: *threo* >93:7)	262
C_6	$CH_3(CH{=}CH)_2CH(OC_2H_5)_2$	$CH_2{=}C(OC_2H_5)CH_3$	$ZnCl_2$	$RCH{=}CHCOCH_3$ (62)[a]	228
	$n\text{-}C_5H_{11}CHO$	$(CH_3)_2C{=}NN(CH_3)_2$	LDA	$RCHOHCH_2C(CH_3){=}NN(CH_3)_2$ (100)	87
		$CH_3COC_3H_7\text{-}n$	$i\text{-}C_4H_9C[Si(CH_3)_3]_2OLi$	$RCHOHCH_2COC_3H_7\text{-}n$ (71)	55
		"	$(n\text{-}C_4H_9)_2BOSO_2CF_3$, $C_2H_5N(C_3H_7\text{-}i)_2$	" (65)	36
		"	9-BBN(OSO_2CF_3), 2,6-lutidine	$RCHOHCH(C_2H_5)COCH_3$ (67)	37
		$CH_3COC_3H_7\text{-}i$	$i\text{-}C_4H_9C[Si(CH_3)_3]_2OLi$	$RCHOHCH_2COC_3H_7\text{-}i$ (81)	55
		$C_2H_5COCH[Si(CH_3)_3]CH_3$	LDA	$RCHOHCH(CH_3)COC_2H_5$ (78)[b]	56
		$CH_3COC_4H_9\text{-}i$	9-BBN(OSO_2CF_3), 2,6-lutidine	$RCHOHCH(C_3H_7\text{-}i)COCH_3$ (57)	37
		"	$i\text{-}C_4H_9C[Si(CH_3)_3]_2OLi$	$RCHOHCH_2COC_4H_9\text{-}i$ (84)	55

TABLE II. REACTION OF KETONES OR THEIR EQUIVALENTS WITH ALDEHYDES OR ACETALS (*Continued*)

No. of C Atoms	Aldehyde or Acetal	Ketone or Equivalent	Reagent	Product(s)[a] and Yield(s) (%)	Refs.
C_6 (*Contd.*)	n-$C_5H_{11}CHO$	$CH_3COC_4H_9$-*sec*	i-$C_4H_9C[Si(CH_3)_3]_2OLi$	$RCHOHCH_2COC_4H_9$-*sec* (78)	55
		$C_2H_5COCH[Si(CH_3)_3]C_2H_5$	LDA	$RCHOHCH(CH_3)COC_3H_7$-n (72)[b]	56
		"	$BF_3 \cdot O(C_2H_5)_2$	$RCHOHCH(C_2H_5)COC_2H_5$ (76)	56
		$C_2H_5COC_4H_9$-t	LDA	$RCHOHCH(CH_3)COC_4H_9$-t (66) (*erythro*)	230
		n-$C_3H_7COC_6H_5$	$CH_3(C_6H_5)NMgBr$	$RCHOHCH(CH_3)COC_6H_5$ (40)	107
		$CH_3COCH_2SOC_6H_5$	NaH, n-C_4H_9Li	$RCHOHCH_2COCH_2SOC_6H_5$ (80)	240
		$CH_3COCH{=}CH$–C_6H_3(3-OCH_3)-4-$OSi(CH_3)_3$	$LiN[Si(CH_3)_3]_2$	$RCHOHCH_2COCH{=}CH$–C_6H_3(3-OCH_3)-4-OH (—)	238
		$CH_3CO(CH_2)_2$–C_6H_3(3-OCH_3)-4-$OSi(CH_3)_3$	"	$RCHOHCH_2CO(CH_2)_2$–C_6H_3(3-OCH_3)-4-OH (57)	238
		$CH_2{=}C(OSi(CH_3)_3)(CH_2)_2$–$C_6H_3$(3-$OCH_3$)-4-$OSi(CH_3)_3$	$TiCl_4$	" (92)	175
		$CH_3COC_3H_7$-n	9-BBN(OSO_2CF_3), 2,6-lutidine	$RCHOHCH(C_2H_5)COCH_3$ (67)	37
		$CH_3COC_3H_7$-i	$(n$-$C_4H_9)_2BOSO_2CF_3$, $C_2H_5N(C_3H_7$-$i)_2$	$RCHOHCH_2COC_3H_7$-i (70)	36
		3,4-$(CH_3O)_2C_6H_3CH_2CH_2C(CH_3){=}N$–N(2-($CH_2OCH_3$)pyrrolidin-1-yl)	n-C_4H_9Li, $(CH_3)_3SiCl$	3,4-$(CH_3O)_2C_6H_3CH_2CH_2C({=}N$–N(2-($CH_2OCH_3$)pyrrolidin-1-yl))$CH_2CH(OSi(CH_3)_3)C_5H_{11}$-$n$ (97)	92
		$CH_3COCH{=}CH$–C_6H_3(3-OCH_3)-4-$OSi(CH_3)_3$	$[(CH_3)_3Si]_2NLi$	$RCHOHCH_2COCH{=}CH$–C_6H_3(3-OCH_3)-4-$OSi(CH_3)_3$ (—)	238

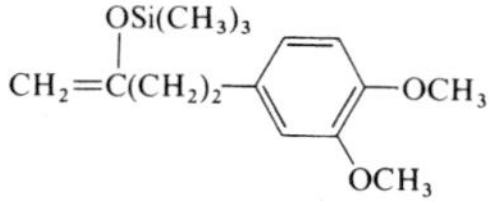

Aldehyde	Enolate precursor	Reagent	Product (Yield %)	Refs.
	$CH_2{=}C(OSi(CH_3)_3)(CH_2)_2$–$C_6H_3$–3,4-$(OCH_3)_2$	$TiCl_4$	$RCHOHCH_2CO(CH_2)_2$–C_6H_3–3,4-$(OCH_3)_2$ (20)	238
	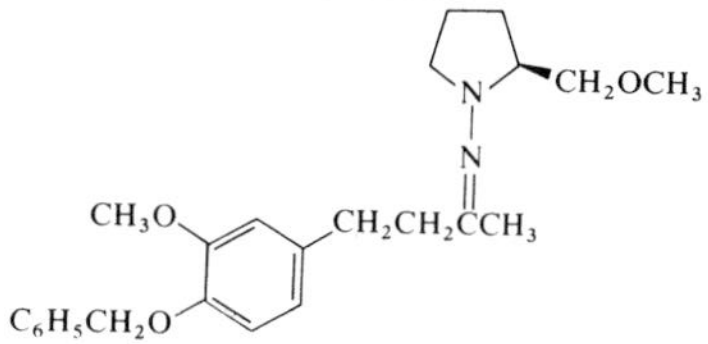 3-CH_3O-4-$C_6H_5CH_2O$-$C_6H_3CH_2CH_2C(CH_3){=}N$–N-pyrrolidine-2-$CH_2OCH_3$	n-C_4H_9Li, $(CH_3)_3SiCl$	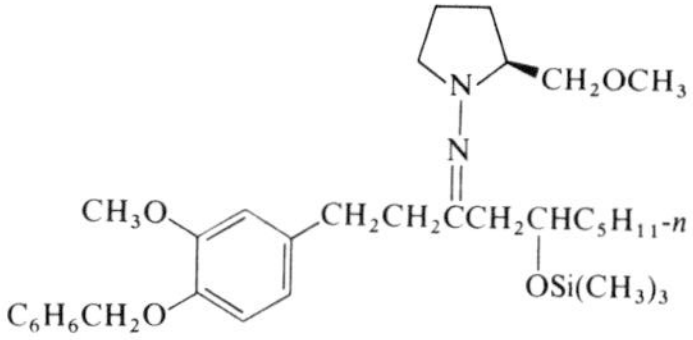 3-CH_3O-4-$C_6H_6CH_2O$-$C_6H_3CH_2CH_2C({=}N$–N-pyrrolidine-2-$CH_2OCH_3)CH_2CH(OSi(CH_3)_3)C_5H_{11}$-$n$ (98)	92
t-$C_4H_9CH_2CHO$	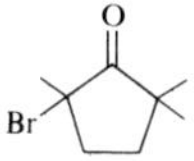 2-Br-2,5,5-trimethylcyclopentanone (O, Br)	Mg	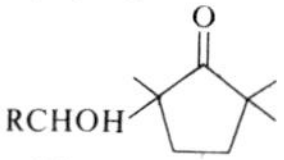 RCHOH–(2,5,5-trimethylcyclopentanone) (67) (*threo* : *erythro* = 57 : 43)	14
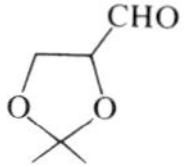 2,2-dimethyl-1,3-dioxolane-4-CHO	$C_2H_5COCH[OSi(CH_3)_3]C_4H_9$-$t$	LDA	$RCHOHCH(CH_3)COCH[OSi(CH_3)_3]C_4H_9$-$t$ (—) (*erythro*)	43
2,2-dimethyl-1,3-dioxolane-4-CHO	COR_1 (di-O-isopropylidene pyranose)	LDA	$COCH(R'')CHOHR$ (di-O-isopropylidene pyranose)	
			$R'' = H$ from $R_1 = CH_3$ (92)	254
			$R'' = CH_3$ from $R_1 = C_2H_5$ (—)	40
			" " (95)	254
	$CH_3COC_4H_9$-t	LDA	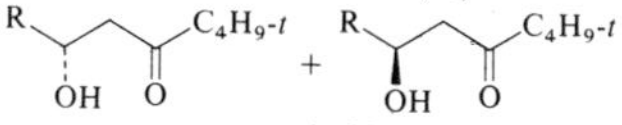 $RCH(OH)CH_2COC_4H_9$-t + $RCH(OH)CH_2COC_4H_9$-t (>95 : 1) (74)	252
	$C_2H_5COC_4H_9$-t	LDA	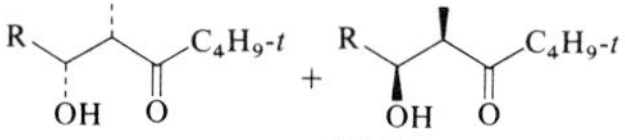 $RCH(OH)CH(CH_3)COC_4H_9$-t + $RCH(OH)CH(CH_3)COC_4H_9$-t (98 : 2) (87)	252
	"	LHMDS	(—)	230

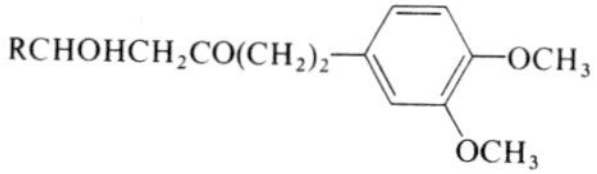

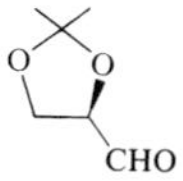

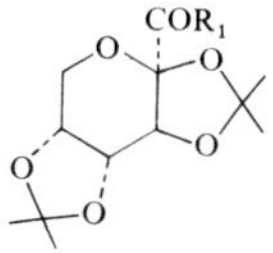

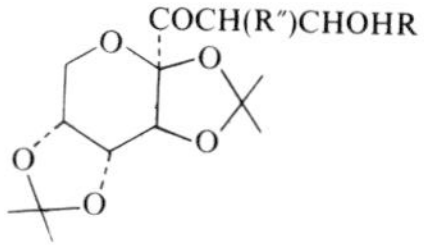

TABLE II. REACTION OF KETONES OR THEIR EQUIVALENTS WITH ALDEHYDES OR ACETALS (*Continued*)

No. of C Atoms	Aldehyde or Acetal	Ketone or Equivalent	Reagent	Product(s)[a] and Yield(s) (%)	Refs.
C_6 (*Contd.*)	n-$C_5H_{11}CHO$	$CH_3COC(CH_3)_2OSi(CH_3)_3$	LDA	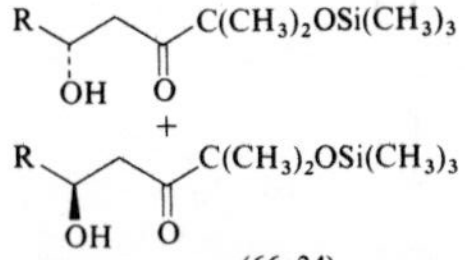 R, OH, O, $C(CH_3)_2OSi(CH_3)_3$ + R, OH, O, $C(CH_3)_2OSi(CH_3)_3$ (72) (66:34)	252
		$C_2H_5COC(CH_3)_2OSi(CH_3)_3$	LTMP	$RCHOHCH(CH_3)COC(CH_3)_2OSi(CH_3)_3$ (−)	254
		$CH_3COCH(C_4H_9$-$t)OSi(CH_3)_3$	LDA	$RCHOHCH_2COCH(C_4H_9$-$t)OSi(CH_3)_3$ (45)	255
		$C_2H_5COC(CH_3)_2OSi(CH_3)_3$	LDA	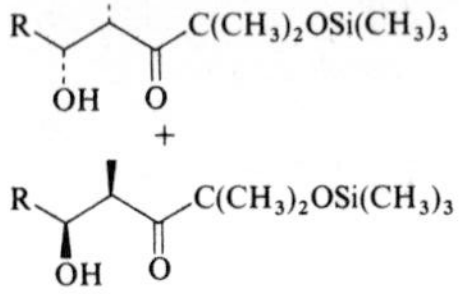 R, OH, O, $C(CH_3)_2OSi(CH_3)_3$ (43) + R, OH, O, $C(CH_3)_2OSi(CH_3)_3$ (85:15)	252
	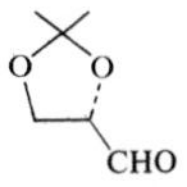 O, O, CHO	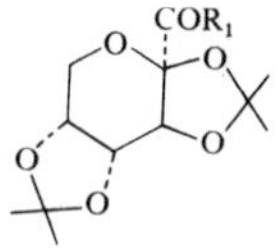 COR_1, O, O, O, O, O	LDA	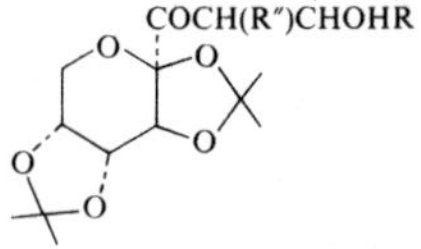 $COCH(R'')CHOHR$, O, O, O, O, O	
				$R'' = H$ from $R_1 = CH_3$ (90)	254
				$R'' = CH_3$ from $R_1 = C_2H_5$ (95)	254
				" " (85–94)	40, 254
C_7	p-ClC_6H_4CHO	$CH_3COCH(C_4H_9$-$n)CO_2CH_2CCl_3$	Zn	$RCHOHCH(C_4H_9$-$n)COCH_3$ (92)	125
		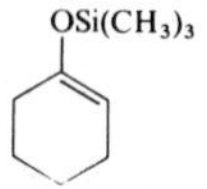 $OSi(CH_3)_3$	$TiCl_4$	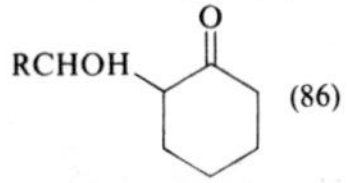 RCHOH, O (86)	177
	p-$O_2NC_6H_4CHO$	$(CH_3)_2CO$	$(C_2H_5)_2NBCl_2$, $(C_2H_5)_3N$	$RCHOHCH_2COCH_3$ (35), $RCH{=}CHCOCH_3$ (4)	149

$C_2H_5C(CH_3){=}NC_6H_{11}$	BCl_3, $N(C_2H_5)_3$	$RCHOHCH_2COCH_2CH_3$ (29), $RCHOHCH(CH_3)COCH_3$ (17), (*threo*: *erythro* = 1:1) $RCHOHCH(CH_3)COCH_2CHOHR$ (10) (*threo*: *erythro* = 1:1)	147
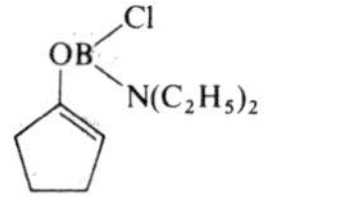 OB(Cl)$N(C_2H_5)_2$	$N(C_2H_5)_3$	 O, CHOHR (73) (*threo*: *erythro* = 1:3.5)	149
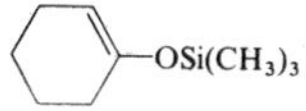 $OSi(CH_3)_3$	$TiCl_4$	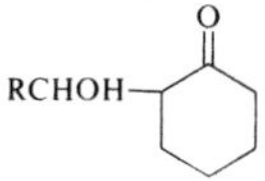 RCHOH, O (85)	177
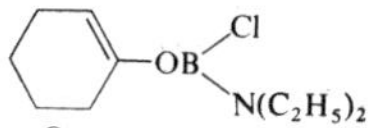 OB(Cl)$N(C_2H_5)_2$	$N(C_2H_5)_3$	" (74) (*threo*: *eryhtro* = 1:2.4)	149
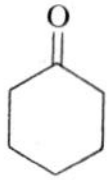 O	$(C_2H_5)_2NBCl_2$,$(C_2H_5)_3N$	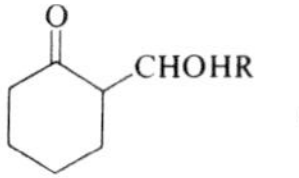 O, CHOHR (71) (*threo*: *erythro* = 1:1) 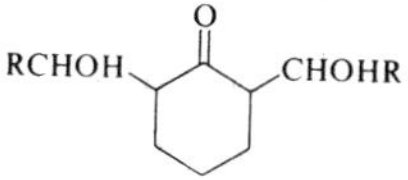 RCHOH, O, CHOHR (< 16)	149
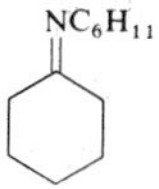 NC_6H_{11}	BCl_3, $N(C_2H_5)_3$	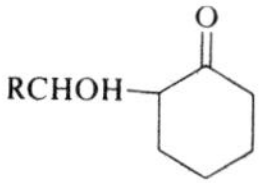 RCHOH, O (76) (*threo*: *erythro* = 1:1.8)	147
$C_6H_5(CH_3)C{=}NC_6H_{11}$	BCl_3, $N(C_2H_5)_3$	$RCHOHCH_2COC_6H_5$ (46) $RCH{=}CHCOC_6H_5$ (2)	147
$C_6H_5(CH_3)C{=}N\overset{*}{C}H(CH_3)C_6H_5$	"	$RCHOHCH_2COC_6H_5$ (42) (optical purity = 34.5%)	148
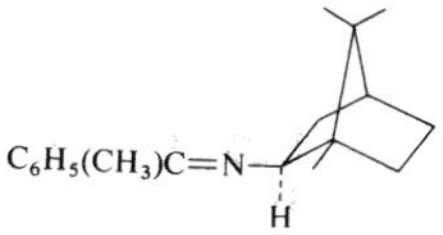 $C_6H_5(CH_3)C{=}N$–, H	"	$RCHOHCH_2COC_6H_5$ (34) (optical purity = 41.5%)	148

TABLE II. REACTION OF KETONES OR THEIR EQUIVALENTS WITH ALDEHYDES OR ACETALS (*Continued*)

No. of C Atoms	Aldehyde or Acetal	Ketone or Equivalent	Reagent	Product(s)[a] and Yield(s) (%)	Refs.
C_7 (*Contd.*)	$p\text{-}O_2NC_6H_4CHO$	$C_6H_5(CH_3)C{=}N\overset{*}{C}H(CH_3)$–(1-naphthyl) 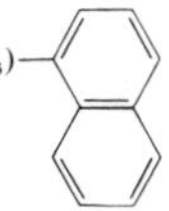	BCl_3, $N(C_2H_5)_3$	$RCHOHCH_2COC_6H_5$ (30) (optical purity = 2.5%)	148
		$C_6H_5COC_2H_5$	$(C_2H_5)_2NBCl_2$, $(C_2H_5)_3N$	$RCHOHCH(CH_3)COC_6H_5$ (37) (*threo*: *erythro* = 1:3)	149
		$CH_3COCH_2(C_4H_9\text{-}n)CO_2CH_2CCl_3$	Zn	$RCHOHCH(C_4H_9\text{-}n)COCH_3$ (70)	125
	C_6H_5CHO	$N_2CHCOCH_3$	$(n\text{-}C_4H_9)_3B$, $LiOC_6H_5$	$RCHOHCH(C_4H_9\text{-}n)COCH_3$ (90) (*erythro*: *threo* > 20:1)	18
		"	$(n\text{-}C_4H_9)_3B$	$RCHOHCH(C_4H_9\text{-}n)COCH_3$ (92) (*threo*: *erythro* = 3:1)	18
		CH_3COCH_3	$NaNH_2$, LiBr	$RCHOHCH_2COCH_3$ (61)	58
		$C_2H_5COSi(CH_3)_3$	$(n\text{-}C_4H_9)_2BOSO_2CF_3$, $(i\text{-}C_3H_7)_2NC_2H_5$	$RCHOHCH(CH_3)COSi(CH_3)_3$ (53) (*erythro*: *threo* = 19:81)	262
		"	 $(C_5H_9)_2BOSO_2CF_3$ (dicyclopentyl), $(i\text{-}C_3H_7)_2NC_2H_5$	" (*erythro*: *threo* = 35:65) (68)	262
		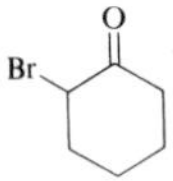 2-Br-cyclohexanone (Br, O)	$(C_2H_5)_2Al$, Zn	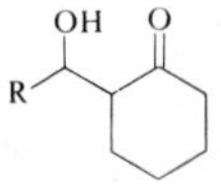 2-(RCH(OH))-cyclohexanone (OH, O, R) (94)	131
		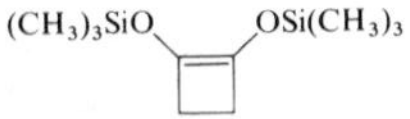 $(CH_3)_3SiO$, $OSi(CH_3)_3$ (cyclobutene)	$TiCl_4$	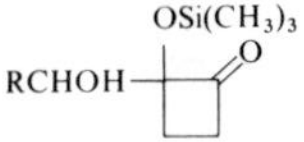 RCHOH–(cyclobutanone, $OSi(CH_3)_3$, O) (78)	185
		$CH_2{=}CHCOCH_3$	$(n\text{-}C_4H_9)_3B$	$RCHOHCH(C_5H_{11}\text{-}n)COCH_3$ (91)	135,137
		"	$(n\text{-}C_4H_9)_2BSC_6H_5$	$RCHOHCH(CH_2SC_6H_5)COCH_3$ (93)	135,137
		"	$(n\text{-}C_4H_9)_2BSCH_2C_6H_5$	$RCHOHCH(CH_2SCH_2C_6H_5)COCH_3$ (72)	137
		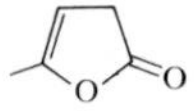 (5-methyl-furan-2(3H)-one: O, O)	$BF_3 \cdot O(C_2H_5)_2$	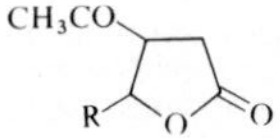 CH_3CO, R (lactone: O, O) (62)	221,222

$OSi(CH_3)_3$ (1-trimethylsilyloxycyclopentene)	$TiCl_4$	RCHOH– (2-oxocyclopentyl) (86)	173
cyclopentanone	$C_2H_5NBCl_2$, $(C_2H_5)_3N$	2-(CHOHR)cyclopentanone (57) (*threo*:*erythro* = 4:5)	149
	LTMP, Cp_2ZrCl_2	" (82) (*erythro*:*threo* = 74:26)	273
	LTMP, $(C_6H_5)_3SnCl$	" (92) (*erythro*:*threo* = 85:15)	271
$CH_3COCH{=}CHCH_3$	$(CH_3)_2CuLi$, $ZnCl_2$	$RCHOHCH(C_3H_7\text{-}i)COCH_3$ (75)	61
$C_2H_5COSi(CH_3)_3$	LHMDS	$RCHOHCH(CH_3)COSi(CH_3)_3$ (–)	251
$C_2H_5COCH[Si(CH_3)_3]CH_3$	LDA	$RCHOHCH(CH_3)COC_2H_5$ (95)[b]	56
$C_2H_5COCH_3$	$(C_2H_5)_2NLi$	$RCHOHCH_2COC_2H_5$ (54)	48
$C_2H_5COC_2H_5$[c]	$(n\text{-}C_4H_9)_2BOSO_2CF_3$, $(C_2H_5)_3N$	$RCHOHCH(CH_3)COC_2H_5$ (92) (*erythro*:*threo* = 96:4)	17,34
"	$(C_2H_5)_3B$, $t\text{-}C_4H_9CO_2B(C_2H_5)_2$	$RCH[OB(C_2H_5)_2]CH(CH_3)COC_2H_5$ (89)	142
"	KH, $n\text{-}C_4H_9Li$	$RCHOHCH(CH_3)COC_2H_5$ (92)	241
$CH_3COC_3H_7\text{-}n$	$(C_2H_5)_2NLi$	$RCHOHCH_2COC_3H_7\text{-}n$ (68)	48
$CH_3COC_3H_7\text{-}i$	$(n\text{-}C_4H_9)_2NLi$	$RCHOHCH_2COC_3H_7\text{-}i$ (48)	48
6-methyl-3,4-dihydro-2H-pyran-2-one	$BF_3\cdot O(C_2H_5)_2$	5-CH_3CO-6-R-tetrahydropyran-2-one (75)	222
2-methylcyclopentanone	$(C_2H_5)_2AlN$ (2,2,6,6-tetramethylpiperidide)	RCHOH– (2-oxo-3-methylcyclopentyl) (72)	132
OCH_3 (1-methoxycyclohexene)	$TiCl_4$, $Ti(OC_3H_7\text{-}i)_4$	2-(RCH=)cyclohexanone (71)	199
2-Br-cyclohexanone	$(C_2H_5)_2AlCl$, Zn	RCHOH– (2-oxocyclohexyl) (97)	131

TABLE II. REACTION OF KETONES OR THEIR EQUIVALENTS WITH ALDEHYDES OR ACETALS (*Continued*)

No. of C Atoms	Aldehyde or Acetal	Ketone or Equivalent	Reagent	Product(s)[a] and Yield(s) (%)	Refs.
C_7 (*Contd.*)	C_6H_5CHO	 1-acetoxycyclohexene ($OCOCH_3$)	CH_3Li, $ZnCl_2$	" (76)	15
		$CH_2{=}C[OSi(CH_3)_3]C_4H_9$-*t*	CH_3Li, $MgBr_2$	$RCHOHCH_2COC_4H_9$-*t* (81)	15
		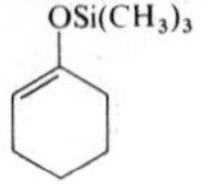 1-trimethylsiloxycyclohexene ($OSi(CH_3)_3$)	$TiCl_4$	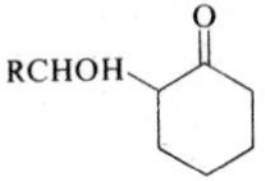 2-(RCHOH)cyclohexanone (92); (90) (*erythro* : *threo* = 36:64)	173, 273
		"	$(n\text{-}C_4H_9)_4NF$	" (80)	70
		"	$(n\text{-}C_4H_9)_2BOSO_2CF_3$	" (90) (*erythro* : *threo* = 17:83)	259
		"	$n\text{-}C_4H_9Li$, Cp_2ZrCl_2	" (98) (*erythro* : *threo* = 72:28)	273
		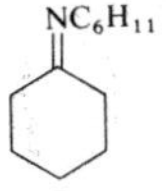 N-cyclohexylcyclohexanimine (NC_6H_{11})	BCl_3, $N(C_2H_5)_3$	" (71) (*threo* : *erythro* = 1:2)	147
		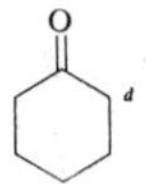 cyclohexanone[d]	$(C_2H_5)_2AlN$(2,2,6,6-tetramethylpiperidide) 	" (78)	132
		"	$(n\text{-}C_4H_9)_2BOSO_2CF_3$, $(i\text{-}C_3H_7)_2NC_2H_5$	" (71) (*erythro* : *threo* = 33:67; 17:83)	34, 262
		"	$C_5H_9(C_6H_{13})BOSO_2CF_3$ $(i\text{-}C_3H_7)_2NC_2H_5$	" (94) (*erythro* : *threo* > 4:96)	34
				" (100) (*erythro* : *threo* = 17:83)	262
		"	$(C_2H_5)_2NBCl_2$, $(C_2H_5)_3N$	" (71) (*threo* : *erythro* = 1:1)	149

$CH_3COCH{=}C(CH_3)_2$	$(CH_3)_2CuLi$, $ZnCl_2$	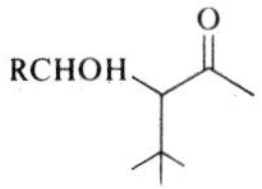 RCHOH (72)	61
$CH_3COCH = C(CH_3)_2$	$(n\text{-}C_4H_9)_2CuLi$, $ZnCl_2$	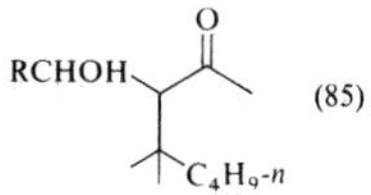 RCHOH, $C_4H_9\text{-}n$ (85)	61
$BrCH_2COC_4H_9\text{-}t$	Mg	$RCHOHCH_2COC_4H_9\text{-}t$ (60)	113
$CH_3COC_4H_9\text{-}n$	LDA	$RCHOHCH_2COC_4H_9\text{-}n$ (75–80)	47
$C_2H_5COC_3H_7\text{-}i^e$	$(n\text{-}C_4H_9)_2BOSO_2CF_3$, $(i\text{-}C_3H_7)_2NC_2H_5$	$RCHOHCH(CH_3)COC_3H_7\text{-}i$ (92) (*erythro*:*threo* = 44:56)	34
"	$(C_5H_9)_2BOSO_2CF_3$, $(i\text{-}C_3H_7)_2NC_2H_5$	" (*erythro*:*threo* = 18:82) (87)	262
$CH_3COC_4H_9\text{-}i$	$(C_2H_5)_2NLi$	$RR'COHCH_2COC_4H_9\text{-}i$ (56)	48
$CH_3COCH_2C_3H_7\text{-}i$	$(C_2H_5)_2NBCl_2$, $(C_2H_5)_3N$	$RCH{=}CHCOCH_2CH(CH_3)_2$ (30), $RCH{=}CHCOCH(CHOHR)CH(CH_3)_2$ (17)	149
(*S*)-$CH_3COCH(CH_3)C_2H_5$	LDA	$RCHOHCH_2COCH(CH_3)C_2H_5$ (78)	54
	$n\text{-}(C_4H_9)_2BOSO_2CF_3$, $(i\text{-}C_3H_7)_2NC_2H_5$	" (–)	
$CH_3COC_4H_9\text{-}t$	LDA	$RCHOHCH_2C_4H_9\text{-}t$ (80)	15
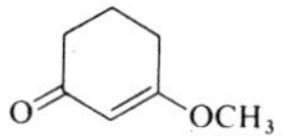 O, OCH_3	"	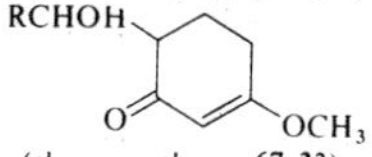 RCHOH, O, OCH_3 (96) (*threo*-*erythro* = 67:33)	53
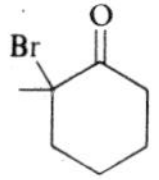 Br, O	$(C_2H_5)_2AlCl$, Zn	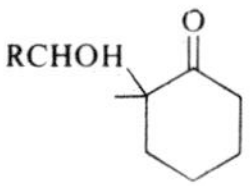 RCHOH, O (100)	131
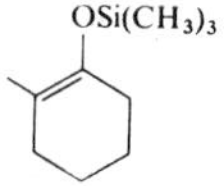 $OSi(CH_3)_3$	$TiCl_4$	" (58)	173
"	$(n\text{-}C_4H_9)_4NF$	" (62)	70
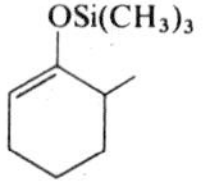 $OSi(CH_3)_3$	$TiCl_4$	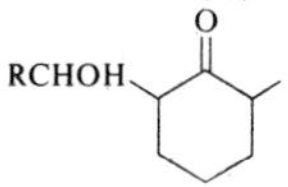 RCHOH, O (81)	173

TABLE II. Reaction of Ketones or Their Equivalents with Aldehydes or Acetals (*Continued*)

No. of C Atoms	Aldehyde or Acetal	Ketone or Equivalent	Reagent	Product(s)[a] and Yield(s) (%)	Refs.
C_7 (*Contd.*)	C_6H_5CHO	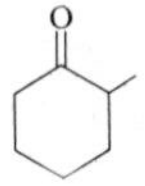 O	$(C_2H_5)_2NBCl_2$, $(C_2H_5)_3N$	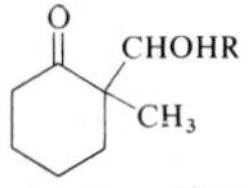 O, CHOHR, CH_3 (40), (*threo*:*erythro* = 10:3) 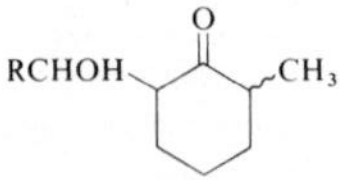	149
				RCHOH, O, CH_3 (10)	
		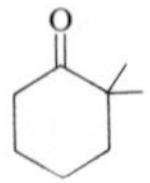 O	LDA	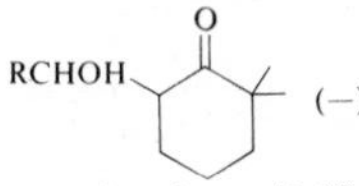 RCHOH, O (—) (*erythro*:*threo* = 13:87)	230
		$C_2H_5COCH_2C_3H_7$-*i*	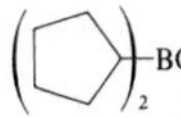 $(C_5H_9)_2BOSO_2CF_3$, (*i*-$C_3H_7)_2NC_2H_5$	$RCHOHCH(CH_3)COCH_2C_3H_7$-*i* (85) (*erythro*:*threo* = 84:16)	262
		$C_2H_5COC(CH_3)_2OH$	LDA	$RCHOHCH(CH_3)COC(CH_3)_2OH$ (88) (*threo*:*erythro* = 1:8)	230
		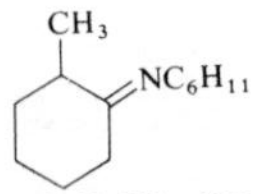 CH_3, NC_6H_{11}	BCl_3, $N(C_2H_5)_3$	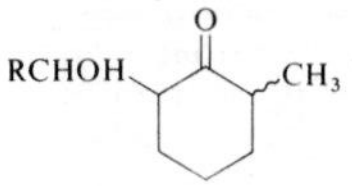 RCHOH, O, CH_3 (37)	149
		n-$C_4H_9CH{=}C(OAc)CH_3$	CH_3Li, $ZnCl_2$	$RCHOHCH(C_4H_9$-*n*$)COCH_3$ (80)	15
		i-$C_4H_9COC_2H_5$	(*n*-$C_4H_9)_2BOSO_2CF_3$, (*i*-$C_3H_7)_2NC_2H_5$	$RCHOHCH(CH_3)COC_4H_9$-*i* (82) (*erythro* *threo* = 97:3)	34
		$C_2H_5COC_4H_9$-*t*	LDA	$RCHOHCH(CH_3)COC_4H_9$-*t* (78) (100% *erythro*)	16, 230
		"	(*n*-$C_4H_9)_2BOSO_2CF_3$, (*i*-$C_3H_7)_2NC_2H_5$	$RCHOHCH(CH_3)COC_4H_9$-*t* (65) (*erythro* *threo* = 97:1)	34
		$CH_3CH{=}C[OSi(CH_3)_3]C_2H_5$	*n*-C_4H_9Li, Cp_2ZrCl_2	$RCHOHCH(CH_3)COC_2H_5$ (90) (*erythro*:*threo* = 83:17)	273

"	$TiCl_4$	" (85) (*erythro* : *threo* = 50:50)	273
"	$(n\text{-}C_4H_9)_2BOSO_2CF_3$	" (63) (*erythro* : *threo* = 95:5)	258
"	$[(C_2H_5)_2N]_3S^+Si(CH_3)_3F_2^-$	" (89) (*erythro* : *threo* = 86:14)	266
"	"	" (84) (*erythro* : *threo* = 63:37)	266
$CH_3CH{=}C[OSi(CH_3)_3]C_3H_7\text{-}i$	CH_3Li	$RCHOHCH(CH_3)COC_3H_7\text{-}i$ (–) (*erythro* : *threo* = 45:55)	230
$CH_3CH{=}C[OSi(CH_3)_3]C_4H_9\text{-}t$	$C_6H_5CH_2N(CH_3)_3F$	$RCHOHCH(CH_3)COC_4H_9\text{-}t$ (55) (*threo*)	230
$CH_3CH{=}C[OSi(CH_3)_3]C_5H_{11}\text{-}n$	$(n\text{-}C_4H_9)_2BOSO_2CF_3$	$RCHOHCH(CH_3)COC_5H_{11}\text{-}n$ (74) (*erythro* : *threo* = 94:6)	258
$CH_3CH{=}C[OSi(CH_3)_3]CH(C_2H_5)_2$	$(n\text{-}C_4H_9)_2BOSO_2CF_3$	$RCHOHCH(CH_3)COCH(C_2H_5)_2$ (73) (*erythro*)	258
$C_2H_5CH{=}C[OSi(CH_3)_3]C_3H_7\text{-}n$	$[(C_2H_5)_2N]_3S^+Si(CH_3)_3F_2^-$	$RCHOHCH(C_2H_5)COC_3H_7\text{-}n$ (65) (*erythro* : *threo* = 86:14)	267
$CH_3CH{=}C[OSi(CH_3)_3]C_6H_5$	$(n\text{-}C_4H_9)_2BOSO_2CF_3$	$RCHOHCH(CH_3)COC_6H_5$ (88) (*erythro* : *threo* = 70:30)	259
"	$(9\text{-BBN})OSO_2CF_3$	" (91) (*erythro* : *threo* = 74:26)	259
"	$[(C_2H_5)_2N]_3S^+Si(CH_3)_3F_2^-$	" (75) (*erythro* : *threo* = 95:5)	267
"	"	" (78) (*erythro* : *threo* = 94:6)	267
$CH_3CH{=}C[OSi(CH_3)_3]C_6H_{11}$	$(n\text{-}C_4H_9)_2BOSO_2CF_3$	$RCHOHCH(CH_3)COC_6H_{11}$ (76) (*erythro*)	258
$CH_3CH{=}C[OSi(CH_3)_3]CH_2C_6H_5$	"	$RCHOHCH(CH_3)COCH_2C_6H_5$ (82) (*erythro* : *threo* = 95:5)	258
$CH_3CH{=}C[OSi(CH_3)_3]C_9H_{19}$	"	$RCHOHCH(CH_3)COC_9H_{19}\text{-}n$ (80) (*erythro* : *threo* = 91:9)	258
$CH_3CH{=}C[OSi(CH_3)_3]C_6H_2(CH_3)_3$	$[(C_2H_5)_2N]_3S^+Si(CH_3)_3F_2^-$	$RCHOHCH(CH_3)COC_6H_2(CH_3)_3$ (77); (65) (*erythro* : *threo* = 95:5)	267
$CH_3CH{=}C[O\text{-}(9\text{-BBN})]CH[OSi(CH_3)_2C_4H_9\text{-}t]C_6H_{11}$		$RCHOHCH(CH_3)COC[OSi(CH_3)_2C_4H_9\text{-}t]C_6H_{11}$ (–)	264
$CH_3CH{=}C[OB(C_4H_9\text{-}n)_2]CH[OSi(CH_3)_2C_4H_9\text{-}t]C_6H_{11}$		" (–)	264
$CH_3CH{=}C[OB(C_5H_9)_2]CH[OSi(CH_3)_2C_4H_9\text{-}t]C_6H_{11}$		" (–)	264
$CH_3(n\text{-}C_5H_{11})C{=}NN(CH_3)_2$	LDA	$RCHOHCH_2C(C_5H_{11}\text{-}n){=}NN(CH_3)_2$ (95)	87
$(i\text{-}C_3H_7)_2CO$	$CH_3(C_6H_5)NMgBr$	$RCHOHC(CH_3)_2COC_3H_7\text{-}i$ (80)	107
$N_2CHCOC_6H_5$	$(n\text{-}C_4H_9)_3B$	$RCHOHCH(C_4H_9\text{-}n)COC_6H_5$ (91) (*erythro* : *threo* = 1:3)	18,135 137
"	$(n\text{-}C_4H_9)_3B$, C_6H_5OLi (catalyst)	" (86) (*erythro* : *threo* > 20:1)	18
$BrCH_2COC_6H_5$	$(C_2H_5)_2AlCl$, Zn	$RCHOHCH_2COC_6H_5$ (95)	131
$C_6H_5COCH_3$	$(C_2H_5)_2NBCl_2$, $(C_2H_5)_3N$	$RCHOHCH_2COC_6H_5$ (26), $RCH{=}CHCOC_6H_5$ (31), $(RCHOH)_2CHCOC_6H_5$ (9)	149

TABLE II. Reaction of Ketones or Their Equivalents with Aldehydes or Acetals (*Continued*)

No. of C Atoms	Aldehyde or Acetal	Ketone or Equivalent	Reagent	Product(s)[a] and Yield(s) (%)	Refs.
C_7 (*Contd.*)	C_6H_5CHO	$C_6H_5(CH_3)C{=}N$– (H)	BCl_3, $N(C_2H_5)_3$	$RCHOHCH_2COC_6H_5$ (30) (optical purity = 47.7%)	148
		$N_2CHCOCH_2C_6H_5$	$(n\text{-}C_4H_9)_3B$	$RCHOHCH(C_4H_9\text{-}n)COCH_2C_6H_5$ (88) (*erythro: threo* = 1:4)	18
		"	$(n\text{-}C_4H_9)_3B$, C_6H_5OLi (catalyst)	" (84) (*erythro: threo* > 20:1)	18
		$CH_3COC(C_6H_5){=}CHN(CH_3)_2$	LTMP	2-C_6H_5-5-C_6H_5-2,3-dihydro-4H-pyran-4-one (64)[b]	237
		$C_2H_5COC_6H_5$	$(n\text{-}C_4H_9)_2BOSO_2CF_3$,	$RCHOHCH(CH_3)COC_6H_5$ (82) (*erythro–threo* = 97:3)	34
		"	LDA	" (96) (*threo–erythro* = 12:88)	230
		"	LCPA	" (–) (*erythro: threo* = 87:13)	230
		"	LHMDS	" (–) (*erythro: threo* = 88:12)	230
		"	LTMP	" (–) (*erythro: threo* = 83:17)	230
		"	LDA, Cp_2ZrCp_2	" (62) (*erythro: threo* = 90:10)	272
		$CH_3COCH[OSi(CH_3)_3]C_4H_9\text{-}t$	LDA	$RCHOHCH_2COCH[OSi(CH_3)_3]C_4H_9\text{-}t$ (71) (diastereomer ratio = 1:1)	255
		$CH_2{=}C[OSi(CH_3)_3]C_6H_5$	CH_3Li, $MgBr_2$	$RCHOHCH_2COC_6H_5$ (81)	15
		$C_2H_5COCH(CH_3)CO_2CH_2CCl_3$	Zn	$RCHOHCH(CH_3)COC_2H_5$ (92)	125
		C_2H_5CO–cyclohexyl	$(\text{cyclopentyl})_2BOSO_2CF_3$, $(i\text{-}C_3H_7)_2NC_2H_5$	$RCHOHCH(CH_3)CO$–cyclohexyl (88) (*erythro: threo* = 14:86)	38
		"	9-BBN(OSO_2CF_3), $(i\text{-}C_3H_7)_2NC_2H_5$	" (79) (*erythro: threo* = 97:3)	38

$C_2H_5COC(CH_3)_2OSi(CH_3)_3$	LDA	$RCHOHCH(CH_3)COC(CH_3)_2OSi(CH_3)_3$ (86)	39
"	"	(*erythro*) (78)	230
$CH_3COCH[OSi(CH_3)_3]CH_2C_6H_5$	LDA	$RCHOHCH_2COCH_2C_6H_5$ (82)[b]	56
"	$BF_3 \cdot O(C_2H_5)_2$	$RCHOHCH(CH_2C_6H_5)COCH_3$ (87)	56
$n\text{-}C_3H_7COC_6H_5$	$(n\text{-}C_4H_9)_2BOSO_2CF_3$, 2,6-lutidine	$RCHOHCH(C_2H_5)COC_6H_5$ (80)	36
$CH_3CO(CH_2)_2C_6H_5$	9-BBN(OSO_2CF_3), 2,6-lutidine	$RCHOHCH(CH_2C_6H_5)COCH_3$ (88) $RCHOHCH_2CO(CH_2)_2C_6H_5$ (2)	37
$CH_3COCH(C_4H_9\text{-}n)CO_2CH_2CCl_3$	Zn	$RCHOHCH(C_4H_9\text{-}n)COCH_3$ (86)	125
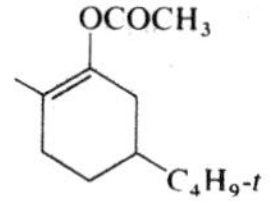 $OCOCH_3$, $C_4H_9\text{-}t$	CH_3Li, $ZnCl_2$	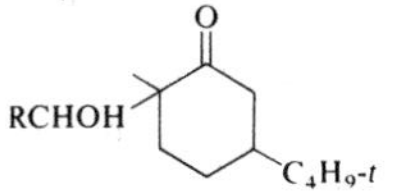 RCHOH, O, $C_4H_9\text{-}t$ (84)	15,120
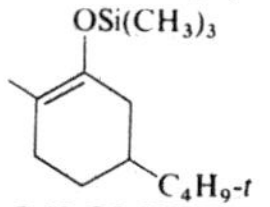 $OSi(CH_3)_3$, $C_4H_9\text{-}t$	$(n\text{-}C_4H_9)_4NF$	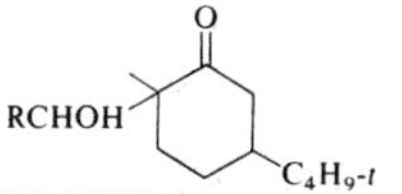 RCHOH, O, $C_4H_9\text{-}t$ (68)	70
$C_2H_5COCH[Si(CH_3)_3]CH_2C_6H_5$	LDA	$RCHOHCH(CH_3)CO(CH_2)_2C_6H_5$ (95)[b]	56
"	$BF_3 \cdot O(C_2H_5)_2$	$RCHOHCH(CH_2C_6H_5)COC_2H_5$ (73)	56
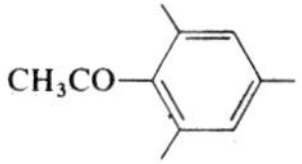 CH_3CO–	CH_3Li, $MgBr_2$	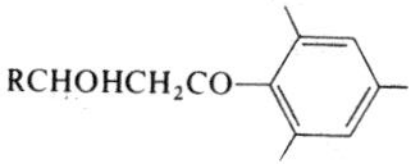 $RCHOHCH_2CO$– (93)	15
"	CH_3MgBr	" (90)	15
$C_2H_5COCH[OSi(CH_3)_3]C_4H_9\text{-}t$	LDA	$RCHOHCH(CH_3)COCH[OSi(CH_3)_3]C_4H_9\text{-}t$ (75) (*threo: erythro* = 1:3)	43, 255
$C_2H_5COC(CH_3)[OSi(CH_3)_3]C_6H_5$	LDA	$RCHOHCH(CH_3)COC(CH_3)[OSi(CH_3)_3]C_6H_5$ (–)	263
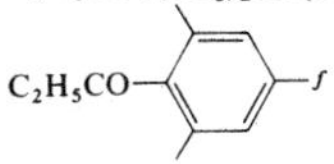 C_2H_5CO–, *f*	LDA	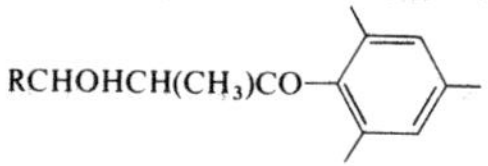 $RCHOHCH(CH_3)CO$– (–) (*threo: erythro* = 92:8)	16
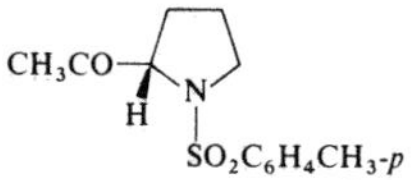 CH_3CO–, H, N, $SO_2C_6H_4CH_3\text{-}p$	LDA	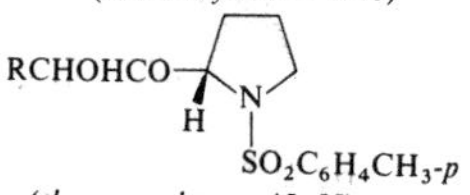 RCHOHCO–, H, N, $SO_2C_6H_4CH_3\text{-}p$ (75) (*threo: erythro* = 45:55)	260
"	$(n\text{-}C_4H_9)_2BOSO_2CF_3$, $(i\text{-}C_3H_7)_2NC_2H_5$	(*threo: erythro* = 83:17) (80)	260

TABLE II. REACTION OF KETONES OR THEIR EQUIVALENTS WITH ALDEHYDES OR ACETALS (*Continued*)

No. of C Atoms	Aldehyde or Acetal	Ketone or Equivalent	Reagent	Product(s)[a] and Yield(s) (%)	Refs.
C_7 (*Contd.*)	C_6H_5CHO	C_2H_5CO-(2-pyrrolyl, N-$CO_2C_4H_9$-*t*)	LDA, $ZnCl_2$	$RCHOHCH(CH_3)CO$-(2-pyrrolyl, N-$CO_2C_4H_9$-*t*) (—) (*threo*:*erythro* = 7:3)	121
		"	$(n\text{-}C_4H_9)_2BOSO_2CF_3$, $(i\text{-}C_3H_7)_2NC_2H_5$	" (70) (*erythro*:*threo* = 97:3)	262
		COR_1 (di-O-isopropylidene pyranose)	LDA	$COCH(R'')CHOHR$ (di-O-isopropylidene pyranose) (*erythro*) $R'' = H$ from $R_1 = CH_3$ (65)	254
				$R'' = CH_3$ from $R_1 = C_2H_5$ (—)	40
				" " (100)	254
		C_2H_5CO(1-adamantyl)	LDA	$RCHOHCH(CH_3)CO$(1-adamantyl) (*threo*:*erythro* = 2:98) (97)	230
		$C_6H_5C(O(9\text{-BBN}))=CHR_1$		$RCHOHCHR_1COC_6H_5$ ($R_1 = n\text{-}C_5H_{11}, n\text{-}C_6H_{13}, n\text{-}C_7H_{15}$) (80–86)	257
	$C_6H_5CH(OCH_3)_2$	$CH_2=C(CH_3)OAc$	$TiCl_4$	$RCH(OR')CH_2COCH_3$ (81) $RCHClCH_2COCH_3$ (14) $RCH=CHCOCH_3$ (5)	214
		4-methylene-2-oxetanone	"	$RCH(OR')CH_2COCH_2CO_2CH_3$ (84)	215
		1-$OSi(CH_3)_3$-cyclohexene	$(CH_3)_3SiO_3SCF_3$	2-[$R(CH_3O)CH$]-cyclohexanone (89) (*threo*:*erythro* = 7:93)	193a
		$CH_3CH=C[OSi(CH_3)_3]C_4H_9\text{-}t$ (*Z*)	"	$RCH(OCH_3)CH(CH_3)COC_4H_9\text{-}t$ (94) (*threo*:*erythro* = 5:95)	193a
		$CH_3CH=C[OSi(CH_3)_3]C_6H_5$ (*E*)	"	$RCH(OCH_3)CH(CH_3)COC_6H_5$ (83) (*threo*:*erythro* = 29:71)	193a

		$CH_3CH{=}C[OSi(CH_3)_3]C_6H_5$ (Z)	"	$RCH(OCH_3)CH(CH_3)COC_6H_5$ (97) (*threo: erythro* = 16:84)	193a
	$C_6H_5CH(OC_2H_5)_2$	$CH_2{=}C(OC_2H_5)CH_3$	$ZnCl_2$	$RCH(OR')CH_2C(OR')_2CH_3$ (81)	242
		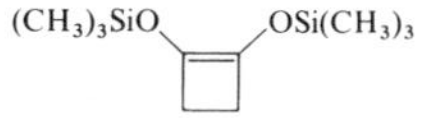 $(CH_3)_3SiO$, $OSi(CH_3)_3$ (cyclobutene)	$BF_3 \cdot O(C_2H_5)_2$	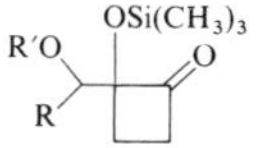 R′O, R, $OSi(CH_3)_3$, O (cyclobutanone) (94)	185
		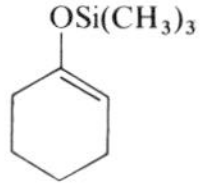 $OSi(CH_3)_3$ (cyclohexene)	$TiCl_4$	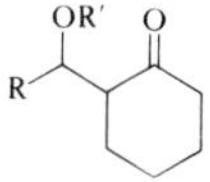 OR′, O, R (cyclohexanone) (95)	180
		$C_6H_5CH{=}C[OSi(CH_3)_3]CH_3$	"	$RCH(OR')CH(C_6H_5)COCH_3$ (95)	180
	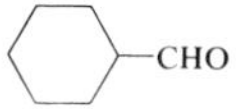 (cyclohexyl)–CHO	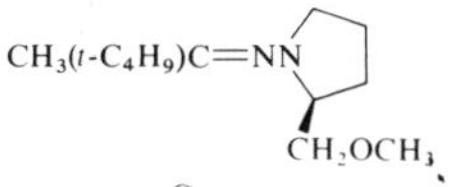 $CH_3(t\text{-}C_4H_9)C{=}NN$ (pyrrolidine) CH_2OCH_3	$n\text{-}C_4H_9Li$	$RCHOHCH_2COC_4H_9\text{-}t$ (60)[b] (< 10% e.e.)	91
	$OHC(CH_2)_4CO_2CH_3$	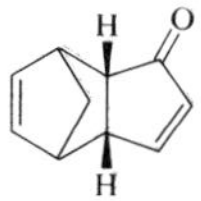 H, O, H (bicyclic enone)	$n\text{-}C_4H_9MgBr$, CuCl	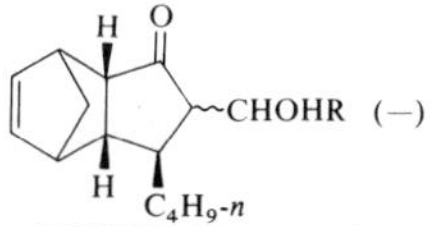 H, O, ~CHOHR, H, $C_4H_9\text{-}n$ (—)	112
	$n\text{-}C_6H_{13}CHO$	$C_2H_5COCH_3$	$(C_2H_5)_2NLi$	$RCHOHCH_2COC_2H_5$ (29)	48
		$CH_3COC_3H_7\text{-}n$	"	$RCHOHCH_2COC_3H_7\text{-}n$ (35)	48
		$CH_3COC_3H_7\text{-}i$	"	$RCHOHCH_2COC_3H_7\text{-}i$ (50)	48
		$CH_3COC_4H_9\text{-}i$	"	$RCHOHCH_2COC_4H_9\text{-}i$ (50)	48
		$i\text{-}C_4H_9COC_4H_9\text{-}i$	$CH_3(C_6H_5)NMgBr$	$RCHOHC(CH_3)_2COC_3H_7\text{-}i$ (88)	107
	$CH_3O_2C(CH_2)_2CH(CH_3)CHO$	$C_2H_5COC_4H_9\text{-}t$	LDA	$RCHOHCH(CH_3)COC_4H_9\text{-}t$ (—)	39
	$CH_3O_2C(CH_2)_2CH(CH_3)CHO$	$C_2H_5COC_4H_9\text{-}t$	1. LDA 2. Ac_2O	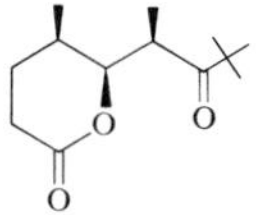 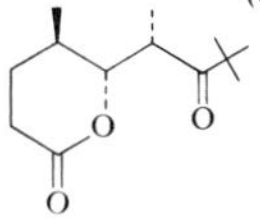 (lactone, O, O, O) (86:14) (86) (lactone, O, O, O)	230
C_8	C_6H_5COCHO	$C_6H_5CH{=}C[OSi(CH_3)_3]CH_3$	$TiCl_4$	$RCHOHCH(C_6H_5)COCH_3$ (83)	173
	$p\text{-}CH_3OC_6H_4CHO$	$CH_3COCH{=}C(CH_3)_2$	LDA	$RCHOHCH_2COCH{=}C(CH_3)_2$ (85)	50

TABLE II. Reaction of Ketones or Their Equivalents with Aldehydes or Acetals (*Continued*)

No. of C Atoms	Aldehyde or Acetal	Ketone or Equivalent	Reagent	Product(s)[a] and Yield(s) (%)	Refs.
C_8 (*Contd.*)	p-$CH_3OC_6H_4CHO$	$CH_3COCH{=}C(CH_3)_2$	$(CH_3)_2CuLi$, $ZnCl_2$	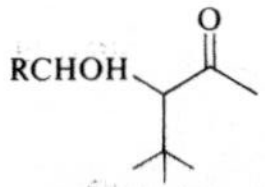 RCHOH … O (46)	61
		$C_2H_5COC_4H_9$-t	LDA	$RCHOHCH(CH_3)COC_4H_9$-t (68) (*erythro*)	230
	m-$CH_3OC_6H_4CHO$	$CH_3CH{=}C(C_2H_5)OSi(CH_3)_3$	$TiCl_4$	$RCHOHCH(CH_3)COC_2H_5$ (88)	177
	p-$CH_3OC_6H_4CHO$	"	"	$RCHOHCH(CH_3)COC_2H_5$ (82)	177
	o-$CH_3CO_2C_6H_4CHO$	"	"	$RCHOHCH(CH_3)COC_2H_5$ (78)	177
	$C_6H_5CH_2CHO$	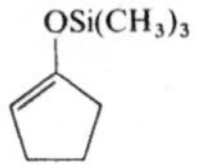 $OSi(CH_3)_3$ (cyclopentenyl)	$TiCl_4$	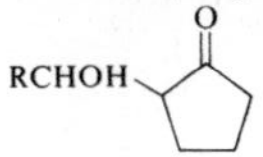 RCHOH … O (95)	173
		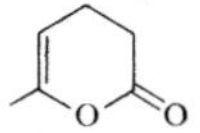 O … O	$BF_3\cdot O(C_2H_5)_2$	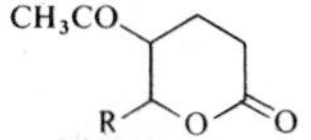 CH_3CO … R … O … O (61)	222
		i-$C_3H_7OC_3H_7$-i	$CH_3(C_6H_5)NMgBr$	$RCHOHC(CH_3)_2COC_3H_7$-i (62)	107
		$C_2H_5COC_4H_9$-t	LDA	$RCHOHCH(CH_3)COC_4H_9$-t (74) (*erythro*)	230
		$C_2H_5COC(CH_3)_2OSi(CH_3)_3$	LDA	$RCHOHCH(CH_3)COC(CH_3)_2OSi(CH_3)_3$ (—) (100% *erythro*)	39
		$CH_2{=}C[OSi(CH_3)_3]C_6H_5$	$TiCl_4$	$RCHOHCH_2COC_6H_5$ (78)	173
		$C_2H_5COCH[OSi(CH_3)_3]C_4H_9$-t	LDA	$RCHOHCH(CH_3)COCH[OSi(CH_3)_3]C_4H_9$-t (75) (*erythro*)	43, 255
		$C_2H_5COC(CH_3)[OSi(CH_3)_3]C_6H_5$	"	$RCHOHCH(CH_3)COC(CH_3)[OSi(CH_3)_3]C_6H_5$ (75)	263
		$C_6H_5CH{=}C[OSi(CH_3)_3]CH_3$	$TiCl_4$	$RCHOHCH(C_6H_5)COCH_3$ (77)	173
		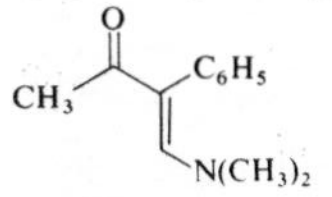 CH_3 … O … C_6H_5 … $N(CH_3)_2$	LTMP	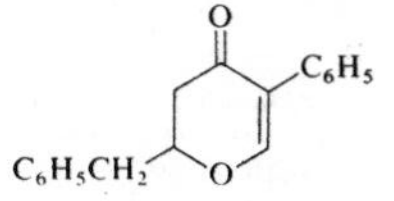 O … C_6H_5 … $C_6H_5CH_2$ … O (37)[b]	237

	$OHC(CH_2)_5CO_2CH_3$	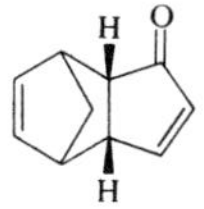	$n-C_8H_{17}MgBr$, CuCl	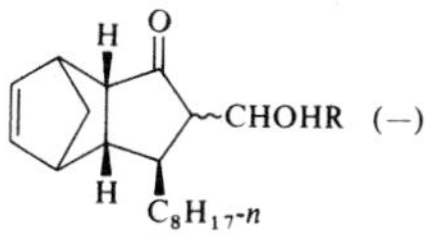 CHOHR, C_8H_{17}-n (—)	112
	$n-C_7H_{15}CHO$	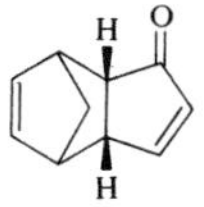	$n-C_8H_{17}MgBr$, CuCl	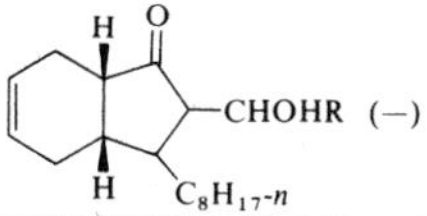 CHOHR, C_8H_{17}-n (—)	112
	$n-C_4H_9CH(C_2H_5)CHO$	$(i-C_3H_7)_2CO$	$C_6H_5NCH_3(MgBr)$	$RCHOHC(CH_3)_2COC_3H_7$-i (67)	107
	$CH_3COCH_2CH(CH_3)(CH_2)_2CH(OCH_3)_2$	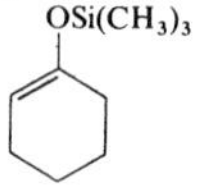 $OSi(CH_3)_3$	$(CH_3)_3SiO_3SCF_3$	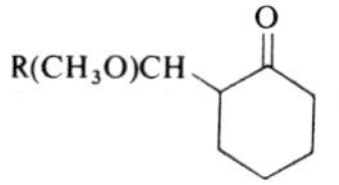 $R(CH_3O)CH$ (92)	193a
	$C_6H_5SeCH_2CHO$[g]	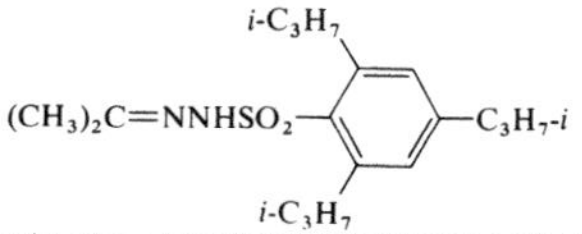 $(CH_3)_2C{=}NNHSO_2$, i-C_3H_7, C_3H_7-i, i-C_3H_7	$n-C_4H_9Li$	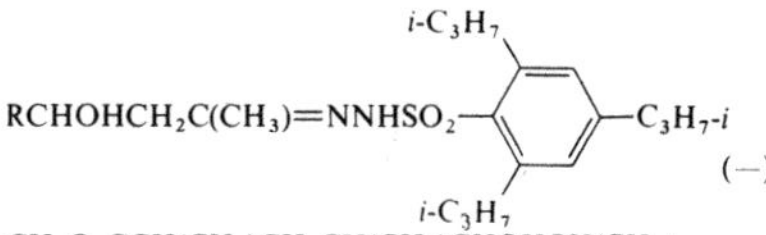 $RCHOHCH_2C(CH_3){=}NNHSO_2$, i-$C_3H_7$, C_3H_7-i, i-C_3H_7 (—)	99
	$CH_3O_2CCH(CH_3)CH_2CH(CH_3)CHO$	$CH_3CH{=}C(OLi)C(CH_3)[OSi(CH_3)_3]C_6H_{11}$		$CH_3O_2CCH(CH_3)CH_2CH(CH_3)CHOHCH(CH_3)$-$COC(CH_3)[OSi(CH_3)_3]C_6H_{11}$ (—)	263
		$CH_3CH{=}C[O(9\text{-}BBN)]$-$CH[OSi(CH_3)_2C_4H_9$-t$]C_6H_{11}$		$CH_3O_2CCH(CH_3)CH_2CH(CH_3)CHOHCH(CH_3)$-$COCH[OSi(CH_3)_2C_4H_9$-t$]C_6H_{11}$ (85)	265
		$CH_3CH{=}C[OB(C_5H_9)_2]$-$CH[OSi(CH_3)_2C_4H_9$-t$]C_6H_{11}$		" (—)	265
C_9	$C_6H_5CH{=}CHCHO$	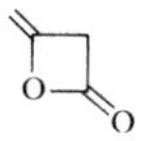	$TiCl_4$	$RCHOHCH_2COCH_2CO_2CH_3$ (45), 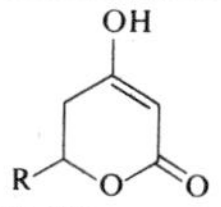 OH, R (7)	216
		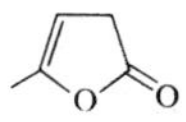	$BF_3\cdot O(C_2H_5)_2$	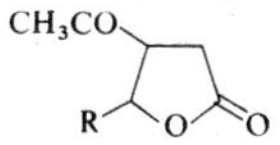 CH_3CO, R (79)	221,222
		$BrCH_2COC_6H_5$	$(C_2H_5)_2AlCl$, Zn	$RCHOHCH_2COC_6H_5$ (92)	131
	$C_6H_5CH{=}CHCH(OCH_3)_2$	$CH_2{=}C(CH_3)OAc$	$TiCl_4$	$RCH(OR')CH_2COCH_3$ (65)	214
	p-$(CH_3)_2NC_6H_4CHO$	$CH_3CH{=}C(C_2H_5)OSi(CH_3)_3$	$TiCl_4$	$RCHOHCH(CH_3)COC_2H_5$ (65)	177
	p-$CH_3O_2CC_6H_4CHO$	$CH_3CH{=}C(C_2H_5)OSi(CH_3)_3$	$TiCl_4$	$RCHOHCH(CH_3)COC_2H_5$ (86)	177

TABLE II. Reaction of Ketones or Their Equivalents with Aldehydes or Acetals (*Continued*)

No. of C Atoms	Aldehyde or Acetal	Ketone or Equivalent	Reagent	Product(s)[a] and Yield(s) (%)	Refs.
C_9 (*Contd.*)	$C_6H_5(CH_2)_2CHO$	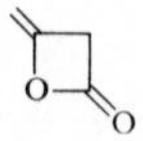	$TiCl_4$	$RCHOHCH_2COCH_2CO_2CH_3$ (80), 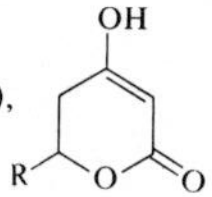 OH, R, O, O (14)	216
		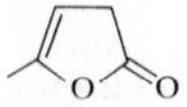 O, O	$BF_3 \cdot O(C_2H_5)_2$	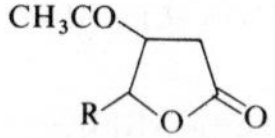 CH_3CO, R, O, O (85)	221,222
		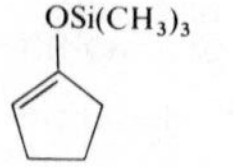 $OSi(CH_3)_3$	$TiCl_4$	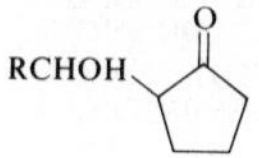 RCHOH, O (94)	173
		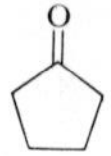 O	$(n\text{-}C_4H_9)_2BOSO_2CF_3$, $C_2H_5N(C_3H_7\text{-}i)_2$	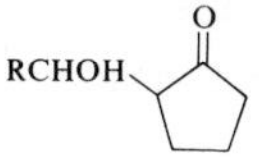 RCHOH, O (80) 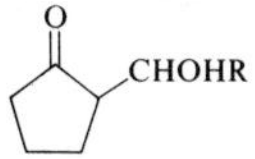 + O, CHOHR (21)	36
		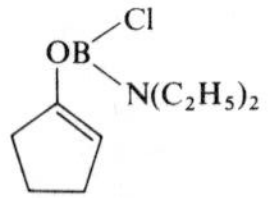 OB, Cl, $N(C_2H_5)_2$	$N(C_2H_5)_3$	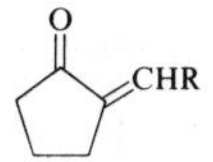 O, CHR (14) 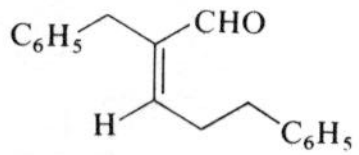 + C_6H_5, CHO, H, C_6H_5 (3)	149
		 O	1. $TiCl_4$, $Ti(OC_3H_7\text{-}i)_4$ 2. CH_3OH	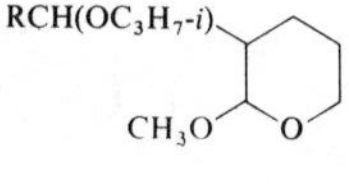 $RCH(OC_3H_7\text{-}i)$, CH_3O, O (100)	199

$CH_3COC_3H_7\text{-}i$	$i\text{-}C_4H_9C[Si(CH_3)_3]_2OLi$	$RCHOHCH_2COC_3H_7\text{-}i$ (83)	55
$CH_3COCH_2CO_2CH_2CCl_3$	Zn	$RCHOHCH_2COCH_3$ (48)	125
6-methyl-3,4-dihydro-2H-pyran-2-one (ring structure)	$BF_3 \cdot O(C_2H_5)_2$	3-CH_3CO-6-R-tetrahydropyran-2-one (ring structure) (84)	222
cyclohexanone (ring structure)	$(n\text{-}C_4H_9)_2BOSO_2CF_3$, $C_2H_5N(C_3H_7\text{-}i)_2$	2-(RCHOH)cyclohexanone (ring structure) (72)	36
cyclohexanone $N\text{-}C_6H_{11}$ imine (ring structure)	BCl_3, $N(C_2H_5)_3$	2-(RCHOH)cyclohexanone (ring structure) (52) (*threo*: *erythro* = 1:1)	147
cyclohexenyl–$OB(Cl)N(C_2H_5)_2$ (ring structure)	$N(C_2H_5)_3$	2-(CHOHR)cyclohexanone (ring structure) (53), 2-(=CHR)cyclohexanone (ring structure) (8) (*threo*: *erythro* = 1:1)	149
$CH_3COC_4H_9\text{-}i$	$(n\text{-}C_4H_9)_2BOSO_2CF_3$, $C_2H_5N(C_3H_7\text{-}i)_2$	$RCHOHCH_2COC_4H_9\text{-}i$ (82)	36
"	$i\text{-}C_4H_9C[Si(CH_3)_3]_2OLi$	" (86)	55
$CH_3COC_4H_9\text{-}sec$	"	$RCHOHCH_2COC_4H_9\text{-}sec$ (80)	55
$CH_3COC_6H_5$	$(n\text{-}C_4H_9)_2BOSO_2CF_3$, $C_2H_5N(C_3H_7\text{-}i)_2$	$RCHOHCH_2COC_6H_5$ (75)	36
"	$C_2H_5NBCl_2$, $(C_2H_5)_3N$	$RCHOHCH_2COC_6H_5$ (32)	149
$C_2H_5COCH(CH_3)CO_2CH_2CCl_3$	Zn	$RCHOHCH(CH_3)COC_2H_5$ (70)	125
$C_2H_5COC_6H_5$	$(n\text{-}C_4H_9)_2BOSO_2CF_3$, $C_2H_5N(C_3H_7\text{-}i)_2$	$RCHOHCH(CH_3)COC_6H_5$ (83)	36
$C_2H_5CO\text{-}C_6H_{11}$ (cyclohexyl, ring structure)	$(C_5H_9)_2BOSO_2CF_3$ (cyclopentyl, ring structure), $(i\text{-}C_3H_7)_2NC_2H_5$	$RCHOHCH(CH_3)CO\text{-}C_6H_{11}$ (cyclohexyl, ring structure) (88) (*erythro*: *threo* = 15:85)	38
"	9-BBN(OSO_2CF_3), $(i\text{-}C_3H_7)_2NC_2H_5$	" (82) (*erythro*: *threo* = 97:3)	38
$CH_3CH{=}C[OSi(CH_3)_3]C_6H_{11}$	$(n\text{-}C_4H_9)_2BOSO_2CF_3$	$RCHOHCH(CH_3)COC_6H_{11}$ (64) (*erythro*: *threo* >98:2)	258
$CH_3COCH_2SOC_6H_5$	NaH, $n\text{-}C_4H_9Li$	$RCHOHCH_2COCH_2SOC_6H_5$ (77)	240
$CH_3COCH(C_4H_9\text{-}n)CO_2CH_2CCl_3$	Zn	$RCHOHCH(C_4H_9\text{-}n)COCH_3$ (82)	125
$CH_3CO(CH_2)_2C_6H_5$	$C_2H_5NBCl_2$, $(C_2H_5)_3N$	$RCHOHCH_2CO(CH_2)_2C_6H_5$ (30)	149
$C_6H_5C[O(9\text{-}BBN)]{=}CHC_7H_{15}\text{-}n$		$RCHOHCH(C_7H_{15}\text{-}n)COC_6H_5$ (77)	257

TABLE II. REACTION OF KETONES OR THEIR EQUIVALENTS WITH ALDEHYDES OR ACETALS (*Continued*)

No. of C Atoms	Aldehyde or Acetal	Ketone or Equivalent	Reagent	Product(s)[a] and Yield(s) (%)	Refs.
C_9 (*Contd.*)	$C_6H_5(CH_2)_2CH(OCH_3)_2$	$CH_2{=}C(CH_3)OAc$	$TiCl_4$	$RCH(OR')CH_2COCH_3$ (89), $RCH{=}CHCOCH_3$ (1)	214
		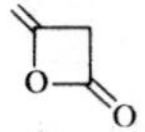 [structure]	"	$RCH(OR')CH_2COCH_2CO_2CH_3$ (87)	215
		$CH_2{=}C[OSi(CH_3)_3]C_6H_5$	"	$RCH(OR')CH_2COC_6H_5$ (73)	180
	$C_6H_5CH(CH_3)CHO$	$C_2H_5COC(CH_3)_2OSi(CH_3)_3$	LDA	$RCHOHCH(CH_3)COC(CH_3)_2OSi(CH_3)_3$ (*erythro*) (—)	39
		$C_2H_5COC(CH_3)_3$	LDA	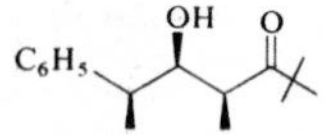 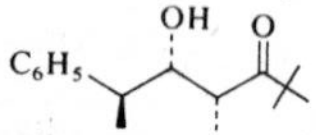 [structure: C_6H_5, OH, O] [structure: C_6H_5, OH, O] (86:14) (80)	230
		$C_2H_5COCH[OSi(CH_3)_3]C_4H_9\text{-}t$	"	$RCHOHCH(CH_3)COCH[OSi(CH_3)_3]C_4H_9\text{-}t$ (—) (*erythro*)	43
	$C_6H_{11}CH(CH_3)CHO$	$C_2H_5COC(CH_3)[OSi(CH_3)_3]C_6H_5$	"	$C_6H_{11}CH(CH_3)CHOHCH(CH_3)COC(CH_3)[OSi(CH_3)_3]C_6H_5$ (72–75)	263
	$C_6H_5CH(CH_3)CH(OCH_3)_2$	$CH_2{=}C(CH_3)OAc$	$TiCl_4$	$RCH(OR')CH_2COCH_3$ (79)	214
	$n\text{-}C_8H_{17}CHO$	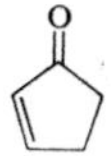 [structure]	$(CH_3)_2AlSC_6H_5$	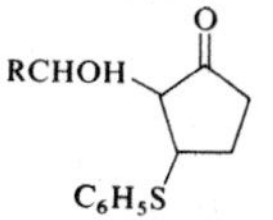 [structure: RCHOH, O, C_6H_5S] (76)	134
		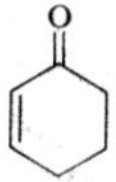 [structure]	$(CH_3)_2AlSC_6H_5$	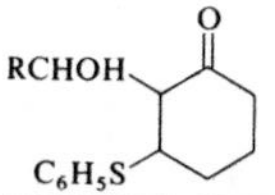 [structure: RCHOH, O, C_6H_5S] (90)	134
		$CH_3COCH[Si(CH_3)_3]CH_2C_6H_5$	LDA	$RCHOHCH_2CO(CH_2)_2C_6H_5$ (78)[b]	56
		"	$BF_3 \cdot O(C_2H_5)_2$	$RCHOHCH(CH_2C_6H_5)COCH_3$ (78)	56
	$THPOCH_2CH(CH_3)CHO$	$(CH_3O)_2CHC(CH_3){=}NC_6H_{11}$	LDA	$RCHOHCH_2COCH(OCH_3)_2$[b] (80)	81
	$CH_3CH[OSi(CH_3)_2C_4H_9\text{-}t]CHO$	$C_2H_5COC(CH_3)_2OSi(CH_3)_3$	"	$CH_3CH[OSi(CH_3)_2C_4H_9\text{-}t]CHOHCH(CH_3)COC(CH_3)_2OSi(CH_3)_3$ (41)	252

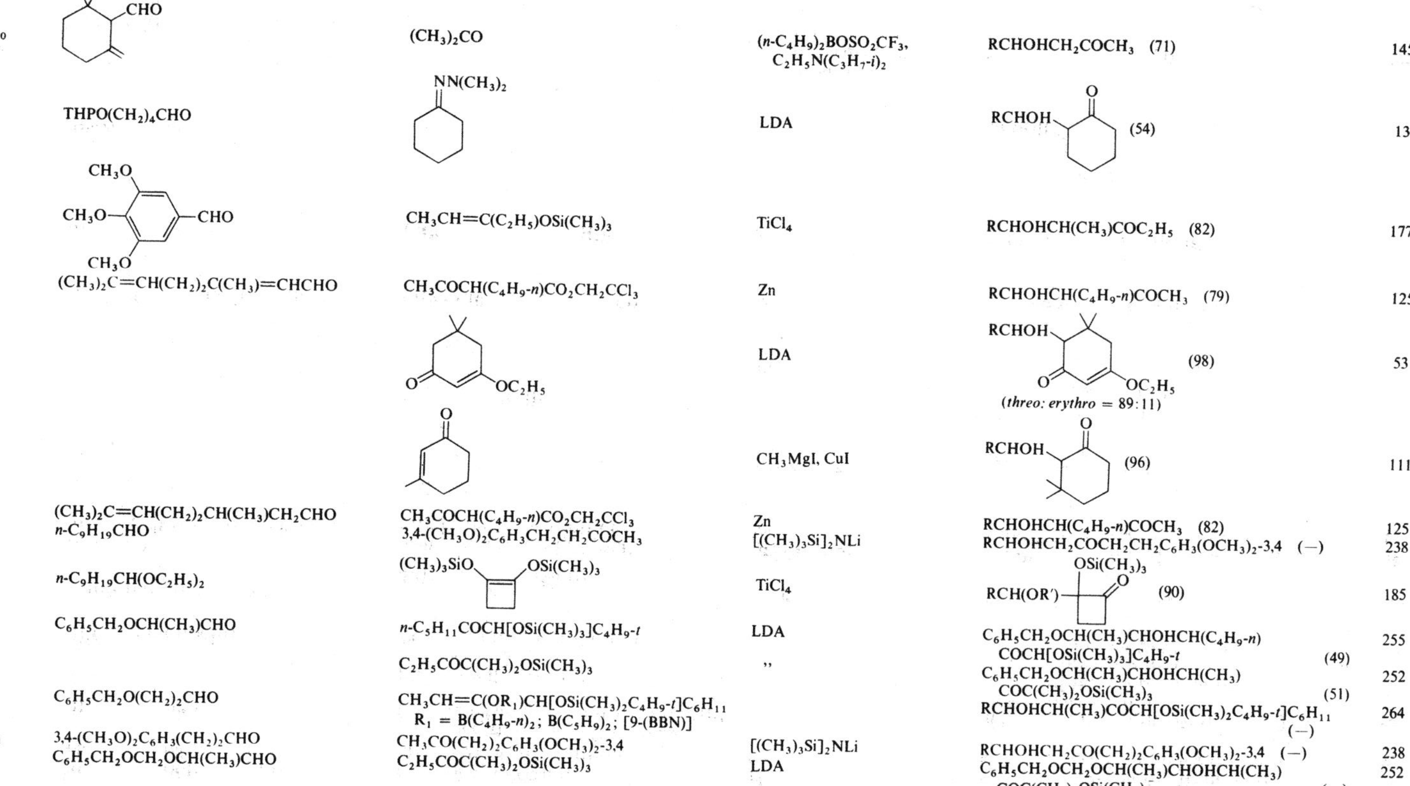

C_{10}	CHO	$(CH_3)_2CO$	$(n\text{-}C_4H_9)_2BOSO_2CF_3$, $C_2H_5N(C_3H_7\text{-}i)_2$	$RCHOHCH_2COCH_3$ (71)	145
	$THPO(CH_2)_4CHO$	$NN(CH_3)_2$	LDA	RCHOH (54)	134
	CH_3O, CH_3O, CHO, CH_3O	$CH_3CH{=}C(C_2H_5)OSi(CH_3)_3$	$TiCl_4$	$RCHOHCH(CH_3)COC_2H_5$ (82)	177
	$(CH_3)_2C{=}CH(CH_2)_2C(CH_3){=}CHCHO$	$CH_3COCH(C_4H_9\text{-}n)CO_2CH_2CCl_3$	Zn	$RCHOHCH(C_4H_9\text{-}n)COCH_3$ (79)	125
		OC_2H_5	LDA	RCHOH, OC_2H_5 (98) (*threo: erythro* = 89:11)	53
			CH_3MgI, CuI	RCHOH (96)	111
	$(CH_3)_2C{=}CH(CH_2)_2CH(CH_3)CH_2CHO$	$CH_3COCH(C_4H_9\text{-}n)CO_2CH_2CCl_3$	Zn	$RCHOHCH(C_4H_9\text{-}n)COCH_3$ (82)	125
	$n\text{-}C_9H_{19}CHO$	$3,4\text{-}(CH_3O)_2C_6H_3CH_2CH_2COCH_3$	$[(CH_3)_3Si]_2NLi$	$RCHOHCH_2COCH_2CH_2C_6H_3(OCH_3)_2\text{-}3,4$ (—)	238
	$n\text{-}C_9H_{19}CH(OC_2H_5)_2$	$(CH_3)_3SiO$, $OSi(CH_3)_3$	$TiCl_4$	$RCH(OR')$, $OSi(CH_3)_3$ (90)	185
	$C_6H_5CH_2OCH(CH_3)CHO$	$n\text{-}C_5H_{11}COCH[OSi(CH_3)_3]C_4H_9\text{-}t$	LDA	$C_6H_5CH_2OCH(CH_3)CHOHCH(C_4H_9\text{-}n)COCH[OSi(CH_3)_3]C_4H_9\text{-}t$ (49)	255
		$C_2H_5COC(CH_3)_2OSi(CH_3)_3$	"	$C_6H_5CH_2OCH(CH_3)CHOHCH(CH_3)COC(CH_3)_2OSi(CH_3)_3$ (51)	252
	$C_6H_5CH_2O(CH_2)_2CHO$	$CH_3CH{=}C(OR_1)CH[OSi(CH_3)_2C_4H_9\text{-}t]C_6H_{11}$ $R_1 = B(C_4H_9\text{-}n)_2$; $B(C_5H_9)_2$; [9-(BBN)]		$RCHOHCH(CH_3)COCH[OSi(CH_3)_2C_4H_9\text{-}t]C_6H_{11}$ (—)	264
C_{11}	$3,4\text{-}(CH_3O)_2C_6H_3(CH_2)_2CHO$	$CH_3CO(CH_2)_2C_6H_3(OCH_3)_2\text{-}3,4$	$[(CH_3)_3Si]_2NLi$	$RCHOHCH_2CO(CH_2)_2C_6H_3(OCH_3)_2\text{-}3,4$ (—)	238
	$C_6H_5CH_2OCH_2OCH(CH_3)CHO$	$C_2H_5COC(CH_3)_2OSi(CH_3)_3$	LDA	$C_6H_5CH_2OCH_2OCH(CH_3)CHOHCH(CH_3)COC(CH_3)_2OSi(CH_3)_3$ (—)	252

TABLE II. REACTION OF KETONES OR THEIR EQUIVALENTS WITH ALDEHYDES OR ACETALS (*Continued*)

No. of C Atoms	Aldehyde or Acetal	Ketone or Equivalent	Reagent	Product(s)[a] and Yield(s) (%)	Refs.
C_{12}	$C_6H_5CH_2OCH_2OCH_2CH(CH_3)CHO$	$(CH_3)_2C[OSi(CH_3)_3]COCH_3$	LDA, $MgBr_2$	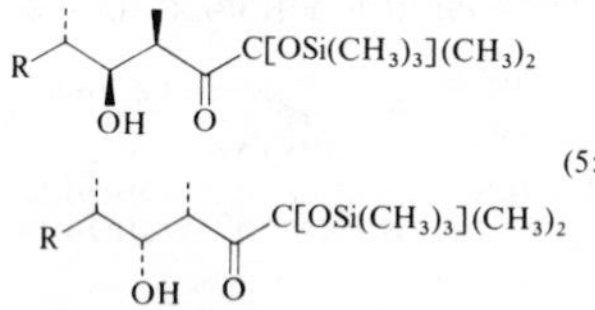 (5:1) (85)	243
C_{13}	n-$C_4H_9CH(OCH_2C_6H_5)CHO$	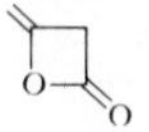	$TiCl_4$	$RCHOHCH_2COCH_2CO_2CH_3$ (—) (*threo–erythro* = 85:15)	217
C_{14}	$(C_6H_5)_2CHCHO$	$C_2H_5COCH[OSi(CH_3)_3]C_4H_9$-$t$	LDA	$RCHOHCH(CH_3)COCH[OSi(CH_3)_3]C_4H_9$-$t$ (*erythro*) (69)	43
	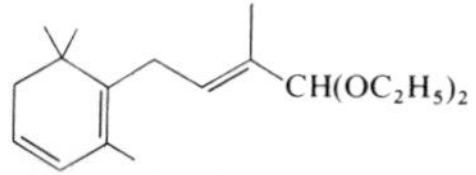	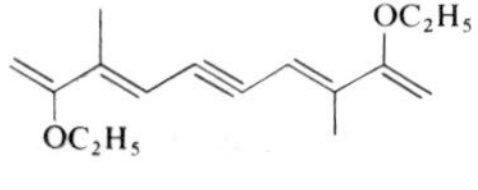	$ZnCl_2$	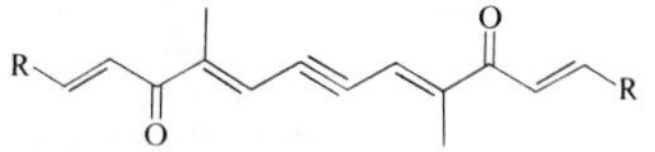 (44)[b]	203
	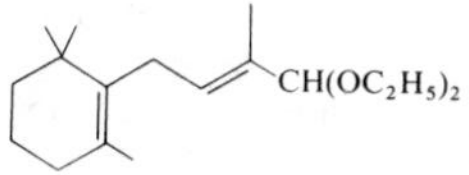	"	$ZnCl_2$	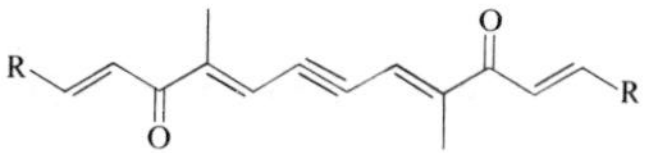 (72)[b]	203
C_{18}	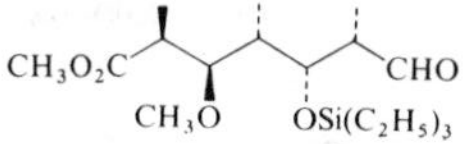	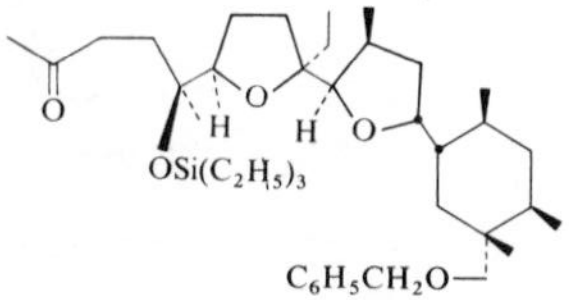	LDA, $MgBr_2$	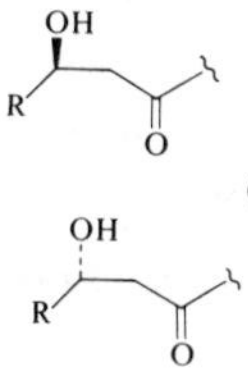 (3:1) (75)	243

	Aldehyde	Ketone/equivalent	Conditions	Product and Yield (%)		Refs.
C_{19}	$CH(OC_2H_5)_2$	OC_2H_5, OC_2H_5	$ZnCl_2$	R, O, O, R	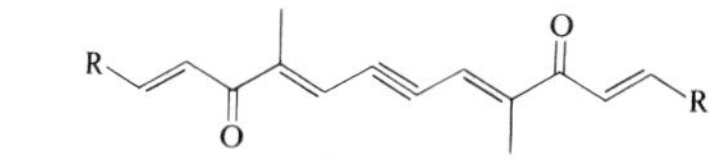 $(40)^b$	203
	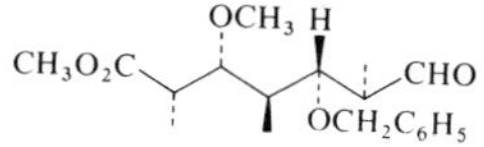 CH_3O_2C, OCH_3 H, CHO, $OCH_2C_6H_5$	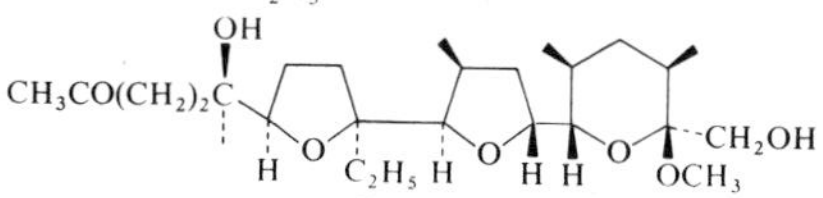 $CH_3CO(CH_2)_2C$, OH, H, C_2H_5 H, H H, CH_2OH, OCH_3	$(i\text{-}C_3H_7)_2NMgBr$	$RCHOHCH_2CO(CH_2)_2C$, OH, H, C_2H_5 H, H H, CH_2OH, OCH_3	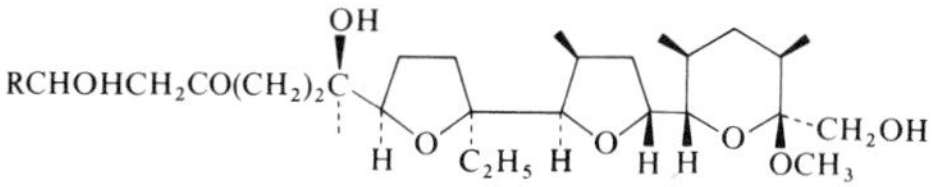 (71)	42
C_{20}	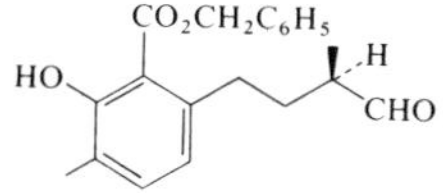 HO, $CO_2CH_2C_6H_5$, H, CHO	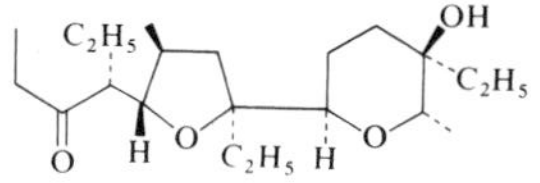 C_2H_5, H, C_2H_5 H, OH, C_2H_5	LDA, $ZnCl_2$	HO, $CO_2CH_2C_6H_5$, OH, O, C_2H_5, H, C_2H_5 H, OH, C_2H_5	(—)	41
C_{23}	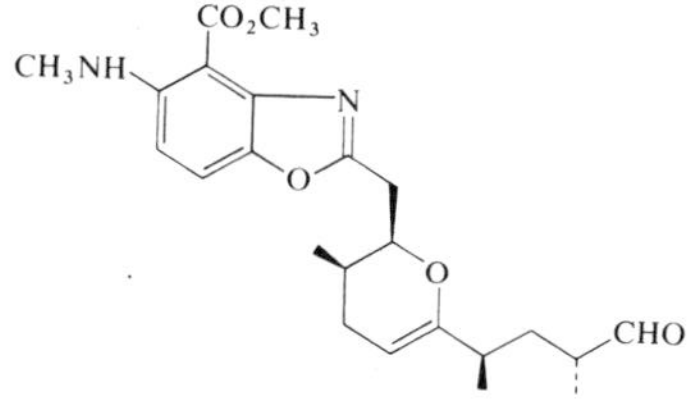 CH_3NH, CO_2CH_3, N, O, O, CHO	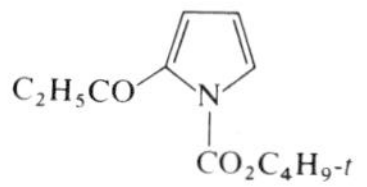 C_2H_5CO, N, $CO_2C_4H_9\text{-}t$	LDA, $ZnCl_2$	$RCHOHCH(CH_3)CO$, N, $CO_2C_4H_9\text{-}t$	(—)	121
	$C_6H_5CO_2$, $O_2CC_6H_5$, $C_6H_5CO_2$, O, OAc	$OSi(CH_3)_3$	$SnCl_4$	$C_6H_5CO_2$, $O_2CC_6H_5$, $C_6H_5CO_2$, O, O	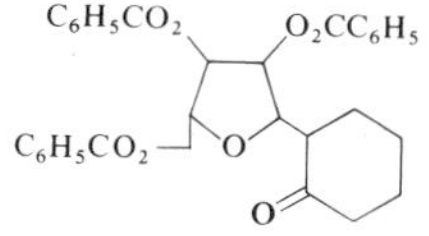 (95)	183

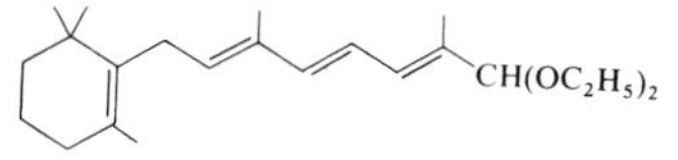

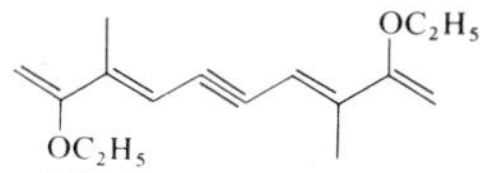

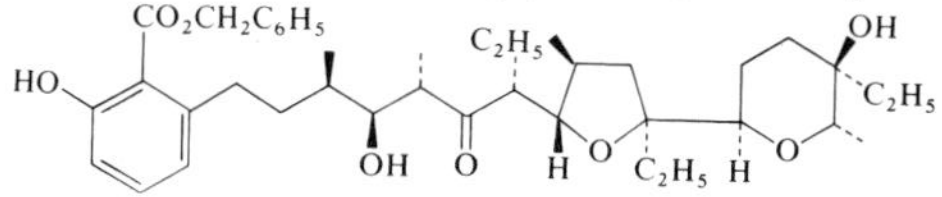

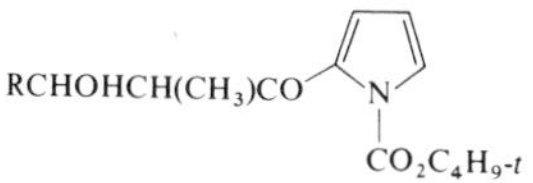

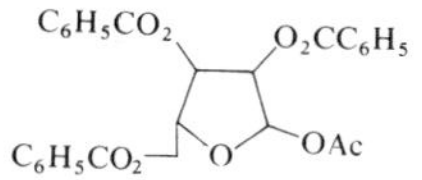

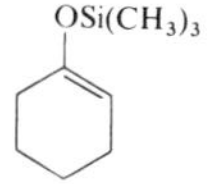

[a] R and R′ in the product are the groups R and R′ in the aldehyde RCHO or in the acetal $RCH(OR')_2$.

[b] The yield given is after hydrolysis.

[c] This reaction has been carried out recently with other reagents, giving yields ranging from 30% (with LTMP) to 92% (with LTMP, $(n\text{-}C_4H_9)_3SnCl$) and *erythro*:*threo* ratios from 30:70 (with LTMP) to >97:3 (with $(n\text{-}C_4H_9)_2BOSO_2CF_3$, $(i\text{-}C_3H_7)_2NC_2H_5$). See refs. 230, 262, 271, and 273.

[d] This reaction has also been carried out with LDA; $C_5H_9B[C(CH_3)_2C_4H_9\text{-}i]OSO_2CF_3$, $(i\text{-}C_3H_7)_2NC_2H_5$; and LTMP, Cp_2ZrCl_2. See refs. 230, 262, and 273, respectively.

[e] This reaction has also been carried out with LDA, LCPA, LHMDS, and LTMP.[230]

[f] This reaction has been carried out recently with LCPA, LHMDS, and LTMP.[230]

[g] For recent reactions with other ketones or ketone equivalents, see ref. 275.

TABLE III. REACTION OF ALDEHYDE EQUIVALENTS WITH KETONES OR KETALS

No. of C Atoms	Ketone or Ketal	Aldehyde Equivalent	Reagent	Product(s)[a] and Yield(s) (%)	Refs.
C_3	CH_3COCH_3	$CH_3CH{=}NC_6H_{11}$	LDA	$RR'C{=}CHCHO$ (51)[b]	75
		$CH_3CH{=}CHOC_2H_5$	$BF_3 \cdot O(C_2H_5)_2$	$RR'C{=}C(CH_3)CHO$ (37)[b]	197
		$C_2H_5CH{=}CHOC_2H_5$	"	$RR'C{=}C(C_2H_5)CHO$ (28)[b]	197
	$(CH_3)_2C(OCH_3)_2$	$(CH_3)_2C{=}CH[OSi(CH_3)_3]$	$(CH_3)_3SiO_3SCF_3$	$RR'C(OCH_3)C(CH_3)_2CHO$ (85)	193a
	$(CH_3)_2C(OC_2H_5)_2$	$CH_2{=}CHOC_2H_5$	$BF_3 \cdot O(C_2H_5)_2$	$RR'C(OR'')CH_2CH(OR'')_2$ (67)	226,232
		$CH_3CH{=}CHOC_2H_5$	"	$RR'C(OR'')CH(CH_3)CH(OR'')_2$ (43)	226
C_4	$CH_3(C_2H_5)CO$	$CH_3CH{=}CHOC_2H_5$	$BF_3 \cdot O(C_2H_5)_2$	$RR'C{=}C(CH_3)CHO$ (33)[b]	197
		$C_2H_5CH{=}CHOC_2H_5$	"	$RR'C{=}C(C_2H_5)CHO$ (29)[b]	197
		$CH_3CH{=}NC_6H_{11}$	LDA	$RR'C(OH)CH_2CH{=}NC_6H_{11}$ (70)	244
		$CH_3CH{=}NC_6H_{11}$	LDA	$RR'C(OH)CH_2CH{=}NC_6H_{11}$ (70)	209
C_6		$CH_3CH{=}NC_6H_{11}$	LDA	$RR'C{=}CHCHO$ (60)[b]	75
C_8	$CH_3COC_6H_5$	$CH_3CH{=}NC_6H_{11}$	LDA	$RR'C(OH)CH_2CH{=}NC_6H_{11}$ (80)	3

	$CH_3(C_6H_5)C(OC_2H_5)_2$	$CH_2{=}CHOC_2H_5$	$BF_3 \cdot O(C_2H_5)_2$	$RR'C(OR'')CH_2CH(OR'')_2$ (27)	212
		$CH_2{=}CHCH{=}CHOC_2H_5$	$ZnCl_2$	$RR'C(OR'')CH_2CH{=}CHCH(OR'')_2$ (13)	212
	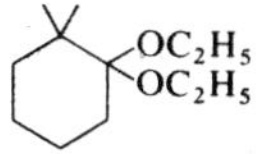 OC_2H_5 OC_2H_5	$CH_2{=}CHOC_2H_5$	$ZnCl_2$	$RR'C(OR'')CH_2CH(OR'')_2$ (94)	245
	$(CH_3)_2C{=}CH(CH_2)_2COCH_3$	$CH_3CH{=}NC_6H_{11}$	LDA	$RR'C{=}CHCHO$ (64)[b]	75
C_{10}	$CH_3CO(CH_2)_2CH(CO_2CH_3)C_3H_7\text{-}i$	$CH_3CH{=}NC_6H_{11}$	LDA	$RR'CHOHCH_2CHO$ (77)[b]	84
C_{13}	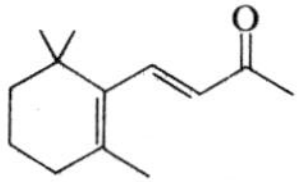 O	$CH_3CH{=}NC_6H_{11}$	LDA	$RR'C(OH)CH_2CH{=}NC_6H_{11}$ (80)	3
		$C_2H_5CH{=}NC_6H_{11}$	"	$RR'C{=}C(CH_3)CHO$ (50)[b]	79
	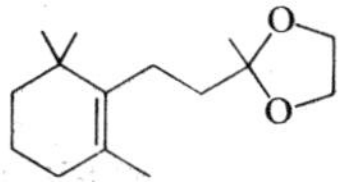 O O	$CH_2{=}CHOC_2H_5$	$BF_3 \cdot O(C_2H_5)_2$	$RR'C{=}CHCH$ 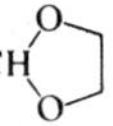 O O (−)	229
	$C_6H_5COC_6H_5$	$CH_3CH{=}NC_6H_{11}$	LDA	$RR'C(OH)CH_2CH{=}NC_6H_{11}$ (91)	3
		"	"	$RR'C{=}CHCHO$ (78–85)[b]	78
		$CH_3CH{=}NN(CH_3)_2$	"	$RR'C(OH)CH_2CH{=}NN(CH_3)_2$ (81)	87
		$CH_3CH{=}CHCH{=}NC_6H_{11}$	LDA	$RR'C(OH)CH_2CH{=}CHCH{=}NC_6H_{11}$ (22)	3
		$C_6H_5CH_2CH{=}NNHC_6H_5$	KNH_2	$RR'C(OH)CH(C_6H_5)CH{=}NNHC_6H_5$ (50–76)	94
C_{19}	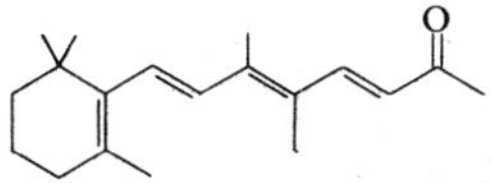 O	$C_2H_5CH{=}NC_6H_{11}$	LDA	$RR'C{=}C(CH_3)CHO$ (36)[b]	79

[a] R, R′, and R″ in the product are the groups R, R′, and R″ in the ketone RR′CO or in the ketal $RR'C(OR'')_2$.

[b] The yield given is after hydrolysis.

TABLE IV. REACTION OF KETONES OR THEIR EQUIVALENTS WITH KETONES OR KETALS

No. of C Atoms	Ketone or Ketal	Ketone or Equivalent	Reagent	Product(s)[a] and Yield(s) (%)	Refs.
C_3	CH_3COCH_3	CH_3COCH_3	$NaNH_2$, LiBr	$RR'COHCH_2COCH_3$ (60)	58
		$CH_2{=}C(OCH_3)CH_3$	$BF_3\cdot O(C_2H_5)_2$	$RR'C(OR'')CH_2COCH_3$ (21)	200
		$CH_2{=}C(OC_2H_5)CH_3$	"	" (45)	200
		$CH_2{=}C(OCH_2C_6H_5)CH_3$	"	" (52)	200
		$(CH_3)_2C{=}NNHSO_2C_6H_4CH_3$-*p*	*n*-C_4H_9Li	$RR'C(OH)CH_2C(CH_3){=}NNHSO_2C_6H_4CH_3$-*p* (57)	97
		$(CH_3)_2C{=}NNHSO_2$-(2,4,6-trimethylphenyl) 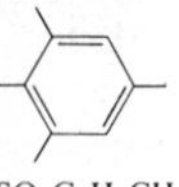	"	$RR'C(OH)CH_2C(CH_3){=}NNHSO_2$-(2,4,6-trimethylphenyl) (95) 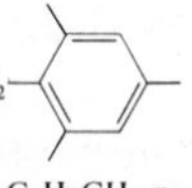	98
		$CH_3(C_2H_5)C{=}NNHSO_2C_6H_4CH_3$-*p*	*n*-C_4H_9Li	$RR'C(OH)CH_2C(CH_3){=}NNHSO_2C_6H_4CH_3$-*p* (61)	97
		$CH_2{=}CHCOCH_3$	$(n\text{-}C_4H_9)_3B$	$RR'C(OH)CH(C_5H_{11}\text{-}n)COCH_3$ (15)	137
		"	$(n\text{-}C_4H_9)_2BSC_6H_5$	$RR'C(OH)CH(CH_2SC_6H_5)COCH_3$ (20)	137
		$C_2H_5COC_2H_5$	$NaNH_2$, LiBr	$RR'C(OH)CH(CH_3)COC_2H_5$ (44)	58
		"	LDA	$RR'C(OH)CH(CH_3)COC_2H_5$ (46)	246
		$CH_3COC_2H_5$	$(C_2H_5)_2NLi$	$RR'C(OH)CH_2COC_2H_5$ (40)	48
		$CH_3COC_3H_7$-*n*	"	$RR'C(OH)CH_2COC_3H_7$-*n* (38)	48
		"	$(i\text{-}C_4H_9)_2NLi$	" (40)	46
		$CH_3COC_3H_7$-*i*	$(C_2H_5)_2NLi$	$RR'C(OH)CH_2COC_3H_7$-*i* (42)	48
		$CH_3COC_4H_9$-*i*	"	$RR'C(OH)CH_2COC_4H_9$-*i* (47)	48
		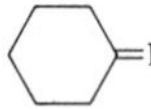 Cyclohexanone ${=}NNHSO_2C_6H_4CH_3$-*p*	*n*-C_4H_9Li	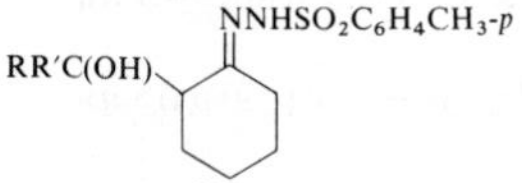 2-[RR'C(OH)]cyclohexanone $NNHSO_2C_6H_4CH_3$-*p* (79)	97
		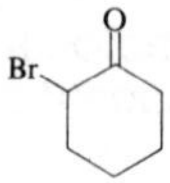 2-Bromocyclohexanone (Br, O)	$(C_2H_5)_2AlCl$, Zn	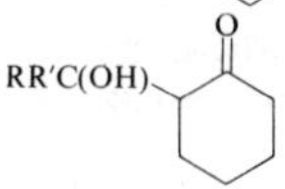 2-[RR'C(OH)]cyclohexanone (O) (75)	131
		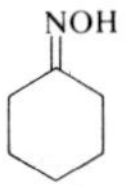 Cyclohexanone oxime (NOH)	*n*-C_4H_9Li	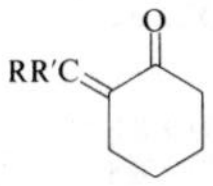 2-(RR'C=)cyclohexanone (O) (48)[b]	101

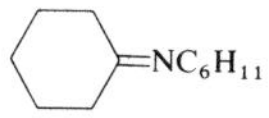	BCl_3, $N(C_2H_5)_3$	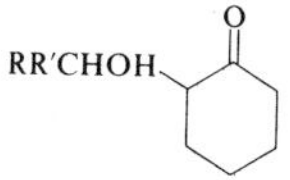 (31)	147
$BrCH_2COC_4H_9$-*t*	Mg	$RR'C(OH)CH_2COC_4H_9$-*t* (58)	114,115
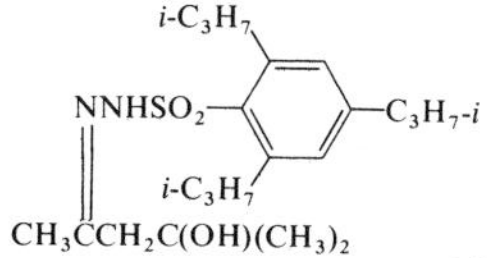	*n*-C_4H_9Li	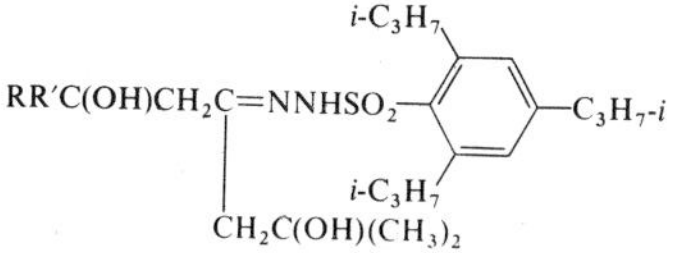 (31)	98
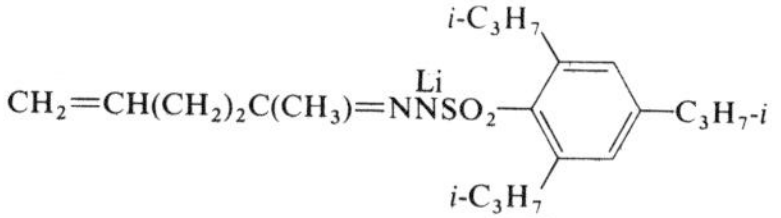	1. *n*-C_4H_9Li 2. $(CH_3)_2CO$, HOAc	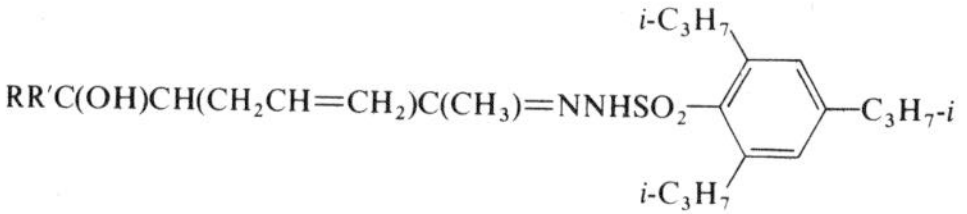 (88)	99
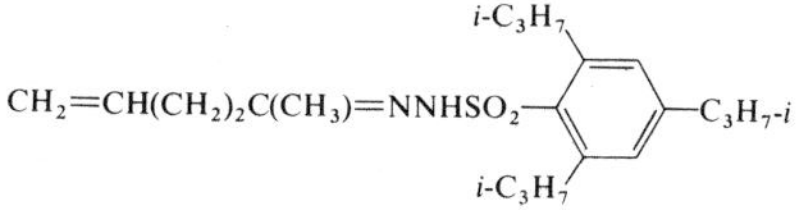	1. *n*-C_4H_9Li 2. HOAc 3. *n*-C_4H_9Li 4. $(CH_3)_2CO$, HOAc	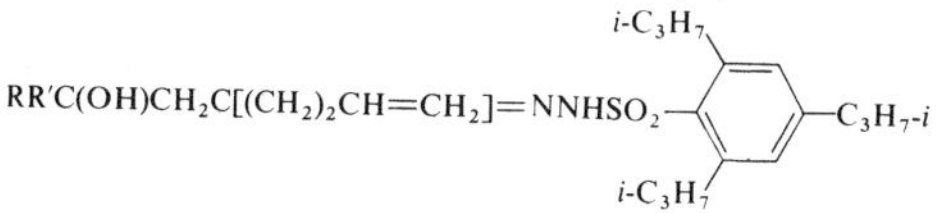 (68)	99
$BrC(CH_3)_2COC_3H_7$-*i*	"	$RR'C(OH)C(CH_3)_2COC_3H_7$-*i* (42)	114
$N_2CHCOC_6H_5$	$(n\text{-}C_4H_9)_3B$	$RR'C(OH)CH(C_4H_9\text{-}n)COC_6H_5$ (42)	136
i-$C_3H_7COC_3H_7$-*i*	LDA	$RR'C(OH)C(CH_3)_2COC_3H_7$-*i* (40)	246
$CH_3(n\text{-}C_5H_{11})C{=}NOTHP$	"	$RR'C(OH)CH_2(n\text{-}C_5H_{11})C{=}NOTHP$ (95)	102
$C_6H_5(CH_3)C{=}NNHSO_2C_6H_4CH_3$-*p*	*n*-C_4H_9Li	$RR'C(OH)CH_2C(C_6H_5){=}NNHSO_2C_6H_4CH_3$-*p* (80)	180
$CH_3COC_6H_5$	$NaNH_2$, LiBr	$RR'C(OH)CH_2COC_6H_5$ (56)	58
"	$(C_2H_5)_2AlN$	" (65)	132
$CH_2{=}C[OSi(CH_3)]C_6H_5$	$TiCl_4$	" (73)	173
$BrCH(C_2H_5)COC_4H_9$-*t*	Mg	$RR'C(OH)CH(C_2H_5)COC_4H_9$-*t* (70)	116
$CH_3(C_6H_5)C{=}NOTHP$	LDA	$RR'C(OH)CH_2(C_6H_5)C{=}NOTHP$ (95)	102
$C_6H_5CH{=}C[OSi(CH_3)_3]CH_3$	$TiCl_4$	$RR'C(OH)CH(C_6H_5)COCH_3$ (60)	173,174
$CH_3CO(CH_2)_2C_6H_5$	9-BBN(OSO_2CF_3), 2,6-lutidine	$RR'C(OH)CH(CH_2C_6H_5)COCH_3$ (63)	37

TABLE IV. REACTION OF KETONES OR THEIR EQUIVALENTS WITH KETONES OR KETALS (*Continued*)

No. of C Atoms	Ketone or Ketal	Ketone or Equivalent	Reagent	Product(s)[a] and Yield(s) (%)	Refs.
C_3 (*Contd.*)	$(CH_3)_2C(OCH_3)_2$ (*Contd.*)	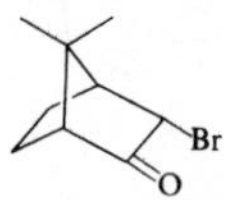 Br, O	$(C_2H_5)_2AlCl$, Zn	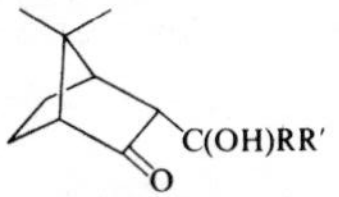 C(OH)RR′, O (79)	131
		p-$CH_3C_6H_4CH(CH_3)CH_2C[OSi(CH_3)_3]{=}CH_2$	$TiCl_4$	$RR'C(OH)CH_2COCH_2CH(CH_3)C_6H_4CH_3$-p (88)	177
		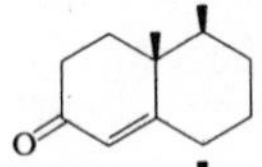 O	LDA, $ZnCl_2$	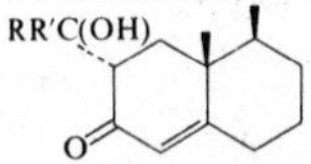 RR′C(OH), O (94)	247
		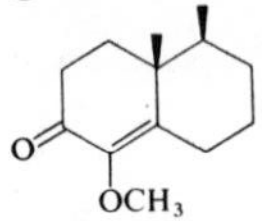 O, OCH_3	LDA, $ZnCl_2$	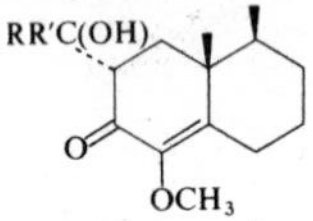 RR′C(OH), O, OCH_3 (86)	247
		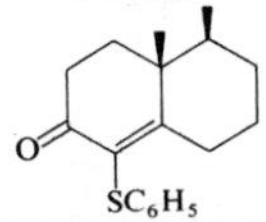 O, SC_6H_5	LDA, $ZnCl_2$	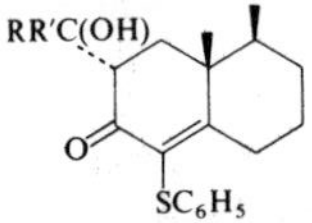 RR′C(OH), O, SC_6H_5 (67)	247
		$(CH_3)_2C{=}C[OSi(CH_3)_3]CH_3$	$(CH_3)_3SiO_3SCF_3$	$RR'C(OCH_3)C(CH_3)_2COCH_3$ (87)	193a
		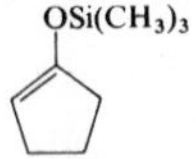 $OSi(CH_3)_3$	"	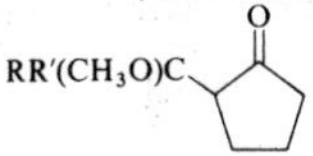 $RR'(CH_3O)C$, O (87)	193a
		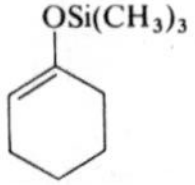 $OSi(CH_3)_3$	"	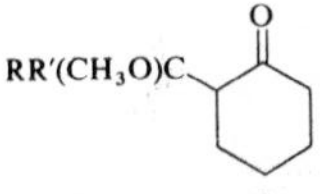 $RR'(CH_3O)C$, O (87)	193a
		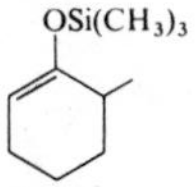 $OSi(CH_3)_3$	"	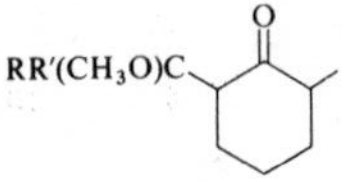 $RR'(CH_3O)C$, O (89)	193a

		$CH_2{=}C[OSi(CH_3)_3]C_6H_5$	$(CH_3)_3SiO_3SCF_3$	$RR'C(OCH_3)CH_2COC_6H_5$ (96)	193a
		$C_6H_5CH{=}C[OSi(CH_3)_3]CH_3$	$TiCl_4$	$RR'C(OR'')CH(C_6H_5)COCH_3$ (66)	180
		$CH_2{=}C[OSi(CH_3)_3]C_6H_5$	"	$RR'C(OR'')CH_2COC_6H_5$ (62)	180
C_4	$CH_3COCH{=}CH_2$	$(CH_3)_2C{=}NOH$	n-C_4H_9Li	$RR'C{=}CHCOCH_3$ (17)[b]	101
		$CH_3COCH{=}C(CH_3)_2$	LDA	$RR'C(OH)CH_2COCH{=}C(CH_3)_2$ (45)	50
		$CH_3COC_4H_9$-i	"	$RR'C{=}CHCOC_4H_9$-i (40)	50
		$CH_3COC_4H_9$-t	LDA, $ZnCl_2$	$RR'C(OH)CH_2COC_4H_9$-t (70)	248
		$CH_3COCH_2SOC_6H_5$	NaH, n-C_4H_9Li	$RR'C(OH)CH_2COCH_2SOC_6H_5$ (65)	240
	$CH_3COC_2H_5$	$CH_3COC_2H_5$	$(i$-$C_4H_9)_2NLi$	$RR'C(OH)CH_2COC_2H_5$ (70)	46
		"	$(C_2H_5)_2NLi$	" (62)	48
C_5	$CH_3COCO_2C_2H_5$	$BrCH_2COC_4H_9$-t	Mg	$RR'C(OH)CH_2COC_4H_9$-t (66)	113,114
		$CH_2{=}C[OSi(CH_3)_3]CH_3$	$TiCl_4$	$RR'C(OH)CH_2COCH_3$ (55)	176,177
		$CH_3CH{=}C[OSi(CH_3)_3]CH_3$	"	$RR'C(OH)CH(CH_3)COCH_3$ (87)	176
		$CH_3CH{=}C[OSi(CH_3)_3]C_2H_5$	"	$RR'C(OH)CH(CH_3)COC_2H_5$ (87)	176
		$OSi(CH_3)_3$ [1-cyclopentenyl]	"	$RR'C(OH)$ [2-cyclopentanonyl] (76)	176,177
		$OSi(CH_3)_3$ [1-cyclohexenyl]	"	$RR'C(OH)$ [2-cyclohexanonyl] (88)	176,177
		$CH_2{=}C[OSi(CH_3)_3]C_6H_5$	"	$RR'C(OH)CH_2C_6H_5$ (63)	176
	$CH_3CO_2CH_2COCH_3$	$OSi(CH_3)_3$ [1-cyclohexenyl]	$TiCl_4$	$RR'C(OH)$ [2-cyclohexanonyl] (34)	177
	$(C_2H_5)_2CO$	CH_3COCH_3	$NaNH_2$, LiBr	$RR'C(OH)CH_2COCH_3$ (55)	58
		"	LDA	" (10)	246
		$BrCH_2COC_4H_9$-t	Mg	$RR'C(OH)CH_2COC_4H_9$-t (48)	116
		$BrCH(CH_3)COC_4H_9$-t	"	$RR'C(OH)CH(CH_3)COC_4H_9$-t (53)	116
		$BrCH(C_2H_5)COC_4H_9$-t	Mg	$RR'C(OH)CH(C_2H_5)COC_4H_9$-$t$ (18)	116
	$(C_2H_5)_2C(OCH_3)_2$	$(CH_3)_3SiO$—[cyclobutene]—$OSi(CH_3)_3$	$BF_3\cdot O(C_2H_5)_2$	$RR'C(OR'')$—[2-$OSi(CH_3)_3$-cyclobutanone] (92)	185
	$CH_3COC_3H_7$-n	$CH_3COC_3H_7$-n	$(i$-$C_4H_9)_2NLi$	$RR'C(OH)CH_2COC_3H_7$-n (87)	46
		$CH_3COC_3H_7$-n	$(C_2H_5)_2NLi$	$RR'C(OH)CH_2COC_3H_7$-n (70)	48
		$CH_3COC_3H_7$-n	$NaNH_2$	" (64)	46
	$CH_3COC_3H_7$-i	$(CH_3)_2C{=}NN$[2-CH_2OCH_3-pyrrolidine]	n-C_4H_9Li	$RR'C^*(OH)CH_2COCH_3$ (58)[b] (47% e.e.)	91

TABLE IV. REACTION OF KETONES OR THEIR EQUIVALENTS WITH KETONES OR KETALS (*Continued*)

No. of C Atoms	Ketone or Ketal	Ketone or Equivalent	Reagent	Product(s)[a] and Yield(s) (%)	Refs.
C_5 (*Contd.*)	$CH_3COC_3H_7$-*i*	$CH_3COC_3H_7$-*i*	$(i\text{-}C_4H_9)_2NLi$	$RR'C(OH)CH_2COC_3H_7$-*i* (73)	46
		"	$(C_2H_5)_2NLi$	" (53)	48
C_6	$CH_3CO(CH_2)_3CN$	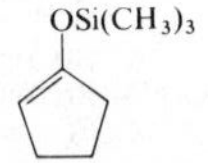 $OSi(CH_3)_3$	$TiCl_4$	$RR'C(OH)$ 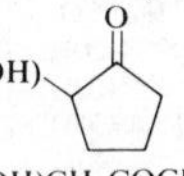 O (73)	177
		$CH_2{=}C(CH_3)OSi(CH_3)_3$	$TiCl_4$	$RR'C(OH)CH_2COCH_3$ (52)	177
	$CH_3O_2C(CH_2)_2C(CH_3)(OCH_3)_2$	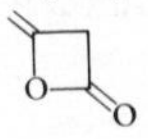 O O	$TiCl_4$	$RR'C(OR'')CH_2COCH_2CO_2R''$ (76)	215
	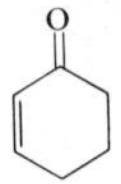 O	$CH_3COC_4H_9$-*t*	LDA	$RR'C(OH)CH_2COC_4H_9$-*t* (93)	249
		$CH_3COCH_2SOC_6H_5$	NaH, *n*-C_4H_9Li	$RR'C(OH)CH_2COCH_2SOC_6H_5$ (75)	240
	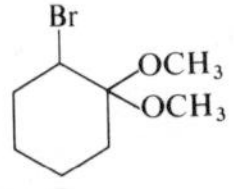 Br OCH_3 OCH_3	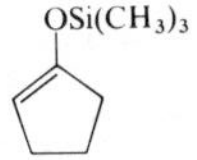 $OSi(CH_3)_3$	$TiCl_4$	$RR'C(OCH_3)$ 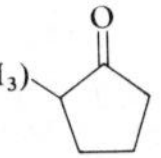 O (—)	182
	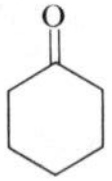 O	$(CH_3)_2C{=}NOH$	*n*-C_4H_9Li	$RR'C{=}CHCOCH_3$ (51)[b]	101
		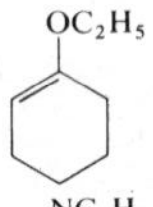 OC_2H_5	$BF_3\cdot(C_2H_5)_2O$	$RR'CHOH$ 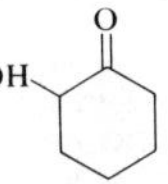 O (35)	200
		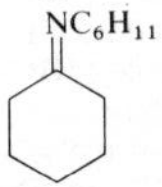 NC_6H_{11}	BCl_3, $N(C_2H_5)_3$	$RR'CHOH$ 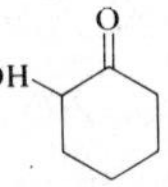 O (28)	147

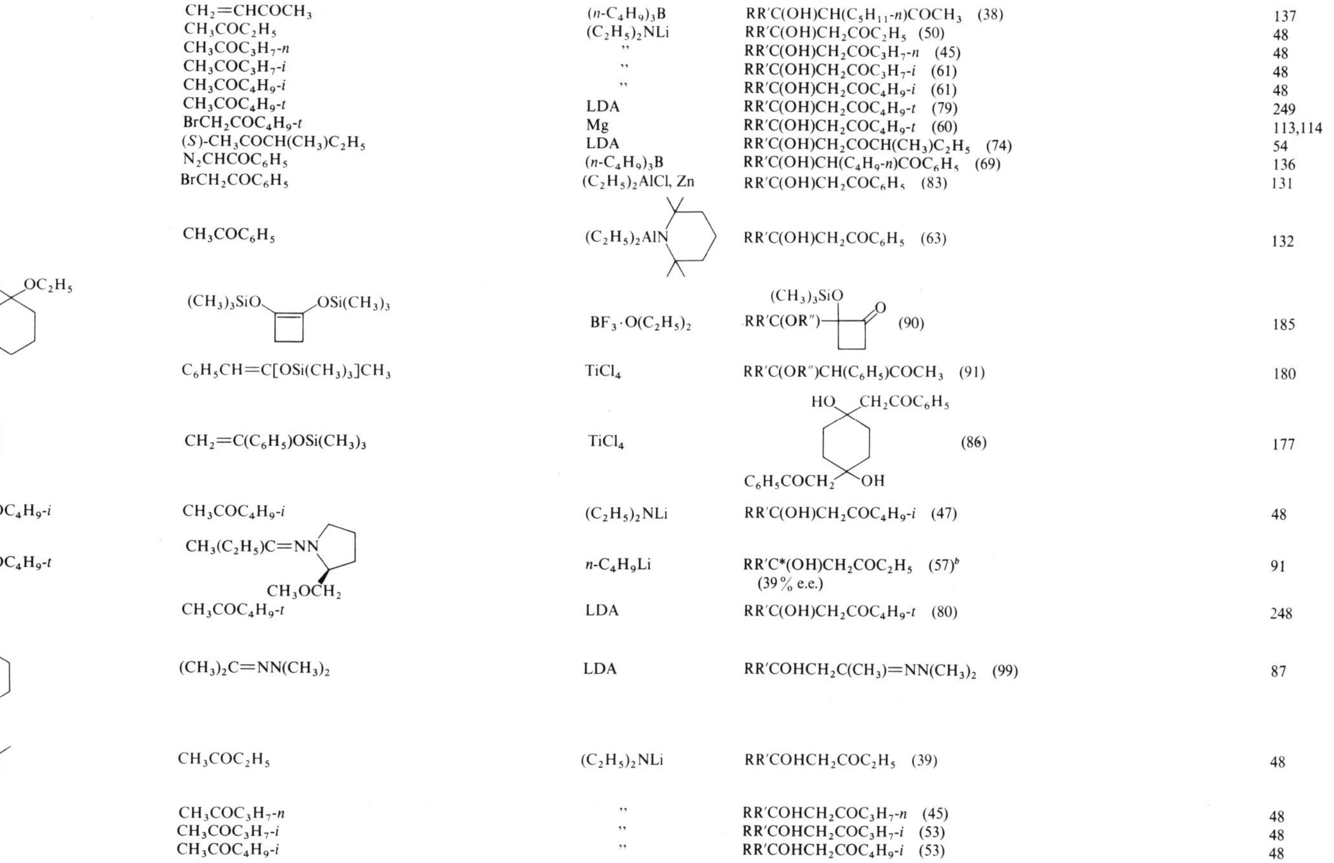

Carbonyl compound	Enolate source	Reagent	Product (Yield, %)	Refs.
	$CH_2{=}CHCOCH_3$	$(n\text{-}C_4H_9)_3B$	$RR'C(OH)CH(C_5H_{11}\text{-}n)COCH_3$ (38)	137
	$CH_3COC_2H_5$	$(C_2H_5)_2NLi$	$RR'C(OH)CH_2COC_2H_5$ (50)	48
	$CH_3COC_3H_7\text{-}n$	"	$RR'C(OH)CH_2COC_3H_7\text{-}n$ (45)	48
	$CH_3COC_3H_7\text{-}i$	"	$RR'C(OH)CH_2COC_3H_7\text{-}i$ (61)	48
	$CH_3COC_4H_9\text{-}i$	"	$RR'C(OH)CH_2COC_4H_9\text{-}i$ (61)	48
	$CH_3COC_4H_9\text{-}t$	LDA	$RR'C(OH)CH_2COC_4H_9\text{-}t$ (79)	249
	$BrCH_2COC_4H_9\text{-}t$	Mg	$RR'C(OH)CH_2COC_4H_9\text{-}t$ (60)	113,114
	$(S)\text{-}CH_3COCH(CH_3)C_2H_5$	LDA	$RR'C(OH)CH_2COCH(CH_3)C_2H_5$ (74)	54
	$N_2CHCOC_6H_5$	$(n\text{-}C_4H_9)_3B$	$RR'C(OH)CH(C_4H_9\text{-}n)COC_6H_5$ (69)	136
	$BrCH_2COC_6H_5$	$(C_2H_5)_2AlCl$, Zn	$RR'C(OH)CH_2COC_6H_5$ (83)	131
	$CH_3COC_6H_5$	$(C_2H_5)_2AlN$ (2,2,6,6-tetramethylpiperidide)	$RR'C(OH)CH_2COC_6H_5$ (63)	132
C_2H_5O OC_2H_5 (cyclohexanone diethyl ketal)	$(CH_3)_3SiO$–cyclobutene–$OSi(CH_3)_3$	$BF_3\cdot O(C_2H_5)_2$	$RR'C(OR'')$–[1-$(CH_3)_3SiO$-cyclobutanone] (90)	185
	$C_6H_5CH{=}C[OSi(CH_3)_3]CH_3$	$TiCl_4$	$RR'C(OR'')CH(C_6H_5)COCH_3$ (91)	180
(1,4-cyclohexanedione)	$CH_2{=}C(C_6H_5)OSi(CH_3)_3$	$TiCl_4$	HO, $CH_2COC_6H_5$ / $C_6H_5COCH_2$, OH (cyclohexane) (86)	177
$CH_3COC_4H_9\text{-}i$	$CH_3COC_4H_9\text{-}i$	$(C_2H_5)_2NLi$	$RR'C(OH)CH_2COC_4H_9\text{-}i$ (47)	48
$CH_3COC_4H_9\text{-}t$	$CH_3(C_2H_5)C{=}NN$(pyrrolidine, 2-CH_3OCH_2)	$n\text{-}C_4H_9Li$	$RR'C^*(OH)CH_2COC_2H_5$ (57)[b] (39% e.e.)	91
	$CH_3COC_4H_9\text{-}t$	LDA	$RR'C(OH)CH_2COC_4H_9\text{-}t$ (80)	248
C_7 (3-methylcyclohex-2-enone)	$(CH_3)_2C{=}NN(CH_3)_2$	LDA	$RR'COHCH_2C(CH_3){=}NN(CH_3)_2$ (99)	87
(2-methylcyclohexanone)	$CH_3COC_2H_5$	$(C_2H_5)_2NLi$	$RR'COHCH_2COC_2H_5$ (39)	48
	$CH_3COC_3H_7\text{-}n$	"	$RR'COHCH_2COC_3H_7\text{-}n$ (45)	48
	$CH_3COC_3H_7\text{-}i$	"	$RR'COHCH_2COC_3H_7\text{-}i$ (53)	48
	$CH_3COC_4H_9\text{-}i$	"	$RR'COHCH_2COC_4H_9\text{-}i$ (53)	48

TABLE IV. Reaction of Ketones or Their Equivalents with Ketones or Ketals (*Continued*)

No. of C Atoms	Ketone or Ketal	Ketone or Equivalent	Reagent	Product(s)[a] and Yield(s) (%)	Refs.
C_7 (*Contd.*)	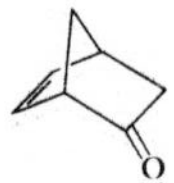	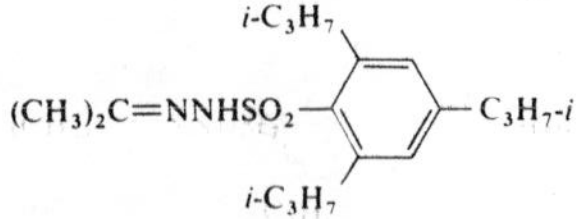 $(CH_3)_2C{=}NNHSO_2C_6H_2(C_3H_7\text{-}i)_3$ (2,4,6)	$n\text{-}C_4H_9Li$	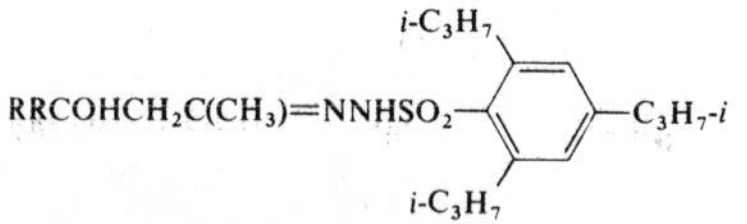 $RRCOHCH_2C(CH_3){=}NNHSO_2C_6H_2(C_3H_7\text{-}i)_3$ (2,4,6) (56)	99
	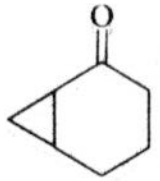	$CH_3COC_4H_9\text{-}t$	LDA	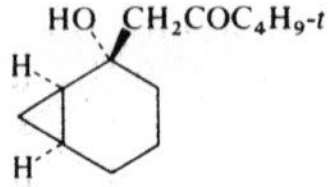 HO, $CH_2COC_4H_9\text{-}t$ (25)	249
	$(C_2H_5O_2C)_2CO$	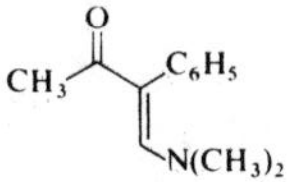 CH_3, C_6H_5, $N(CH_3)_2$	LTMP	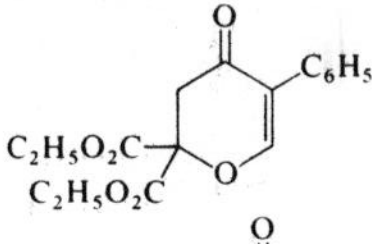 $C_2H_5O_2C$, $C_2H_5O_2C$, C_6H_5 (50)	237
	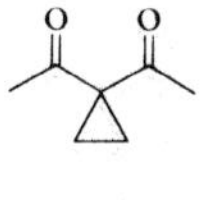	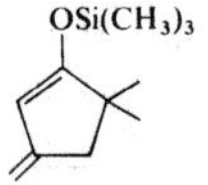 $OSi(CH_3)_3$	$TiCl_4$	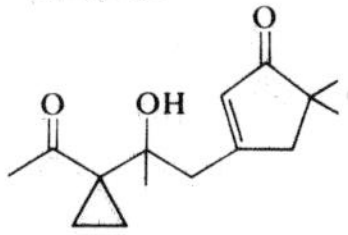 OH (30)	192
	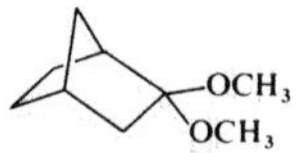 OCH_3, OCH_3	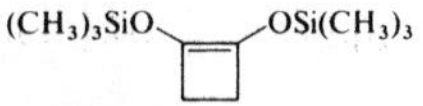 $(CH_3)_3SiO$, $OSi(CH_3)_3$	$BF_3\cdot O(C_2H_5)_2$	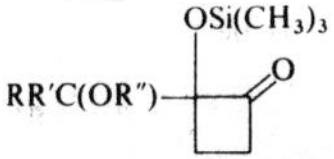 $RR'C(OR'')$, $OSi(CH_3)_3$ (60)	185
	$CH_3CO(CH_2)_2CO_2C_2H_5$	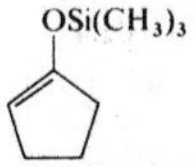 $OSi(CH_3)_3$	$TiCl_4$	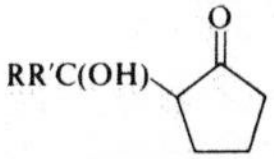 $RR'C(OH)$ (73)	176,177
		$CH_3CH{=}C[OSi(CH_3)_3]C_2H_5$	"	$RR'C(OH)CH(CH_3)COC_2H_5$ (86)	176,177
		$CH_2{=}C(C_6H_5)OSi(CH_3)_3$	"	$RR'C(OH)CH_2COC_6H_5$ (41)	177
	$CH_3CO(CH_2)_3CO_2CH_3$	$CH_3CH{=}C[OSi(CH_3)_3]C_2H_5$	$TiCl_4$	$RR'C(OH)CH(CH_3)COC_2H_5$ (71)	176,177

		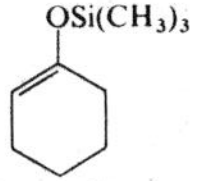 $OSi(CH_3)_3$ (structure)	"	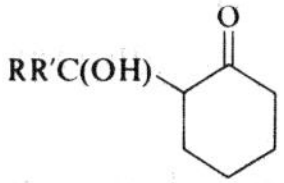 $RR'C(OH)$ (structure), O (69)	176,177
C_8	$CH_3COC_6H_5$	CH_3COCH_3	$NaNH_2$, LiBr	$RR'C(OH)CH_2COCH_3$ (40)	58
		$CH_2{=}C(OC_2H_5)C_6H_5$	$BF_3 \cdot O(C_2H_5)_2$	$RR'C(OR'')CH_2COC_6H_5$ (37)	200
		$CH_3COC_2H_5$	$(C_2H_5)_2NLi$	$RR'C(OH)CH_2COC_2H_5$ (45)	48
		$CH_3COC_3H_7$-*n*	"	$RR'C(OH)CH_2COC_3H_7$-*n* (45)	48
		$CH_3COC_3H_7$-*i*	"	$RR'C(OH)CH_2COC_3H_7$-*i* (42)	48
		$CH_3COC_4H_9$-*t*	"	$RR'C(OH)CH_2COC_4H_9$-*t* (55)	48
		$CH_3(n\text{-}C_5H_{11})C{=}NN(CH_3)_2$	LDA	$RR'C(OH)CH_2C(C_5H_{11}\text{-}n){=}NN(CH_3)_2$ (80)	87
		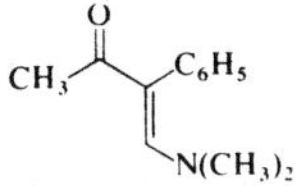 O, CH_3, C_6H_5, $N(CH_3)_2$ (structure)	LTMP	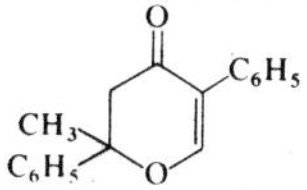 O, C_6H_5, CH_3, C_6H_5, O (structure) (51)[b]	237
	p-$O_2NC_6H_4COCH_3$	$CH_2{=}C[OSi(CH_3)_3]C_6H_5$	$TiCl_4$	*p*-$O_2NC_6H_4C(CH_3)(OH)CH_2COC_6H_5$ (52)	177
	$CH_3CO(CH_2)_3CO_2C_2H_5$	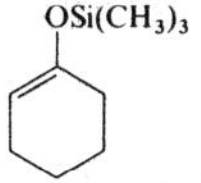 $OSi(CH_3)_3$ (structure)	$TiCl_4$	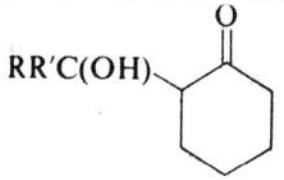 $RR'C(OH)$ (structure), O (52)	176,177
	O, C_2H_5 (structure)	$CH_2{=}C[OSi(CH_3)_3]C_6H_5$	"	$RR'C(OH)CH_2COC_6H_5$ (38)	176,177
	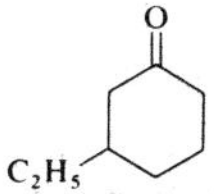	$CH_3COC_4H_9$-*t*	LDA	$RR'C(OH)CH_2COC_4H_9$-*t* (—)	249
	$CH_3COC_6H_{13}$-*n*	$CH_3COC_6H_{13}$-*n*	$(i\text{-}C_4H_9)_2AlOC_6H_5$, C_5H_5N	$RR'C{=}CHCOC_6H_{13}$-*n* (85)	133
	CH_3COCH_2OTHP	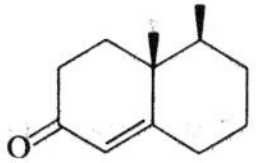 O (structure)	LDA, $ZnCl_2$	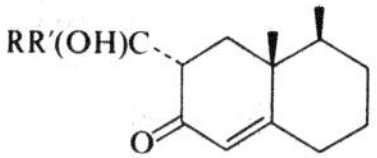 $RR'(OH)C$ (structure), O (86)	247
		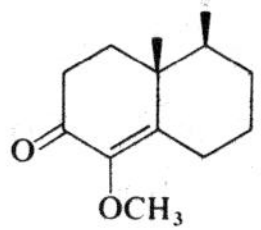 O, OCH_3 (structure)	"	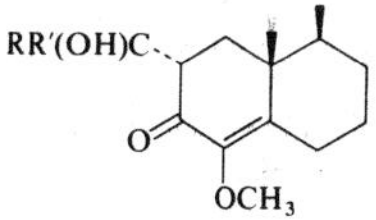 $RR'(OH)C$ (structure), O, OCH_3 (65)	247

TABLE IV. REACTION OF KETONES OR THEIR EQUIVALENTS WITH KETONES OR KETALS (*Continued*)

No. of C Atoms	Ketone or Ketal	Ketone or Equivalent	Reagent	Product(s)[a] and Yield(s) (%)	Refs.
C_8 (*Contd.*)	CH_3COCH_2OTHP	SC_6H_5	LDA, $ZnCl_2$	$RR'(OH)C$ … SC_6H_5 (62)	247
C_9	$CH_3COCH_2C_6H_5$	$CH_3COCH_2C_6H_5$	LDA	$RR'C(OH)CH_2COC_6H_5$ (52)	46
		$CH_3COC_2H_5$	$(C_2H_5)_2NLi$	$RR'C(OH)CH_2COC_2H_5$ (21)	48
		$CH_3COC_3H_7$-*n*	"	$RR'C(OH)CH_2COC_3H_7$-*n* (27)	48
		$CH_3COC_3H_7$-*i*	"	$RR'C(OH)CH_2COC_3H_7$-*i* (31)	48
		$CH_3COC_4H_9$-*i*	LDA	$RR'C(OH)CH_2COC_4H_9$-*i* (44)	48
	p-$CH_3C_6H_4COCH_3$.	$CH_2{=}C(CH_3)OSi(CH_3)_3$	$TiCl_4$	$RR'C{=}CHCOCH_3$ (53)	177
		$CH_3COCH{=}C(CH_3)_2$	LDA	$RR'C(OH)CH_2COCH{=}C(CH_3)_2$ (60)	50
	$C_6H_5SeCH_2COCH_3$	Cyclohexanone	LDA, $ZnCl_2$	C(OH)RR′ (67)	272
C_{10}	$CH_3COCH{=}CHC_6H_5$	CH_3COCH_3	$NaNH_2$, LiBr	$RR'C(OH)CH_2COCH_3$ (98)	58
	t-C_4H_9–(cyclohexanone)	$CH_3COC_4H_9$-*t*	*i*-C_3H_7MgCl	$RR'C(OH)CH_2COC_4H_9$-*t* (—)	108
		$CH_3COCH_2C_3H_7$-*i*	"	$RR'C(OH)CH_2COCH_2C_3H_7$-*i* (—)	108
		$CH_3COCH_2C_4H_9$-*t*	"	$RR'C(OH)CH_2COCH_2C_4H_9$-*t* (—)	108
		$(i\text{-}C_3H_7)_2CO$	"	$RR'C(OH)C(CH_3)_2COC_3H_7$-*i* (—)	108
		$C_2H_5COC_4H_9$-*t*	"	$RR'C(OH)CH(CH_3)COC_4H_9$-*t* (—)	108
		n-$C_3H_7COC_4H_9$-*t*	"	$RR'C(OH)CH(C_2H_5)COC_4H_9$-*t* (—)	108
		$(CH_3)_2CHCH_2COC_4H_9$-*t*	"	$RR'C(OH)CH[CH(CH_3)_2]COC_4H_9$-*t* (—)	108
		i-$C_3H_7CH_2COCH_2C_3H_7$-*i*	"	$RR'C(OH)CH[CH(CH_3)_2]COCH_2$ -)	108
	p-$CH_3CO_2C_6H_4COCH_3$	$CH_3CH{=}C(C_2H_5)OSi(CH_3)_3$	$TiCl_4$	$RR'C(OH)CH(CH_3)COC_2H_5$ (36)	177
		$CH_2{=}C(C_6H_5)OSi(CH_3)_3$	"	$RR'C(OH)CH_2COC_6H_5$ (11)	177

	$C_6H_5(CH_2)_2C(CH_3)(OCH_3)_2$	[structure: O, =O]	$TiCl_4$	$RR'C(OR'')CH_2COCH_2CO_2R''$ (87)	215
	[structure: CH_3O, SO_3CH_3, $C(OCH_3)_2CH_3$]	[structure: $(CH_3)_3SiO$, $OSi(CH_3)_3$]	$BF_3 \cdot O(C_2H_5)_2$	[structure: OCH_3, CH_3O_3S, CH_3, O, O] (50)	250
C_{11}	p-$C_2H_5OCOC_6H_4COCH_3$	$CH_2{=}C(C_2H_5)OSi(CH_3)_3$	$TiCl_4$	$RR'C(OH)CH(CH_3)COC_2H_5$ (68)	177
	[structure: S, COC_6H_5]	$CH_3(C_6H_5)C{=}NOH$	n-C_4H_9Li	$RR'C(OH)CH_2C({=}NOH)C_6H_5$ (40)	100
		p-$CH_3C_6H_4(CH_3)C{=}NOH$	"	$RR'C(OH)CH_2C({=}NOH)C_6H_4CH_3$-$p$ (16)	100
C_{12}	[structure: OCH_3, OCH_3]	[structure: $(CH_3)_3SiO$, $OSi(CH_3)_3$]	$BF_3 \cdot O(C_2H_5)_2$	[structure: $RR'C(OR'')$, $OSi(CH_3)_3$, =O] (92)	185
C_{13}	p-$ClC_6H_4COC_6H_5$	$CH_3(C_6H_5)C{=}NNHCOC_6H_5$	n-C_4H_9Li	$RR'C(OH)CH_2C({=}NNHCOC_6H_5)C_6H_5$ (84)	95
	$C_6H_5COC_6H_5$	$CH_3(p$-$CH_3C_6H_4)C{=}NNHSO_2C_6H_4CH_3$-$p$	"	$RR'C(OH)CH_2C({=}NNHSO_2C_6H_4CH_3$-$p)C_6H_4CH_3$-$p$ (61)	95
		$(CH_3)_2C{=}NOH$	n-C_4H_9Li	$RR'C{=}CHCOCH_B$ (55)[b]	101
		[structure: O, CH_3, CH_3, $N(CH_3)_2$]	LTMP	[structure: O, CH_3, C_6H_5, O, C_6H_5] (55)[b]	237
		[structure: $NN(CH_3)_2$]	LDA	[structure: $RR'C(OR'')$, $NN(CH_3)_2$] (>95)	87
		$CH_3(n$-$C_5H_{11})C{=}NN(CH_3)_2$	"	$RR'C(OH)CH_2C(C_5H_{11}$-$n){=}NN(CH_3)_2$ (95)	87
		$CH_3COCH_2SOC_6H_5$	NaH, n-C_4H_9Li	$RR'C(OH)CH_2COCH_2SOC_6H_5$ (91)	240
		[structure: O, C_2H_5, C_6H_5, $N(CH_3)_2$]	LTMP	[structure: O, CH_3, C_6H_5, C_6H_5, O, C_6H_5] (51)[b]	237
		$CH_3(C_6H_5)C{=}NNHC_6H_5$	n-C_4H_9Li	$RR'C(OH)CH_2C({=}NNHC_6H_5)C_6H_5$ (75)	93
		$CH_3(C_6H_5)C{=}NNHSO_2C_6H_5$	"	$RR'C(OH)CH_2C({=}NNHSO_2C_6H_5)C_6H_5$ (78)	95

TABLE IV. REACTION OF KETONES OR THEIR EQUIVALENTS WITH KETONES OR KETALS (*Continued*)

No. of C Atoms	Ketone or Ketal	Ketone or Equivalent	Reagent	Product(s)[a] and Yield(s) (%)	Refs.
C_{13} (*Contd.*)	$C_6H_5COC_6H_5$	$CH_3(p\text{-}CH_3C_6H_4)C{=}NOH$	$n\text{-}C_4H_9Li$	$RR'C(OH)CH_2C({=}NOH)C_6H_4CH_3\text{-}p$ (74)	100
		$C_6H_5CH_2(C_6H_5)C{=}NNHCOC_6H_5$	"	$RR''C(OH)CH(C_6H_5)C({=}NNHCOC_6H_5)C_6H_5$ (47)	95
	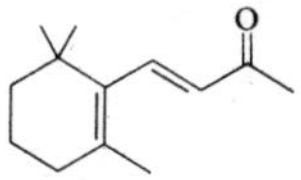	$CH_3COCH{=}C(OCH_3)CH_3$	LDA	$RR'C(OH)CH_2COCH{=}C(OCH_3)CH_3$ (—)	51
	(−)-CH_3COCO_2– 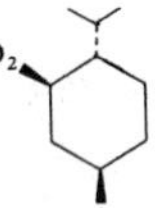	$(CH_3)_2C{=}C[OSi(C_2H_5)_3]CH_3$	$TiCl_4$	$RR'C(OH)C(CH_3)_2COCH_3$ (100)	179
		$CH_2{=}C[OSi(CH_3)_3]C_4H_9\text{-}t$	"	$RR'C(OH)CH_2COC_4H_9\text{-}t$ (77)	179
		$CH_2{=}C[OSi(CH_3)_3]C_6H_5$	"	$RR'C(OH)CH_2COC_6H_5$ (75)	179
C_{14}	$p\text{-}CH_3C_6H_4COC_6H_5$	$CH_3(C_6H_5)C{=}NNHCOC_6H_5$	$n\text{-}C_4H_9Li$	$RR'C(OH)CH_2C({=}NNHCOC_6H_5)C_6H_5$ (100)	95
		$CH_3(C_6H_5)C{=}NNHSO_2C_6H_5$	"	$RR'C(OH)CH_2C({=}NNHSO_2C_6H_5)C_6H_5$ (72)	95
		$CH_3(p\text{-}CH_3C_6H_4)C{=}NOH$	"	$RR'C(OH)CH_2C({=}NOH)C_6H_4CH_3\text{-}p$ (17)	100
		$CH_3(p\text{-}CH_3C_6H_4)C{=}NNHSO_2C_6H_4CH_3\text{-}p$	"	$RR'C(OH)CH_2C({=}NNHSO_2C_6H_4CH_3\text{-}p)C_6H_4CH_3\text{-}p$ (54)	95
C_{15}	$C_6H_5CH{=}CHCOC_6H_5$	$CH_3COC_4H_9\text{-}t$	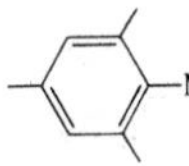 –MgBr	$RR'C(OH)CH_2COC_4H_9\text{-}t$ (—)	109

		$C_2H_5COC_4H_9$-*t*	—MgBr	$RR'COHCH(CH_3)COC_4H_9$-*t* (—)	109
	$C_6H_5CH_2COCH_2C_6H_5$	$OSi(CH_3)_3$	$TiCl_4$	$RR'C(OH)$ (61)	173,174
		$OSi(CH_3)_3$	"	$RR'C(OH)$ (64)	173,174
		$CH_2{=}C[OSi(CH_3)_3]C_6H_5$	"	$RR'C(OH)CH_2COC_6H_5$ (39)	174
	$CH_3C(CH_2)_2CO(CH_2)_2C_6H_5$ (S, S)	$CH_2{=}C(CH_3)OSi(CH_3)_3$	$TiCl_4$	$RR'C(OH)CH_2COCH_3$ (76)	177
C_{18}	$C_6H_5COCO_2$—	$CH_2{=}C[OSi(CH_3)_3]C_4H_9$-*t*	$TiCl_4$	$RR'C(OH)CH_2COC_4H_9$-*t* (84)	179
		$CH_2{=}C[OSi(CH_3)_3]C_6H_5$	"	$RR'C(OH)CH_2COC_6H_5$ (85)	179
		$(CH_3)_2C{=}C[OSi(C_2H_5)_3]CH_3$	"	$RR'C(OH)CH_2COCH_3$ (95)	179

[a] R, R′, and R″ in the product are the groups R, R′ and R″ in the ketone RR′CO or in the ketal RRC(OR″)$_2$.
[b] The yield given is after hydrolysis.

TABLE V. INTRAMOLECULAR REACTIONS

No. of C Atoms	Diketone or Equivalent	Reagent	Product and Yield (%)	Refs.
C_{10}	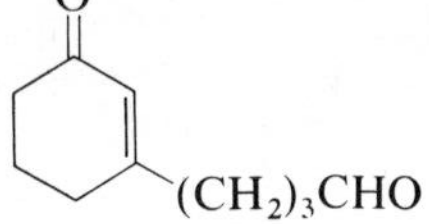 $(CH_2)_3CHO$	$(CH_3)_2AlSC_6H_5$	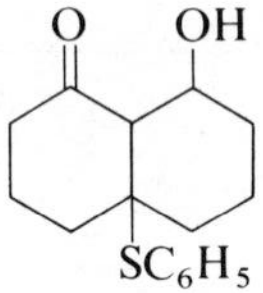 O, OH, SC_6H_5 (60)	134
C_{11}	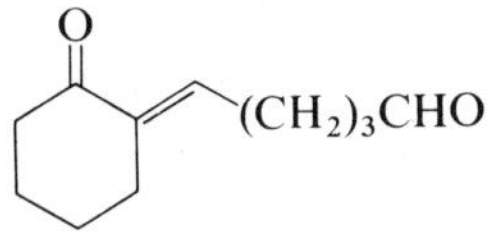 $(CH_2)_3CHO$	$(CH_3)_2AlSC_6H_5$	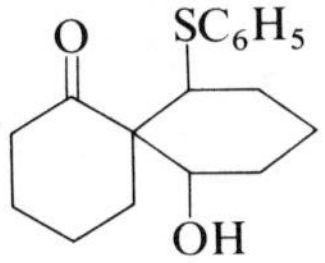 O, SC_6H_5, OH (94)	134
	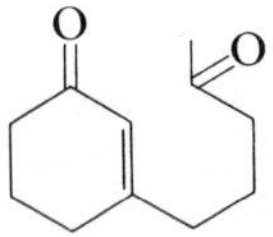 O, O	$(CH_3)_2CuLi$	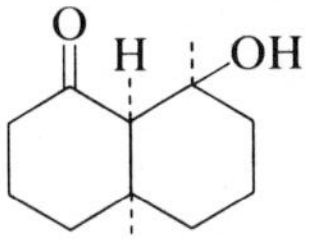 O, H, OH (63)	64
	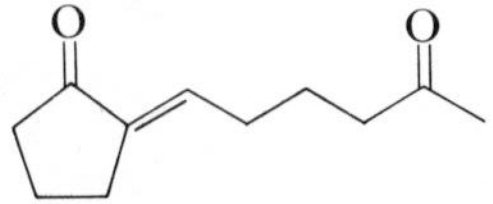 O, O	$(CH_3)_2CuLi$	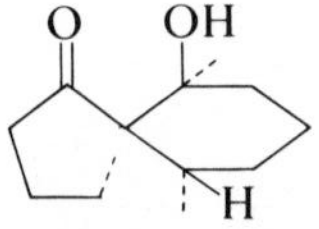 O, OH, H (—)	64

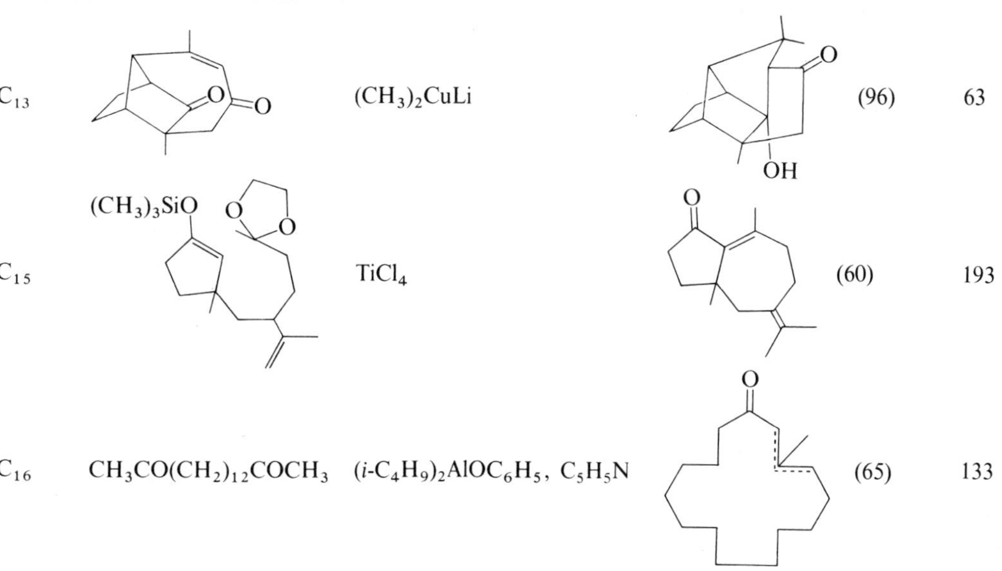

C_{13}	(structure)	$(CH_3)_2CuLi$	(structure)	(96)	63
C_{15}	(structure, $(CH_3)_3SiO$)	$TiCl_4$	(structure)	(60)	193
C_{16}	$CH_3CO(CH_2)_{12}COCH_3$	$(i\text{-}C_4H_9)_2AlOC_6H_5$, C_5H_5N	(structure)	(65)	133

REFERENCES

[1] A. T. Nielsen and W. J. Houlihan, *Org. Reactions*, **16**, 1 (1968).

[2] H. O. House, *Modern Synthetic Reactions*, 2nd ed., Benjamin, New York, 1972, Chap. 10.

[3] G. Wittig and H. D. Frommeld, *Angew. Chem., Int. Ed. Engl.*, **2**, 683 (1963).

[4] G. Wittig and H. Reiff, *Angew. Chem., Int. Ed. Engl.*, **7**, 7 (1968).

[5] H. Reiff, *Newer Methods Prep. Org. Chem.*, **6**, 48 (1971).

[6] G. Wittig, *Top. Curr. Chem.*, **67**, 1 (1976).

[7] G. Wittig, *Fortschr. Chem. Forsch.*, **67**, 1 (1976).

[8] T. Mukaiyama and K. Narasaka, *Chemistry*, **32**, 576 (1977).

[9] E. Pohjala, *Taydennyskoulutuskurssi-Suom. Kem. Seura*, **1978**, 20 (Synt. Org. Kem. Uusia Menetelmia), 149 [*C.A.*, **91**, 107627r (1979)].

[10] K. Narasaka, *Yuki Gosei Kagaku Kyokai Shi*, **37**, 307 (1979) [*C.A.*, **91**, 55592d (1979)].

[11] Z. G. Hajos, in *Carbon-Carbon Bond Formation*, R. L. Augustine, Ed., Vol. I, Marcel Dekker, New York, 1979, Chap. 1, pp. 1–84.

[12] D. Caine, *ibid.*, Vol. I, Chap. 2, pp. 264–276.

[13] P. A. Bartlett, *Tetrahedron*, **36**, 2 (1980).

[14] P. Fellmann and J. E. Dubois, *ibid.*, **34**, 1349 (1978).

[15] H. O. House, D. S. Crumrine, A. Y. Teranishi, and H. D. Olmstead, *J. Am. Chem. Soc.*, **95**, 3310 (1973).

[16] W. A. Kleschick, C. T. Buse, and C. H. Heathcock, *ibid.*, **99**, 247 (1977).

[17] T. Inoue and T. Mukaiyama, *Bull. Chem. Soc. Jpn.*, **53**, 174 (1980).

[18] S. Masamune, S. Mori, D. V. Horn, and D. W. Brooks, *Tetrahedron Lett.*, **1979**, 1665.

[19] E. A. Jeffery, A. Meisters, and T. Mole, *J. Organomet. Chem.*, **74**, 373 (1974).

[20] R. E. Ireland and A. K. Willard, *Tetrahedron Lett.*, **1975**, 3975.

[21] J. Mulzer, J. Segner, and G. Brüntrup, *ibid.*, **1977**, 4651.

[22] J. Mulzer, A. Pointner, A. Chucholowski, and G. Brüntrup, *Chem. Commun.*, **1979**, 52.

[23] J. Mulzer, G. Brüntrup, J. Finke, and M. Zippel, *J. Am. Chem. Soc.*, **101**, 7723 (1979).

[24] T. H. Chan, T. Aida, P. W. K. Lau, V. Gorys, and D. N. Harpp, *Tetrahedron Lett.*, **1979**, 4029.

[25] A. I. Meyers and P. J. Reider, *J. Am. Chem. Soc.*, **101**, 2501 (1979).

[26] F. DiNinno, T. R. Beattie, and B. G. Christensen, *J. Org. Chem.*, **42**, 2960 (1977).

[27] J. A. Aimetti and M. S. Kellogg, *Tetrahedron Lett.*, **1979**, 3805.

[28] J. E. Dubois and M. Dubois, *ibid.*, **1967**, 4215.

[29] J. E. Dubois and M. Dubois, *Chem. Commun.*, **1968**, 1567.

[30] J. E. Dubois and J. F. Fort, *Tetrahedron*, **28**, 1653 (1972).

[31] J. E. Dubois and J. F. Fort, *ibid.*, **28**, 1665 (1972).

[32] J. E. Dubois and P. Fellmann, *C.R. Hebd. Seances Acad. Sci., Ser. C*, **274**, 1307 (1972).

[33] J. E. Dubois and P. Fellmann, *Tetrahedron Lett.*, **1975**, 1225.

[34] D. A. Evans, E. Vogel, and J. V. Nelson, *J. Am. Chem. Soc.*, **101**, 6120 (1979).

[35] M. Hirama and S. Masamune, *Tetrahedron Lett.*, **1979**, 2225.

[36] T. Mukaiyama and T. Inoue, *Chem. Lett.*, **1976**, 559.

[37] T. Inoue, T. Uchimaru, and T. Mukaiyama, *ibid.*, **1977**, 153.

[38] D. E. Van Horn and S. Masamune, *Tetrahedron Lett.*, **1979**, 2229.

[39] C. T. Buse and C. H. Heathcock, *J. Am. Chem. Soc.*, **99**, 8109 (1977).

[40] C. H. Heathcock and C. T. White, *J. Am. Chem. Soc.*, **101**, 7076 (1979).

[41] T. Nakata, G. Schmid, B. Vranesic, M. Okigawa, T. Smith-Palmer, and Y. Kishi, *ibid.*, **100**, 2933 (1978).

[42] T. Fukuyama, K. Akasaka, D. S. Karanewsky, C.-L. J. Wang, G. Schmid, and Y. Kishi, *ibid.*, **101**, 262 (1979).

[43] C. H. Heathcock, M. C. Pirrung, C. T. Buse, J. P. Hagen, S. D. Young, and J. E. Sohn, *ibid.*, **101**, 7077 (1979).

[43a] D. A. Evans and L. R. McGee, *Tetrahedron Lett.*, **21**, 3975 (1980).

[44] J. d'Angelo, *Tetrahedron*, **32**, 2979 (1976).

[45] G. Stork, *Pure Appl. Chem.*, **43**, 553 (1975).

[46] M. Gaudemar, *C.R. Hebd. Seances Acad. Sci., Ser. C*, **279**, 961 (1974).

[47] G. Stork, G. A. Kraus, and G. A. Garcia, *J. Org. Chem.*, **39**, 3459 (1974).

[48] F. Gaudemar-Bardone and M. Gaudemar, *J. Organomet. Chem.*, **104**, 281 (1976).

[49] M. Matsui, T. Kitahara, and K. Takagi, *Jpn. Kokai Tokkyo Koho*, 79, 48,737 [*C.A.*, **91**, 157942k (1979)].

[50] O. S. Park, Y. Grillasca, G. A. Garcia, and L. A. Maldonado, *Synth. Commun.*, **7**, 345 (1977).

[51] G. Stork and G. A. Kraus, *J. Am. Chem. Soc.*, **98**, 2351 (1976).

[52] S. Torii, T. Okamoto, and S. Kadono, *Chem. Lett.*, **1977**, 495.

[53] S. Torii, T. Inokuchi, and H. Ogawa, *Bull. Chem. Soc. Jpn.*, **52**, 1233 (1979).

[54] D. Seebach, V. Ehrig, and M. Teschner, *Justus Liebigs Ann. Chem.*, **1976**, 1357.

[55] I. Kuwajima, T. Sato, M. Arai, and N. Minami, *Tetrahedron Lett.*, **1976**, 1817.

[56] I. Kuwajima, T. Inoue, and T. Sato, *ibid.*, **1978**, 4887.

[57] G. Stork and J. d'Angelo, *J. Am. Chem. Soc.*, **96**, 7114 (1974).

[58] M. Gaudemar, *C.R. Hebd. Seances Acad. Sci., Ser. C*, **278**, 533 (1974).

[59] H. O. House and J. M. Wilkins, *J. Org. Chem.*, **41**, 4031 (1976), and references cited therein.

[60] K. K. Heng and R. A. J. Smith, *Tetrahedron Lett.*, **1975**, 589.

[61] K. K. Heng and R. A. J. Smith, *Tetrahedron*, **35**, 425 (1979).

[62] G. Stork and M. Isobe, *J. Am. Chem. Soc.*, **97**, 6260 (1975).

[63] J. E. McMurry and S. J. Isser, *ibid.*, **94**, 7132 (1972).

[64] F. Näf, R. Decorzant, and W. Thommen, *Helv. Chim. Acta*, **58**, 1808 (1975).

[65] G. Stork and P. F. Hudrlik, *J. Am. Chem. Soc.*, **90**, 4462 (1968).

[66] G. Stork and P. F. Hudrlik, *ibid.*, **90**, 4464 (1968).

[67] G. H. Posner, J. J. Sterling, C. E. Whitten, C. M. Lentz, and D. J. Brunelle, *ibid.*, **97**, 107 (1975).

[68] J. E. Dubois and C. Lion, *C.R. Hebd. Seances Acad. Sci., Ser. C*, **280**, 217 (1975).

[69] I. Kuwajima and E. Nakamura, *J. Am. Chem. Soc.*, **97**, 3257 (1975).

[70] R. Noyori, K. Yokoyama, J. Sakata, I. Kuwajima, E. Nakamura, and M. Shimizu, *ibid.*, **99**, 1265 (1977).

[71] E. Nakamura, K. Hashimoto, and I. Kuwajima, *J. Org. Chem.*, **42**, 4166 (1977).

[72] G. Wittig, H. J. Schmidt, and H. Renner, *Chem. Ber.*, **95**, 2377 (1962).

[73] G. Wittig and H. D. Frommeld, *Chem. Ber.*, **97**, 3541 (1964).

[74] G. Wittig and H. D. Frommeld, *Chem. Ber.*, **97**, 3548 (1964).

[75] G. Wittig and P. Suchanek, *Tetrahedron* (*Suppl.*), No. 8, 347 (1966).

[76] G. Büchi and H. Wüest, *Helv. Chim. Acta*, **50**, 2440 (1967).

[77] N. Y. Grigor'eva and A. V. Semenovskii, *Izv. Akad. Nauk SSSR, Ser. Khim*, **1976**, 2644 [*C.A.*, **86**, 89115n (1977)].

[78] G. Wittig and A. Hesse, *Org. Synth.*, **50**, 66 (1970).

[79] S. P. Tanis, R. H. Brown, and K. Nakanishi, *Tetrahedron Lett.*, **1978**, 869.

[80] G. Büchi and H. Wüest, *J. Org. Chem.*, **34**, 1122 (1969).

[81] A. I. Meyers and C. C. Shaw, *Tetrahedron Lett.*, **1974**, 717.

[82] A. I. Meyers, C. C. Shaw, and D. Horne, *ibid.*, **1975**, 1745.

[83] A. I. Meyers and R. S. Brinkmeyer, *ibid.*, **1975**, 1749.

[84] W. G. Dauben, G. H. Beasley, M. D. Broadhurst, B. Muller, D. J. Peppard, P. Pesnelle, and C. Suter, *J. Am. Chem. Soc.*, **97**, 4973 (1975).

[85] R. F. Borch, A. J. Evans, and J. J. Wade, *ibid.*, **99**, 1612 (1977).

[86] E. J. Corey and D. Enders, *Chem. Ber.*, **111**, 1337 (1978).

[87] E. J. Corey and D. Enders, *Chem. Ber.*, **111**, 1362 (1978).

[88] E. J. Corey and D. Enders, *Tetrahedron Lett.*, **1976**, 73.

[89] E. J. Corey, D. Enders, and M. G. Bock, *ibid.*, **1976**, 7.

[90] E. J. Corey and D. Enders, *ibid.*, **1976**, 11.

[91] H. Eichenauer, E. Friedrich, W. Lutz, and D. Enders, *Angew. Chem., Int. Ed. Engl.*, **17**, 206 (1978).

[92] D. Enders, H. Eichenauer, and R. Pieter, *Chem. Ber.*, **112**, 3703 (1979).

[93] D. C. Reames, C. E. Harris, L. W. Dasher, R. M. Sandifer, W. M. Hollinger, and C. F. Beam, *J. Heterocycl. Chem.*, **12**, 779 (1975).

[94] F. E. Henoch, K. G. Hampton, and C. R. Hauser, *J. Am. Chem. Soc.*, **89**, 463 (1967).

[95] R. M. Sandifer, S. E. Davis, and C. F. Beam, *Synth. Commun.*, **6**, 339 (1976).

[96] R. H. Shapiro, M. F. Lipton, K. J. Kolonko, R. L. Buswell, and L. A. Capuano, *Tetrahedron Lett.*, **1975**, 1811.

[97] M. F. Lipton and R. H. Shapiro, *J. Org. Chem.*, **43**, 1409 (1978).

[98] R. M. Adlington and A. G. M. Barrett, *Chem. Commun.*, **1978**, 1071.

[99] R. M. Adlington and A. G. M. Barrett, *ibid.*, **1979**, 1122.

[100] C. F. Beam, C. A. Park, D. C. Reames, S. A. Miller, and C. R. Hauser, *J. Chem. Eng. Data*, **23**, 183 (1978).

[101] W. G. Kofron and M. K. Yeh, *J. Org. Chem.*, **41**, 439 (1976).

[102] H. E. Hensley and R. Lohr, *Tetrahedron Lett.*, **1978**, 1415.

[103] J. Colonge, *Bull. Soc. Chim. Fr.*, [5] **1**, 1101 (1934).

[104] J. Colonge, *ibid.*, [5] **5**, 98 (1938).

[105] V. V. Chelintsev and A. V. Pataraya, *J. Gen. Chem.* (*USSR*), **11**, 461 (1941) [*C.A.*, **35**, 6571 (1941)].

[106] V. I. Aksenova, *Uch. Zap. Saratov. Gos. Univ. N. G. Chernyschevskogo, Sb. Nauch. Rabot Studentov*, No. 2, 92 (1939) [*C.A.*, **35**, 6238 (1941)].

[107] A. T. Nielsen, C. Gibbons, and C. A. Zimmerman, *J. Am. Chem. Soc.*, **73**, 4696 (1951).

[108] Y. Maroni-Barnaud, P. Maroni, and R. Cantagrel, *Bull. Soc. Chim. Fr.*, **1971**, 4051.

[109] J. Bertrand, N. Cabrol, L. Gorrichon-Guigon, and Y. Maroni-Barnaud, *Tetrahedron Lett.*, **1973**, 4683.

[110] Y. Koudsi and Y. Maroni-Barnaud, *ibid.*, **1975**, 2525.

[111] F. Näf and R. Decorzant, *Helv. Chim. Acta*, **57**, 1317 (1974).

[112] J-B. Wiel and F. Rouessac, *Chem. Commun.*, **1976**, 446.

[113] J. Colonge and J. Grenet, *C.R. Hebd. Seances Acad. Sci.*, **234**, 1181 (1952).

[114] J. Colonge and S. Grenet, *Bull. Soc. Chim. Fr.*, **20**, c 41 (1953).

[115] J. Colonge and J. Grenet, *ibid.*, **1954**, 1304.

[116] J. E. Dubois, G. Schutz, and J. M. Normant, *ibid.*, **1966**, 3578.

[117] J. A. Miller, M. H. Durand, and J. E. Dubois, *Tetrahedron Lett.*, **1965**, 2381.

[118] J. E. Dubois and J. Itzkowitch, *ibid.*, **1965**, 2839.

[119] T. Nakata and Y. Kishi, *ibid.*, **1978**, 2745.

[120] R. A. Auerbach, D. S. Crumrine, D. L. Ellison, and H. O. House, *Org. Synth.*, **54**, 49 (1974).

[121] D. A. Evans, C. E. Sacks, W. A. Kleschick, and T. R. Taber, *J. Am. Chem. Soc.*, **101**, 6789 (1979).

[122] I. J. Borowitz, E. W. R. Casper, R. K. Crouch, and K. C. Yee, *J. Org. Chem.*, **37**, 3873 (1972).

[123] G. Stork and M. Isobe, *J. Am. Chem. Soc.*, **97**, 4745 (1975).

[124] T. A. Spencer, R. W. Britton, and D. S. Watt, *ibid.*, **89**, 5727 (1967).

[125] T. Mukaiyama, T. Sato, S. Suzuki, T. Inoue, and H. Nakamura, *Chem. Lett.*, **1976**, 95.

[126] C. Schöpf and K. Thierfelder, *Justus Liebigs Ann. Chem.*, **518**, 127 (1935).

[127] M. Stiles, D. Wolf, and G. V. Hudson, *J. Am. Chem. Soc.*, **81**, 628 (1959).

[128] E. A. Jeffery, A. Meisters, and T. Mole, *J. Organomet. Chem.*, **74**, 365 (1974).

[129] E. A. Jeffery and A. Meisters, *ibid.*, **82**, 307 (1974).

[130] E. A. Jeffery and A. Meisters, *ibid.*, **82**, 315 (1974).

[131] K. Maruoka, S. Hashimoto, Y. Kitagawa, H. Yamamoto, and H. Nozaki, *J. Am. Chem. Soc.*, **99**, 7705 (1977).

[132] H. Nozaki, K. Oshima, K. Takai, and S. Ogawa, *Chem. Lett.*, **1979**, 379.

[133] J. Tsuji, T. Yamada, M. Kaito, and T. Mandai, *Tetrahedron Lett.*, **1979**, 2257.

[134] A. Itoh, S. Ozawa, K. Oshima, and H. Nozaki, *ibid.*, **1980**, 361.

[135] T. Mukaiyama, K. Inomata, and M. Muraki, *J. Am. Chem. Soc.*, **95**, 967 (1973).

[136] K. Inomata, M. Muraki, and T. Mukaiyama, *Bull. Chem. Soc. Jpn.*, **46**, 1807 (1973).

[137] M. Muraki, K. Inomata, and T. Mukaiyama, *ibid.*, **48**, 3200 (1975).

[138] D. J. Pasto and P. W. Wojtkowski, *Tetrahedron Lett.*, **1970**, 215.

[139] J. Hooz and J. N. Bridson, *Can. J. Chem.*, **50**, 2387 (1972).

[140] W. Fenzl, R. Köster, and H-J. Zimmermann, *Justus Liebigs Ann. Chem.*, **1975**, 2201.

[141] R. Köster, H.-J. Zimmermann, and W. Fenzl, *ibid.*, **1976**, 1116.

[142] W. Fenzl and R. Köster, *ibid.*, **1975**, 1322.

[143] W. Fenzl and R. Köster, *Angew. Chem., Int. Ed. Engl.*, **10**, 750 (1971).

[144] W. Fenzl, H. Kosfeld, and R. Köster, *Justus Liebigs Ann. Chem.*, **1976**, 1370.

[145] T. Mukaiyama, K. Saigo, and O. Takazawa, *Chem. Lett.*, **1976**, 1033.

[145a] I. Kuwajima, M. Kato, and A. Mori, *Tetrahedron Lett.*, **21**, 2745 (1980).

[146] I. Kuwajima, M. Kato, and A. Mori, *ibid.*, **21**, 4291 (1980).

[147] T. Sugasawa, T. Toyoda, and K. Sasakura, *Synth. Commun.*, **1979**, 515.

[148] T. Sugasawa and T. Toyoda, *Tetrahedron Lett.*, **1979**, 1423.

[149] T. Sugasawa, T. Toyoda, and K. Sasakura, *Synth. Commun.*, **1979**, 583.
[150] O. Isler and P. Schudel, *Adv. Org. Chem.*, **14**, 115 (1963).
[151] F. Effenberger, *Angew. Chem., Int. Ed. Engl.*, **8**, 295 (1969).
[152] T. Mukaiyama, *ibid.*, **16**, 817 (1977).
[153] H. O. House, L. J. Czuba, M. Gall, and H. D. Olmstead, *J. Org. Chem.*, **34**, 2324 (1969).
[154] H. O. House, M. Gall, and H. D. Olmstead, *ibid.*, **36**, 2361 (1971).
[155] R. E. Donaldson and P. L. Fuchs, *ibid.*, **42**, 2032 (1977).
[156] J. K. Rasmussen, *Synthesis*, **1977**, 91.
[157] S. Torkelson and C. Ainsworth, *ibid.*, **1976**, 722.
[158] S. Torkelson and C. Ainsworth, *ibid.*, **1977**, 431.
[159] G. Simchen and W. Kober, *ibid.*, **1976**, 259.
[160] H. Sakurai, K. Miyoshi, and Y. Nakadaira, *Tetrahedron Lett.*, **1977**, 2671.
[161] Y. Seki, A. Hidaka, S. Murai, and N. Sonoda, *Angew. Chem., Int. Ed. Engl.*, **16**, 174 (1977).
[162] E. Nakamura, K. Hashimoto, and I. Kuwajima, *Tetrahedron Lett.*, **1978**, 2079.
[163] E. W. Colvin, *Chem. Soc. Rev.*, **7**, 15 (1978).
[164] E. Nakamura, T. Murofushi, M. Shimizu, and I. Kuwajima, *J. Am. Chem. Soc.*, **98**, 2346 (1976).
[165] G. Stork and J. Singh, *ibid.*, **96**, 6181 (1974).
[166] R. K. Boeckman, Jr., *ibid.*, **95**, 6867 (1973).
[167] R. K. Boeckman, Jr., *J. Org. Chem.*, **38**, 4450 (1973).
[168] R. K. Boeckman, Jr., *J. Am. Chem. Soc.*, **96**, 6179 (1974).
[169] G. M. Rubottom, R. C. Mott, and D. S. Krueger, *Synth. Commun.*, **7**, 327 (1977).
[170] R. M. Coates, L. O. Sandefur, and R. D. Smillie, *J. Am. Chem. Soc.*, **97**, 1619 (1975).
[171] B. M. Trost and M. J. Bogdanowicz, *ibid.*, **95**, 5311 (1973).
[172] C. Girard, P. Amice, J. P. Barnier, and J. M. Conia, *Tetrahedron Lett.*, **1974**, 3329.
[173] T. Mukaiyama, K. Banno, and K. Narasaka, *J. Am. Chem. Soc.*, **96**, 7503 (1974).
[174] T. Mukaiyama, K. Narasaka, and K. Banno, *Chem. Lett.*, **1973**, 1011.
[175] K. Banno and T. Mukaiyama, *Bull. Chem. Soc. Jpn.*, **49**, 1453 (1976).
[176] K. Banno and T. Mukaiyama, *Chem. Lett.*, **1975**, 741.
[177] K. Banno, *Bull. Chem. Soc. Jpn.*, **49**, 2284 (1976).
[178] K. Banno and T. Mukaiyama, *Chem. Lett.*, **1976**, 279.
[179] I. Ojima, K. Yoshida, and S. Inaba, *Chem. Lett.*, **1977**, 429.
[180] T. Mukaiyama and M. Hayashi, *Chem. Lett.*, **1974**, 15.
[181] T. Mukaiyama, H. Ishihara, and K. Inomata, *Chem. Lett.*, **1975**, 527.
[182] H. Ishihara, K. Inomata, and T. Mukaiyama, *Chem. Lett.*, **1975**, 531.
[183] T. Ogawa, A. G. Pernet, and S. Hanessian, *Tetrahedron Lett.*, **1973**, 3543.
[184] T. Sato, M. Arai, and I. Kuwajima, *J. Am. Chem. Soc.*, **99**, 5827 (1977).
[185] E. Nakamura and I. Kuwajima, *ibid.*, **99**, 961 (1977).
[186] T. Mukaiyama and A. Ishida, *Chem. Lett.*, **1975**, 319.
[187] A. Ishida and T. Mukaiyama, *Bull. Chem. Soc. Jpn.*, **50**, 1161 (1977).
[188] A. Ishida and T. Mukaiyama, *Chem. Lett.*, **1975**, 1167.
[189] T. Mukaiyama and A. Ishida, *ibid.*, **1975**, 1201.
[190] A. Ishida and T. Mukaiyama, *ibid.*, **1977**, 467.
[191] A. Ishida and T. Mukaiyama, *Bull. Chem. Soc. Jpn.*, **51**, 2077 (1978).
[192] Y. Hayashi, M. Nishizawa, and T. Sakan, *Chem. Lett.*, **1975**, 387.
[193] A. Alexakis, M. J. Chapdelaine, G. H. Posner, and A. W. Runquist, *Tetrahedron Lett.*,**1978**, 4205.
[193a] S. Murata, M. Suzuki, and R. Noyori, *J. Am. Chem. Soc.*, **102**, 3248 (1980).
[194] J. W. Copenhaver, U.S. Pat. 2,543,312 (1951) [*C.A.*, **45**, 5447d (1951)].
[195] R. I. Hoaglin and D. H. Hirsh, U.S. Pat. 2,628,257 (1953) [*C.A.*, **48**, 1423d (1954)].
[196] R. I. Hoaglin, D. G. Kubler, and R. E. Leech, *J. Am. Chem. Soc.*, **80**, 3069 (1958).
[197] S. Satsumabayashi, K. Nakajo, and R. Soneda, *Bull. Chem. Soc. Jpn.*, **43**, 1586 (1970).
[198] H. Saikachi and H. Ogawa, *J. Am. Chem. Soc.*, **80**, 3652 (1958).
[199] E. Kitazawa, T. Imamura, K. Saigo, and T. Mukaiyama, *Chem. Lett.*, **1975**, 569.
[200] J. Albaigés, F. Camps, J. Castells, J. Fernàndez, and A. Guerrero, *Synthesis*, **1972**, 378.
[201] M. M-Cunradi and K. Pieroh, *U.S. Pat.*, 2,165,962 (1939) [*C.A.*, **33**, 3210^2 (1939)].
[202] L. S. Povarov, *Usp. Khim.*, **34**, 1489 (1965) [*C.A.*, **63**, 16145b (1965)].
[203] O. Isler, N. Montavon, R. Rüegg, and P. Zeller, *Justus Liebigs Ann. Chem.*, **603**, 129 (1957).

[204] O. Isler, H. Lindlar, M. Montavon, R. Rüegg, and P. Zeller, *Helv. Chim. Acta*, **39**, 249 (1956).
[205] O. Isler, M. Montavon, R. Rüegg, and P. Zeller, *ibid.*, **39**, 259 (1956).
[206] O. Isler, H. Gutmann, H. Lindlar, M. Montavon, R. Rüegg, G. Ryser, and P. Zeller, *ibid.*, **39**, 463 (1956).
[207] O. Isler, H. Lindlar, M. Montavon, R. Rüegg, G. Saucy, and P. Zeller, *ibid.*, **39**, 2041 (1956).
[208] O. Isler, H. Lindlar, M. Montavon, R. Rüegg, G. Sauchy, and P. Zeller, *ibid.*, **40**, 456 (1957).
[209] R. Rüegg, M. Montavon, G. Ryser, G. Saucy, U. Schwieter, and O. Isler, *ibid.*, **42**, 854 (1959).
[210] I. N. Nazarov and Zh. A. Krasnaya, *Dokl. Akad. Nauk SSSR*, **118**, 716 (1958) [*C.A.*, **52**, 11737e (1958)].
[211] I. N. Nazarov and Zh. A. Krasnaya, *ibid.*, **121**, 1034 (1958) [*C.A.*, **53**, 1183i (1959)].
[212] B. M. Mikhailov and G. S. T-Sarkisyan, *Zh. Obsch. Khim.*, **29**, 1642 (1959) [*C.A.*, **54**, 10952f (1960)].
[213] Zh. A. Krasnaya and V. F. Kucherov, *ibid.*, **32**, 64 (1962) [*C.A.*, **57**, 16671c (1962)].
[214] T. Mukaiyama, T. Izawa, and K. Saigo, *Chem. Lett.*, **1974**, 323.
[215] T. Izawa and T. Mukaiyama, *Chem. Lett.*, **1974**, 1189.
[216] T. Izawa and T. Mukaiyama, *Chem. Lett.*, **1975**, 161.
[217] T. Izawa and T. Mukaiyama, *Chem. Lett.*, **1978**, 409.
[218] S. N. Huckin and L. Weiler, *Tetrahedron Lett.*, **1971**, 4835.
[219] S. N. Huckin and L. Weiler, *Can. J. Chem.*, **52**, 2157 (1974).
[220] S. S. Yufit and V. F. Kucherov, *Izv. Akad. Nauk SSSR, Otd. Khim. Nauk*, **1960**, 1658 [*C.A.*, **55**, 9273h (1961)].
[221] T. Mukaiyama, J. Hanna, T. Inoue, and T. Sato, *Chem. Lett.*, **1974**, 381.
[222] T. Sato, J. Hanna, H. Nakamura, and T. Mukaiyama, *Bull. Chem. Soc. Jpn.*, **49**, 1055 (1976).
[223] T. Mukaiyama, M. Wada, and J. Hanna, *Chem. Lett.*, **1974**, 1181.
[224] H. C. Brown and G. W. Kabalka, *J. Am. Chem. Soc.*, **92**, 714 (1970); A. Suzuki, A. Arase, H. Matsumoto, M. Itoh, H. C. Brown, M. Rogic, and M. W. Rathke, *ibid.*, **89**, 5708 (1967).
[225] R. I. Hoaglin and D. H. Hirsh, *ibid.*, **71**, 3468 (1949).
[226] H. Normant and G. Martin, *Bull. Soc. Chim. Fr.*, **1963**, 1646.
[227] B. M. Mikhailov and L. S. Povarov, *Izv. Akad. Nauk SSSR, Otd. Khim. Nauk*, **1960**, 1903; [*C.A.*, **55**, 13409f (1961)].
[228] I. N. Nazarov, I. T. Nazarova, and T. V. Torgov, *Dokl. Akad. Nauk SSSR*, **122**, 82 (1958) [*C.A.*, **53**, 1123f (1959)].
[229] H. Pommer, German Pat. 1,031,301 (1958) [*C.A.*, **54**, 22712d (1960)].
[230] C. H. Heathcock, C. T. Buse, W. A. Kleschick, M. C. Pirrung, J. E. Sohn, and J. Lampe, *J. Org. Chem.*, **45**, 1066 (1980).
[231] B. M. Mikhailov and L. S. Povarov, *Izv. Akad. Nauk SSSR, Otd. Khim. Nauk*, **1957**, 1239 [*C.A.*, **52**, 6253f (1958)].
[232] B. M. Mikhailov and L. S. Povarov, *Zh. Obsch. Khim.*, **29**, 2079 (1959) [*C.A.*, **54**, 10851e (1960)].
[233] O. Isler, M. Montavon, R. Rüegg, G. Saucy, and P. Zeller, U.S. Pat. 2,827,481 (1958) [*C.A.*, **52**, 12904d (1958)].
[234] O. Isler, M. Montavon, R. Rüegg, G. Saucy, and P. Zeller, U.S. Pat. 2,827,482 (1958) [*C.A.*, **52**, 12906c (1958)].
[235] S. Danishefsky, K. Vaughan, R. C. Gadwood, and K. Tsuzuki, *J. Am. Chem. Soc.*, **102**, 4262 (1980).
[236] E. E. van Tamelen, J. P. Demers, E. G. Taylor, and K. Koller, *ibid.*, **102**, 5425 (1980).
[237] F. J. Vinick and H. E. Gschwend, *Tetrahedron Lett.*, **1978**, 315.
[238] P. Denniff and D. A. Whiting, *Chem. Commun.*, **1976**, 712.
[239] C. Byon, G. Büyüktür, P. Choay, and M. Gut, *J. Org. Chem.*, **42**, 3619 (1977).
[240] I. Kuwajima and H. Iwasawa, *Tetrahedron Lett.*, **1974**, 107.
[241] J. S. Hubbard and T. M. Harris, *J. Am. Chem. Soc.*, **102**, 2110 (1980).
[242] B. M. Mikhailov and L. S. Povarov, *Izv. Akad. Nauk SSSR, Otd. Khim. Nauk*, **1959**, 1948 [*C.A.*, **54**, 10952c (1960)].
[243] D. B. Collum, J. H. McDonald, III, and W. C. Still, *J. Am. Chem. Soc.*, **102**, 2120 (1980).
[244] J. C. Gilbert and K. R. Smith, *J. Org. Chem.*, **41**, 3883 (1976).
[245] J. P. de Sauza and A. M. R. Gonçalves, *ibid.*, **43**, 2068 (1978).
[246] F. Gaudemar-Bardone and M. Gaudemar, *Synthesis*, **1979**, 463.
[247] H. Hagiwara, H. Uda, and T. Kodama, *J. Chem. Soc., Perkin Trans. I*, **1980**, 963.
[248] H. O. House and M. J. Lusch, *J. Org. Chem.*, **42**, 183 (1977).

[249] H. O. House and K. A. J. Snoble, *ibid.*, **41**, 3076 (1976).
[250] W. K. Anderson and G. E. Lee, *Synth. Commun.*, **10**, 351 (1980).
[251] M. C. Pirrung and C. H. Heathcock, *J. Org. Chem.*, **45**, 1727 (1980).
[252] C. H. Heathcock, S. D. Young, J. P. Hagen, M. C. Pirrung, C. T. White, and D. VanDerveer, *J. Org. Chem.*, **45**, 3846 (1980).
[253] C. T. White and C. H. Heathcock, *J. Org. Chem.*, **46**, 191 (1981).
[254] C. H. Heathcock, C. T. White, J. J. Morrison, and D. VanDerveer, *J. Org. Chem.*, **46**, 1296 (1981).
[255] C. H. Heathcock, M. C. Pirrung, J. Lampe, C. T. Buse, and S. D. Young, *J. Org. Chem.*, **46**, 2290 (1981).
[256] C. H. Heathcock, in "*Comprehensive Carbanion Chemistry*," T. Durst and E. Buncel, Eds., Vol. II, Elsevier, Amsterdam, 1981.
[257] T. Mukaiyama, M. Murakami, T. Oriyama, and M. Yamaguchi, *Chem. Lett.*, **1981**, 1193.
[258] I. Kuwajima, M. Kato, and A. Mori, *Tetrahedron Lett.*, **21**, 4291 (1980).
[259] M. Wada, *Chem. Lett.*, **1981**, 153.
[260] D. A. Evans and T. R. Taber, *Tetrahedron Lett.*, **21**, 4675 (1980).
[261] D. A. Evans, J. Bartroli, and T. L. Shih, *J. Am. Chem. Soc.*, **103**, 2127 (1981).
[262] D. A. Evans, J. V. Nelson, E. Vogel, and T. R. Taber, *J. Am. Chem. Soc.*, **103**, 3099 (1981).
[263] S. Masamune, Sk. A. Ali, D. L. Snitman, and D. S. Garvey, *Angew. Chem., Int. Ed. Engl.*, **19**, 557 (1980).
[264] S. Masamune, W. Choy, F. A., J. Kerdesky, and B. Imperiali, *J. Am. Chem. Soc.*, **103**, 1566 (1981).
[265] S. Masamune, M. Hirama, S. Mori, Sk. A. Ali, and D. S. Garvey, *J. Am. Chem. Soc.*, **103**, 1568 (1981).
[266] R. Noyori, I. Nishida, J. Sakata, and M. Nishizawa, *J. Am. Chem. Soc.*, **102**, 1223 (1980).
[267] R. Noyori, I. Nishida, and J. Sakata, *J. Am. Chem. Soc.*, **103**, 2106 (1981).
[268] S. Shoda and T. Mukaiyama, *Chem. Lett.*, **1981**, 723.
[269] T. Mukaiyama and T. Harada, *Chem. Lett.*, **1982**, 467.
[270] S. Shenvi and J. K. Stille, *Tetrahedron Lett.*, **23**, 627 (1982).
[271] Y. Yamamoto, H. Yatagai, and K. Maruyama, *J. Chem. Soc., Chem. Commun.*, **1981**, 162.
[272] D. A. Evans and L. R. McGee, *Tetrahedron Lett.*, **21**, 3975 (1980).
[273] Y. Yamamoto and K. Maruyama, *Tetrahedron Lett.*, **21**, 4607 (1980).
[274] D. A. Evans and L. R. McGee, *J. Am. Chem. Soc.*, **103**, 2876 (1981).
[275] D. L. J. Clive and C. G. Russell, *J. Chem. Soc., Chem. Commun.*, **1981**, 434.
[276] D. B. Collum, J. H. McDonald, III, and W. C. Still, *J. Am. Chem. Soc.*, **102**, 2120 (1980).
[277] M. Koreeda and Y. P. Liang Chen, *Tetrahedron Lett.*, **22**, 15 (1981).
[278] R. F. Newton, S. M. Roberts, B. J. Wakefield, and G. T. Woolley, *J. Chem. Soc., Chem. Commun.*, **1981**, 922.
[279] T. Mukaiyama, R. W. Stevens, and N. Iwasawa, *Chem. Lett.*, **1982**, 353.
[280] T. Mukaiyama, R. W. Stevens, and N. Iwasawa, *Chem. Lett.*, submitted.

AUTHOR INDEX, VOLUMES 1–28

CHAPTER AND TOPIC INDEX, VOLUMES 1–28

Many chapters contain brief discussions of reactions and comparisons of alternative synthetic methods which are related to the reaction that is the subject of the chapter. These related reactions and alternative methods are not usually listed in this index. In this index the volume number is in BOLDFACE, the chapter number in ordinary type.

SUBJECT INDEX, VOLUME 28

Since the table of contents provides a quite complex index, only those items not readily found from the contents pages are listed here. Numbers in BOLDFACE refer to experimental procedures.